THE LIBRARY
ST. MARY'S COLLEGE OF MARYLAND
ST. MARY'S CITY, MARYLAND 20686

D1567126

MOLECULAR INTERACTIONS

Wiley Tutorial Series in Theoretical Chemistry

Series Editors

D. Clary, University College London, London, UK
A. Hinchliffe, UMIST, Manchester, UK
D. S. Urch, Queen Mary and Westfield College, London, UK

Previously Published Titles

Thermodynamics of Irreversible Processes: Applications to Diffusion and Rheology

Gerard D. C. Kuiken, Delft University of Technology, The Netherlands

Published 1994, ISBN 0471 948446

Modelling Molecular Structures

Alan Hinchliffe, UMIST, Manchester, UK

Published 1995, ISBN 0471 959219 (cloth), 0471 959235 (paper)

MOLECULAR INTERACTIONS

From van der Waals to Strongly Bound Complexes

Edited by
Steve Scheiner
*Department of Chemistry,
Southern Illinois University, USA*

JOHN WILEY & SONS
Chichester · New York · Weinheim · Brisbane · Singapore · Toronto

Copyright © 1997 by John Wiley & Sons Ltd,
Baffins Lane, Chichester,
West Sussex PO19 1UD, England

National   01243 779777
International (+44) 1243 779777
e-mail (for orders and customer service enquiries): cs-books@wiley.co.uk
Visit our Home Page on http://www.wiley.co.uk
or http://www.wiley.com

All Rights Reserved. No part of this book may be reproduced, stored in a retrieval system, or transmitted, in any form or by any means, electronic, mechanical, photocopying, recording or otherwise, except under the terms of the Copyright, Designs and Patents Act 1988 or under the terms of a licence issued by the Copyright Licensing Agency, 90 Tottenham Court Road, London, UK W1P 9HE, without the permission in writing of the publisher.

Other Wiley Editorial Offices

John Wiley & Sons, Inc., 605 Third Avenue,
New York, NY 10158-0012, USA

VCH Verlagsgesellschaft mbH,
Pappelallee 3, D-69 469 Weinheim, Germany

Jacaranda Wiley Ltd, 33 Park Road, Milton,
Queensland 4064, Australia

John Wiley & Sons (Asia) Pte Ltd, 2 Clementi Loop #02-01,
Jin Xing Distripark, Singapore 129809

John Wiley & Sons (Canada) Ltd, 22 Worcester Road,
Rexdale, Ontario M9W 1L1, Canada

Library of Congress Cataloging-in-Publication Data

Molecular interactions : from van der Waals to strongly bound
    complexes / edited by Steve Scheiner.
        p.   cm. – (Wiley tutorial series in theoretical chemistry)
    Includes bibliographical references and index.
    ISBN 0-471-97154-5 (alk. paper)
     1. Molecular association.   2. Molecular dynamics.   I. Scheiner,
Steve.   II. Series.
QD461.M616   1997
541.2'26–dc20   96-43509
                                                                  CIP

British Library Cataloguing in Publication Data

A catalogue record for this book is available from the British Library

ISBN 0-471-97154 5

Typeset in 10/12 pt Times by Keytec Typesetting Ltd, Bridport, Dorset.
Printed and bound in Great Britain by Bookcraft (Bath) Ltd, Midsomer-Norton, Somerset
This book is printed on acid-free paper responsibly manufactured from sustainable forestation, for which at least two trees are planted for each one used for paper production.

Contents

List of contributors.. xi
Series preface.. xiii
Introduction.. xv

SECTION I THEORETICAL FRAMEWORK 1

1 Symmetry-adapted perturbation theory of intermolecular interactions 3

K. Szalewicz and B. Jeziorski

1.1 Introduction 3
1.2 Fundamental problems in perturbation theory of intermolecular interactions   7
    1.2.1 Polarization expansion.................................... 9
    1.2.2 Symmetry-adapted perturbation theory..................... 12
1.3 Intramonomer correlation problem 15
    1.3.1 Double-perturbation theory approach 15
        1.3.1.1 Relation to Hartree–Fock theory 16
        1.3.1.2 Dispersion interaction........................... 17
        1.3.1.3 Electrostatic correlation 18
        1.3.1.4 Induction correlation............................ 19
        1.3.1.5 Exchange correlation 20
    1.3.2 Accuracy of SAPT and comparison to the supermolecular approach... 21
1.4 SAPT treatment of three-body nonadditivities 25
1.5 Computational aspects.. 30
1.6 Review of recent applications 31
References.. 38

2 *Ab initio* studies of nonadditive interactions........................... 45

M. M. Szczęśniak and G. Chałasiński

2.1 Introduction 45
2.2 *Ab initio* theory of nonadditive effects 48
    2.2.1 Classification.. 48
    2.2.2 Convergence of SAPT.................................... 50
    2.2.3 Intrasystem correlation effects............................. 51

		2.2.4	Combination of supermolecular and PT treatments of nonadditive effects	51
	2.3	Physical interpretation		53
		2.3.1	Heitler–London exchange	53
		2.3.2	Induction nonadditivity	53
		2.3.3	Dispersion nonadditivity	55
	2.4	Applications		56
		2.4.1	Rare-gas trimers	56
			2.4.1.1 Dispersion nonadditivity in supermolecular calculations	56
			2.4.1.2 Analysis of three-body effects in the equilateral triangle Ar_3	58
			2.4.1.3 Some best results for RG cyclic trimers	60
		2.4.2	RG_2-chromophore clusters	61
		2.4.3	Polar trimers	68
			2.4.3.1 Water trimer	68
			2.4.3.2 HF trimer	72
		2.4.4	Ion-(molecule)$_2$ clusters	73
	2.5	Conclusion		75
	References			76

3 Basis set superposition error — 81

F. B. van Duijneveldt

3.1	BSSE: a short history		81
3.2	How BSSE affects the calculated properties of molecular complexes		84
	3.2.1	Role of basis incompleteness in deciding the size of BSSE	84
	3.2.2	Equilibrium distances and binding energies	86
	3.2.3	Vibrational properties	89
	3.2.4	Electrical properties	89
3.3	Removal of BSSE: full counterpoise and virtuals-only counterpoise		90
	3.3.1	Problems with virtual spaces?	93
	3.3.2	Is it safe to use CP if BSSE is large?	94
	3.3.3	DCBS versus MCBS, and 'secondary BSSE'	94
3.4	Removal of BSSE: local counterpoise, alternative methods		96
	3.4.1	Local counterpoise	96
	3.4.2	Alternative methods: larger basis sets, R-12 methods, geminals	98
3.5	Geometry optimization. Non-uniqueness of calculated interaction energies		99
	3.5.1	Algorithm for geometry optimization	99
	3.5.2	Ambiguities for defining the fragments	100
3.6	Conclusion		101
References			102

4 Theory and computation of vibration, rotation and tunneling motions of Van der Waals complexes and their spectra — 105

A. van der Avoird, P. E. S. Wormer and R. Moszynski

4.1	Introduction	105
4.2	Choice of coordinates	106
4.3	Kinetic energy expressions	107
4.4	(*Ab initio*) intermolecular potentials; representations	110
4.5	Methods for the calculation of VRT states	113
4.6	Properties, transitions, intensities	120
4.7	Symmetry aspects	122
4.8	Comparison with experimental spectra, test and improvement of *ab initio* potentials, semi-empirical potential fits.	124
4.8.1	He—HF	124
4.8.2	He—CO	131
4.8.3	NH_3—NH_3	136
4.8.4	CO in C_{60}	142
4.9	Conclusion	148
	References	148

SECTION II APPLICATIONS 155

5 The quest for reliability in calculated properties of hydrogen-bonded complexes 157

J. E. Del Bene and I. Shavitt

5.1	Introduction	157
5.2	Methodological considerations	159
5.2.1	Basis sets	159
5.2.2	Wavefunction model	161
5.3	Structures and vibrational spectra of hydrogen-bonded complexes	162
5.3.1	Intermolecular distances	162
5.3.2	A–H vibrational frequency shifts in complexes with A—H...B hydrogen bonds	165
5.4	Binding energies and enthalpies	165
5.4.1	Basis set effects	168
5.4.2	Electron correlation effects	172
5.4.3	Comparisons with experimental data	174
5.5	Conclusions	176
	References	177

6 *Ab initio* predictions of the vibrational spectra of some molecular complexes: comparison with experiment 181

T. A. Ford

6.1	Introduction	181
6.2	Computational details	182
6.3	Computed vibrational spectra	183
6.3.1	The water–ammonia–hydroxylamine system	183
6.3.2	The complexes of water with hydrogen cyanide, acetonitrile and cyanogen	192

	6.3.3	The methanol–dimethylamine–dimethyl ether–trimethylamine system	196
	6.3.4	The hydrogen fluoride–hydrogen chloride–methane–silane system	204
	6.3.5	The complexes of water with carbon dioxide and nitrous oxide	208
6.4	Summary	212
References	213

7 Conventional and unconventional hydrogen-bonded ionic clusters 217

C. A. Deakyne

7.1	Introduction	217
7.2	Computational details	219
	7.2.1 General aspects	219
	7.2.2 Calculational protocols	221
7.3	Conventionally hydrogen-bonded ionic systems	224
	7.3.1 Proton-bound clusters	224
	7.3.2 Intracluster reactions	232
	7.3.3 Polyethers	234
	7.3.4 Negatively charged clusters	237
7.4	Unconventionally hydrogen-bonded ionic systems	239
	7.4.1 CH...Y and XH...C hydrogen bonds	240
	7.4.2 XH...π hydrogen bonds	245
7.5	Summary	253
References	258

8 Case studies in cooperativity in hydrogen-bonded clusters and polymers .. 265

A. Karpfen

8.1	Introduction	265
8.2	Weak cooperativity: the case of hydrogen cyanide	268
8.3	Medium cooperativity: the case of hydrogen fluoride	274
8.4	The transition to strong ionic hydrogen bonds: protonated and deprotonated chain-like clusters of hydrogen fluoride	287
8.5	Conclusions and outlook	290
References	292

9 Electrostatic forces in molecular interactions 297

S. L. Price

9.1	Introduction	297
9.2	Representations of the electrostatic interaction	298
	9.2.1 The central multipole expansion of the electrostatic energy	298
	9.2.2 The atomic point charge model	304
	9.2.3 Distributed multipole models	306
	9.2.4 Limitations to the accuracy of distributed multipole electrostatic energies	308

9.3		The importance of the electrostatic energy in intermolecular interactions	309
9.4		Molecular crystal structure modelling	311
9.5		Rationalising intermolecular preferences	316
	9.5.1	'Lone-pair directionality'—electrostatic or 'specific interactions'?	317
9.6		Biological modelling	321
	9.6.1	Van der Waals complexes as models for interactions within proteins	322
	9.6.2	Electrostatic modelling of nucleic acid bases—the variability of hydrogen bonds with environment	323
	9.6.3	Electrostatic similarity—a tool for protein–ligand design	326
9.7		Future outlook for biological modelling—the issue of transferability	327
	9.7.1	Variation with conformation	327
	9.7.2	Transferability between molecules	328
9.8		Overview and future directions	329
		References	331

10 Protein–ligand interactions ... 335

G. Náray-Szabó

10.1		Introduction	335
10.2		Principles	335
10.3		Methodology	338
	10.3.1	Molecular force fields	338
	10.3.2	Active site–environment partition	339
	10.3.3	Reaction field theories	340
	10.3.4	The Poisson–Boltzmann equation	343
	10.3.5	Thermodynamic cycle—perturbation method	344
10.4		Applications	345
	10.4.1	Quantitative structure–activity relationships	345
	10.4.2	Computer-aided drug design	347
		References	348

Index of Complexes ... 351

Subject Index ... 353

List of Contributors

G. Chalasinski	*Department of Chemistry, Hannah Hall, Room 357, Oakland University, Rochester, MI 48309, USA*
C. A. Deakyne	*Chemistry Department, Eastern Illinois University, Charleston, IL 61920, USA*
J. E. Del Bene	*Department of Chemistry, Youngstown State University, Youngstown, OH 44555, USA*
T. A. Ford	*Department of Chemistry and Applied Chemistry, University of Natal, Durban, Private Bag X10, Dalbridge 4014, South Africa*
B. Jeziorski	*Department of Physics and Astronomy, University of Delaware, Newark, DE 19716, USA*
A. Karpfen	*Institut für Theoretische Chemie und Strahlenchemie der Universität Wien, Währingerstrasse 17, A-1090 Wien, Austria*
R. Moszynski	*Department of Chemistry, University of Warsaw, Pasteura 1, 02-093 Warsaw, Poland*
G. Náray-Szabó	*Department of Chemistry, Eötvös University Budapest, PO Box 32, H-1518 Budapest 112, Hungary*
S. L. Price	*Department of Chemistry, University College London, 20 Gordon Street, London, WC1H 0AJ, UK*
S. Scheiner	*Department of Chemistry, Southern Illinois University, Carbondale, IL 62901-4409, USA*
I. Shavitt	*Youngstown State University, Youngstown, OH 44555-3663, USA*
K. Szalewicz	*Department of Physics and Astronomy, University of Delaware, Newark, DE 19716, USA*
M. Szczesniak	*Department of Chemistry, Hannah Hall, Room 357, Oakland University, Rochester, MI 48309, USA*
A. van der Avoird	*Institute of Theoretical Chemistry, University of Nijmegen, Toernooiveld, 6525 ED Nijmegen, The Netherlands*
F. B. van Duijneveldt	*Rijkuniversiteit Utrecht, Vakgroep Theoretische Chemie, Padualaan 8, De Uithof, Utrecht, The Netherlands*
P. Wormer	*Institute of Theoretical Chemistry, University of Nijmegen, Toernooiveld, 6525 ED Nijmegen, The Netherlands*

Series Preface

Theoretical chemistry is one of the most rapidly advancing and exciting fields in the natural sciences today. This series is designed to show how the results of theoretical chemistry permeate and enlighten the whole of chemistry together with the multifarious applications of chemistry in modern technology. This is a series designed for those who are engaged in practical research, in teaching and for those who wish to learn about the role of theory in chemistry today. It will provide the foundation for all subjects which have their roots in the field of theoretical chemistry.

How does the materials scientist interpret the properties of a novel doped-fullerene superconductor or a solid-state semiconductor? How do we model a peptide and understand how it docks? How does an astrophysicist explain the components of the interstellar medium? Where does the industrial chemist turn when he wants to understand the catalytic properties of a zeolite or a surface layer? What is the meaning of 'far-from-equilibrium' and what is its significance in chemistry and in natural systems? How can we design the reaction pathway leading to the synthesis of a pharmaceutical compound? How does our modelling of intermolecular forces and potential energy surfaces yield a powerful understanding of natural systems at the molecular and ionic level? All these questions will be answered within our series which covers the broad range of endeavour referred to as 'theoretical chemistry'.

The aim of the series is to present the latest fundamental material for research chemists, lecturers and students across the breadth of the subject, reaching into the various applications of theoretical techniques and modelling. The series concentrates on teaching the fundamentals of chemical structure, symmetry, bonding, reactivity, reaction mechanism, solid-state chemistry and applications in molecular modelling. It will emphasise the transfer of theoretical ideas and results to practical situations so as to demonstrate the role of theory in the solution of chemical problems in the laboratory and in industry.

D. Clary, A. Hinchliffe and D. S. Urch
June 1994

Introduction

S. Scheiner

Southern Illinois University, USA

Much of chemistry is, of course, concerned with the covalent bonds that hold molecules together. Most chemical reactions involve the breaking of such bonds and the reforming of new ones. But before any such chemical reaction can begin, the two moieties must approach within a critical distance of one another. This approach is controlled by the forces between them, known as molecular interactions. These interactions are important not only as a precursor to a chemical reaction between the species but also with respect to the clustering of molecules about one another in the liquid and solid state, for example, or as determining factors in the structure of biomolecules like DNA and proteins.

Experimental measurements have made many important contributions over the years to our present understanding of molecular interactions. For instance, spectra have been analyzed to arrive at empirical force fields that cover a wide area of configurational space and are consistent with the measurements. But it is a theoretical attack that has the greatest potential to analyze the fundamental nature of these interactions. Such theoretical calculations can be used to improve upon the potentials generated from spectral analysis. From a more fundamental perspective, one can separate the total interaction energy into its constituent forces, and analyze each separately.

It is the intent of this volume to present the reader with an up-to-date account of the means by which theoretical methods can be brought to bear on the problem of molecular interactions. The startling rate of advances in computer technology and algorithms have greatly broadened the scope of systems which are amenable to theoretical analysis and sharpened the accuracy which can be expected from such calculations.

Intermolecular forces vary over a wide spectrum in terms of strength and underlying causes. For this reason, it is common to take a somewhat different tack in studying systems of each sort. For example, whereas a strong hydrogen bond can be investigated to some degree of reliability with even relatively crude methods, the much weaker forces holding together rare gas atoms require much more sophisticated approaches. As there is no single method which is suited to all sorts of interactions, this volume attempts to sweep the field from very weak van der Waals complexes to the strong forces of attraction between ions.

Another dimension of the problem is the size of the system of interest. Although it might be possible to apply a high-level of theory to a system as small as the helium dimer, such a calculation becomes unthinkable for a larger complex like an aggregate of numerous water molecules or a pair of DNA bases. A range of different methods has developed over the years, each best suited to systems of a certain level of complexity.

The first Section of this text presents a theoretical framework by which one might understand the fundamental nature of molecular interactions. One can dissect the total force into its constituent elements, each of which is associated with a clear physical phenomenon. Szalewicz and Jeziorski describe in the opening chapter a formal framework by which the total interaction energy can be partitioned into such components and provides a common language for the remainder of the text. Of course, there are various means by which this interaction energy can be broken into pieces. The perturbation treatment adopted by Szalewicz and Jeziorski has certain advantages over some others, including rigorous definitions and clear connections to physically meaningful concepts. Work such as this provides a framework for the development of improved analytical functions for molecular dynamics and Monte Carlo simulations. The authors provide in their chapter a detailed analysis of the accuracy of computations on weakly bound systems such as He_2 and Ar··HF, and hydrogen bonds as might occur in the water dimer.

Szalewicz and Jeziorski introduce the notion of nonadditivity in the interactions among a cluster of species, which concerns the perturbation of a particular interaction induced by the introduction of a third molecule. This topic is further elaborated and forms the heart of Chapter 2, authored by Szczesniak and Chalasinski, who continue the language of symmetry-adapted perturbation theory in order to elucidate the various contributing terms to each interaction. The authors go to great pains to describe the physical interpretations of the formal mathematical constructs that go into the nonadditivity for three- and four-body terms and higher. Their work thus constitutes a framework by which the nonadditivity which is so integral a part of condensed phases may be understood. Examples provided by Szczesniak and Chalasinski span a range of interaction strength from rare gas atoms to polar molecules such as water and HF, to systems where the nonadditivity is magnified by the presence of an ion. It thus becomes possible for the reader to watch certain contributing terms which might be dominant for one sort of interaction fade into virtual insignificance for others.

The computation of intermolecular potentials by *ab initio* means has become progressively more accurate over the years as it has become feasible to apply basis sets with greater flexibility and make use of higher levels of electron correlation. Yet one intrinsic limitation of any finite basis set arises from the variational principle which states that the 'trial' electronic energy of any system is lowered by any improvement in the basis set used for its description. Since the basis set of the full complex by definition contains more basis functions than does the set of each of its constituent subunits, there is an inherent discrepancy which can bias the computed interaction energy of any system. In fact, this bias, which has come to be known as the 'basis set superposition error' or BSSE, can be so profound as to alter the very qualitative character of certain computed interaction potentials. The fundamental nature of the basis set superposition error has stimulated a great deal of debate among theorists over the years. Van Duijneveldt outlines the nature of this controversy in Chapter 3 and offers insights into its resolution. He highlights the central issues of which one should be aware when perusing the literature, and places BSSE in perspective with other errors that are likely to occur in typical calculations.

Not all information about molecular interactions comes from theoretical calculations of course. There is a wealth of experimental data, much of which is derived from gas-phase spectra. Unfortunately, there is no clear and error-free path to convert an experimental spectrum into a function representing the interaction potential. On the other hand, if one could start with a computed potential and transform it into a spectrum, this could be compared directly with the experimental spectrum so as to check the validity of the former, and could suggest refinements in the theoretical potential to improve its accuracy. Van der Avoird, Wormer and Moszynski describe in Chapter 4 the procedures for obtaining a spectrum directly from a potential energy surface. Their methods involve a solution of the equations of nuclear motion which first separate inter and intramolecular types of motion. Because of the large amplitudes of the intermolecular motions, the vibrations, rotations, and translations (VRT) cannot be rigorously separated from one another but must be treated together to obtain VRT states. The examples in this chapter consist of the generally weak interactions involving such molecules as He, CO, NH_3 and C_{60}.

Following the preceding chapters which provide a theoretical context, the chapters in the second Section are devoted to applications. Del Bene and Shavitt turn the reader's attention to hydrogen bonds in Chapter 5. Their emphasis is placed on the limits of accuracy that may be expected at a given level of theory, and how well theory does in reproducing experiment. Of particular use as yardsticks of accuracy are the length of the H-bond, the interaction energy of the equilibrium structure, and the shift undergone by a key vibrational frequency upon formation of the complex. In addition to the particular vibrational mode examined by Del Bene and Shavitt, there are a great many other spectral lines that are related to the nature of the molecular interaction. Ford devotes Chapter 6 to a detailed analysis of the vibrational spectra of H-bonded complexes, with particular emphasis on a comparison between theory and experiment. He describes the utility of theory in assigning certain bands in the experimental spectrum. This chapter is not limited exclusively to hydrogen-bonded systems but goes on to consider electron donor–acceptor complexes so as to highlight the differences between the two.

Hydrogen bonds can form not only between neutral molecules but can also involve a molecular ion. As described by Deakyne in Chapter 7, these bonds can be particularly strong. In fact, the presence of an ion is conducive to the formation of a 'cluster' of molecules around it, forming shells of varied structure. Such clusters can be considered as the precursors to the solvent state where the ion is immersed in a liquid containing numerous solvent molecules. One topic of especial interest is the solvated proton. Does the structure resemble a proton surrounded by solvent molecules or does the proton attach itself to one solvent molecule in particular? When there are more than one type of molecule present, with which is the proton more likely to be associated? This chapter also concerns itself with the clustering of a proton with molecules that contain more than one site with which the proton might simultaneously interact. Deakyne goes on to consider interactions with molecules that would not ordinarily be considered as proton acceptors when the complex is uncharged and a special case with biological significance.

In its description of the formation of ionic and neutral clusters, Chapter 7 touches on some of the aspects of nonadditivity, i.e. how the various H-bonding interactions affect one another. This idea serves as the central basis of Chapter 8 in which Karpfen examines clusters of H-bonded molecules that form long chains, and compares them to cyclic clusters in which the chain curves back on itself. Attempts are made to extrapolate the results for $(HCN)_n$ and $(HF)_n$ all the way up to $n = \infty$ so as to ascertain the limits of cooperativity in such H-bonded chains. The author focuses in particular on certain markers of H-bond strength: the interaction energy, the distance between subunits, the stretch in the bond that links the hydrogen to the proton-donor atom, the shift in the stretching frequency that corresponds to the latter bond, and the NMR shielding constant for the bridging proton. By comparing these neutral systems with those in which an excess proton has been added, or one removed, Karpfen characterizes the cooperativity intrinsic to strong ionic H-bonds.

Coulombic interactions are an important part of molecular interactions, and in many cases a dominating force. For that reason, a great deal of effort has been expended in formulating efficient means for its evaluation. Price summarizes some of these efforts in Chapter 9 and describes in detail one of these methods in particular. This procedure treats each subunit as a collection of points, generally the atomic nuclei. To each point is assigned a set of multipoles so as to mimic the overall charge density of the entire molecule and the electrostatic potential generated by it. After describing the essence of this technique, Price goes on to summarize the sorts of results that are achieved in its use with particular emphasis on the correct reproduction of angular preferences in hydrogen bonding. Model systems range from van der Waals dimers, to crystal structures, to biological systems such as nucleic acid base pairs and protein–ligand interactions.

The topic of the interactions between a protein and its ligands is the focus of Chapter 10 where Náray-Szabó describes some of the modern and efficient methodologies that have been developed for such large systems. In addition to exploring the steric aspects of the interaction, the author discusses the complementarity that is essential in other aspects such as electrostatics and hydrophobicity. The latter concept takes the reader into the realm of competitive binding of ligand versus solvent, as well as the interactions between the solvent molecules themselves. The treatment of this problem involves a separation of the entire system into various domains where the 'central' system, which may be thought of as an active site, is surrounded first by a polarizable region, which is itself embedded in a nonpolarizable area, containing ionized groups and counterions. Finally, the solvent in which the complex is immersed is treated as an overall biophase.

A last order of business concerns the units in the various chapters. It is not unusual for researchers in different areas to employ different units as well, particularly for energy. Whereas kcal/mol (or kJ/mole) is the common energy denomination for chemists working with strong interactions or covalent bonds, the reciprocal centimeter or electron volt is perhaps more natural for the spectroscopist. Computational chemists who deal with weak van der Waals clusters find it useful to talk in terms of Hartree, or more commonly microhartree, μH. In the case of particularly weak interactions, the energy is sometimes converted into degrees Kelvin. The reader may find the following set of conversion factors useful:

$$1 \text{ kcal/mol} = 4.184 \text{ kJ/mol} = 350 \text{ cm}^{-1} = 0.0434 \text{ eV} = 1594 \, \mu H(\mu E_H) = 503 \text{ K}$$

Section I

THEORETICAL FRAMEWORK

1 Symmetry-adapted Perturbation Theory of Intermolecular Interactions

K. SZALEWICZ[a] and B. JEZIORSKI[b]

[a]*University of Delaware, USA*
[b]*University of Warsaw, Poland*

1.1 INTRODUCTION

The perturbation theory of intermolecular interactions has been extensively reviewed by Jeziorski *et al.* (1994). That work contains an exhaustive list of literature references. The present chapter is intended to be more didactic, thus, it will also be narrower in scope. However, nonadditive effects not included in Jeziorski *et al.* (1994) will be discussed. We will review in some detail the new work published in the past two years but will not repeat discussion of older literature.

The chemical bond is at the present a well-understood phenomenon. Several high-quality quantum chemistry programs exist which can be used to predict properties of systems formed by chemical forces. However, a vast area of molecular phenomena are determined by much weaker intermolecular forces, of quite different nature. Included in these phenomena are bulk properties of gases, liquids, solutions, and molecular solid phase, as well as a large part of biological processes. For example, we know that the biological pattern recognition or catalytic activity of enzymes are all due to weak intermolecular interactions. Theoretical methods describing such systems always utilize some kinds of intermolecular potentials. The fields of molecular science where the use of these potentials is most critical are theories of condensed phase, the molecular dynamics (MD) and Monte Carlo (MC) simulations of condensed phase and of biological aggregates, molecular scattering calculations, and predictions of rovibrational spectra of atomic and molecular clusters. Since until recently no reliable potentials have been available, the common practice was to use model potentials with parameters adjusted to recover existing experimental information in a given field. Clearly, the first principles approach

Molecular Interactions. Edited by S. Scheiner
© 1997 John Wiley & Sons Ltd

which would utilize an independently derived information about intermolecular potentials would be preferred.

Intermolecular interaction energy (or intermolecular potential energy surface) is defined as

$$E_{\text{int}}(R, \zeta_A, \zeta_B; \Xi_A, \Xi_B) = E_{AB}(R, \zeta_A, \zeta_B; \Xi_A, \Xi_B) - E_A(\Xi_A) - E_B(\Xi_B) \quad (1.1)$$

where E_{AB} is the total energy of the dimer, E_X is the total energy of monomer X, R is the distance between the centers of mass of the monomers, ζ_X denotes the Euler angles defining orientation of monomer X in space, and Ξ_X is the set of internal coordinates of each monomer. Notice that the same values of internal coordinates Ξ_X are used in calculating all energies in equation (1.1). The energies entering equation (1.1) are usually defined by the clamped nuclei (electronic) Schrödinger equation, although the diagonal adiabatic corrections to these energies (Kolos, 1970) could be added if necessary. Thus, the concept of the interaction potential appears in the framework of the Born–Oppenheimer approximation and loses its usefulness when the nonadiabatic effects are significant. In a great majority of chemical physics problems the nonadiabatic effects are negligible and the potential energy surface defined in equation (1.1) is sufficient to predict spectra, scattering cross-sections, and bulk properties of investigated system. It should also be emphasized that the interaction energy itself is not a directly measurable quantity since in no experiment can we fix the distance R, the orientation angles ζ_A, ζ_B, and the internal coordinates Ξ_A, Ξ_B. The so-called 'experimental' interaction potentials are not defined by equation (1.1) but are rather *effective* potentials obtained by a fit to a set of experimental data. These potentials always include in an averaged way the adiabatic and nonadiabatic effects and in many cases also the effects of monomer non-rigidity and even the pair-wise nonadditivity of intermolecular forces. The theoretical calculations provide thus, in principle, the most straightforward and unambiguous way to obtain information about intermolecular potential energy surfaces.

It should be noted that in some works the interaction energy E_{int} is defined in such a way that different values of internal coordinates Ξ_A (Ξ_B) are used in the first and in the second (third) terms on the right-hand side of equation (1.1) (for example, equilibrium monomers' geometries are used to calculate E_X and internal coordinates which minimize the total energy of the dimer are used to obtain E_{AB}). The interaction energies defined in such a way are of no use in dynamical calculations and in fact it is not clear how they can be employed to calculate any measurable quantity, except perhaps for stabilization energies of very strongly bound complexes involving ions.

For many applications, especially involving low-energy processes, the monomers can be assumed to be rigid to a very good approximation. The intermolecular potential depends then only on R and on the Euler angles ζ_A, ζ_B and is usually calculated from equation (1.1) for some fixed values of the internal coordinates Ξ_A and Ξ_B. This simplification dramatically reduces the number of points which have to be calculated and therefore is often applied. One may mention that a more appropriate method of obtaining the interaction energy of rigid molecules is to average the complete Ξ_A, Ξ_B dependent potential over internal coordinates using suitable vibrational wave functions of monomers, but this procedure is rarely employed in dynamics calculations (Moszynski *et al.*, 1994e).

In a majority of *ab initio* calculations of interaction energies of rigid molecules the internal coordinates are fixed at the equilibrium geometry of the monomers (we will refer to it as the r_e geometry). This choice may be not the best since in most experiments the monomers are in some well-defined vibrational state (usually the ground state), their nuclei moving in a potential energy well with an amplitude of 5–10 per cent of the bond length. This means that the mean monomer geometry, obtained by averaging over the vth vibrational wave function (we will refer to this geometry as the r_v geometry) would be a better choice for internal coordinates. The intermolecular potentials obtained in this way depend then on the vibrational state of monomers. Such potentials can be used to interpret spectroscopic measurements providing vibrational state specific information about interaction energies. Transition energies obtained from potentials computed assuming monomers frozen in r_v geometries are usually very good approximations to the measured quantities (van der Avoird et al., 1994).

Early experimental investigations of intermolecular forces were dominated by studies of bulk gas properties and molecular beam scattering (Maitland et al., 1981). The bulk studies in general do not provide enough information to determine more than a one-dimensional potential function, although measurements of such properties as sound absorption and rotational line broadening do give some information about the anisotropy of the potential. The scattering experiments, in principle, can provide enough information to allow determination of a complete potential energy surface. Such determination has so far been possible only for a few very favorable cases such as Ne–H_2 (Anders et al., 1980). However, the recently developed technique of direct absorption IR laser probe measurements of the scattering cross-sections (Schiffman et al., 1996) may change this situation. The last two decades have seen the emergence of high-resolution spectroscopic methods (McKellar and Welsh, 1971; Dyke et al., 1972; Dyke and Muenter, 1975; Klemperer, 1980; Harris et al., 1974; Keenan et al., 1981; Dixon et al., 1981; Lovejoy et al. 1986a,b; Saykally, 1989; Cohen and Saykally, 1992; Nesbitt, 1988; Weber, 1987; Halberstadt and Janda, 1990; Leopold et al., 1994; Nesbitt, 1994). The new spectroscopic techniques provided a wealth of accurate data on interaction potentials for atom–atom and atom–diatom systems (Weber, 1987; Hutson, 1990), as well as on some more complicated systems such as the HF dimer (Dayton et al., 1989; Barton and Howard, 1982; Pine and Howard, 1986), Ar–H_2O (Cohen et al., 1988; Cohen and Saykally, 1991; Fraser et al., 1990; Cohen et al., 1990; Suzuki et al., 1991; Lascola and Nesbitt, 1991; Zwart and Meerts, 1991; Cohen and Saykally, 1990, 1993), Ar–NH_3 (Fraser et al., 1985; Nelson et al., 1986; Bizarri et al., 1988; Gwo et al., 1990; Zwart et al., 1991; Fraser et al., 1991; Schmuttenmaer et al., 1991, 1994), water dimer (Fraser, 1991; Nelson and Klemperer, 1987; Pugliano et al., 1993), DF trimer (Suhm et al., 1993), and water trimer (Liu et al., 1994), with the number of systems studied rising almost exponentially in recent years (Leopold et al., 1994).

Due to the above experimental progress, the past few years have been a period when our knowledge of intermolecular interactions increased dramatically. It will be fair to say that prior to 1986 there existed no quantitatively correct intermolecular potentials (some interatomic potentials were relatively accurate by that time: see Chalasinski and Gutowski, 1998). In the late 1980s the first accurate empirical intermolecular potentials appeared. The important factor enabling construction of such potentials was the development of computational techniques for calculating the rovibrational levels of van der Waals complexes (Hutson, 1990; Law et al., 1993). These techniques are now advanced enough to allow accurate calculations of rovibrational states of systems containing three or four atoms (five or even six in special cases). Such calculations could repeatedly be performed on an analytical potential energy surface dependent on some number of adjustable parameters, which allowed fitting to the observed spectra. In this way several accurate potential energy surfaces have been produced in the last decade. Most of them are for complexes of atoms with molecules containing up to four atoms. As we know from very recent *ab initio* calculations, these potentials are usually quantitatively accurate, i.e. differ from exact potentials on the order of 10 per cent or less in most regions of configuration space. The empirical potentials obtained from spectroscopic data are most accurate in the region of the potential well. Other regions of the surfaces are very difficult to determine without an input from electronic structure calculations. Although the best empirical potentials (Le Roy and Hutson, 1987; Hutson, 1992) are very accurate, the individual physical components typically differ by up to about 20 per cent from those computed *ab initio* (Williams *et al.*, 1993; Lotrich *et al.*, 1995).

For larger systems the situation is much less satisfactory. Even if high-resolution spectra of many complexes involving polyatomic monomers are available (for example, for the water dimer), dynamical calculations are not yet feasible for general case of a van der Waals molecule containing more than four nuclei (Klemperer and Yaron, 1990). Several empirical potentials of a very simple analytical form have been obtained for medium-size molecules by fitting results of MD/MC calculations on large ensembles of such molecules to various bulk properties, mostly the density and the enthalpy of vaporization of liquids. These potentials cannot be considered quantitatively accurate as the interaction energies predicted by them can easily be a few times smaller or larger than the true values. Even less accurate intermolecular potentials are used in MD/MC simulations of very large molecules. These are built from atom–atom contributions determined by simple formulas on the basis of atomic properties such as ionization potentials and polarizabilities. An additional complication in this case results from the three- and higher many-body effects (see discussion below). Almost all empirical potentials used in molecular simulations are *effective* two-body potentials which means that they incorporate the three-body effects by a distortion of the two-body potentials. The many-body effects are often invoked to explain very significant deviations of those potentials from *ab initio* pair potentials.

The advances in construction of empirical potentials for small dimers were initially not accompanied by a parallel progress in theory. Theory was especially in trouble in the regions around the van der Waals minima. This situation made Buckingham *et al.* (1988) refer to this part of the potential as to the *theoretically intractable 'intermediate region'*. Within the past few years several accurate *ab initio* potentials have been obtained (Williams *et al.*, 1993; Lotrich *et al.*, 1995; Moszynski *et al.*, 1994a,b, 1995a,b; Tao and Klemperer, 1994, 1995; Tao *et al.*, 1994, 1995; Huang *et al.*, 1995; Kukawska-Tarnawska *et al.*, 1994) which match the accuracy of the empirical potentials. At present such calculations are being extended to somewhat larger molecules (for example, the water dimer, Mas *et al.*, to be published) for which empirical potentials fitted to spectroscopic data are not yet available.

The currently available computational methods can be divided into supermolecular and those which compute the interaction energies directly. We will call the latter methods perturbational theories of intermolecular interactions since the interaction energy is expanded in powers of intermolecular interaction operator. The present chapter will mainly be devoted to discussion of such theories and their applications. Notice, however, that the supermolecular approach can also employ perturbation theory as a computational technique, e.g. a Møller–Plesset type expansion of the dimer energy. In that case the perturbation operator (Møller–Plesset or fluctuation potential for the dimer) is not directly related to the strength of the intermolecular interaction.

The majority of efforts in the field of intermolecular interactions have been devoted to studying dimers. Structure of clusters containing more than two molecules and the properties of condensed phase are usually described assuming the pair-wise additivity of intermolecular interactions. However, in many cases this approximation is insufficient and to obtain quantitative (or even qualitative) agreement with measurements an inclusion of 'pair-wise nonadditive' interactions is necessary. The most important of those are the three-body interactions. Our present theoretical understanding of three-body interactions is rather unsatisfactory, in particular in view of recent experimental activity in this field (Elrod and Saykally, 1994; Suhm and Nesbitt, 1995).

1.2 FUNDAMENTAL PROBLEMS IN PERTURBATION THEORY OF INTERMOLECULAR INTERACTIONS

Generation of accurate potential energy surfaces is only one of the aims of the theory of intermolecular forces. Perhaps even more important is to extend our understanding of the underlying physical mechanisms. This aim cannot be achieved by the supermolecular method since it can only provide a single number representing the total interaction energy. In contrast, the symmetry-adapted perturbation theory (SAPT) developed during the last decade into an efficient computational tool allows us to understand the structure and energetics of van der Waals complexes in terms of the four *fundamental* intermolecular interactions: *electrostatic*, *exchange*, *induction*, and *dispersion*. Perturbational studies can help to understand in detail what the real physical mechanisms of the formation of the van der Waals bond are and which physical interactions are responsible for the minima, tunneling barriers, and anisotropies of the potentials. This is important not only from the point of view of elucidating mechanisms of intermolecular interactions but also because modern empirical potentials are usually formulated in terms of those four interactions. Comparison of the potentials obtained from SAPT with those derived from the high-resolution spectroscopy data will be helpful in the construction of intermolecular potentials for larger polyatomic systems, such as those currently investigated using the MD/MC simulations. The SAPT results can also provide necessary benchmarks and reference data for various electrostatic and polarization models frequently used to interpret the structure and energetics of van der Waals and hydrogen-bonded complexes.

The SAPT approach is asymptotically related to the classical London theory, based on the multipole expansion of the interaction operator V and on the Rayleigh–Schrödinger perturbation expansion in powers of V. Specifically, the $1/R$ asymptotic expansion of the nth-order interaction energy in SAPT is given precisely by the London theory. It is known (Ahlrichs, 1976; Morgan and Simon, 1980) that the London theory provides the correct values of the constants C_n defining the asymptotic expansion of the interaction energy in the sense that

$$\left| E_{\text{int}} - \sum_{n=1}^{N} \frac{C_n}{R^n} \right| = O(R^{-N-1}) \qquad (1.2)$$

Thus, the correct large R asymptotics is built into the SAPT theory as well.

Although the London theory has been widely used for describing long-range interactions, i.e. for large intermolecular distances R, its applicability at shorter distances (e.g. at the van der Waals minimum) is severely limited. There are two reasons for that. First, the series on the r.h.s. of equation (1.2) is divergent for any finite value of R and cannot be used to reconstruct the exact value of interaction energy at finite intermolecular distances (Cizek *et al.*, 1986). Second, when this series is truncated (arbitrarily) at finite values of N, a qualitatively incorrect result is obtained—the interaction potential does not become repulsive at small intermolecular separations since the needed repulsive components, resulting from the exchange and penetration (charge-overlap) effects, are neglected by the London theory. There have been several methods proposed which account for the neglected effects in some approximate way. For example, one can utilize the long-range information available from the London theory and short-range data from (scaled) SCF supermolecular calculations to construct a hybrid potential, called the Hartree–Fock plus dispersion (HFD) model (Hepburn *et al.*, 1975; Ahlrichs *et al.*, 1977; Douketis *et al.*, 1982; Tang and Toennies, 1978). Another example is the exchange-Coulomb (XC) approach of Meath *et al.* (1981) (see also Ng *et al.*, 1978, 1979a; Wheatley and Meath, 1993). To obtain better values of the electrostatic (Coulomb) energy for large, polyatomic molecules several methods based on the concept of distributed multipoles have been proposed (Stone, 1991; Dykstra, 1993). Recently, this concept was extended to the induction component using both distributed multipoles and polarizabilities (Stone, 1985; Le Sueur and Stone, 1993; Angyan *et al.*, 1994).

The SAPT approach is a rigorous method of including the penetration and exchange effects. In SAPT, as in the London method, the dimer is described in zeroth-order by the product of the wave functions of unperturbed monomers. The penetration effects are included in SAPT by keeping the interaction potential in the nonexpanded form. The exchange effects are included by acting on the wave function with operators interchanging electrons between the interacting systems. An early version of SAPT was proposed by Eisenschitz and London (1930). Various forms of SAPT were developed and successfully applied to one- and two-electron systems in the late 1960s and in the 1970s (Hirschfelder and Silbey, 1966; Murrell and Shaw, 1967; Musher and Amos, 1967; Chipman, 1977; Adams and Polymeropoulos, 1978; Chalasinski *et al.*, 1977; Jeziorski *et al.*, 1978, 1979, 1980; Chalasinski and Szalewicz, 1980; Klein, 1987). See Arrighini (1981), Jeziorski and Kolos (1982), and Kaplan (1987) for reviews of these developments. The extension of the method to many-electron systems required a solution of the intramonomer correlation problem. This solution has recently been achieved in a series of papers (Szalewicz and Jeziorski, 1979; Rybak *et al.*, 1987; Jeziorski *et al.*, 1989; Jankowski *et al.*, 1990; Rybak *et al.*, 1991; Moszynski *et al.*, 1993a,b, 1994c,d,f; Williams *et al.*, 1995b). To account for the intramonomer correlation effects in a size-extensive manner SAPT uses a methodology similar to the standard many-body perturbation theory (MBPT)/coupled cluster (CC) approach. The resulting theory is referred to as the *many-body* (or *many-electron*) SAPT. Since the interaction energy in SAPT is obtained directly and not as a difference, it is completely free from the basis set superposition error. This energy is a sum of contributions corresponding to the four fundamental intermolecular interactions: electrostatic, exchange, induction, and dispersion. At large intermolecular distances, where the exchange and penetration effects are negligible, SAPT reduces to the classical London theory. The individual corrections in SAPT, each with a unique radial and angular dependence, as well as with distinct basis set requirements, can be examined separately to achieve state-of-the-art values and to find reliable approximations for very large systems. Each term can be represented in the form of an analytical potential with physically interpretable parameters. SAPT calculations can provide accurate *ab initio* values for all the terms included in the empirical and semiempirical potentials, as well as precise assessment of the components neglected by those approaches.

1.2.1 Polarization Expansion

The polarization expansion represents the simplest perturbation approach to intermolecular interactions. This name denotes just the standard Rayleigh–Schrödinger (RS) perturbation theory based on the splitting of the total Hamiltonian H into an unperturbed Hamiltonian $H_0 = H_A + H_B$, being the sum of monomers' Hamiltonians H_A and H_B, and the interaction operator $V = H - H_0$. Unlike in the London theory, V is taken in a complete, nonexpanded form. We will restrict our attention to interactions of closed-shell atoms or molecules in their ground states described by wave functions Ψ_X, $X = A$ or B. The product of such wave functions $\Psi_0 = \Psi_A \Psi_B$ is an eigenfunction of H_0, satisfying the zeroth-order equation $H_0 \Psi_0 = E_0 \Psi_0$, where $E_0 = E_A + E_B$ is the sum of ground-state energies. The well-known equations of the RS perturbation expansion define then the polarization energies and wave functions of arbitrary order:

$$E_{\text{int}} = \sum_{n=1}^{\infty} E_{\text{pol}}^{(n)} \tag{1.3}$$

$$\Psi = \sum_{n=0}^{\infty} \Psi_{\text{pol}}^{(n)} \tag{1.4}$$

$$E_{\text{pol}}^{(n)} = \langle \Psi_0 | V \Psi_{\text{pol}}^{(n-1)} \rangle \tag{1.5}$$

$$\Psi_{\text{pol}}^{(n)} = -\hat{R}_0 V \Psi_{\text{pol}}^{(n-1)} + \sum_{k=1}^{n-1} E_{\text{pol}}^{(k)} \hat{R}_0 \Psi_{\text{pol}}^{(n-k)} \tag{1.6}$$

where the reduced resolvent $\hat{R}_0$ obeys the relation $\hat{R}_0 (H_0 - E_0) = 1 - |\Psi_0\rangle\langle\Psi_0|$ and can be conveniently represented in the form of the following spectral sum:

$$\hat{R}_0 = \sum_{m \neq 0} \frac{|\Psi_m\rangle\langle\Psi_m|}{E_m - E_0} \tag{1.7}$$

The quantities denoted by subscript m are excited eigenenergies and eigenfunctions of H_0.

The polarization energies of lowest order have clear physical interpretation. The first-order polarization energy $E_{\text{pol}}^{(1)}$ represents the energy of electrostatic interaction of unperturbed charge distributions of the monomers. It is usually referred to as the *electrostatic energy* and denoted by $E_{\text{elst}}^{(1)}$. The second-order polarization energy is a sum of the well-known *induction* and *dispersion* energies. Note that in older literature the induction contribution is sometimes referred to as the polarization energy. In the third order one can identify the induction, dispersion, and also the mixed, *induction–dispersion* contributions (Jeziorski *et al.*, 1994).

The polarization expansion has very unusual convergence properties which have been thoroughly investigated for small systems such as H_2^+, H_2, and He_2 (Chalasinski *et al.*, 1977; Cwiok *et al.*, 1992a; Korona *et al.*, in press). For intermolecular distances R which correspond to van der Waals minima the partial sum of the series very rapidly (practically in second order) reaches the value of the so-called *Coulomb energy*, defined as the average of the energies of all states corresponding to the same dissociation limit (this Coulomb energy should be distinguished from the first-order, electrostatic energy which is sometimes also referred to as the Coulomb energy). Then the convergence rate drops dramatically, i.e. the consecutive corrections cease to decrease although they are so small that it might appear that a limit value of the series has been reached. This unusual behavior has caused recently some controversy as to the actual limit of the polarization expansion (Tang and Toennies, 1990). Presently, there is no doubt (Cwiok *et al.*, 1992a; Kutzelnigg, 1992; Scott *et al.*, 1993) that in reality the series converges (although extremely slowly) to the energy of the ground, fully symmetric state of the considered systems (for He_2 this state corresponds to a nonphysical, Pauli-forbidden solution of the Schrödinger equation). The pathologically slow convergence of the series is now well understood as a consequence of the singular behavior of the analytic function defined by the perturbation series (Cwiok *et al.*, 1992a; Kutzelnigg, 1992). The convergence to the eigenenergy can be seen clearly for smaller values of R since when R decreases, the high-order convergence rate becomes faster.

When one of the monomers has three or more electrons, the situation is quite different. Group-theoretical arguments, given by Kutzelnigg (1980), show that the polarization expansion cannot converge in such a case, at least for sufficiently large distances R. These theoretical arguments were recently confirmed by Adams in numerical calculations of the interaction energy between lithium and hydrogen atoms (Adams, 1991, 1992, 1994). The divergence of the polarization series has few consequences in applications since for many-electron systems the high-order polarization corrections cannot be calculated in practice anyway. Moreover, after the exchange corrections (see Section 1.3) are added to the polarization ones, a low-order perturbation treatment turns out to be entirely satisfactory. The divergence of the polarization series shows, however, that the concepts of infinite-order polarization or exchange energy are not well defined and that the separate polarization and exchange contributions to the interaction energy can be given a rigorous definition only in a finite order of perturbation theory.

For large intermolecular distances, independently of its divergence or pathologically slow convergence, the polarization expansion provides already through the second-order a very good approximation to the interaction energy. The reason is that several (three to six, depending on the interacting system) leading terms in the $1/R$ asymptotic expansion of the exact interaction energy (equation (1.2)) are completely recovered in the first and second order of the polarization theory. In general, each term in equation (1.2) (i.e. each constant C_n) can be obtained by an asymptotic expansion of a finite-order polarization energy (Jeziorski and Kolos, 1982). In practice this is done by evaluating equations (1.5) and (1.6) using the multipole expanded form of V:

$$V = \sum_{l_1=0}^{\infty} \sum_{l_2=0}^{\infty} V_{l_1 l_2} R^{-l_1-l_2-1} \qquad (1.8)$$

The operators $V_{l_1 l_2}$ are expressed in the most transparent way using the spherical form of the multipole moment operators:

$$Q_m^l = \sum_{p \in X} Z_p r_p^l C_m^l(\theta_p, \phi_p) \tag{1.9}$$

where the summation index p runs over all particles, both nuclei and electrons, of molecule X (X = A or X = B), Z_p are the charges of those particles, $C_m^l(\theta, \phi) = (4\pi)^{1/2}(2l+1)^{-1/2} Y_{lm}(\theta, \phi)$ are the spherical harmonics in the Racah normalization and with the Condon and Shortley phase (Brink and Satchler, 1975) and r_p, θ_p, ϕ_p are the spherical components of the vector $\mathbf{r}_p$ pointing from the center of mass of the molecule X to the particle p ($\mathbf{r}_p = \mathbf{R}_p - \mathbf{R}_X$, where $\mathbf{R}_p$ and $\mathbf{R}_X$ denote the position vectors of particle p and the center of mass of the molecule X, respectively, expressed in a space-fixed coordinate system). All vector coordinates are measured here in an arbitrary space-fixed coordinate system the same for molecules A and B. The operators Q_m^l, $m = -l, \ldots, l$ form the components of the irreducible spherical tensor $\mathbf{Q}^l$ of the rotation group SO(3). The operators V_{l_1, l_2} can be written in the form explicitly invariant under arbitrary rotations the coordinate system (Wormer 1975, Wormer et al., 1977; van der Avoird et al., 1980):

$$V_{l_1,l_2} = (-1)^{l_2} \binom{2l_1 + 2l_2}{2l_1}^{1/2} \sum_{m=-l_1-l_2}^{l_1+l_2} (-1)^m C_{-m}^{l_1+l_2}(\Theta, \Phi) [\mathbf{Q}^{l_1} \otimes \mathbf{Q}^{l_2}]_m^{l_1+l_2} \tag{1.10}$$

where R, Θ, Φ are the spherical components of the vector $\mathbf{R} = \mathbf{R}_B - \mathbf{R}_A$. By $[\mathbf{Q}^{l_1} \otimes \mathbf{Q}^{l_2}]^{l_1+l_2}$ we denote the irreducible tensor product of the 2^{l_1} pole moment tensor $\mathbf{Q}^{l_1}$ for molecule A and the 2^{l_2} pole moment tensor $\mathbf{Q}^{l_2}$ for molecule B:

$$[\mathbf{Q}^{l_1} \otimes \mathbf{Q}^{l_2}]_m^l = \sum_{m_1=-l_1}^{l_1} \sum_{m_2=-l_2}^{l_2} \langle l_1, m_1; l_2, m_2 | l, m \rangle Q_{m_1}^{l_1} Q_{m_2}^{l_2} \tag{1.11}$$

where $\langle l_1, m_1; l_2, m_2 | l, m \rangle$ are the Clebsch–Gordan coefficients (Brink and Satchler, 1975). Note that the summation in equation (1.10) represents the Clebsch–Gordan coupling of the tensors $\mathbf{C}^{l_1+l_2}(\Theta, \Phi)$ and $[\mathbf{Q}^{l_1} \otimes \mathbf{Q}^{l_2}]^{l_1+l_2}$ to a scalar. The consecutive terms in equation (1.8) can be interpreted as interactions of permanent and instantaneous 2^{l_1} moments on monomer A with the permanent and instantaneous 2^{l_2} moments on monomer B. The multipole expansion of V converges pointwise when $R > r_1 + r_2$ (for necessary convergence conditions see Stolarczyk and Piela, 1979) but its possible divergence in other parts of the configuration space is irrelevant for the asymptotic analysis of polarization energies (Ahlrichs, 1976). The resulting asymptotic expansion of a polarization correction

$$E_{\text{pol}}^{(k)} \sim \sum_{n=1}^{\infty} \frac{C_n^{(k)}}{R^n} \tag{1.12}$$

is usually referred to as its multipole expansion. Note that for interactions of molecules the coefficients $C_n^{(k)}$ depend on the set of angles determining the relative orientation of the interacting systems. This angular dependence can always be expressed in a closed form (van der Avoird et al., 1980; Jeziorski et al., 1994—note that for nonlinear monomers the formulas for $C_n^{(2)}$ given in these references contain an erroneous phase factor corrected for recently by Heijmen et al., 1996). For low values of k the coefficients

$C_n^{(k)}$ are easier to compute than the complete correction $E_{\text{pol}}^{(k)}$. They can also be related to a monomer's properties such as multipole moments and static and dynamic polarizabilities, enabling a further insight into the physics of the interaction and providing a guidance in construction of approximate potentials for larger systems. Efficient methods of determining the coefficients $C_n^{(k)}$ have been developed (see Jeziorski et al., 1994, for references) and for low values of n and k these coefficients are available either from *ab initio* calculations or from semiempirical work (Maitland et al., 1981).

The transition from the left- to the right-hand side of equation (1.12) is usually referred to as the neglect of the penetration (or charge-overlap) effects. The multipole expansion for $E_{\text{pol}}^{(k)}$ is divergent for any R (except possibly for $k=1$). Therefore, the role of the penetration effects is not only to give some additive short-range (exponentially decaying) correction to the series on the right-hand side of equation (1.12) but first of all to provide damping factors $f_n^{(k)}$, which after multiplication by $C_n^{(k)}$ would transform the divergent series into a convergent one. This additive short-range correction, referred sometimes to as 'spherical term', and the damping factors $f_n^{(k)}$ can be computed directly using the so-called bipolar expansion of V (Ng et al., 1976, 1977, 1979b; Knowles and Meath, 1986, 1987). If the polarization energies are computed without the multipole expansion of V, then approximate values of the damping factors and the additive short-range correction ('spherical term') can be obtained by fitting. A theoretically well-motivated and practically very efficient form for the damping factors has been proposed by Tang and Toennies (1984):

$$f_n^{(k)} = 1 - \exp(-\beta^{(k)} R) \sum_{m=0}^{n} \frac{(\beta^{(k)} R)^m}{m!} \tag{1.13}$$

In general, the fitting parameters $\beta^{(k)}$ do depend on the mutual orientations and internal geometries of the monomers, although this dependence is often weak and can be neglected. The 'spherical term' is usually assumed in the Born–Mayer form Ae^{-bR}.

The damped multipole expansion supplemented with the exponentially decaying 'spherical term' is very useful as an analytical representation of the polarization energies calculated with the complete form of V. The knowledge of coefficients $C_n^{(k)}$ from an independent direct calculation is important for the construction of analytical potential energy surfaces since it reduces the number of fitted parameters. The knowledge of a few first constants (usually through C_{10}) is sufficient for most purposes. The role of the penetration effects grows as R decreases and at the van der Waals minimum their neglect would lead in most cases to quantitatively incorrect representation of the polarization contributions to the interaction energy.

1.2.2 Symmetry-adapted Perturbation Theory

Although low-order calculations using the polarization expansion are very accurate for large intermolecular distances (much larger than the van der Waals minimum distance), at shorter distances the results of such calculations are qualitatively incorrect since no minimum and no repulsive wall appear. This is due to the fact that the polarization series is not able (in low order) to recover the repulsive contributions to the intermolecular potential. These contributions are due to the electron exchange, i.e. to the physical process of the (resonance) tunneling of electrons between interacting systems. The unperturbed function $\Psi_0 = \Psi_A \Psi_B$ as well as higher-order polarization functions describe the situation when the electrons stay within their monomers but the permanent and instantaneous polarization is allowed (therefore the name 'polarization expansion'). The true wave function contains also components corresponding to tunneling of one, two or more electron pairs between interacting units. These additional components are neglected in the polarization approximation.

Another point of view is that the zeroth-order function, written more explicitly as

$$\Psi_0 = \Psi_A(1, \ldots, N_A)\Psi_B(N_A + 1, \ldots, N_A + N_B) \qquad (1.14)$$

where N_A (N_B) is the number of electrons in system A (B), does not possess the correct permutational symmetry, i.e. does not fulfill the Pauli principle. This function is clearly not antisymmetric under the exchange of electrons between systems A and B. The zeroth-order function of correct symmetry is, of course, $\mathscr{A} \Psi_0$, where $\mathscr{A}$ is the total N-electron antisymmetrization operator (antisymmetrizer). The use of such a zeroth-order function in Rayleigh–Schrödinger perturbation theory is not possible since it is not an eigenfunction of H_0. The perturbation formalisms aimed at resolving this difficulty are collectively referred to as the symmetry-adapted perturbation theory. Numerous such formalisms have been proposed thus far (see Jeziorski *et al.*, 1994; Jeziorski and Kolos, 1982; Kaplan, 1987, for more comprehensive reviews). The simplest of these is the so-called Symmetrized Rayleigh–Schrödinger (SRS) theory (Jeziorski *et al.*, 1978; Cwiok *et al.*, 1992b) which extends the idea of symmetrization of Ψ_0 to higher-order functions of the polarization expansion (see also Chipman and Hirschfelder, 1973, and Chipman *et al.*, 1973, for an early formulation of a similar idea). This means that the wave function corrections in SRS are exactly the same as in the polarization theory. The energy corrections are computed from the following expression (Jeziorski *et al.*, 1978):

$$E_{SRS}^{(n)} = N_0 \left(\langle \Psi_0 | V \mathscr{A} \Psi_{pol}^{(n-1)} \rangle - \sum_{k=1}^{n-1} E_{SRS}^{(k)} \langle \Psi_0 | \mathscr{A} \Psi_{pol}^{(n-k)} \rangle \right) \qquad (1.15)$$

where $N_0 = \langle \Psi_0 | \mathscr{A} \Psi_0 \rangle$. In other, more complicated, formulations of SAPT the symmetry projection operators appear in the perturbation equations. The perturbed wave functions contain then some exchange contributions which reflect the so-called exchange deformation of the wave function. The simplest of such formulations, asymptotically consistent with London theory, has been proposed by Hirschfelder and Silbey (1966). Calculations for one- and two-electron systems have shown (Chalasinski *et al.*, 1977; Cwiok *et al.*, 1992b, 1994) that the high-order convergence of the Hirschfelder–Silbey (HS) expansion is considerably faster than the convergence of the SRS method. Unfortunately, without further simplifications the HS theory cannot be applied in practice to many-electron systems. Recently, Korona *et al.* (in press) compared the SRS and HS approaches in large-order calculations for the interaction of two helium atoms. It turned out that the low-order results (up to the tenth order) are virtually identical in both methods. The convergence of the HS theory becomes superior only in higher orders and at very short interatomic separations. This means that the effects of the exchange deformation are negligible in low orders of the HS theory. Thus, despite a dramatic increase in complexity, the HS theory is not an essential improvement over the SRS approach.

When one of the monomers has more than two electrons, the SRS series, like the polarization expansion, cannot be convergent in the conventional sense (Adams, 1992,

1994, 1996). We believe, however, that as for the He$_2$ system, the low-order SRS corrections are excellent approximations to corrections of a much more complicated, convergent expansion, accounting for the exchange deformation. One may also note that the antisymmetrized finite-order polarization function used in the SRS theory provides the correct large R asymptotic expansion for the exact wave function (Ahlrichs, 1976). Thus, this wave function includes asymptotically all components due to the electron exchange and, when used in an appropriate energy expression, can give a very good approximation for the exchange contribution to the interaction energy. On the basis of a large body of numerical evidence, gathered also for large, many-electron systems, we believe that the SRS theory through second order correctly accounts for the main exchange contributions to the interaction energy and gives results which are accurate to within a few percent in the van der Waals minimum region. Since the only SAPT approach we will consider from now on will be the SRS expansion, the SRS subscript will be dropped in further discussion.

Because of the close relation to the polarization theory, SAPT provides a transparent physical interpretation of the interaction energy. In each order the SRS correction $E^{(n)}$ can be separated into the polarization and exchange parts:

$$E^{(n)} = E^{(n)}_{\text{pol}} + E^{(n)}_{\text{exch}} \tag{1.16}$$

In low-order we get the familiar components: $E^{(1)}_{\text{pol}}$—the electrostatic energy; $E^{(1)}_{\text{exch}}$—leading exchange contribution (practically the same as predicted by the familiar Heitler–London theory); $E^{(2)}_{\text{pol}} = E^{(2)}_{\text{ind}} + E^{(2)}_{\text{disp}}$—induction and dispersion energies; $E^{(2)}_{\text{exch}} = E^{(2)}_{\text{exch-ind}} + E^{(2)}_{\text{exch-disp}}$—*exchange–induction* and *exchange–dispersion* components.

For intermonomer separations corresponding to van der Waals minima the evaluation of the exchange corrections can be dramatically simplified by neglecting higher than single-electron exchanges between monomers. The simplified, approximate expressions for $E^{(n)}_{\text{exch}}$ are then quadratic in the intermolecular overlap densities $\rho_{\lambda\mu}(\mathbf{r}) = \phi_\lambda(\mathbf{r})\phi_\mu(\mathbf{r})$ (and intermolecular overlap integrals $S_{\lambda\mu} = \langle\phi_\lambda|\phi_\mu\rangle$) and are accordingly denoted by $E^{(n)}_{\text{exch}}(S^2)$ (ϕ_λ and ϕ_μ stand for the orbitals used to build the wave functions for monomers A and B, respectively). In the first and second order these approximate expressions take the form

$$E^{(1)}_{\text{exch}}(S^2) = -\langle\Psi_0|V - \langle V\rangle)\mathscr{P}\Psi_0\rangle \tag{1.17}$$

$$E^{(2)}_{\text{exch}}(S^2) = -\langle\Psi_0|(V - \langle V\rangle)(\mathscr{P} - \langle\mathscr{P}\rangle)\Psi^{(1)}_{\text{pol}}\rangle \tag{1.18}$$

where $\langle V\rangle = \langle\Psi_0|V\Psi_0\rangle$, $\langle\mathscr{P}\rangle = \langle\Psi_0|\mathscr{P}\Psi_0\rangle$, and $\mathscr{P}$ denotes the sum of all $N_A N_B$ operators performing transpositions of electrons between monomers. Equations (1.17) and (1.18) represent a very good approximation to the accurate values of $E^{(1)}_{\text{exch}}$ and $E^{(2)}_{\text{exch}}$ since their error is of the fourth order in the intermonomer overlap densities. Since the first-order polarization function $\Psi^{(1)}_{\text{pol}}$ is the sum of the induction and dispersion components, $\Psi^{(1)}_{\text{pol}} = \Psi^{(1)}_{\text{ind}} + \Psi^{(1)}_{\text{disp}}$, the expressions for the exchange–induction and exchange–dispersion contributions (in the S^2 approximations) can be obtained from equation (1.18) by replacing $\Psi^{(1)}_{\text{pol}}$ by $\Psi^{(1)}_{\text{ind}}$ and $\Psi^{(1)}_{\text{disp}}$, respectively.

The exchange–induction and the exchange–dispersion energies appear since the induction and dispersion contributions to the wave function are antisymmetrized in the second-order energy expression. In the short-range part of the intermolecular potential the exchange–induction energy quenches a substantial part of the induction contribution.

The exchange–dispersion energy, which is a pure correlation effect, is less important but cannot be neglected in any quantitatively accurate calculation.

1.3 INTRAMONOMER CORRELATION PROBLEM

1.3.1 Double-perturbation Theory Approach

Since for many-electron systems monomers' wave functions are not known (i.e. Ψ_0 not available), we have to use Hartree–Fock determinants Φ_X as the best practical approximations, and account for the intramonomer correlation contribution in a perturbative way, i.e. to use a double-perturbation approach. To this end Møller–Plesset partitioning is used to decompose the monomer Hamiltonians H_A and H_B into the Fock operator F_X, $X = A$ or $X = B$, and the intramonomer correlation operator $W_X = H_X - F_X$. The total Hamiltonian decomposes then as $H = F + V + W$, where $F = F_A + F_B$ is the new unperturbed operator, $W = W_A + W_B$ is the operator accounting for intramonomer correlation effects, and V is the (nonexpanded) intermolecular interaction operator considered in the previous sections. The zeroth-order function is now given by the product of the Hartree–Fock determinants:

$$\Phi_0 = \Phi_A \Phi_B \tag{1.19}$$

The double-polarization expansion for the wave function and the interaction energy is obtained by Taylor expansion around $\lambda = 0$ and $\mu = 0$ of the functions $E(\lambda, \mu)$ and $\Phi(\lambda, \mu)$ defined for each λ and μ by the equation

$$(F + \lambda V + \mu W)\Phi(\lambda, \mu) = E(\lambda, \mu)\Phi(\lambda, \mu) \tag{1.20}$$

The corrections of the double SRS theory are defined by an analogous Taylor expansion of the function

$$\mathscr{E}(\lambda, \mu) = \frac{\langle \Phi(0, \mu) | \lambda V \mathscr{A} \Phi(\lambda, \mu) \rangle}{\langle \Phi(0, \mu) | \mathscr{A} \Phi(\lambda, \mu) \rangle} \tag{1.21}$$

The resulting double SAPT series for the interaction energy can be written as

$$E_{\text{int}} = \sum_{n=1}^{\infty} \sum_{i=0}^{\infty} [E_{\text{pol}}^{(ni)} + E_{\text{exch}}^{(ni)}] \tag{1.22}$$

where n denotes the order with respect to V and i the order with respect to W.

If a finite orbital basis set is used in the calculations, the double SAPT corrections of equation (1.22) can be expressed in terms of one- and two-electron integrals and the orbital energies. The resulting expressions are to a large degree similar to the expressions of the standard many-body perturbation theory (MBPT). If they are derived using the familiar Slater–Condon rules to evaluate many-electron matrix elements then numerous disconnected (not size-extensive) terms appear that must be cancelled out. In standard MBPT these cancellations are taken care of automatically by using diagrammatic techniques. Since no diagrammatic approach is available for the evaluation of exchange corrections, purely algebraic techniques have been developed and applied to derive directly the explicitly connected expressions for SAPT corrections of both the polarization and exchange type. These techniques, described in detail by Jeziorski *et al.* (1989),

Rybak *et al.* (1991) and Moszynski *et al.* (1993a, 1994c), are based on the ideas of the coupled cluster theory. Their main advantage is that they lead to very efficient coupled-cluster-type infinite-order partial resummations of the perturbation expansion in W (Moszynski *et al.*, 1993a; Williams *et al.*, 1995b; Williams, 1995). Due to the explicitly connected nature of the developed formalism, the resulting approximations are always size extensive.

1.3.1.1 Relation to Hartree–Fock Theory

Double SAPT corrections of lowest-order can be easily identified as describing either the interactions at the Hartree–Fock level or as representing an effect of the electron correlation. The relation between the Hartree–Fock interaction energy $E_{\text{int}}^{\text{HF}}$ (defined using the supermolecular method) and the SAPT theory is not straightforward and has been the subject of extensive investigations (Jeziorski and van Hemert, 1976; Chalasinski and Jeziorski, 1976; Chalasinski, 1983; Jaszunski, 1980; Gutowski *et al.*, 1983, Jeziorska *et al.*, 1987, Moszynski *et al.*, in press). Initially, it has been found that the corrections $E_{\text{pol}}^{(10)}$, $E_{\text{exch}}^{(10)}$, $E_{\text{ind}}^{(20)}$, and $E_{\text{exch-ind}}^{(20)}$ contribute to the Hartree–Fock interaction energy $E_{\text{int}}^{\text{HF}}$ (Jeziorski and van Hemert, 1976, Chalasinski and Jeziorski, 1976; Chalasinski, 1983). Sadlej (1980) pointed out, however, that the induction and exchange–induction contributions at the Hartree–Fock level should include the coupled Hartree–Fock (CHF) response of the perturbed system. These contributions, denoted by $E_{\text{ind,resp}}^{(20)}$ and $E_{\text{exch-ind,resp}}^{(20)}$ and defined explicitly by Jaszunski (1980), include entirely the correction $E_{\text{ind}}^{(21)}$ and partially all higher-order (in W) induction and exchange–induction corrections. The terms included in $E_{\text{ind,resp}}^{(20)}$ and $E_{\text{exch-ind,resp}}^{(20)}$ which are not of the zeroth order in W are sometimes referred to as the *apparent* correlation terms. The computation of the induction contributions with the inclusion of the CHF response corresponds to an infinite-order partial resummation of the series (22).

In some cases, especially for interactions involving ions, the sum of the first- and second-order contributions $E_{\text{pol}}^{(10)} + E_{\text{exch}}^{(10)} + E_{\text{ind,resp}}^{(20)} + E_{\text{exch-ind,resp}}^{(20)}$ deviates by 10 per cent or more from the accurate value of $E_{\text{int}}^{\text{HF}}$. As has been observed by Gutowski *et al.* (1983), this effect can be attributed to the fact that the exchange corrections considered before ($E_{\text{exch}}^{(10)}$ and $E_{\text{exch-ind,resp}}^{(20)}$) do not take account of the exchange deformation of orbitals. In fact, a perturbation expansion can converge to the Hartree–Fock interaction energy $E_{\text{int}}^{\text{HF}}$ only if the wave function corrections are computed with a full account of the exchange effects (notice that SAPT expansions of the Hartree–Fock interaction energy are distinct from expansions of the total interaction energy even if these expansions bear the same name). This has been demonstrated initially for He$_2$ (Jeziorska *et al.*, 1987). Very recently the method of Jeziorska *et al.* (1987) has been extended to many-electron monomers (Moszynski *et al.*, in press) and it has been shown that for a wide range of investigated systems the SAPT expansion through the second (third) order with respect to V recovers the HF interaction energy to within about 2–15 per cent (1–5 per cent).

Up to the second-order in V one can write then

$$E_{\text{int}}^{\text{HF}} = E_{\text{pol}}^{(10)} + E_{\text{exch}}^{(10)} + E_{\text{ind,resp}}^{(20)} + E_{\text{exch-ind,resp}}^{(20)} + E_{\text{exch-def,HF}}^{(20)} + \cdots \quad (1.23)$$

where $E_{\text{exch-def,HF}}^{(20)}$ denotes the second-order contribution from the Hartree–Fock exchange deformation and the dots stand for the third- and higher-order induction, exchange–

induction and exchange deformation terms. (Equation (1.23) neglects some very small terms of the order of S^4—these are terms of the zeroth-order in V appearing in the Heitler–London interaction energy, Jeziorski et al., 1976.) Nowadays the Hartree–Fock interaction energy can be computed very accurately and inexpensively so a calculation of all SAPT corrections shown on the right-hand side of equation (1.23) is not really necessary and one can represent the interaction energy as

$$E_{\text{int}} = E_{\text{int}}^{\text{HF}} + E_{\text{int}}^{\text{CORR}} \quad (1.24)$$

and approximate $E_{\text{int}}^{\text{CORR}}$ by SAPT components

$$E_{\text{int}}^{\text{CORR}} = E_{\text{disp}} + E_{\text{elst}}^{\text{CORR}} + E_{\text{ind}}^{\text{CORR}} + E_{\text{exch}}^{\text{CORR}} \quad (1.25)$$

where E_{disp}, $E_{\text{elst}}^{\text{CORR}}$, $E_{\text{ind}}^{\text{CORR}}$, and $E_{\text{exch}}^{\text{CORR}}$ stand for the sum of SAPT contributions representing the dispersion interaction and the correlation contributions to the electrostatic, exchange, and induction energies, respectively. The dispersion contribution E_{disp} does not have to carry the CORR superscript since there are no dispersion contributions at the Hartree–Fock level.

1.3.1.2 Dispersion Interaction

We shall now briefly discuss the present state of the theory and computer codes available for the calculation of the successive terms in equation (1.25). By far the most important term on the right-hand side of equation (1.25) is the dispersion contribution E_{disp}. The simplest possible approximation to E_{disp} corresponds to the complete neglect of the effects of intramonomer electron correlation and is given by the correction of the zeroth-order in W: $E_{\text{disp}}^{(20)}$. After expansion in terms of two-electron integrals and orbital energies this correction takes the form (Jeziorski and van Hemert, 1976)

$$E_{\text{disp}}^{(20)} = 4 v_{ab}^{rs} v_{rs}^{ab} / \varepsilon_{rs}^{ab} \quad (1.26)$$

where v_{ij}^{kl} are two-electron integrals over monomer (spin-free) orbitals ϕ_k

$$v_{ij}^{kl} = \langle \phi_i(1)\phi_j(2) | r_{12}^{-1} | \phi_k(1)\phi_l(2) \rangle \quad (1.27)$$

and $\varepsilon_{ij}^{kl} \equiv \varepsilon_k + \varepsilon_l - \varepsilon_i - \varepsilon_j$ are differences of the appropriate orbital energies. In equation (1.26) and in subsequent equations unrestricted summation over repeated upper and low indices will be implied. We shall also always assume that the indices a and r (b and s) refer to occupied and virtual orbitals of system A (B), respectively. This convention includes all prime superscripting and numerical subscripting of these labels.

The correction $E_{\text{disp}}^{(20)}$, corresponding to the dispersion interaction of monomers described at the Hartree–Fock level of approximation and sometimes referred to as the 'Hartree–Fock dispersion energy', gives usually quite good, semi-quantitative representation of the total dispersion interaction. As shown by Jeziorski and van Hemert (1976), expression (1.26) results from the minimization of an appropriate functional. Due to this variational property, approximate, finite basis set computations converge from above to the accurate value of $E_{\text{disp}}^{(20)}$. The variational nature of $E_{\text{disp}}^{(20)}$ enables a dispersion-based optimization of orbital exponents, which is very important in view of relatively slow convergence of this correction with the increase of orbital basis sets (Williams et al., 1995a).

The leading intramonomer correlation correction to $E_{\text{disp}}^{(20)}$ is linear in W and is given by

$E_{\text{disp}}^{(21)}$. The explicit orbital expression for $E_{\text{disp}}^{(21)}$ has been derived by Jeziorski et al. (1989) and takes the form

$$E_{\text{disp}}^{(21)} = 4 t_{ab}^{rs} g_{ra'}^{ar'} t_{r's}^{a'b} + 8 \theta_{rr'}^{aa'} v_{a's}^{r'b} t_{ab}^{rs} + (A \leftrightarrow B) \tag{1.28}$$

where $t_{rs}^{ab} \equiv v_{rs}^{ab}/\varepsilon_{rs}^{ab}$, $t_{rr'}^{aa'} \equiv v_{rr'}^{aa'}/\varepsilon_{rr'}^{aa'}$, $t_{ab}^{rs} \equiv t_{rs}^{ab}$, $g_{ij}^{kl} \equiv 2v_{ij}^{kl} - v_{ij}^{lk}$, $\theta_{rr'}^{aa'} \equiv 2t_{rr'}^{aa'} - t_{rr'}^{a'a}$, and (A $\leftrightarrow$ B) denotes the preceding expression in which orbitals of monomers A and B are interchanged. In equation (1.28) and in further equations we assume that the orbitals are real. As shown by Chalasinski and Szczesniak (1988), the dispersion corrections $E_{\text{disp}}^{(20)}$ and $E_{\text{disp}}^{(21)}$ are taken into account in supermolecular calculations at the level of the second and third order, respectively, of the supermolecular (i.e. conventional) many-body perturbation theory. The inclusion of the correction $E_{\text{disp}}^{(21)}$ is usually not a major improvement over $E_{\text{disp}}^{(20)} + E_{\text{disp}}^{(21)}$ calculations. It is thus recommended that the intramonomer correlation contributions to the dispersion energy are calculated at least through the second order in W. The explicit orbital expression for $E_{\text{disp}}^{(22)}$ has been derived by Rybak et al. (1991) but is too complicated to be reported here.

If Feynman-type diagrams are used to represent individual perturbation theory terms, the corrections $E_{\text{disp}}^{(20)}$ and $E_{\text{disp}}^{(21)}$ correspond to simplest ring diagrams. By performing an infinite order summation of selected (Rybak et al., 1987) or all ring diagrams (Moszynski et al., 1993a) one obtains an improved approximation to the dispersion energy. Still better approximation is obtained by summing all coupled-pair (CCD) diagrams (Williams et al., 1995b). The resulting approximation to $E_{\text{disp}}^{(2)}$, denoted by $E_{\text{disp}}^{(2)}(\text{CCD})$, does not take account of the contributions from single and triple excitations. These contributions, denoted by $E_{\text{disp}}^{(22)}(S)$ and $E_{\text{disp}}^{(22)}(T)$, respectively, are included in $E_{\text{disp}}^{(22)}$ and should be added to $E_{\text{disp}}^{(2)}(\text{CCD})$. An even better approach is to calculate the expressions for $E_{\text{disp}}^{(22)}(S)$ and $E_{\text{disp}}^{(22)}(T)$ with the converged CCD amplitudes in the spirit of the well-known CCSD(T) method (Williams et al., 1995b). The resulting infinite-order (in W) contribution from single and triple excitations is denoted by $E_{\text{disp}}^{(2)}[\text{ST(CCD)}]$. The best currently coded approximation to the energy of the dispersion interaction is given by the sum

$$E_{\text{disp}}^{(2)} \approx E_{\text{disp}}^{(2)}(\text{CCD}) + E_{\text{disp}}^{(2)}[\text{ST(CCD)}]$$

It should be noted, however, that the effect of the infinite order summation (beyond second order in W) of intramonomer correlation contributions included in the above equation is not large. While the contribution of the infinite summation at the CCD level is quite substantial, this effect cancels to a large extent with the change produced by ST(CCD) and the final result is very close to the perturbation expansion through the second order in W, i.e. to $E_{\text{disp}}^{(20)} + E_{\text{disp}}^{(21)} + E_{\text{disp}}^{(22)}$.

For the interaction of two two-electron system a specific code has been developed (Korona et al., in press) which computes $E_{\text{disp}}^{(2)}$ exactly at the FCI level (without an expansion in powers of W). This code has been used to test the approximations discussed above and to provide the best current value of the dispersion energy for the helium dimer (Williams et al., 1996; Korona et al., submitted).

1.3.1.3 Electrostatic Correlation

The simplest possible approximation to the electrostatic correlation term $E_{\text{elst}}^{\text{CORR}}$ is given by the correction $E_{\text{pol}}^{(12)}$ (the contribution of the first order in W vanishes on account of

Brillouin theorem). This correction, derived by Jeziorski et al. (1989), can be expressed as

$$E_{\text{pol}}^{(12)} = 4(\omega_B)_a^r t_r^a + 2\theta_{rr'}^{aa'}(\omega_B)_{r'}^{r'} t_{aa'}^{rr''} - 2\theta_{rr'}^{aa'}(\omega_B)_{a'}^{a''} t_{aa''}^{rr'} + (A \leftrightarrow B) \qquad (1.29)$$

where $(\omega_B)_k^l = (v_B)_k^l + 2v_{bk}^{bl}$ are matrix elements of the complete electrostatic potential of molecule B calculated in the Hartree–Fock approximation ($(v_B)_k^l$ denotes matrix elements of the electrostatic potential of nuclei from molecule B), $t_{aa'}^{rr'} \equiv t_{rr'}^{aa'}$, and t_r^a are the second-order single excitation amplitudes defined by

$$t_r^a = [\theta_{r''r'}^{a'a} v_{a'r}^{r''r'} - \theta_{r'r}^{a''a'} v_{a''a'}^{r'a}]/\epsilon_r^a \qquad (1.30)$$

It appears that the convergence of the electrostatic correlation energy $E_{\text{elst}}^{\text{CORR}}$ in powers of W is rather slow and the leading term in this expansion, given by $E_{\text{pol}}^{(12)}$, provides only qualitative approximation to $E_{\text{elst}}^{\text{CORR}}$. To obtain quantitative accuracy for this (usually small) correlation term one has to include at least the third-order contribution $E_{\text{pol}}^{(13)}$ and in some cases also the fourth-order one $E_{\text{pol}}^{(14)}$ (Moszynski et al., 1993b). The explicit orbital expressions for these corrections have been derived by Moszynski et al. (1993b). Significantly faster convergence is obtained if the successive intramonomer correlation corrections are calculated with the inclusion of the coupled-Hartree–Fock response of the monomers (Moszynski et al., 1991, 1993b). Calculation of such corrections, denoted by $E_{\text{pol,resp}}^{(1i)}$, may be viewed as a partial infinite-order resummation of the series (22) for $n = 1$. It has been shown (Chalasinski et al., 1990, 1994; Chalasinski and Szczesniak, 1994) that the corrections $E_{\text{pol,resp}}^{(1i)}$, $i = 2, 3$, are included in the supermolecular MBPT interaction energy of ith order while the correction $E_{\text{pol,resp}}^{(14)}$ contributes both to the fourth- and the fifth-order MBPT energy. When the electrostatic effects dominate the interaction energy asymptotically then the corrections $E_{\text{pol,resp}}^{(1i)}$, $i = 2, 3$, and the main part of the correction $E_{\text{pol,resp}}^{(14)}$ determine the asymptotic behavior of the supermolecular MBPT energy through the fourth order.

Explicit expressions for the corrections $E_{\text{pol,resp}}^{(1i)}$, $i = 2, 3, 4$, have been given by Moszynski et al. (1993b). The corrections $E_{\text{pol,resp}}^{(14)}$ and $E_{\text{pol}}^{(14)}$ include the effect of triple excitations and are very time-consuming to evaluate. SAPT codes currently available for distribution do not include these corrections and approximate the electrostatic correlation as

$$E_{\text{elst}}^{\text{CORR}} = E_{\text{pol,resp}}^{(12)} + E_{\text{pol,resp}}^{(13)} \qquad (1.31)$$

1.3.1.4 Induction Correlation

Since the induction correction of the first order in W is entirely contained in $E_{\text{ind,resp}}^{(20)}$ (and consequently in the Hartree–Fock interaction energy), the leading intramonomer correlation contribution to the induction energy is included in the correction $E_{\text{ind}}^{(22)}$. The expression for the latter correction contains also some terms (diagrams) present in $E_{\text{ind,resp}}^{(20)}$ (so-called apparent correlation terms). These terms have to be eliminated if the true correlation contribution ${}^t E_{\text{ind}}^{(22)}$ is to be found. This elimination can be achieved either diagrammatically or by removing response terms from the induction energy $E_{\text{ind,resp}}^{(22)}$ calculated with response effects (Moszynski et al., 1994f), since the latter by definition is free from a Hartree–Fock component. The resulting expression for ${}^t E_{\text{ind}}^{(22)}$ (Jeziorski et al., unpublished research report) can most conveniently be written as ${}^t E_{\text{ind}}^{(22)} = {}^t E_{\text{ind}}^{(220)} + {}^t E_{\text{ind}}^{(202)}$, where the second and third superscripts denote the orders with respect to the operators W_A and W_B, respectively. The component ${}^tE_{\text{ind}}^{(220)}$ can be split into terms corresponding to the electrostatic potential of monomer B inducing moments on A and vice versa:

$${}^tE_{\text{ind}}^{(220)} = {}^tE_{\text{ind}}^{(220)}(A \leftarrow B) + {}^tE_{\text{ind}}^{(220)}(A \rightarrow B)$$

which are given by

$${}^tE_{\text{ind}}^{(220)}(A \leftarrow B) = (2X_{aa'}^{rr'} - X_{aa'}^{r'r})X_{rr'}^{aa'}/\varepsilon_{rr'}^{aa'} + 4t_a^r z_r^a - 2\theta_{rr'}^{aa'}I_{r''}^{a''}(\omega_{a''}^{r'}t_{a'a}^{r''r} + \omega_{a'}^{r''}t_{a''a}^{r'r})$$
$$+ 2I_a^r I_{a'}^{r'}(\theta_{rr'}^{aa'}[2]\varepsilon_{rr'}^{aa'} - 2g_{ra}^{ar''}\theta_{r''r'}^{a''a'}) - 2\theta_{aa'}^{rr'}I_{a''}^{r''}(I_{r'}^{a'}v_{r''r}^{a''a} + I_{r'}^{a''}v_{r''r}^{a'a}) \quad (1.32)$$

where

$$X_{rr'}^{aa'} = \omega_r^{r''}t_{r''r'}^{aa'} + \omega_{r'}^{r''}t_{rr''}^{aa'} - \omega_{a''}^a t_{rr'}^{a''a'} - \omega_{a''}^{a'}t_{rr'}^{aa''} + v_{rr'}^{ar'}I_{r''}^{a'} + v_{rr'}^{r''a'}I_a^{a''} - v_{ra''}^{aa'}I_{r'}^{a''} - v_{a''r'}^{aa'}I_r^{a''} \quad (1.33)$$

$$z_r^a = \omega_r^{r'}I_{r'}^a - \omega_{a'}^a I_r^{a'} \quad (1.34)$$

$X_{aa'}^{rr'} \equiv X_{rr'}^{aa'}, t_a^r \equiv t_r^a, I_a^r \equiv (\omega_B)_r^a/\varepsilon_r^a, I_a^r \equiv I_r^a, \theta_{rr'}^{aa'}[2] \equiv 2t_{rr'}^{aa'}[2] - t_{rr'}^{a'a}[2]$ and $t_{rr'}^{aa'}[2]$ are the second-order cluster amplitudes defined as in Moszynski *et al.* (1993b) or in Williams *et al.* (1995b):

$$t_{rr'}^{aa'}[2] = (v_{rr'}^{r''r'''}t_{r''r'''}^{aa'} + v_{a''a'''}^{aa'}t_{rr'}^{a''a'''} + v_{r'a''}^{a'r'''}\theta_{rr''}^{aa''} + v_{ra''}^{ar'''}\theta_{r'r''}^{a'a''}$$
$$- v_{a''r'}^{a'r'''}t_{rr''}^{aa''} - v_{a''r}^{ar'''}t_{r'r''}^{a'a''} - v_{a''r'}^{ar'''}t_{rr''}^{a'a''} - v_{a''r}^{a'r'''}t_{r'r''}^{aa''})/\varepsilon_{rr'}^{aa'}$$

and

$${}^tE_{\text{ind}}^{(220)}(A \rightarrow B) = -8\theta_{rr'}^{a'a'}t_{a''a}^{rr'}v_{a'b}^{as}I_s^b + 8\theta_{r''r}^{aa'}t_{aa'}^{rr''r'}v_{r'b}^{rs}I_s^b + 16t_a^r v_{rb}^{as}I_s^b \quad (1.35)$$

The potential ω in equations (1.32)–(1.34) refers to molecule B and we dropped the subscript B only to simplify the notation. The corrections ${}^tE_{\text{ind}}^{(202)}(A \leftarrow B)$ and ${}^tE_{\text{ind}}^{(202)}(A \rightarrow B)$ can be obtained from the expressions given above by interchanging orbitals of monomer A with those of monomer B. The induction correlation correction $E_{\text{ind}}^{\text{CORR}}$ is usually quite small (Moszynski *et al.*, 1994f) and it is expected that it can be approximated with sufficient accuracy by ${}^tE_{\text{ind}}^{(22)}$.

1.3.1.5 Exchange Correlation

The theory of the intramonomer correlation effects in the first-order exchange interaction has been worked out by Moszynski *et al.* (1994c,d). The correlation effect appears already in the first order in W but the correction $E_{\text{exch}}^{(11)}$ is usually small and the main intramonomer correlation effect is of the second order in W. The explicit orbital expressions for $E_{\text{exch}}^{(11)}$ and $E_{\text{exch}}^{(12)}$ have been given by Moszynski *et al.* (1994c,d). Two completely different, though numerically equivalent, forms of these expressions have been derived and coded to assure the correctness of the results. The comparison with full CI results has shown that the convergence of the exchange correlation as a series in W is rather slow (Moszynski *et al.*, 1994c). This difficulty can be circumvented if the first- and second-order perturbation theory amplitudes entering the expression for $E_{\text{exch}}^{(12)}$ are replaced by the converged coupled cluster amplitudes of the coupled cluster singles and doubles (CCSD) theory (Moszynski *et al.*, 1994c). The resulting approximation to $E_{\text{exch}}^{\text{CORR}}$ will be denoted by $\epsilon_{\text{exch}}^{(1)}(\text{CCSD})$. This approximation represents the most advanced currently available method of calculating the first-order exchange repulsion energy.

The second-order exchange–dispersion energy is usually small and can be approximated with the neglect of intramonomer correlation, i.e. by the correction $E^{(20)}_{\text{exch-disp}}$. The explicit orbital expressions $E^{(20)}_{\text{exch-disp}}$ can be found in the paper by Rybak *et al.* (1991). For interactions of two-electron monomers a method to calculate the exchange-dispersion energy at the FCI level has been developed and applied to obtain highly accurate interatomic potential for the helium dimer (Williams *et al.*, 1996; Korona *et al.*, submitted).

As already mentioned, at shorter intermolecular distances the induction energy is largely quenched by the exchange–induction effect. In some situations when the induction correlation is significant (Lotrich *et al.*, 1995) it may be necessary to include the exchange quenching of $^t E^{(22)}_{\text{ind}}$. The theoretical framework to compute this correction has not been developed thus far and one may estimate its value by scaling $E^{(20)}_{\text{exch-ind,resp}}$ using the known values of $^t E^{(22)}_{\text{ind}}$ and $E^{(20)}_{\text{ind,resp}}$:

$$^t E^{(22)}_{\text{exch-ind}} \approx E^{(20)}_{\text{exch-ind,resp}} \frac{^t E^{(22)}_{\text{ind}}}{E^{(20)}_{\text{ind,resp}}} \quad (1.36)$$

The best presently available approximation to the exchange correlation contribution to the SAPT interaction energy is thus given by the formula

$$E^{\text{CORR}}_{\text{exch}} = \epsilon^{(1)}_{\text{exch}}(\text{CCSD}) + E^{(20)}_{\text{exch-disp}} + {}^t E^{(22)}_{\text{exch-ind}} \quad (1.37)$$

1.3.2 Accuracy of SAPT and Comparison to the Supermolecular Approach

Recent work provided precise estimates of the accuracy attainable with the presently developed level of SAPT. For the helium dimer comparisons can be made using almost exact values of energy components computed in bases close to saturation (using a mixed geminal plus orbital approach). Such comparisons show (see Table 1.1) that near the van der Waals minimum and for larger separations SAPT (at the currently available level) recovers about 98 per cent of the exact interaction energy (Williams *et al.*, 1996; Korona *et al.*, submitted). For the strongly repulsive separation of 4 bohr SAPT gives the error of 6.2 per cent. The reason for the increase of error is the deterioration of the convergence

Table 1.1 Comparison of SAPT with supermolecular MBPT and CCSD(T) approaches for He$_2$

R (bohr)	4.0	5.6	7.0
E_{int}	291.64	−11.059	−4.629
Coded-level SAPT	309.58	−10.843	−4.557
Difference	−17.94	−0.216	−0.072
MBPT4	296.24	−10.247	−4.424
Difference	−4.60	−0.812	−0.205
CCSD(T)	294.91	−10.668	−4.540
Difference	−3.27	−0.371	−0.089

Interaction energies are taken from Williams *et al.* (1996), Korona *et al.* (submitted), Bukowski *et al.* (1996), and from unpublished results of Bukowski *et al.* (1996) and are given in kelvins, 1 Hartree = 315 733 K. All quantities were computed using mixed orbital/geminal basis sets and are close to saturation. Rows denoted as 'difference' are differences between the total interaction energy E_{int} and a given approximation to it.

rate, which is to be expected since at smaller R the perturbation due to the interaction is becoming stronger. At $R = 4$ bohr the contribution of the corrections beyond the second order of SAPT amounts to 7.8 per cent of the interaction energy while the part of the second-order component neglected in the coded SAPT contributes 1.6 per cent (of opposite sign). It should be pointed out that for He_2 the $R = 4$ bohr point is located rather high on the repulsive wall—the interaction energy there is 26 times larger than the interaction at the van der Waals minimum. Typically, calculations of intermolecular potentials are restricted to the region where the above ratio is not larger than about 2.

Basis set saturated values of several leading SAPT corrections have been computed using Gaussian geminals (Rybak et al., 1989; Bukowski et al., 1996). Orbital calculations were able to reproduce these values to about 0.5 per cent. Thus, one may expect that the current level of SAPT may provide interaction energies with about 98 per cent accuracy. This may be difficult to accomplish for larger systems since at the present time the largest practically usable basis sets will have to be relatively less complete than those used for the helium dimer. Nevertheless, for $Ar-H_2$ and for $Ar-HF$ agreement to within 5 per cent or so with the very accurate empirical potentials has been achieved (Williams et al., 1993; Lotrich et al., 1995).

SAPT results have often been compared with $MBPTn/CC$ supermolecular energies computed in the same (or very similar) basis set ($MBPTn$ denotes results through nth order). In the region of the van der Waals minimum and for larger separations the agreement of the presently coded level of SAPT with MBPT4 or CCSD(T) is excellent provided that the supermolecular calculations are corrected for the basis set superposition error (BSSE). For example, at the van der Waals minimum of the water dimer SAPT differs from MBPT4 and CCSD(T) by 0.09 kcal/mol (1.8 per cent) and 0.07 kcal/mol (1.4 per cent), respectively (Mas and Szalewicz, 1996, and unpublished results from that work). The difference between SAPT and MBPT4 usually becomes larger (up to about 10 per cent) in the repulsive region.

A very detailed comparison of SAPT and supermolecular results can be made for the helium dimer. The relevant values taken from Williams et al. (1996) and Bukowski et al. (1996) are collected in Table 1.1. Both the SAPT and supermolecular MBPT/CC interaction energies were computed using mixed GTG/orbital basis sets discussed elsewhere in this chapter. Thus, this comparison can be viewed as a genuine comparison of methods, not biased by basis set incompleteness effects. As one can see from Table 1.1, for He_2 the CCSD(T) approach provides a significant improvement over the MBPT4 one, particularly for larger R. The errors committed using the two approaches are 1.1 per cent versus 1.6 per cent, 3.4 per cent versus 7.4 per cent and 1.9 per cent versus 4.4 per cent for $R = 4$, 5.6, and 7 bohr, respectively. For larger systems, like the water dimer example quoted above, such differences are usually less pronounced. We further observe that the presently coded level of SAPT is significantly more accurate than MBPT4 or even CCSD(T) for the van der Waals minimum and larger distances as it is in error of only 2.0 per cent and 1.6 per cent for $R = 5.6$ and 7 bohr, respectively. In the very strongly repulsive region MBPT/CC results become eventually more accurate than SAPT since there are no particular reasons why the accuracy of the former approaches should deteriorate there as is the case with SAPT. We can conclude that in the region of interest the presently coded level of SAPT should compare favorably with the supermolecular CCSD(T) calculations.

Table 1.2 shows some details of the low-order convergence of the presently coded

SAPT for He$_2$, where comparisons can be made with corrections computed without using the expansion in powers of W. All the quantities were computed in the same 147-term basis (Korona *et al.*, submitted). As one can see, at least for this system both the electrostatic and exchange first-order components are very well converged, with residual error of the total first-order energy smaller than 1 per cent. The largest truncation error appears in the dispersion component, where it amounts to 2 per cent. The induction, exchange–induction, and exchange–dispersion energies exhibit somewhat larger relative errors but the absolute value of the neglected contributions is smaller than for dispersion energy. The total error of the result obtained with the current level of the general SAPT code is 2.1 per cent compared to the nonexpanded sum of $E^{(1)}$ and $E^{(2)}$. For He$_2$ the neglected terms in the first and second order in V dominate the error, since the components of the higher orders contribute only 0.52 per cent.

The SAPT method has played a substantial role in development of the supermolecular approach. The major accomplishment was a demonstration (Korona *et al.*, in press; Williams *et al.*, 1996) that the supermolecular full configuration interaction (FCI)

Table 1.2 First- and second-order double SAPT energies (in K) for He$_2$ and their exact (nonexpanded) counterparts

$E^{(10)}_{pol}$	-1.56
$E^{(12)}_{pol,resp} + E^{(13)}_{pol,resp}$	-0.13
$E^{(10)}_{pol} + E^{(12)}_{pol,resp} + E^{(13)}_{pol,resp}$	-1.69
$E^{(1)}_{pol}$	-1.72
$E^{(10)}_{exch}$	11.25
$\epsilon^{(1)}_{exch}(CCSD)$	1.06
$E^{(10)}_{exch} + \epsilon^{(1)}_{exch}(CCSD)$	12.31
$E^{(1)}_{exch}$	12.33
$E^{(20)}_{disp}$	-17.07
$E^{(2)}_{disp}(CCD) + E^{(2)}_{disp}[ST(CCD)]$	-21.82
$E^{(2)}_{disp}$	-22.28
$E^{(20)}_{ind,resp}$	-0.25
$^tE^{(22)}_{ind}$	-0.01
$E^{(20)}_{ind,resp} + {}^tE^{(22)}_{ind}$	-0.26
$E^{(2)}_{ind}$	-0.28
$E^{(20)}_{exch-ind,resp}$	0.21
$E^{(2)}_{exch-ind}$	0.25
$E^{(20)}_{exch-disp}$	0.51
$E^{(2)}_{exch-disp}$	0.71
Double SAPT sum[a]	$-10.75\ (-1.63)$[b]
$E^{(1)} + E^{(2)}$	$-10.98\ (0.52)$[b]

Internuclear distance is 5.6 bohr. Basis set is $7s5p3d1f$ on each atom and $3s2p1d$ on bond (102 functions). The total FCI interaction energy in this basis is equal to -10.86 K.
[a] Sum of all double SAPT corrections listed above.
[b] Percentage error with respect to E^{FCI}_{int}.

calculations should remove BSSE using the Boys–Bernardi (1970) counterpoise (CP) scheme. The supermolecular method computes interaction energies by performing a calculation of the energy for the 'supermolecule', i.e. the dimer, and subtracting from it the sum of the energies of the monomers. The natural approach would be to compute the monomers' energies in their own orbital basis sets. However, in the dimer calculation the energy of each monomer is additionally lowered since they may utilize the basis of the interacting partner. This nonphysical effect is called BSSE. Boys and Bernardi (1970) proposed a solution to this problem consisting of calculating monomer energies in the complete dimer's basis set. Various other CP schemes were proposed later and until recently there was no universal agreement on which of them, if any, should be used (for extensive discussions of this subject we refer to van Duijneveldt *et al.*, 1994, and Chalasinski and Szczesniak, 1994). This resulted in difficulties in assessing the accuracy of supermolecular work: a single publication could contain several significantly different values of the interaction energy (for the same molecular geometry) differing by the applied CP, as well as basis set choices and level of theory—without a clear indication which of the numbers should be considered most reliable. Comparison of results from different papers could, of course, be even more tricky.

The numerical proof of the validity of the Boys–Bernardi CP follows from the convergence of the SAPT expansion for He_2 to the FCI interaction energy computed in the full dimer basis set, while the CP-uncorrected FCI energy is a factor of two or so larger (depending on the basis set, Korona *et al.*, in press). Since, by definition, SAPT does not contain BSSE, the CP-corrected energy does not contain it either. This numerical demonstration agrees with SAPT-based formal arguments given by van Duijneveldt *et al.* (1994). There are no rigorous proofs for other theory levels, like MBPT2 or MBPT4, but vast numerical experience indicates that the same statement holds in those cases (van Duijneveldt *et al.*, 1994; Chalasinski and Szczesniak, 1994; Mas and Szalewicz, 1996). Thus, the supermolecular studies should recommend only the CP-corrected values of the interaction energy. The size of the BSSE does not even need to be given since it does not really matter. In fact, some of the best bases exhibit very large BSSE.

A related issue is the use of automatic geometry optimizations techniques which became a part of most of quantum chemistry packages. These optimizations will always lead to a structure biased by BSSE. Automatic minimization scheme with CP correction included is possible but has not been implemented yet. Thus, to obtain the true minimum unbiased by BSSE one has to compute the potential at several points in the minimum region.

The major disadvantage of using the supermolecular method is the low interpretative value of the results. The single number produced in such calculations represents a sum of various effects with diverse radial and angular dependencies, often of opposite signs. This single number allows no physical insight into the nature of intermolecular interactions. Therefore, one of major applications of SAPT, in particular when only low-order SAPT corrections are available, is to interpret the supermolecular potentials in physical terms. In the case of the MBPT approach the basis for this interpretation was an analysis by Chalasinski and Szczesniak (1988, 1994). In general, one should apply a supermolecular method only if it adequately accounts for the essential physical components of the interaction energy in the complex considered. The comparison with SAPT allows to determine the adequate level of supermolecular theory for a given system. For example, for systems where the dispersion interaction plays a major role, one should use at least the MBPT4 theory since it contains the important triple excitation contribution to this component. In no case could SCF-level calculations be considered adequate, i.e. the MBPT2 should be the minimal acceptable level of supermolecular theory.

Another important contribution of SAPT to supermolecular calculations is the analysis of the construction of optimal basis sets. It has been clear for a long time that off-the-shelf correlation optimized bases are not adequate for computations of interaction energies (although plenty of papers applying such bases still appear each year). Investigations of basis set convergence is very difficult in the supermolecular approach due to the significant cancellation of various effects with dramatically different basis set dependencies. In contrast, individual energy components computed in SAPT describe identifiable physical interactions which in most cases require specific bases. For more discussion see Section 1.5.

1.4 SAPT TREATMENT OF THREE-BODY NONADDITIVITIES

Properties of non-metallic condensed phase are determined by intermolecular interactions. In the first approximation this phase, as well as clusters consisting of three or more molecules, can be described by the pair-wise additive interactions or 'two-body' terms. The total interaction energy may be viewed as a many-body expansion beginning with the two-body contribution (notice that the term 'many-body' has been applied so far with a different meaning, to denote methods used for many-electron systems). In general, this expansion is assumed to be rapidly convergent, but in many cases the two-body approximation is insufficient to obtain quantitative (or even qualitative) agreement with measurements and inclusion of 'pair-wise nonadditive'—first of all the three-body—interactions are necessary. Since the three-body effects are so small, in order to investigate them one has to first know very precisely the two-body effects, which has not been possible until recently.

As stated in the Introduction, our present theoretical understanding of three-body interactions is rather unsatisfactory, in particular in view of recent experimental activity in this field (Elrod and Saykally, 1994; Suhm and Nesbitt, 1995). High-resolution spectroscopy techniques are now able to identify and precisely measure transitions corresponding to trimers or tetramers. In particular studies of the Ar_2–HF and Ar_2–HCl clusters (Elrod *et al.*, 1991, 1993; Hutson *et al.*, 1990; Cooper and Hutson, 1993; Gutowsky *et al.*, 1987; McIlroy *et al.*, 1991; McIlroy and Nesbitt, 1992; Ernesti and Hutson, 1994, 1995) have shown that the three-body effects are clearly contributing to the observed transitions. At the same time, as stated by Ernesti and Hutson (1995), *for systems containing molecules there are no reliable models of nonadditive forces*. Only the three-body implementation of SAPT can provide such models. One may hope that such implementation will allow an understanding of nonadditive forces in terms of fundamental physical interactions as well as provide their accurate values.

Hutson *et al.* (1990) and Ernesti and Hutson (1994, 1995) performed dynamical calculations of the lowest rovibrational levels of Ar_2–HF and Ar_2–HCl. For these systems very accurate empirical two-body potentials exist. Hutson *et al.* have found significant discrepancies between predictions from two-body potentials and experimental results. They have also used model three-body potentials to show that their inclusion improves the agreement in most cases. Notice that in contrast to work for dimers, empirical adjustments of parameters in such three-body potentials have not yet been possible.

A large number of transitions have been measured also for some other trimers (Elrod and Saykally, 1994; Suhm and Nesbitt, 1995; Leopold *et al.*, 1994), in particular for the water trimer (Liu *et al.*, 1994) which has at the same time been the subject of very active theoretical research (see Sabo *et al.*, 1995, and references therein). There are some systems where three-body effects can show up in a more direct way than by a small distortion of dimer spectra. One example is Ar_3, which should exhibit transitions forbidden in the absence of three-body forces, but no data have been yet obtained for this system (Suhm and Nesbitt, 1995). Another example is Ar_2-CO_2 where the deviation of the red shift from the linear dependence on the number of Ar atoms in Ar_n-CO_2 is directly related to the three-body nonadditivity (Sperhac *et al.*, 1996).

There are various other phenomena exhibiting nonadditive forces. The three-body interactions affect significantly the structure and binding energies of rare gas crystals and in fact until recently this field was the major playground for the many-body work (Elrod and Saykally, 1994). The classical challenge for theory is the prediction of the observed fcc structure of the argon crystal. Until recently the most reliable theory consistently predicted the hcp structure. An implementation of the three-body SAPT described in this section shows how various interaction energy components influence the energetical ordering of those structures (Lotrich and Szalewicz, work in progress). Some bulk properties like the third virial coefficient are directly related to nonadditive effects. In general, the effective potentials fitted in MD/MC calculations of liquid phase imply very significant nonadditive component, in disagreement with existing *ab initio* data which predict the ratio of three- to two-body effects of the order of only 10 per cent.

The total interaction energy for a trimer can be divided into a sum of two-body (additive) and three-body nonadditive contributions:

$$E_{int}^{ABC} = E_{int}[2,3] + E_{int}[3,3] \quad (1.38)$$

where in general $E_{int}[K, M]$ denotes the K-body contribution to the interaction energy of M-bodies. Supermolecular methods compute $E_{int}[2, 3]$ and $E_{int}[3, 3]$ by subtraction of independently computed total energies, three in the case of the additive energies and seven in the case of the nonadditive energies. As in the dimer case, the nonadditivity computed by the supermolecular method cannot be rigorously separated into components with a clear physical interpretation, in contrast to the SAPT approach, where $E_{int}[3, 3]$ is computed directly and is naturally decomposed into physically interpretable terms.

The SAPT expansion of the three-body interaction energy is based on the same partition of the total Hamiltonian: $H = F + V + W$, as for dimers, except that now $F = F_A + F_B + F_C$ is the sum of all three Fock Hamiltonians for the monomers, $V = V_{AB} + V_{BC} + V_{AC}$ is the sum of three binary interaction potentials, and $W = W_A + W_B + W_C$. In the actual derivation of SAPT one has to consider all six perturbations separately, which leads to a sextuple perturbation expansion. The global order in V or W is just the sum of orders of the constituent operators. The total nonadditivity can be now expanded as

$$E_{int}[3,3] = \sum_{i=1, j=0} [E_{pol}^{(i,j)}[3,3] + E_{exch}^{(i,j)}[3,3]] \quad (1.39)$$

where, as before, i (j) refer to the order with respect to V (W). One can also consider quantities with the complete account of intramonomer correlation effects defined as

$$E_{\text{pol}}^{(i)}[3,3] = \sum_{j=0}^{\infty} E_{\text{pol}}^{(i,j)}[3,3] \qquad (1.40)$$

and similarly for the exchange corrections. The first-order polarization energy is additive. In the second-order in V the polarization component of $E_{\text{int}}[3,3]$ contains only a pure induction interaction, i.e. the dispersion terms are zero in this order. In the third and higher orders the polarization nonadditivity can be divided into the induction, dispersion, and mixed induction–dispersion parts. All these contributions also have their exchange counterparts. Moreover, the exchange nonadditivities are not zero even if their polarization counterparts are.

The earliest applications of SAPT to three-body problems date back about three decades ago, but only very recently this approach has been developed to the level comparable to that for the dimer case. The first-order exchange correction with complete neglect of intramonomer correlation, $E_{\text{exch}}^{(1,0)}[3,3]$, was developed by Jeziorski et al. (1976). This component has been computed for several atomic trimers (Bulski, 1975; Bulski and Chalasinski, 1980, 1982) but no molecular applications have been made so far. The important exchange effect has instead been computed using the Heitler–London expression. This expression is closely related (Jeziorski et al., 1976) to $E_{\text{exch}}^{(1,0)}[3,3]$ as it differs only by the 'zeroth-order' exchange energy which is proportional to S^4, where S denotes, as before, an overlap integral between orbitals of different monomers. The first non-vanishing polarization nonadditivity is $E_{\text{ind}}^{(2,0)}[3,3]$. This component has been derived by Cybulski (1995) and computed for various trimers by Chalasinski, Szczesniak, and collaborators (Chalasinski et al., 1991; Szczesniak et al., 1991, 1993). Cybulski (1995) has also derived the third-order induction contribution $E_{\text{ind}}^{(3,0)}[3,3]$. This term has recently been computed for Ar_2–HF and Ar_2–HCl trimers by Cybulski et al. (1994). In all those papers, as well as in some earlier work (Chalasinski et al., 1989b, 1990; Szczesniak et al., 1990), the same group computed the leading dispersion nonadditivity $E_{\text{disp}}^{(3,0)}[3,3]$. For large separations this component decays as the famous Axilrod–Teller–Muto (Axilrod and Teller, 1943; Muto, 1943) triple-dipole term. To our knowledge the detailed formula for this correction has not been published by this group. The $E_{\text{exch}}^{(2,0)}[3,3]$ component was first analyzed by Jeziorski (1974). Following this work Bulski and Chalasinski (1987) obtained explicit expressions for the $E_{\text{exch-disp}}^{(2,0)}[3,3]$ component (including terms proportional to S^2 and S^3), but only for the specific case of the He trimer. Recently Tachikawa and Iguchi (1994) investigated the components $E_{\text{exch}}^{(1,0)}[3,3]$, $E_{\text{ind}}^{(2,0)}[3,3]$, $E_{\text{exch-ind}}^{(2,0)}[3,3]$, and $E_{\text{exch-disp}}^{(2,0)}[3,3]$, but only through terms proportional to S^2 which are known to be insufficient for studies of nonadditivities (Bulski and Chalasinski, 1980, 1987).

The references listed above include almost all applications of (low-order) SAPT to three-body problems. There have been a larger number of *ab initio* supermolecular calculations (see Elrod and Saykally, 1994, and Chapter 2, this volume, for references), in particular at the self-consistent field (SCF) level. Most of the correlated calculations applied MBPTn, with n ranging from two to four. Chalasinski, Szczesniak, and collaborators computed MBPTn nonadditivities for several trimers (Chalasinski et al., 1989a,b, 1990, 1991, 1994; Szczesniak et al., 1990, 1991, 1993; Cybulski et al., 1994). These were usually interpreted in terms of SAPT corrections computed in the same basis set (Chalasinski and Szczesniak, 1988, 1994; Szczesniak and Chalasinski, 1992), providing in this way some physical insight into their results.

The polarization nonadditivities in the long-range region, where the multipole expansion of the interaction potential can be applied, have been investigated since the 1940s when Axilrod and Teller (1943) and Muto (1943) derived the leading triple-dipole term of $E_{\text{disp}}^{(3)}[3,3]$ for the atomic case. Higher multipole contributions were obtained by Bell (1970). Extensions to nonspherical systems and to other components of $E_{\text{pol}}^{(3)}[3,3]$ are due to Stogryn (1970, 1971, 1972). Piecuch (1986, 1988a,b, 1990, 1993) derived very general formulas for these nonadditivities (as well as for $E_{\text{ind}}^{(4)}[3,3]$) using spherical tensor formalism. A number of methods for calculation of the long-range nonadditive coefficients have been proposed (Meath and Aziz, 1984; Meath and Koulis, 1991).

Recently several three-body SAPT components of nonadditivity have been developed (Moszynski et al., 1995c; Lotrich and Szalewicz, submitted). Both papers rederived (in some cases in a new or more general form) the expressions considered previously: $E_{\text{exch}}^{(1,0)}[3,3]$ (not included in Moszynski et al., 1995c), $E_{\text{ind}}^{(2,0)}[3,3]$, $E_{\text{exch-ind}}^{(2,0)}[3,3]$, $E_{\text{exch-disp}}^{(2,0)}[3,3]$, $E_{\text{ind}}^{(3,0)}[3,3]$, and $E_{\text{disp}}^{(3,0)}[3,3]$. For the exchange corrections terms proportional to S^2 and S^3 have been included. Additionally, the first-order S^4 terms were taken into account by Lotrich and Szalewicz (submitted). Both papers developed the previously neglected $E_{\text{ind-disp}}^{(3,0)}[3,3]$ component. In addition, Lotrich and Szalewicz (submitted) derived $E_{\text{disp}}^{(3,1)}[3,3]$ and $E_{\text{disp}}^{(4,0)}[3,3]$. The first of the corrections listed above, $E_{\text{ind-disp}}^{(3,0)}[3,3]$, should be included to complete the third-order nonadditivity with neglect of intramonomer correlation. This term certainly will be important for systems where both induction and dispersion nonadditivities are large. The next term is the leading intramonomer correlation contribution to the third-order nonadditivity. The fourth-order dispersion energy is the level where the single and triple excitations occur for the first time. Near the minimum and for somewhat larger intersystem distances $E_{\text{disp}}^{(3,1)}[3,3]$ and $E_{\text{disp}}^{(4,0)}[3,3]$ are expected to reproduce the nonadditivity at the fourth order of MBPT (Chalasinski and Szczesniak, 1994). Asymptotically the former component will prevail due to its R^{-9} decay versus R^{-12} decay for the latter one. Specific expressions in terms of one- and two-electron integrals and orbital energies were given (Moszynski et al., 1995c; Lotrich and Szalewicz, submitted) for all the considered corrections. The third-order nonadditivities have been developed by Moszynski et al. (1995c) using the polarization propagator approach in the random-phase approximation (RPA) and therefore the resulting expressions are significantly different from those of Lotrich and Szalewicz (submitted). It is possible to obtain the third-order expressions of zeroth order in W presented in Lotrich and Szalewicz (submitted) from those of Moszynski et al. (1995c) by replacing the RPA and coupled Hartree–Fock amplitudes and denominators by the uncoupled quantities and neglecting certain terms. Although the RPA approach contains $E_{\text{disp}}^{(3,1)}[3,3]$ plus some higher-order terms, the RPA nonadditivities do not need to be closer to the true values than those obtained in the first order in W. As found by Moszynski et al. (1993a) in the dimer case the RPA dispersion energy is not a considerable improvement over $E_{\text{disp}}^{(20)} + E_{\text{disp}}^{(21)}$. The paper by Moszynski et al. (1995c) contains also an unwarranted criticism of the earlier work of Piecuch (1986, 1988a). One more recently completed development (Lotrich et al., in preparation) has been the derivation of the leading correlation effects in the first-order exchange, i.e. the $E_{\text{exch}}^{(1,1)}[3,3]$ and $E_{\text{exch}}^{(1,2)}[3,3]$ corrections.

A short summary of major components of three-body nonadditivity is given in Table 1.3. Again, as in the case of dimer interactions, a hybrid approach consisting of adding to

Table 1.3 Major components of three-body nonadditivity

$E_{\text{pol}}^{(1,j)}[3,3] = 0$	No polarization nonadditivity
$E_{\text{exch}}^{(1,0)}[3,3]$	Always important
$E_{\text{ind}}^{(2,0)}[3,3]$	Dominates for polar systems
$E_{\text{exch-ind}}^{(2,0)}[3,3]$	Significant for polar systems
$E_{\text{disp}}^{(2,j)}[3,3] = 0$	Dispersion additive in second order
$E_{\text{exch-disp}}^{(2,0)}[3,3]$	Nonnegligible
$E_{\text{ind}}^{(3,0)}[3,3]$	Important for polar systems
$E_{\text{disp}}^{(3,0)}[3,3]$	Contains Axilrod–Teller–Muto nonadditivity
$E_{\text{ind-disp}}^{(3,0)}[3,3]$	Importance to be determined
$E_{\text{disp}}^{(3,1)}[3,3]$	Leading intramonomer correlation effect
$E_{\text{disp}}^{(4,0)}[3,3]$	Small

the Hartree–Fock nonadditivity the correlation components from SAPT can be used. The leading correlated terms will then be the induction–dispersion and dispersion energies of zeroth order with respect to W:

$$E_{\text{ind-disp}}^{(3,0)}[3,3] = 8[t_{ab}^{rs} v_{rs}^{as'}(I_C)_{s'}^{b} - t_{ab}^{rs} v_{rb'}^{ab}(I_C)_{s}^{b'}] + 4[t_{ab}^{rs} t_{rs}^{ab}(\omega_C)_{s}^{s'} - t_{ab}^{rs} t_{rs}^{ab'}(\omega_C)_{b'}^{b}] + \cdots \quad (1.41)$$

$$E_{\text{disp}}^{(3,0)}[3,3] = 16(t_{ab}^{rs} t_{rt}^{ac} v_{sc}^{bt} + t_{ab}^{rs} t_{st}^{bc} v_{rc}^{at} + t_{bc}^{st} t_{rt}^{ac} v_{as}^{rb}) \quad (1.42)$$

where the indices ct denote occupied (virtual) orbitals of monomer C and $(I_C)_s^b = (\gamma_C)_s^b(\omega_C)_s^b/\varepsilon_s^b$ are induction amplitudes (analogous to amplitudes I_r^a). The dots in equation (1.41) denote terms obtained by all possible permutations of monomers: P_{AB}, P_{AC}, P_{BC}, P_{ABC}, and P_{ACB}.

The convergence of the expansion of the exchange corrections in powers of S is an important question in the theory of three-body nonadditivities. It has been investigated only for some rare-gas trimers (Bulski, 1975; Bulski and Chalasinski, 1980). More thorough analysis at the first-order level can be performed now since terms proportional up to S^4 have been derived and the exact expression for trimer's $E_{\text{exch}}^{(1,0)}$ is available (according to the formula of Jeziorski et al., 1976). Instead of $E_{\text{exch}}^{(1,0)}$ the Heitler–London expression is often used. It has been shown (Jeziorski et al., 1976) that the difference between the two formulas is proportional to S^4 and the explicit expression for this difference has been given. The Heitler–London energy can be easily obtained from the first iteration of a standard SCF program. Thus, one can compare the two approaches and establish the range of applicability of the expansion in powers of S. Such work may also shed light on the functional form used to approximate the exchange part of three-body nonadditivity. The most used form is the so-called exchange quadrupole–dipole approximation (Jansen, 1962). This form can be related to the S^2 terms in the expansion of $E_{\text{exch}}^{(1,0)}[3,3]$. One of the aims of future ab initio work should be determination of the class of systems for which it dominates the nonadditivity. For example, the S^2 terms are small for atomic trimers where the S^3 terms give major contributions. Further, the importance of the second-order exchange components should be evaluated and possibilities of fitting them to various analytical expressions should be explored. These components have been found to be quite significant in Ar$_3$ (Lotrich and Szalewicz, submitted).

1.5 COMPUTATIONAL ASPECTS

The physically transparent interaction mechanisms related to each SAPT component make it possible to analyze which characteristics of basis sets are most important for optimal description of a given effect. Notice that such separation of effects is not possible in the supermolecular approach. It has been known for some time that standard, correlation energy optimized basis sets perform poorly in calculations of interaction energies and use of more diffuse functions has been advocated (van Duijneveldt-van de Rijdt and van Duijneveldt, 1982; Szalewicz et al., 1988). Important contributions to our understanding of those problems have been made by Gutowski et al. (1986, 1987, 1993) in investigations of the helium dimer. Recently use of functions located on intermolecular bond became popular (Gutowski et al., 1987; Tao and Pan, 1992). Burcl et al. (1995) and Williams et al. (1995a) have shown that the dispersion energy requires basis functions located on the intermolecular bond since this is the region where intermolecular electron correlation takes place. The exchange and induction components require some small number of basis functions to be placed at the other nuclear center to allow for the resonance tunneling of electrons and charge transfer between systems. Finally, intramolecular correlation optimized exponents must be included in the monomer basis set to correctly describe the coupling of intra- and intermonomer correlation effects. Using this guidance, Williams et al. (1995a) developed a new method of performing SAPT calculations. So far most of such calculations have been performed using dimer-centered orbital basis sets. It is possible to reduce the time of calculations by about an order of magnitude by using what is called monomer-centered 'plus' basis sets (MC$^+$BS), i.e. basis sets in which most of the functions are located on the considered monomer and just a few are located on the other one. Such bases may also incorporate the bond functions and it has been argued (Williams et al., 1995a) that these functions are very effective in interaction energy calculations. Several possibilities of constructing MC$^+$BS have been investigated and the following simple rule has been recommended: an MC$^+$BS should be obtained from the corresponding DC$^+$BS by removing the polarization functions on the interacting partner. The MC$^+$BS approach allows us to decrease the size of basis sets by about 30 per cent without any significant effect on the calculated interaction energies. The MC$^+$BS method has been applied to the helium and water dimers providing benchmark energies for both systems (Williams et al., 1996; Mas and Szalewicz, 1996).

The conclusions from these investigations are applicable to both SAPT and supermolecular calculations. Clearly, to account for the major components of interaction energy one has to use bases which are flexible enough. Bases without polarization functions lead to severely distorted potentials and should not be used in any circumstances. If only a single polarization shell is present, it should consist of diffuse functions, with exponents obtained by minimization of $E_{\text{disp}}^{(20)}$ (these diffuse functions help also to correctly reproduce the multipole moments and polarizabilities of monomers). As more polarization functions are added, these should gradually include more and more intramonomer correlation optimized exponents to properly take into account effects of this correlation on the interaction energies. The dispersion energies are best served by a relatively small number of functions placed on the midbond.

The SAPT codes are a part of the METECC-94 package (Jeziorski *et al.*, 1993). A chapter in the METECC book describes some details of the program. It contains an introduction to the many-body SAPT, a summary of the final programmed formulas, and some examples of applications. A manual for the SAPT programs is distributed with the package. A newer version of the program is available via anonymous ftp from the authors. Recent optimizations of SAPT programs resulted in speedups of some major parts of those codes by a factor of about 20 in bases of the order of 100 orbitals (even more for larger calculations). The overall speedup of a complete calculation (including integral evaluations, SCF, transformation, monomer coupled cluster, and SAPT) in a basis of this size was close to a factor of 5. At the same time, the disk requirements have been decreased. A complete SAPT calculation involving all the developed corrections is now somewhat faster than a supermolecular MBPT4 calculation employing the basis set of the same size. Since SAPT can use the MC$^+$BS-type bases, which give results practically identical to those obtained in the dimer-centered bases, SAPT calculations of intermolecular potential energy surfaces are now a couple times faster than using the supermolecular approach at a similar level of theory.

1.6 REVIEW OF RECENT APPLICATIONS

The SAPT calculations can provide a complete potential energy surface for any weakly bound molecular system. These potentials can then be used to compute various observables (rovibrational quanta, scattering cross-sections, second virial coefficient, etc.), as has already been done for several systems (Williams *et al.*, 1993; Lotrich *et al.*, 1995; Moszynski *et al.*, 1994a,b,e, 1995a,b). Thus, this work enables one to understand the structure and dynamics of van der Waals clusters from first-principle calculations involving accurate and physically meaningful potentials. It leads to direct comparisons with measured quantities and therefore can test experimental findings and guide experimental work. This is particularly important for systems where the number of measured transitions is small. One should stress that calculations including only a few points or selected cuts through the potential are at present of limited utility. The interplay with experimental effort requires the knowledge of full, multidimensional intermolecular potentials, often even including a dependence on intramonomer vibrational coordinates.

We will describe in this section only the most recent applications of SAPT, which have not been discussed by Jeziorski *et al.* (1994). These include an application to the helium dimer, the benchmark system in the field of van der Waals interactions. Because of its simplicity, theory can achieve here accuracy rivaling or even surpassing the experimental accuracy. Recently van Mourik and van Lenthe computed FCI interaction energies for this system using a large 155-term interaction optimized basis set (van Mourik, 1994; van Mourik and van Lenthe, 1995). They obtained -10.947 K for the interactions energy at the minimum of 5.6 bohr and estimated that it is accurate to 0.01 K. Aziz *et al.* (1995) fitted a potential to these FCI results as well as to quantum Monte Carlo points at very small R computed by Ceperley and Partridge (1986). Notice that only *ab initio* information was used in the fit. The transport properties and virial coefficients of helium computed with this potential were found to agree very well with experiment. It has been proposed that the *ab initio* potential should set the standard for calibrating apparatus used to measure these quantities. This procedure would be an unusual example of an inversion of traditional practices in the field. However, when the relativistic retardation corrections were later added to the potential of Aziz *et al.* (1995), the agreement with experiment deteriorated, particularly for the low-temperature ^{3}He second virial coefficients (Janzen and Aziz, to be published). It is rather unusual that retardation effects influence the bulk properties since, typically, such effects make contributions many orders of magnitude smaller than measured quantities (Spruch, 1993).

Another important question is the existence of bound helium dimer. Modern interaction potentials support a bound vibrational state of ^{4}He$_2$. A few years ago this dimer has been indirectly detected by Luo *et al.* (1993, 1994). The results of this experiment have been disputed by Meyer *et al.* (1994). Recently the existence of bound ^{4}He$_2$ has been confirmed in a different measurement by Schöllkopf and Toennies (1996). At the same time, Luo *et al.* (1996) have measured the size of the dimer, determining the mean internuclear distance $\langle R \rangle$ to be 62 ± 10 Å. Since the vibrational wave function of He$_2$ extends to such large distances, the retardation effects have a significant impact on the properties of this molecule.

For He$_2$ it has been possible to extend the SAPT approach to an arbitrary (in practice, infinite) order. The lowest-order components have been calculated almost exactly applying the basis of explicitly correlated Gaussian-type geminals (GTG), following the methodology developed for He$_2$ by Szalewicz and Jeziorski (1979) and later improved upon by Jankowski *et al.* (1990) and Rybak *et al.* (1987, 1989). The intermediate-order components have been computed using the standard SAPT codes and very large orbital basis sets. Finally, the high-order components have been computed using a He$_2$-specific program calculating the SRS corrections to an arbitrary order. For the van der Waals minimum distance of 5.6 bohr these three levels of theory contributed -9.205, -1.679, and -0.157 K, respectively, resulting in the total interaction energy of -11.059 K. The basis-set saturated values of SAPT corrections are listed in Table 1.4.

The new He$_2$ interaction potential for the first time contains major components truly converged to 0.005 K. It has been estimated that the total error of the SAPT interaction energy at $R = 5.6$ bohr is smaller than 0.03 K. Comparison with the recent very accurate MBPT/CC calculations (Klopper and Noga, 1995; Bukowski *et al.*, 1996) indicates that

Table 1.4 SAPT calculation of the helium dimer potential at the equilibrium distance of 5.6 bohr

	Energy	Method
Electrostatic	-1.71	Geminals
First-order exchange	12.27	Geminals
Dispersion	-22.39	Geminals + orbitals
Induction	-0.28	Orbitals
Exchange–dispersion	0.74	Geminals + orbitals
Exchange–induction	0.25	Orbitals
SRS(2)	-11.12	
Higher-order terms	0.06	Orbitals
Total	-11.06	Infinite-order SAPT
van Mourik and van Lenthe (1995)	-10.95	Orbital FCI

All energies are in kelvin (K). The reported components represent the best presently available values obtained using a mixture of orbital and geminal techniques (Williams *et al.*, 1996; Korona *et al.*, submitted).

this error most likely makes SAPT interaction energies slightly too attractive. The SAPT potential is significantly deeper at the minimum (by about 0.1 K) than the recent FCI potential of van Mourik and van Lenthe (1995). Knowledge of the limit values of the two major interaction energy components shows that the error estimates of the latter potential, given as 0.01 K, are definitely too tight. The SAPT potential with retardation effects added recovers properties of bulk helium very accurately (see Janzen and Aziz, to be published)—everywhere within experimental error bars, including the very sensitive ^{3}He virials at very low temperatures. The retardation correction distinctly improves the agreement. The dissociation energy of the He$_2$ molecule is 1.878 mK without retardation and 1.713 mK with the retardation damping, while the corresponding values of $\langle R \rangle$ are 43.9 and 45.8 Å. The computed $\langle R \rangle$ is significantly smaller than measured in Luo et al. (1996), suggesting a possible need for reinterpretation of the experimental data.

Recently Huiszoon and Caffarel (1996) computed some SAPT components for He$_2$ using a quantum MC approach. Their value of $E_{\text{pol}}^{(2)}$ equal to -23.4 ± 0.3 K at $R = 5.6$ bohr can be compared to the result of Williams et al. (1996) and Korona et al. (submitted), equal to -22.67 K and believed to be accurate to about 0.03 K. While at this point there is no explanation for the discrepancy, one may note that lowering of the potential by 0.7 K would completely destroy the present excellent agreement with experiment.

The accuracy achieved by SAPT for the helium dimer required refinements of the methods which utilize the computed interaction energies. The form of the analytical fitting function previously used in He$_2$ work turned out to be inadequate to represent the computed points and a more elaborate and physically better justified form has been introduced (Williams et al., 1996). The scattering calculation had to distinguish properly between the nuclear and atomic masses of He. Most importantly, this work pointed out to the need of a more advanced treatment of relativistic effects.

The first accurate SAPT potential has been obtained for Ar–H$_2$ (Williams et al., 1993). It agreed very well with the empirical potential of Le Roy and Hutson (1987), which was at that time the most accurate of all empirical potentials for systems involving molecules. The rovibrational quanta generated using this potential (Moszynski et al., 1994e) agree to about 0.1 cm^{-1} with measured values (see Jeziorski et al., 1994, for a more extensive review). Recently, Bissonnette et al. (1996) developed a new empirical potential for this system which is even more accurate than the potential of Le Roy and Hutson (1987). Commenting on the SAPT potential of Williams et al. (1993), they wrote 'In view of the fact that no experimental data was used to refine it, the agreement of the SAPT potential with the fitted potentials, and the quality of predictions it yields, are really remarkably good'. They also pointed out to the areas of the SAPT potential which are most deficient. With the new SAPT corrections developed after the paper by Williams et al. (1993) was published and with the present capability of using larger basis sets, the quality of the SAPT potential could indeed be improved.

Lotrich et al. (1995) recently computed a SAPT potential for Ar–HF. All the corrections have been obtained in a 95-term *spdfg* basis while converged results for the dispersion energy have been obtained using a large 212-term set containing *spdfgh*-symmetry orbitals. The SAPT potential agrees well with the empirical H6(4,3,2) potential of Hutson (1992), including a reasonably similar account of the anisotropy. It predicts an absolute minimum of -207.4 cm^{-1} for the linear Ar–HF geometry at an intermolecular separation of 6.53 bohr and a secondary minimum of -111.0 cm^{-1} for the linear Ar–FH geometry at an intermolecular separation of 6.36 bohr. The corresponding values for the

H6(4.3.2) potential are -211.1 cm^{-1} at an intermolecular separation of 6.50 bohr and -108.8 cm^{-1} at an intermolecular separation of 6.38 bohr. Despite this agreement in the overall potentials, the individual components describing different physical effects are quite different in the SAPT and H6(4,3,2) potentials. The SAPT potential has been used to generate rovibrational levels of the complex which were compared to the levels predicted by H6(4,3,2) at the equilibrium separation. These results are quoted in Table 1.5. The agreement is excellent for stretch-type states (to within one cm^{-1}), while states corresponding to bending vibrations agree to a few cm^{-1}. The latter discrepancies are consistent with the differences in anisotropies of the two potentials. The SAPT potential is somewhat closer (by about 15–25 cm^{-1} near the minimum) to the empirical one than the recent MBPT4 potential of Tao and Klemperer (1994). SAPT potential for Ar–HF was fitted separately for the major interaction energy components. This fit allowed quite precise rationalization of the shape of the potential energy surface in terms of long-range coefficients and their anisotropies and anisotropies of exchange terms which follow the overlap of electron clouds. The shapes result from a delicate balance between all four major components (even the electrostatic interaction which results purely from overlap contributes quite significantly for smaller R). It has been shown that near the van der Waals minimum and for closer separations damping of the dispersion energy is similar in magnitude to the value of the purely exponential component of this energy.

A SAPT potential for He–CO has been computed by Moszynski et al. (1995b). They found that the interaction in this system is dominated by first-order exchange effects and by the dispersion energy. The potential has a single minimum for the linear He–OC geometry ($\theta = 180°$) at $R_m = 3.62$ Å with depth of 24.9 cm^{-1}. These findings are in some disagreement with MBPT4 studies of Tao et al. (1994) and Kukawska-Tarnawska et al. (1994). The former work found a minimum near a T-shaped configuration at $R_m = 3.49$ Å and $\theta_m = 120°$ with a depth of 20.32 cm^{-1}, while for the latter the corresponding values are 3.4 Å, 110°, and 21.95 cm^{-1}. The empirical potential $V_{(3,3,3)}$ from Chuaqui et al. (1994) has a depth of 22.91 cm^{-1}, a minimum distance $R_m = 3.39$ Å, and a minimum angle $\theta_m = 121.3°$. Another empirical potential dubbed XC(fit) was obtained by Le Roy

Table 1.5 Comparison of the spectroscopic properties of Ar–HF computed from the SAPT (Lotrich et al., 1995) and H6(4,3,2) (Hutson, 1992) potentials

	SAPT $r = r_e$	H6(4,3,2) $r = r_e$	H6(4,3,2) $r = r_0$	Observed
Ground state (D_0)	101.28	97.35	101.26	101.7
$E(J=1) - E(J=0)$	0.203 28	0.204 28	0.204 56	0.204 51
$n = 1$ stretch	38.40	37.61	38.70	38.69
$E(J=1) - E(J=0)$	0.184 17	0.184 38	0.184 94	0.184 05
Σ bend	44.50	49.82	52.05	52.06
$E(J=1) - E(J=0)$[a]	0.205 77	0.203 41	0.203 70	0.204 49
Π bend	60.05	63.47	65.81	65.81
$E(J=2^-) - E(J=1^-)$[a]	0.402 81	0.400 98	0.401 14	0.399 99
$n = 2$ stretch	67.01	65.98	66.59	66.60
$E(J=1) - E(J=0)$[a]	0.166 08	0.164 55	0.162 11	0.162 00

[a]Coriolis coupling neglected.
All energies are in cm^{-1} and levels are given as distances above the ground state. The observed values are from Fraser and Pine (1986), Lovejoy and Nesbitt (1989), and Dvorak et al. (1991).

et al. (1994). This potential has two minima with almost the same depth, about 22.5 cm^{-1}, one at $R_m = 3.62$ Å and $\theta_m = 180°$ (global minimum) and another at $R_m = 3.39$ Å and an almost T-shaped orientation. The former minimum agrees very well with SAPT predictions. The somewhat confusing lack of agreement on the parameters of potential minimum does not indicate significant differences between various theoretical/empirical potentials but rather results from the specific shape of the bottom of the well, which contains an almost flat valley between points (3.6, 180°) and (3.4, 90°). The potentials fitted to the computed SAPT (Moszynski *et al.*, 1995b) and MBPT4 (Tao *et al.*, 1994) points have been used to generate bound rovibrational states of the system. The predicted positions (and intensities in Moszynski *et al.*, 1995b) of lines in the infrared spectrum are in good agreement (to within about 0.5 cm^{-1}) with the experimental results from Chuaqui *et al.* (1994). The SAPT potential was also used to compute second virial coefficients which agreed with experimental data to within experimental error bars for a wide range of temperatures.

Moszynski *et al.* (1995a) computed a SAPT potential for the He–C$_2$H$_2$ dimer. This complex has been extensively investigated in scattering experiments (Danielson *et al.*, 1987, 1988; Buck *et al.*, 1993) and recently a part of near infrared spectrum has been measured (Miller, unpublished results). The SAPT calculations have been performed for 55 dimer geometries assuming acetylene in a linear rigid configuration corresponding to the equilibrium structure. All SAPT corrections were computed in a 107-term *spdf* DCBS and, in addition, $E_{disp}^{(20)}$ was computed in a 164-term basis. The SAPT potential exhibits a minimum depth of 22.3 cm^{-1} at a nearly linear configuration ($R_m = 8.20$ bohr and $\theta_m = 14.16°$). The depth of the minimum is in a good agreement with empirical potentials based on scattering work (Danielson *et al.*, 1987, 1988; Buck *et al.*, 1993) but disagrees significantly with earlier supermolecular MBPT2 potential (Slee *et al.*, 1992), which predicted a depth of only 8 cm^{-1}. This failure of MBPT2 theory can be associated with the triple bond in acetylene, which requires a higher-order treatment including coupled-cluster summations. The SAPT potential predicts another minimum of 22.0 cm^{-1} at a skew geometry. The potential energy surface is characterized by a very dramatic increase of anisotropy with decrease of R. Moszynski *et al.* (1995a) performed dynamical calculations of the bound states of He–C$_2$H$_2$ using the SAPT potential and found 34 of such states. The authors wrote that their spectrum agrees qualitatively with the data of Miller (unpublished). It is assumed that for quantitative agreement one would have to take into account dependence of the intermolecular potential on the acetylene internal coordinate.

All the systems considered above contained two degrees of freedom in intermolecular motion. Theoretical methods of computing dynamics of nuclear motion are strongly dependent on the number of degrees of freedom, therefore accurate potentials are needed to test methods which can deal with more than two degrees. An important system for three-dimensional dynamics is Ar–H$_2$O. This system is also a useful model of the forces associated with hydrophobic effects. High-resolution spectra in the microwave, near-infrared, and far-infrared regions have been measured for this system (Cohen *et al.*, 1988, 1990; Cohen and Saykally, 1991; Fraser *et al.*, 1990; Suzuki *et al.*, 1991; Lascola and Nesbitt, 1991; Zwart and Meerts, 1991). See Saykally (1989) and Cohen and Saykally (1992) for a review. These data have been used to construct empirical potential energy surfaces. Cohen and Saykally (1993) determined their first three-dimensional potential in 1990, later improving upon this work. The latter potential energy surface (denoted by

AW2) has been obtained (Cohen and Saykally, 1993) by a direct nonlinear least square fit of 12 parameters to 37 spectroscopic measurements. Some parameters in the latter potential were fixed at their *ab initio* values computed by Bulski *et al.* (1991). The most accurate *ab initio* calculation for this system has been performed by Tao and Klemperer (1994). The agreement of MBPT4 interaction energies with AW2 is to within about 20 cm^{-1} for R close to the minimum value. The position of the global minimum is to within 0.04 Å of AW2 and the minimum depth is 130 cm^{-1} compared to 143 cm^{-1} in AW2. This agreement is significantly better than that obtained by earlier *ab initio* MBPT2 calculation of Chalasinski *et al.* (1989a) and approximate SAPT calculations (with model damping functions) of Bulski *et al.* (1991). Unfortunately, the number of points computed at the MBPT4 level was not sufficient to fit a three-dimensional potential. Obtaining a reliable *ab initio* potential becomes now very important in view of recent measurement of inelastic cross-sections for rotational excitations in Ar–H$_2$O scattering (Chapman *et al.*, to be published). The measured cross-sections are in a dramatic disagreement with those calculated from the AW2 potential. Therefore, further theoretical effort is warranted. A new SAPT potential which has recently been obtained for this system (Szalewicz *et al.*, to be published) is significantly different from AW2 in the repulsive region. This potential is being used in dynamical calculations to predict rovibrational levels and low-energy cross-sections for direct comparison with measured quantities.

A recent SAPT application (Mas and Szalewicz, 1996) sheds light on the long-standing problem of the discrepancy between the theoretical and empirical values of the potential depth for the water dimer. Empirical potential depth is 5.4 ± 0.7 kcal/mol (Curtiss *et al.*, 1979) while the best previous theory predicted values close to 4.8 kcal/mol (Wang *et al.*, 1994). It has been shown (Mas and Szalewicz, 1996) that the use of the vibrationally averaged monomer geometry (r_0) rather than the equilibrium monomer geometry (r_e) and employment of a large, 152-orbital MC$^+$BS which balances the description of the intramonomer correlation and dispersion effects leads to an increase of the minimum depth by about 0.2 kcal/mol compared to the previously best theoretical results. This SAPT calculation provides a new *ab initio* well depth of 5.05 kcal/mol. The supermolecular MBPT4 gives a similar value of about 4.95 kcal/mol. On the other hand, replacement of the harmonic zero-point energy in the empirical estimate by a more accurate one that includes effects of anharmonicity decreases the empirical depth by at least 0.2 kcal/mol (Althorpe and Clary, 1994; Franken and Dykstra, 1994), to the value of about 5.2 kcal/mol. A comparison of the parameters of the minimum from SAPT calculations with the experimental values is presented in Table 1.6. Notice the near-

Table 1.6 Potential minimum parameters for water dimer

	ϵ_m (kcal/mol)	R_m (Å)	θ_m	α_m
SAPT computed	5.05	2.953	124	6.8
Experiment	5.4 ± 0.7[a]	2.952 ± 0.004[b]	123 ± 10[b]	5 ± 10[b]
SAPT/MBPT estimate	5.1 ± 0.1			
Experiment corrected for anharmonicity	5.2			

SAPT and MBPT results are from Mas and Szalewicz (1996).
[a]Curtiss *et al.* (1979).
[b]Odutola and Dyke (1980).

perfect agreement of SAPT-optimized dimer geometry with the measurements of Odutola and Dyke (1980).

The water dimer is an example of a relatively large system for which it has been possible to saturate the interaction energy to within about 0.05 kcal/mol. For r_e monomer geometry and for the often used for comparisons dimer's geometry listed in Feller (1992) the largest 243-term DC$^+$BS from Mas and Szalewicz (1996) gave potential depth of 4.77 kcal/mol at the MBPT2 level. The previously best calculation of Wang et al. (1994) in a 187-term DC$^+$BS gave a very similar value of 4.75 kcal/mol. Recently, Feyereisen et al. (1996) used even larger bases. Unfortunately, for the largest basis (containing 574 orbitals) they have utilized an approximate MBPT2 method called the 'resolution of identity' MBPT2 (RI-MBPT2) which introduces an error of the order of 0.05 kcal/mol—significant at this accuracy level—and they have not computed the CP correction. The next to the largest basis from that work (denoted by aug-cc-pVQZ) contained 344 orbitals. The frozen-core (FC) RI-MBPT2 result computed with the CP correction was 4.82 kcal/mol and the corresponding regular MBPT2(FC) was estimated to be 4.79 kcal/mol. However, since the difference between the respective *computed* CP-uncorrected results in the same basis is 0.06 kcal/mol, it seems that this estimate might be somewhat too large and the true value in this basis might be closer to 4.76 kcal/mol. Assuming that the frozen-core approximation misses about 0.02 kcal/mol, our estimate of the MBPT2 (all electrons) CP corrected depth in the aug-cc-pVQZ basis would be 4.78 kcal/mol, very close to the value of 4.77 kcal/mol computed in Mas and Szalewicz (1996). This very small change under the basis size increase from 243 to 344 orbitals most likely does not indicate that the result is converged to 0.01 kcal/mol but rather follows from the fact that aug-cc-pVQZ is not an interaction-optimized basis (it is a correlation-optimized one with an additional diffused set of functions in each symmetry) and, in particular, it does not include functions on bond which are critical for converging the dispersion energy. Feyereisen et al. (1996) have also made extrapolations to the complete basis set limit. Unfortunately, depending on the use of the CP correction and inclusion of the diffuse functions in the basis the extrapolated results for the MBPT(FC) depth range from 4.53 to 4.97 kcal/mol. However, the most dependable extrapolation from CP corrected energies in bases with diffuse functions is 4.83 kcal/mol. After correcting for the FC approximation this becomes 4.85 kcal/mol, quite close to the estimate of the limit value of 4.82 kcal/mol from Mas and Szalewicz (1996). This discussion would indicate that the true basis set limit is most likely 4.85 to within a few hundredth of kcal/mol. However, Klopper et al. (1995) computed (at the discussed geometry) a value of 4.92 kcal/mol using the so-called MP2-R12 approach, i.e. an MBPT2 energy in a basis containing explicitly interelectronic distances. A rather erratic pattern of convergence with basis set size observed by Klopper et al. (1995) suggests caution in accepting this value as the true limit.

Chalasinski et al. (1991) have computed the MBPT3 interaction energies for the water dimer and trimer. They also computed interaction energies predicted by several popular empirical potentials from MD/MC work. Since the empirical potentials are effectively including the many-body effects, the comparison was made using the energy of the trimer. Near the minimum of the cyclic planar configuration the empirical potentials were from about 2 to 4 kcal/mol (20–40 per cent) deeper than the *ab initio* result. The differences were even larger for other values of the angular coordinate (corresponding to simultaneous rotation of all monomers around the positions of oxygens), approaching 20 kcal/mol.

Recent two- and three-body SAPT interaction energies computed by Mas *et al.* (to be published) do not change this picture. The new *ab initio* SAPT potential based on more than 1000 points will enable more extensive comparisons with empirical potentials.

Langlet *et al.* (1995) performed low-level SAPT calculations (without including intramonomer correlation effects) for several hydrogen bonded dimers consisting of water, ammonia, and hydrogen fluoride. They discussed the relative importance of the electrostatic, exchange, induction, and dispersion interactions for creation of the hydrogen bond as well as on its anisotropy. They found that already at this level of theory the qualitative description of the interaction is correct and independent of minor changes in methodology.

ACKNOWLEDGMENTS

This work was partly supported by the NSF grant CHE-9626739, by the Polish Research Council grant KBN 3 T09A 072 09, and by the NSF through a grant to the Institute for Theoretical Atomic and Molecular Physics at the Harvard University and Smithsonian Astrophysical Observatory.

REFERENCES

Adams, W. H. (1991) *Int. J. Quantum Chem.*, S **25**, 165.
Adams, W. H. (1992) *J. Math. Chem.*, **10**, 1.
Adams, W. H. (1994) *Chem. Phys. Lett.*, **229**, 472.
Adams, W. H. (1996) *Int. J. Quantum Chem.*, **60**, 273.
Adams, W. H. and Polymeropoulos, E. E. (1978) *Phys. Rev.*, **A17**, 11, 18, 24.
Ahlrichs, R. (1976) *Theor. Chim. Acta.*, **41**, 7.
Ahlrichs, R., Penco, R. and Scoles, G. (1977) *Chem. Phys.*, **19**, 119.
Althorpe, S. C. and Clary, D. C. (1994) *J. Chem. Phys.*, **101**, 3603.
Anders, J., Buck, U., Huisken, F., Schluesener, J. and Torello, F. (1980) *J. Chem. Phys.*, **73**, 5620.
Angyan, J. G., Jansen, G., Loos, M., Hättig, C. and Hess, B. A. (1994) *Chem. Phys. Lett.*, **219**, 267.
Arrighini, G. P. (1981) *Intermolecular Forces and Their Evaluation by Perturbation Theory*, Springer, Berlin.
van der Avoird, A., Wormer, P. E. S., Mulder, F. and Berns, R. M. (1980) *Top. Curr. Chem.*, **93**, 1.
van der Avoird, A., Wormer, P. E. S. and Moszynski, R. (1994) *Chem. Rev.*, **94**, 1931.
Axilrod, B. M. and Teller, E. (1943) *J. Chem. Phys.*, **11**, 299.
Aziz, R. A., Janzen, A. R. and Moldover, M. R. (1995) *Phys. Rev. Lett.*, **74**, 1586.
Barton, A. E. and Howard, B. J. (1982) *Faraday Discuss. Chem. Soc.*, **73**, 45.
Bell, R. J. (1970) *J. Phys. B*, **3**, 751.
Bissonnette, C., Crowell, K. G., Le Roy, R. J., Wheatley, R. J. and Meath, W. J., (1996) *J. Chem. Phys.*, **105**, 2639.
Bizarri, A., Heijmen, B., Stolte, S. and Reuss, J. (1988) *Z. Phys. D*, **10**, 291.
Boys, F. S. and Bernardi, D. (1970) *Mol. Phys.*, **19**, 553.
Brink, D. M. and Satchler, G. R. (1975) *Angular Momentum*, Clarendon Press, Oxford.
Buck, U., Ettischer, I., Schlemmer, S., Yang, M., Vohralik, P. and Watts, R. O. (1993) *J. Chem. Phys.*, **99**, 3494.
Buckingham, A. D., Fowler, P. W. and Hutson, J. M. (1988) *Chem. Rev.*, **88**, 963.
Bukowski, R., Jeziorski, B. and Szalewicz, K. (1996) *J. Chem. Phys.*, **104**, 3306.
Bulski, M. (1975) *Mol. Phys.*, **29**, 1171.
Bulski, M. and Chalasinski, G. (1980) *Theor. Chim. Acta*, **56**, 199.

REFERENCES

Bulski, M. and Chalasinski, G. (1982) *Chem. Phys. Lett.*, **89**, 450.
Bulski, M. and Chalasinski, G. (1987) *J. Chem. Phys.*, **86**, 937.
Bulski, M., Wormer, P. E. S. and van der Avoird, A. (1991) *J. Chem. Phys.*, **94**, 8096.
Burcl, R., Chalasinski, G., Bukowski, R. and Szczesniak, M. (1995) *J. Chem. Phys.*, **103**, 1498.
Ceperley, D. M. and Partridge, H. J. (1986) *J. Chem. Phys.*, **84**, 820.
Chalasinski, G. (1983) *Mol. Phys.*, **49**, 1353; *Chem. Phys.*, **82**, 207.
Chalasinski, G. and Gutowski, M. (1988) *Chem. Rev.*, **88**, 943.
Chalasinski, G. and Jeziorski, B. (1976) *Mol. Phys.*, **32**, 81.
Chalasinski, G. and Szalewicz, K. (1980) *Int. J. Quantum Chem.*, **18**, 1071.
Chalasinski, G. and Szczesniak, M. M. (1988) *Mol. Phys.*, **63**, 205.
Chalasinski, G. and Szczesniak, M. M. (1994) *Chem. Rev.*, **94**, 1723.
Chalasinski, G., Jeziorski, B. and Szalewicz, K. (1977) *Int. J. Quantum Chem.*, **11**, 247.
Chalasinski, G., Szczesniak, M. M. and Scheiner, S. (1989a) *J. Chem. Phys.*, **94**, 2807.
Chalasinski, G., Cybulski, S. M., Szczesniak, M. M. and Scheiner, S. (1989b) *J. Chem. Phys.*, **91**, 7048.
Chalasinski, G., Szczesniak, M. M. and Cybulski, S. M. (1990) *J. Chem. Phys.*, **92**, 2481.
Chalasinski, G., Szczesniak, M. M., Cieplak, P. and Scheiner, S. (1991) *J. Chem. Phys.*, **94**, 2873.
Chalasinski, G., Szczesniak, M. M. and Kendall, R. (1994) *J. Chem. Phys.*, **101**, 8860.
Chapman, W. B., Kulcke, A., Schiffman, A. and Nesbitt, D. J., to be published.
Chipman, D. M. (1977) *J. Chem. Phys.*, **66**, 1830.
Chipman, D. M. and Hirschfelder, J. O. (1973) *J. Chem. Phys.*, **59**, 2838.
Chipman, D. M., Bowman, J. D. and Hirschfelder, J. O. (1973) *J. Chem. Phys.*, **59**, 2830.
Chuaqui, C. E., Le Roy, R. J. and McKellar, A. R. W. (1994) *J. Chem. Phys.*, **101**, 39.
Cizek, J., Damburg, R. J., Graffi, S., Grechi, V., Hassels, E. M., Harris, J. G., Nakai, S., Paldus, J., Propin, R. K. and Silverstone, H. J. (1986) *Phys. Rev.*, **33**, 12.
Cohen, R. C. and Saykally, R. J. (1990) *J. Phys. Chem.*, **94**, 7991.
Cohen, R. C. and Saykally, R. J. (1991) *J. Chem. Phys.*, **95**, 7891.
Cohen, R. C. and Saykally, R. J. (1992) *J. Phys. Chem.*, **96**, 1024.
Cohen, R. C. and Saykally, R. J. (1993) *J. Chem. Phys.*, **98**, 6007.
Cohen, R. C., Busarow, K. L., Laughlin, K. B., Blake, G. A., Havenith, M., Lee, Y. T. and Saykally, R. J. (1988) *J. Chem. Phys.*, **89**, 4494.
Cohen, R. C., Busarow, K. L., Lee, Y. T. and Saykally, R. J. (1990) *J. Chem. Phys.*, **92**, 169.
Cooper, A. R. and Hutson, J. M. (1993) *J. Chem. Phys.*, **98**, 5337.
Curtiss, L. A., Frurip, D. J. and Blander, M. (1979) *J. Chem. Phys.*, **71**, 2703.
Cwiok, T., Jeziorski, B., Kolos, W., Moszynski, R., Rychlewski, J. and Szalewicz, K. (1992a) *Chem. Phys. Lett.*, **195**, 67.
Cwiok, T., Jeziorski, B., Kolos, W., Moszynski, R. and Szalewicz, K. (1992b) *J. Chem. Phys.*, **97**, 7555.
Cwiok, T., Jeziorski, B., Kolos, W., Moszynski, R. and Szalewicz, K. (1994) *J. Mol. Struct. (Theochem.)*, **307**, 135.
Cybulski, S. M. (1995) *Chem. Phys. Lett.*, **238**, 261.
Cybulski, S. M., Szczesniak, M. M. and Chalasinski, G. (1994) *J. Chem. Phys.*, **101**, 10708.
Danielson, L. J., McLeod, K. M. and Keil, M. (1987) *J. Chem. Phys.*, **87**, 239.
Danielson, L. J., Keil, M. and Dunlop, P. J. (1988) *J. Chem. Phys.*, **88**, 4218.
Dayton, D. C., Jucks, K. W. and Miller, R. E. (1989) *J. Chem. Phys.*, **90**, 2631.
Dixon, T. A., Joyner, C. H., Baiocchi, F. A. and Klemperer, W. (1981) *J. Chem. Phys.*, **74**, 6539.
Douketis, C., Scoles, G., Marchetti, S., Zen, M. and Thakkar, A. J. (1982) *J. Chem. Phys.*, **76**, 3057.
van Duijneveldt-van de Rijdt, J. G. C. M. and van Duijneveldt, F. B. (1982) *J. Mol. Struct. (Theochem.)*, **89**, 185.
van Duijneveldt, F. B., van Duijneveldt-van de Rijdt, J. G. C. M. and van Lenthe, J. H. (1994) *Chem. Rev.*, **94**, 1873.
Dvorak, M. A., Reeve, S. W., Burns, W. A., Grushow, A. and Leopold, K. R. (1991) *Chem. Phys. Lett.*, **185**, 399.
Dyke, T. R. and Muenter, J. S. (1975) in *Molecular Structure and Properties*, edited by Buckingham, A. D. (MTP Int. Rev. Sci., Phys. Chem. Ser. 2, vol. 2), Butterworths, London, p. 27.

Dyke, T. R., Howard, B. and Klemperer, W. (1972) *J. Chem. Phys.*, **56**, 2442.
Dykstra, C. E. (1993) *Chem. Rev.*, **93**, 2339.
Eisenschitz, R. and London, F. (1930) *Z. Phys.*, **60**, 491.
Elrod, M. J. and Saykally, R. J. (1994) *Chem. Rev.*, **94**, 1975.
Elrod, M. J., Steyert, D. W. and Saykally, R. J. (1991) *J. Chem. Phys.*, **94**, 58; **93**, 3182.
Elrod, M. J., Loeser, J. G. and Saykally, R. J. (1993) *J. Chem. Phys.*, **98**, 5352.
Ernesti, A. and Hutson, J. M. (1994) *Faraday Discuss.*, **97**, 119.
Ernesti, A. and Hutson, J. M. (1995) *Phys. Rev. A*, **51**, 239.
Feller, D. (1992) *J. Chem. Phys.*, **96**, 6104.
Feyereisen, M. W., Feller, D. and Dixon, D. A. (1996) *J. Phys. Chem.*, **100**, 2993.
Franken, K. A. and Dykstra, C. E. (1994) *J. Chem. Phys.*, **100**, 2865.
Fraser, G. T. (1991) *Int. Rev. Phys. Chem.*, **10**, 189.
Fraser, G. T. and Pine, A. S. (1986) *J. Chem. Phys.*, **85**, 2502.
Fraser, G. T., Nelson, D. D., Charo, A. and Klemperer, W. (1985) *J. Chem. Phys.*, **82**, 2535.
Fraser, G. T., Lovas, F. J., Suenram, R. D. and Matsumura, K. (1990) *J. Mol. Spectrosc.*, **144**, 97.
Fraser, G. T., Pine, A. S. and Kreiner, W. A. (1991) *J. Chem. Phys.*, **94**, 7061.
Gutowski, M., Kakol, M. and Piela, L. (1983) *Int. J. Quantum Chem.*, **23**, 1843.
Gutowski, M., van Lenthe, J. H., Verbeek, J., van Duijneveldt, F. B. and Chalasinski, G. (1986) *Chem. Phys. Lett.*, **124**, 370.
Gutowski, M., Verbeek, J., van Lenthe, J. H. and Chalasinski, G. (1987) *Chem. Phys.*, **111**, 271.
Gutowski, M., van Duijneveldt-van de Rijdt, J. G. C. M., van Lenthe, J. H. and van Duijneveldt, F. B. (1993) *J. Chem. Phys.*, **98**, 4728.
Gutowsky, H. S., Klots, T. D., Chuang, C., Schmuttenmaer, C. A. and Emilson, T. (1987) *J. Chem. Phys.*, **86**, 569.
Gwo, J., Havenith, M., Busarow, K. L., Cohen, R. C., Schmuttenmaer, C. A. and Saykally, R. J. (1990) *Mol. Phys.*, **71**, 453.
Halberstadt, N. and Janda, K. C. (eds) (1990) *Dynamics of Polyatomic van der Waals Complexes*, NATO ASI Series B, Vol. 227, Plenum, New York.
Harris, S. J., Novick, S. E. and Klemperer, W. (1974) *J. Chem. Phys.*, **60**, 3208.
Heijmen, T. G. A., Moszynski, R., Wormer, P. E. S. and van der Avoird, A. (1996) *Mol. Phys.*, **89**, 81.
Hepburn, J., Scoles, G. and Penco, R. (1975) *Chem. Phys. Lett.*, **36**, 451.
Hirschfelder, J. O. and Silbey, R. (1966) *J. Chem. Phys.*, **45**, 2188.
Huang, S. S., Bieler, C. R., Janda, K. C., Tao, F.-M., Klemperer, W., Casavecchia, P., Volpi, G. G. and Halberstadt, N. (1995) *J. Chem. Phys.*, **102**, 8846.
Huiszoon, C. and Caffarel, M. (1996) *J. Chem. Phys.*, **104**, 4621.
Hutson, J. M. (1990) *Ann. Rev. Phys. Chem.*, **41**, 123.
Hutson, J. M. (1992) *J. Chem. Phys.*, **96**, 6752.
Hutson, J. M., Beswick, J. A. and Halberstadt, N. (1990) *J. Chem. Phys.*, **90**, 1337.
Jankowski, P., Jeziorski, B., Rybak, S. and Szalewicz, K. (1990) *J. Chem. Phys.*, **92**, 7441.
Jansen, L. (1962) *Phys. Rev.*, **135**, 1798.
Janzen, A. R. and Aziz, R. A. to be published.
Jaszunski, M. (1980) *Mol. Phys.*, **39**, 777.
Jeziorska, M., Jeziorski, B. and Cizek, J. (1987) *Int. J. Quantum Chem.*, **32**, 149.
Jeziorski, B. (1974) PhD thesis, University of Warsaw.
Jeziorski, B. and van Hemert, M. C. (1976) *Mol. Phys.*, **31**, 713.
Jeziorski, B. and Kolos, W. (1982) in *Molecular Interactions*, edited by Ratajczak, H. and Orville-Thomas, W. J., Wiley, New York, Vol. 3, p. 1.
Jeziorski, B., Bulski, M. and Piela, L. (1976) *Int. J. Quantum Chem.*, **10**, 281.
Jeziorski, B., Szalewicz, K. and Chalasinski, G. (1978) *Int. J. Quantum Chem.*, **14**, 271.
Jeziorski, B., Szalewicz, K. and Jaszunski, M. (1979) *Chem. Phys. Lett.*, **61**, 391.
Jeziorski, B., Schwalm, W. A. and Szalewicz, K. (1980) *J. Chem. Phys.*, **73**, 6215.
Jeziorski, B., Moszynski, R., Rybak, S. and Szalewicz, K. (1989) in *Many-Body Methods in Quantum Chemistry*, edited by Kaldor, U., Lecture Notes in Chemistry, Vol. 52, Springer-Verlag, New York, p. 65.
Jeziorski, B., Moszynski, R., Ratkiewicz, A., Rybak, S., Szalewicz, K. and Williams, H. L. (1993) in

Methods and Techniques in Computational Chemistry: METECC94, Vol. B Medium-Size Systems, edited by Clementi, E., STEF, Cagliari, p. 79.
Jeziorski, B., Moszynski, R. and Szalewicz, K. (1994) *Chem. Rev.*, **94**, 1887.
Jeziorski, B., Moszynski, R. and Szalewicz, K. Unpublished research report.
Kaplan, I. G. (1987) *Theory of Intermolecular Interactions*, Elsevier, Amsterdam.
Keenan, M. R., Buxton, L. W., Campbell, E. J., Legon, A. C. and Flygare, W. H. (1981) *J. Chem. Phys.*, **74**, 2133.
Klein, D. J. (1987) *Int. J. Quantum Chem.*, **32**, 377.
Klemperer, W. (1980) *J. Mol. Str.*, **59**, 161.
Klemperer, W. and Yaron, D. (1990) in *Dynamics of Polyatomic van der Waals Complexes*, edited by Halberstadt, N. and Janda, K. C., NATO ASI Series B, Vol. 227, Plenum, 1990.
Klopper, W. and Noga, J. (1995) *J. Chem. Phys.*, **103**, 6127.
Klopper, W., Schütz, M., Luthi, H. P. and Leutwyler, S. (1995) *J. Chem. Phys.*, **103**, 1085.
Knowles, P. J. and Meath, W. J. (1986) *Mol. Phys.*, **59**, 965.
Knowles, P. J. and Meath, W. J. (1987) *Mol. Phys.*, **60**, 1143.
Kolos, W. (1970) *Adv. Quantum Chem.*, **5**, 99.
Korona, T., Williams, H. L., Bukowski, R., Jeziorski, B. and Szalewicz, K. *J. Chem. Phys.*, submitted.
Korona, T., Moszynski, R. and Jeziorski, B. *Adv. Quantum Chem.*, in press.
Kukawska-Tarnawska, B., Chalasinski, G. and Olszewski, K. (1994) *J. Chem. Phys.*, **101**, 4964.
Kutzelnigg, W. (1980) *J. Chem. Phys.*, **73**, 343.
Kutzelnigg, W. (1992) *Chem. Phys. Lett.*, **195**, 77.
Langlet, J., Caillet, J. and Caffarel, M. (1995) *J. Chem. Phys.*, **103**, 8043.
Lascola, R. and Nesbitt, D. J. (1991) *J. Chem. Phys.*, **95**, 7917.
Law, M. M., Hutson, J. M. and Ernesti, A. (eds) (1993) *Fitting Molecular Potential Energy Surfaces*, Collaborative Computational Project on Heavy Particle Dynamics CCP6, Daresbury.
Leopold, K. R., Fraser, G. T., Novick, S. E. and Klemperer, W. (1994) *Chem. Rev.*, **94**, 1807.
Le Roy, R. J. and Hutson, J. M. (1987) *J. Chem. Phys.*, **86**, 837.
Le Roy, R. J., Bissonnette, C., Wu, T. H., Dham, A. K. and Meath, W. J. (1994) *Faraday Disc.*, **97**, 81.
Le Sueur, C. R. and Stone, A. J. (1993) *Mol. Phys.*, **78**, 1267.
Liu, K., Loeser, J. G., Elrod, M. J., Host, B. C., Rzepiela, J. A. and Saykally, R. J. (1994) *J. Am. Chem. Soc.*, **116**, 3507.
Lotrich, V. F. and Szalewicz, K., *J. Chem. Phys.*, submitted.
Lotrich, V. F. and Szalewicz, K. Work in progress.
Lotrich, V. F., Williams, H. L., Szalewicz, K., Jeziorski, B., Moszynski, R., Wormer, P. E. S. and van der Avoird, A. (1995) *J. Chem. Phys.*, **103**, 6076.
Lotrich, V. F., Jeziorski, B. and Szalewicz, K. (1996) Manuscript in preparation.
Lovejoy, C. M. and Nesbitt, D. J. (1989) *J. Chem. Phys.*, **91**, 2790.
Lovejoy, C. M., Schuder, M. D. and Nesbitt, D. J. (1986a) *Chem. Phys. Lett.*, **127**, 374.
Lovejoy, C. M., Schuder, M. D. and Nesbitt, D. J. (1986b) *J. Chem. Phys.*, **85**, 4890.
Luo, F., McBane, C., Kim, G., Giese, C. F. and Gentry, W. R. (1993) *J. Chem. Phys.*, **98**, 3564.
Luo, F., McBane, C., Kim, G., Giese, C. F. and Gentry, W. R. (1994) *J. Chem. Phys.*, **100**, 4023.
Luo, F., Giese, C. F. and Gentry, W. R. (1996) *J. Chem. Phys.*, **104**, 1151.
Maitland, G. C., Rigby, M., Smith, E. B. and Wakeham, W. A. (1981) *Intermolecular Forces*, Clarendon Press, Oxford.
Mas, E. M. and Szalewicz, K. (1996) *J. Chem. Phys.*, **104**, 7606.
Mas, E. M. *et al.*, to be published.
McIlroy, A. and Nesbitt, D. J. (1992) *J. Chem. Phys.*, **97**, 6044.
McIlroy, A., Lascola, R., Lovejoy, C. M. and Nesbitt, D. J. (1991) *J. Chem. Phys.*, **95**, 2636.
McKellar, A. R. W. and Welsh, H. L. (1971) *J. Chem. Phys.*, **55**, 595.
Meath, W. J. and Aziz, R. A. (1984) *Mol. Phys.*, **52**, 225.
Meath, W. J. and Koulis, M. (1991) *J. Mol. Struct. (Theochem.)*, **226**, 1.
Meath, W. J., Margoliash, D. J., Jhanwar, B. L., Koide, A. and Zeiss, G. D. (1981) in *Intermolecular Forces*, edited by Pullman, B., Reidel, Dordrecht, p. 101.
Meyer, E. S., Mester, J. C. and Silvera, I. F. (1994) *J. Chem. Phys.*, **100**, 4021.

Miller, R. E., unpublished results.
Morgan, III, J. D. and Simon, B. (1980) *Int. J. Quantum Chem.*, **17**, 1143.
Moszynski, R., Rybak, S., Cybulski, S. M. and Chalasinski, G. (1991) *Chem. Phys. Lett.*, **166**, 609.
Moszynski, R., Jeziorski, B. and Szalewicz, K. (1993a) *Int. J. Quantum Chem.*, **45**, 409.
Moszynski, R., Jeziorski, B., Ratkiewicz, A. and Rybak, S. (1993b) *J. Chem. Phys.*, **99**, 8856.
Moszynski, R., Wormer, P. E. S., Jeziorski, B. and van der Avoird, A. (1994a) *J. Chem. Phys.*, **101**, 2811.
Moszynski, R., Jeziorski, B., van der Avoird, A. and Wormer, P. E. S. (1994b) *J. Chem. Phys.*, **101**, 2825.
Moszynski, R., Jeziorski, B. and Szalewicz, K. (1994c) *J. Chem. Phys.*, **100**, 1312.
Moszynski, R., Jeziorski, B., Rybak, S., Szalewicz, K. and Williams, H. L. (1994d) *J. Chem. Phys.*, **100**, 5080.
Moszynski, R., Jeziorski, B., Wormer, P. E. S. and van der Avoird, A. (1994e) *Chem. Phys. Lett.*, **221**, 161.
Moszynski, R., Cybulski, S. M. and Chalasinski, G. (1994f) *J. Chem. Phys.*, **100**, 4998.
Moszynski, R., Wormer, P. E. S. and van der Avoird, A. (1995a) *J. Chem. Phys.*, **102**, 8385.
Moszynski, R., Korona, T., Wormer, P. E. S. and van der Avoird, A. (1995b) *J. Chem. Phys.*, **103**, 321.
Moszynski, R., Wormer, P. E. S., Jeziorski, B. and van der Avoird, A. (1995c) *J. Chem. Phys.*, **103**, 8058.
Moszynski, R., Heijmen, T. G. A. and Jeziorski, B., *Mol. Phys.*, in press.
van Mourik, T. (1994) PhD thesis, The University of Utrecht, The Netherlands.
van Mourik, T. and van Lenthe, J. H. (1995) *J. Chem. Phys.*, **102**, 7479.
Murrell, J. N. and Shaw, G. (1967) *J. Chem. Phys.*, **46**, 46.
Musher, J. I. and Amos, A. T. (1967) *Phys. Rev.*, **164**, 31.
Muto, Y. (1943) *Proc. Phys. Math. Soc. Jpn*, **17**, 629.
Nelson, D. D., Jr and Klemperer, W. (1987) *J. Chem. Phys.*, **87**, 139.
Nelson, D. D., Fraser, G. T., Peterson, K. I., Klemperer, W., Lovas, F. J. and Suenram, R. D. (1986) *J. Chem. Phys.*, **85**, 5512.
Nesbitt, D. J. (1988) *Chem. Rev.*, **88**, 843.
Nesbitt, D. J. (1994) *Ann. Rev. Phys. Chem.*, **45**, 367.
Ng, K.-C., Meath, W. J. and Allnatt, A. R. (1976) *Mol. Phys.*, **32**, 177.
Ng, K.-C., Meath, W. J. and Allnatt, A. R. (1977) *Mol. Phys.*, **33**, 699.
Ng, K.-C., Meath, W. J. and Allnatt, A. R. (1978) *Chem. Phys.*, **32**, 175.
Ng, K.-C., Meath, W. J. and Allnatt, A. R. (1979a) *Mol. Phys.*, **37**, 237.
Ng, K.-C., Meath, W. J. and Allnatt, A. R. (1979b) *Mol. Phys.*, **38**, 449.
Odutola, J. A. and Dyke, T. R. (1980) *J. Chem. Phys.*, **72**, 5062.
Piecuch, P. (1986) *Mol. Phys.*, **59**, 1067, 1085, 1097.
Piecuch, P. (1988a) in *Molecules in Physics, Chemistry and Biology,* vol. 2 *Physical Aspects of Molecular Systems*, edited by Maruani, J., Kluwer, Dordrecht, p. 417.
Piecuch, P. (1988b) *Acta Phys. Polon. A*, **74**, 563.
Piecuch, P. (1990) *Acta Phys. Polon. A*, **77**, 453.
Piecuch, P. (1993) *Int. J. Quantum Chem.*, **47**, 261.
Pine, A. S. and Howard, B. J. (1986) *J. Chem. Phys.*, **84**, 590.
Pugliano, N., Cruzan J. D., Loeser, J. G. and Saykally, R. J. (1993) *J. Chem. Phys.*, **98**, 6600.
Rybak, S., Szalewicz, K., Jeziorski, B. and Jaszunski, M. (1987) *J. Chem. Phys.*, **86**, 5652.
Rybak, S., Szalewicz, K. and Jeziorski, B. (1989) *J. Chem. Phys.*, **91**, 4779.
Rybak, S., Jeziorski, B. and Szalewicz, K. (1991) *J. Chem. Phys.*, **95**, 6576.
Sabo, D., Bacic, Z., Burgi, T. and Leutwyler, S. (1995) *Chem. Phys. Lett.*, **244**, 283.
Sadlej, A. J. (1980) *Mol. Phys.*, **39**, 1249.
Saykally, R. J. (1989) *Acc. Chem. Res.*, **22**, 295.
Schiffman, A., Chapman, W. B. and Nesbitt, D. J. (1996) *J. Chem. Phys.*, **100**, 3402.
Schmuttenmaer, C. A., Cohen, R. C., Loeser, J. G. and Saykally, R. J. (1991) *J. Chem. Phys.*, **95**, 9.
Schmuttenmaer, C. A., Cohen, R. C. and Saykally, R. J. (1994) *J. Chem. Phys.*, **101**, 146.
Schöllkopf, W. and Toennies, J. P. (1996) *J. Chem. Phys.*, **104**, 1155.
Scott, T. C., Babb, J. F., Dalgarno, A. and Morgan III, J. D. (1993) *Chem. Phys. Lett.*, **203**, 175; *J. Chem. Phys.*, **99**, 2841.

Slee, T., Le Roy, R. J. and Chuaqui, C. E. (1992) *Mol. Phys.*, **77**, 111.
Sperhac, J. M., Weida, M. J. and Nesbitt, D. J. (1996) *J. Chem. Phys.*, **104**, 2202.
Spruch, L. (1993) in *Long-Range Casimir Forces. Theory and Recent Experiments in Atomic Systems*, edited by Levine, F. S. and Micha, D. Plenum, New York, p. 1.
Stogryn, D. E. (1970) *Phys. Rev. Lett.*, **24**, 971.
Stogryn, D. E. (1971) *Mol. Phys.*, **22**, 81.
Stogryn, D. E. (1972) *Mol. Phys.*, **23**, 897.
Stolarczyk, L. Z. and Piela, L. (1979) *Int. J. Quantum Chem.*, **15**, 701.
Stone, A. J. (1985) *Mol. Phys.*, **56**, 1065.
Stone, A. J. (1991) in *Theoretical Models of Chemical Bonding*, edited by Maksic, Z. B., Springer, New York, vol. 4, p. 103.
Suhm, M. A. and Nesbitt, D. J. (1995) *Chem. Soc. Rev.*, **24**, 45.
Suhm, M. A., Farrell, Jr, J. T., Ashworth, S. H. and Nesbitt, D. J. (1993) *J. Chem. Phys.*, **98**, 5985.
Suzuki, S., Bumgarner, R. E., Stockman, P. A., Green, P. G. and Blake, G. A. (1991) *J. Chem. Phys.*, **94**, 824.
Szalewicz, K. and Jeziorski, B. (1979) *Mol. Phys.*; **38**, 191.
Szalewicz, K., Cole, S. J., Kolos, W. and Bartlett, R. J. (1988) *J. Chem. Phys.*, **89**, 3662.
Szalewicz, K. *et al.*, to be published.
Szczesniak, M. M. and Chalasinski, G. (1992) *J. Mol. Struct. (Theochem.)*, **261**, 37.
Szczesniak, M. M., Chalasinski, G., Cybulski, S. M. and Scheiner, S. (1990) *J. Chem. Phys.*, **93**, 4243.
Szczesniak, M. M., Kendall, R. and Chalasinski, G. (1991) *J. Chem. Phys.*, **95**, 5196.
Szczesniak, M. M., Chalasinski, G. and Piecuch, P. (1993) *J. Chem. Phys.*, **99**, 6732.
Tachikawa, M. and Iguchi, K. (1994) *J. Chem. Phys.*, **101**, 3062.
Tang, K. T. and Toennies, J. P. (1978) *J. Chem. Phys.*, **68**, 5501.
Tang, K. T. and Toennies, J. P. (1984) *J. Chem. Phys.*, **80**, 3726.
Tang, K. T. and Toennies, J. P. (1990) *Chem. Phys. Letters*, **175**, 511.
Tao, F.-M. and Klemperer, W. (1994) *J. Chem. Phys.*, **101**, 1129.
Tao, F.-M. and Klemperer, W. (1995) *J. Chem. Phys.*, **103**, 950.
Tao, F.-M. and Pan, Y.-K. (1992) *J. Chem. Phys.*, **97**, 4989.
Tao, F.-M., Drucker, S., Cohen, R. C. and Klemperer, W. (1994) *J. Chem. Phys.*, **101**, 8680.
Tao, F.-M., Drucker, S. and Klemperer, W. (1995) *J. Chem. Phys.*, **102**, 7289.
Wang, Y.-B., Tao, F.-M. and Pan, Y.-K. (1994) *J. Mol. Struct. (Theochem.)*, **309**, 235.
Weber, A. (ed.) (1987) *Structure and Dynamics of Weakly Bound Molecular Complexes*, NATO ASI Series C, Vol. 212, Reidel, Dordrecht.
Wheatley, R. J. and Meath, W. J. (1993) *Mol. Phys.*, **79**, 253.
Williams, H. L. (1995) PhD dissertation, University of Delaware, Newark, DE.
Williams, H. L., Szalewicz, K., Jeziorski, B., Moszynski, R. and Rybak, S. (1993) *J. Chem. Phys.*, **98**, 1279.
Williams, H. L., Mas, E. M., Szalewicz, K. and Jeziorski, B. (1995a) *J. Chem. Phys.*, **103**, 7374.
Williams, H. L., Szalewicz, K., Moszynski, R. and Jeziorski, B. (1995b) *J. Chem. Phys.*, **103**, 4586.
Williams, H. L., Korona, T., Bukowski, R., Jeziorski, B. and Szalewicz, K., (1996) Chem. Phys. Lett., **262**, 431.
Wormer, P. E. S. (1975) PhD thesis, University of Nijmegen.
Wormer, P. E. S., Mulder, F. and van der Avoird, A. (1977) *Int. J. Quantum Chem.*, **11**, 959.
Zwart, E. and Meerts, W. L. (1991) *Chem. Phys.*, **151**, 407.
Zwart, E., Linnartz, H., Meerts, W. L., Fraser, G. T., Nelson, D. D. and Klemperer, W. (1991) *J. Chem. Phys.*, **96**, 793.

2 *Ab Initio* Studies of Nonadditive Interactions

M. M. SZCZĘŚNIAK

Oakland University, USA

G. CHAŁASIŃSKI

University of Warsaw, Poland

2.1 INTRODUCTION

Weak intermolecular forces are of key importance in studies of the properties of matter in condensed phases. A molecular description of condensed-phase behavior relies on the full characterization of pairwise interactions. Over the past decade, great progress has been achieved in understanding pair interactions at both experimental and theoretical levels. As a result of recent advances in high-resolution spectroscopic techniques coupled with new mathematical approaches which allow the translation of the spectra into multidimensional surfaces, several very accurate potential energy surfaces (PESs) have become available (Cohen and Saykally, 1991, 1992; Hutson, 1990). A great deal of progress has also been achieved in the *ab initio* techniques which can now calculate PESs with almost spectroscopic accuracy (Chałasiński and Szczęśniak, 1994; Jeziorski et al., 1994). However, the description of interactions involving several atomic or molecular species requires not only the knowledge of pair interaction but also information on the so-called many-body forces:

$$V = \Sigma V_{ij} + \Sigma V_{ijk} + \Sigma V_{ijkl} + \ldots \quad (2.1)$$

where the terms on the right-hand side describe, respectively, two-, three-, four-, etc. body interactions.

Many-body interactions are known to affect directly and indirectly many properties of

Molecular Interactions. Edited by S. Scheiner
© 1997 John Wiley & Sons Ltd

solids and liquids (Barker, 1976; Klein and Venables, 1976). One of the first straightforward indications that these effects are important was the so-called 'paradox of rare gases'. The observed lattices of Ne, Ar, Kr, etc. were of the face-centered cubic type, while the predictions of lattice structures by the summation over the pair interactions predicted the hexagonal ones (Jansen, 1962; Niebel and Venables, 1976). Many-body effects strongly affect the properties of the most common substance: water. Cieplak *et al.* (1990) and Corongiu (1992) found it impossible to simultaneously determine the thermodynamic and structural properties of the gas, liquid and solid phases of water without the explicit consideration of many-body interactions. In nonpure systems such as solutions where the assumption of additivity is intrinsically questionable, many-body effects are known to dramatically alter the coordination numbers of solvated ions (Cieplak *et al.*, 1987), structures of solvation shells (Perera and Berkowitz, 1991), etc. Since all living processes take place in an aqueous solution the full understanding of these processes cannot be accomplished without the inclusion of many-body forces.

One way of bypassing the nonadditivity problems is to represent the total potential energy of N interacting molecules as a sum of two-body terms, the so-called effective pair interactions. Such effective potentials have been proposed for many substances by simulations of liquid-state properties (Jorgensen, 1986; Jorgensen and Tirado-Rives, 1988). It should be emphasized, however, that the effective potentials are 'effective' only for a limited range of properties and do not represent the true interactions within the pairs, but instead have folded into them various effects, such as nonadditive interactions, polarization of the surrounding, nonrigidity of monomers, effects due to oversimplified anisotropy, etc. Not surprisingly, these potentials are generally nontransferable from phase to phase.

Although the evidence for the departure from pairwise additivity has existed for quite some time, experimental studies were long hindered by the fact that there were no techniques available which could probe three-body forces without the complicating effects of higher many-body forces. This situation radically changed with recent advances in the spectroscopic techniques which are able to probe small van der Waals clusters (Cohen and Saykally, 1991; Nesbitt, 1994; Miller, 1990; Leopold *et al.*, 1994). These spectroscopic techniques coupled with advances in the theoretical treatments of vibrational dynamics (Hutson, 1991; Cooper and Hutson, 1993, Cohen and Saykally, 1992; van der Avoird *et al.*, 1994) moved investigations of many-body interactions to the forefront of the studies of intermolecular forces. Microwave studies of gas-phase trimers provide evidence of nonpairwise behavior via the observation of rotational constants. The high-resolution laser IR spectroscopy in slit jets allows for the observation of intermolecular combination bands built on top of intramolecular transitions in the near-IR region (Nesbitt, 1994). These techniques provide unique information on the n-body dependence of red-shifts in clusters of varying size. They also allow for studies of energy transfer dynamics, such as vibrational predissociation, intramolecular vibrational redistribution, etc. Another powerful technique involves the direct observation of intermolecular vibrations in the far-IR region (Cohen and Saykally, 1991, 1992, Saykally and Blake, 1993) via the so-called vibration-rotation-tunelling spectra (VRT). These vibrations serve as an extremely sensitive probe of the shape of intermolecular potential energy surfaces. However, the extraction of PESs from spectral features also requires a means of solving the ro-vibrational Schrödinger equation in many degrees of freedom, which is in itself a formidable task. These solutions are first carried out on the PESs obtained from the accurate pair-potentials; the three-body interaction potentials are then chosen so as to reproduce the observed spectral features. It should be stressed that accurate pair potentials are seldom available. Another problem with this procedure is the need for *a priori* postulating the form of the three-body potential. Very little is known on how to do it. Despite the difficulties, this approach proved very successful in elucidating three-body potentials in Ar_n-chromophore systems.

Generally, in these techniques three-body interactions are extracted as a very small effect comparable in magnitude to the errors in two-body potentials. Simultaneously with perfecting them, there are also experimental efforts directed toward finding such experiments which would measure the three-body effects directly. One interesting example is provided by Ar_nCO_2 clusters. In these clusters the nonlinearity of the n-dependence of red-shift of an asymmetric CO_2 stretch has been linked directly to the three-body interactions (Sperhac *et al.*, 1996). Another possible candidate is Ar_3. Ar_2 by symmetry cannot generate even a temporary dipole and therefore, if a collision-induced spectrum is ever observed for this species it would result directly from the three-body forces (for more discussion see Suhm and Nesbitt, 1995; Cooper *et al.*, 1993).

Despite great experimental progress, our fundamental understanding of nonadditivity is still incomplete. Part of the problem stems from the fact that many-body interactions behave differently from their two-body counterparts. For example, the three-body exchange term can be either attractive or repulsive, while the two-body exchange can be only repulsive. Furthermore, the three-body interactions can no longer be described in terms of only the monomer properties, and may require certain 'interaction-induced dimer properties' which cannot be obtained from classical considerations.

The theoretical modeling of nonadditive phenomena must take into account that there are numerous physical sources of nonadditivity, some of them classical in nature, and some of them resulting from the quantum behavior of electrons. The *ab initio* calculations can provide much needed insights by identifying the dominant effects and by examining their properties and behavior with respect to geometrical degrees of freedom (Szczęśniak and Chałasiński, 1992; Chałasiński and Szczęśniak, 1994).

There are two types of quantum-chemical approaches which could be used to evaluate the many-body potentials: supermolecular and perturbational. In the supermolecular approach the many-body terms are obtained by successive subtractions of energies of n-mers and their various subsets up through the monomers. For this procedure to be meaningful it must provide the energies of these n-mers with similar accuracy; otherwise the errors will magnify. So for obvious reasons only methods which are size-consistent can be used. The accuracy of energy evaluation also depends in a major way upon the basis set. For this reason the treatment of interaction energies should also be basis-set consistent, i.e. the interaction energies should be evaluated in the same basis set of the whole cluster. Yet another requirement is that a method must describe all the effects which give rise to nonadditivity, such as intersystem and intrasystem electron correlation, intermolecular exchange of electrons, etc. The electron-correlation components prove to be particularly demanding in the choice of applicable methods. Finally, a successful computational scheme must provide numercial values which, besides reliability, should be physically interpretable.

Most insights on how to use the supermolecular treatment ultimately come from the perturbation theory of intermolecular forces. This theory treats the many-body interaction as a small perturbation of the energies of monomers. It offers a physical interpretation to the nonadditive contributions by expressing them in the form of a sum of physically sound components. However, this approach is not free from drawbacks. Some of the problems stem from the need to satisfy the antisymmetry requirement for the trimer wavefunction. Modern approaches have largely overcome these difficulties, and the symmetry-adapted treatments are now available which can handle both the permutational symmetry of the wavefunction and the intramolecular correlation effects (Szalewicz and Jeziorski, 1979; Jeziorski et al., 1994). The most useful in practical calculations of the many-body interactions is the treatment combining the supermolecular and perturbational approaches. This treatment allows for a dissection of the supermolecular nonadditive terms into perturbation components which have a clear physical sense (Chałasiński and Szczęśniak, 1988, 1994).

Many-body effects have been reviewed in the past on many occasions. Early model considerations were discussed in the review by Margenau and Stamper (1967) and in the monograph by Margenau and Kestner (1967). The early *ab initio* results have been summarized by Schuster *et al.* (1980). Many issues pertaining to nonadditive properties of rare gases have been discussed in the monograph *Rare gas solids* (1976) and in reviews by Meath and Aziz (1984) and Meath and Koulis (1991). *Ab initio* and perturbational results have been reviewed by Szczęśniak and Chałasiński (1992). More recently, Elrod and Saykally (1994) have presented a comprehensive overview of the experimental and theoretical approaches to many-body effects. The spectroscopy and calculations on HF-containing clusters have been surveyed by Suhm and Nesbitt (1995), while H_2O clusters have been reviewed by Liu *et al.* (1996).

In this chapter we will address some recent developments in the theoretical treatment of many-body interactions. For a number of model systems which span a wide range of intermolecular interactions we will discuss the *ab initio* calculations of many-body effects and their interpretation in terms of the perturbation theory of intermolecular forces. We will focus primarily on the physical insights which such an approach can provide.

The structure of this article is as follows. In Section 2.2 we describe the *ab initio* theory of many-body effects. In Section 2.3 the physical interpretation of various many-body terms will be discussed. Section 2.4 will include the results for the selected types of clusters: rare gases, RG_2-chromophore, trimers of polar molecules, and ion–molecule–molecule interactions. For selected clusters from each group we will discuss the origins of nonadditive interactions, the dependence of these interactions on geometrical variables, and the analytical modeling of the three-body potentials.

2.2 *AB INITIO* THEORY OF NONADDITIVE EFFECTS

2.2.1 Classification

Since the seminal monographs of Hirschfelder *et al.* (1954) and Margenau and Kestner (1967) (see also Hirschfelder and Meath, 1967; Kaplan 1986; Arrighini, 1981; Claverie, 1978; Jeziorski and Kolos, 1982), it has been known that the best framework to analyse the origins of nonadditive interactions is within the perturbation theory of intermolecular forces. The classic Rayleigh–Schrödinger (RS) perturbation theory formalism predicts that there are three fundamental long-range interaction energy components: electrostatic, induction and dispersion. The theory correctly predicts that two of them, induction and dispersion, are nonadditive. In the intermediate and short range the exchange effects, which are neglected by the RS formalism, appear and they are also nonadditive. Thus, the treatment of nonadditive effects requires theories which can deal with the intermolecular permutational symmetry of the wavefunction. One such theory is the symmetry adapted perturbation theory (SAPT), which provides a rigorous formal framework for including permutational symmetry in the perturbation expansion (Jeziorski and Kolos, 1982; Jeziorski *et al.*, 1993, 1994). Since it is applied to many-electron problems through the Møller–Plesset's choice of the zero-order hamiltonian and many-body perturbation theory expansion, it is often referred to as the intermolecular Møller–Plesset perturbation theory (I-MPPT) (Chałasiński and Szczęśniak, 1994).

The only nonadditive effect which can be interpreted within classical physics is the induction interaction. The induction effect originates from a classical electric polarization of interacting species. Two other mechanisms of nonadditivity arise only because of the quantum-mechanical character of the molecular world. The dispersion energy results from a quantum-mechanical Coulomb correlation of electrons belonging to different monomers. The exchange energy is also a quantum effect caused by the Fermi correlation of electrons from different monomers, as imposed by the Pauli exclusion principle. The induction and dispersion nonadditives can be described by classical or semi-classical models related to the properties of interacting monomers such as static and dynamic polarizabilities and charge distributions. The exchange nonadditivity cannot be directly related to monomer properties and classical models of interactions.

A classification and analytical description of two- and many-body effects in the framework of the RS perturbation theory has been summarized by Piecuch (1988). The two-body induction and dispersion energies appear in every order of perturbation expansion with respect to the interaction operator V, beginning with the second order. In addition, beginning with the third order in V, they can couple with each other. The corrections, which include three-body terms are listed in Table 2.1. The three-body induction terms appear in the second and higher orders. The three-body dispersion terms appear in the third and higher orders. In the third order we can also distinguish the dispersion–induction three-body terms. Inclusion of exchange effects introduces the nonadditive exchange corrections of the two following types: (1) the exchange counterparts of the RS nonadditive contributions (such as second-order exchange–induction or

Table 2.1 Nonadditive terms predicted by perturbation theory

'Classic' Rayleigh–Schrödinger	
Induction	$\varepsilon_{ind}^{(2)}, \varepsilon_{ind}^{(3)}, \ldots$
Dispersion	$\varepsilon_{disp}^{(3)}, \varepsilon_{disp}^{(4)}, \ldots$
Induction–Dispersion	$\varepsilon_{ind-disp}^{(3)}, \ldots$
Exchange	
Exchange counterparts of 'classic' terms	
	$\varepsilon_{exch-disp}^{(3)}, \ldots$
	$\varepsilon_{exch-ind}^{(2)}, \ldots$
Nonadditive counterparts of additive terms	
	$\varepsilon_{es}^{(1)} - \varepsilon_{exch}^{HL} \ (\approx \varepsilon_{exch}^{(1)})$
	$\varepsilon_{disp}^{(2)} - \varepsilon_{exch-disp}^{(2)}, \ldots$

third-order exchange–dispersion), and (2) exchange counterparts of the additive RS terms (such as second-order exchange–dispersion). This crude classification may help the reader to better appreciate a vast variety of different terms which formally must be considered.

Some initial ideas underlying the SAPT approach to many-body forces have been set forth by Löwdin (1956) in his study of elastic constants in ionic solids. A thorough conceptual framework for the perturbation treatment of exchange effects was laid out by Jansen (1962, 1965). Although the results of the various interaction energies obtained by Jansen and coworkers proved to be much too large (Meath and Aziz, 1984; Bulski and Chałasiński, 1987; Jeziorski, 1974), nevertheless the ideas such as the Gaussian effective electron model and exchange multipole interactions are very useful and have lately enjoyed renewed interest (Dotelli and Jansen, 1996; Cooper and Hutson, 1993; Ernesti and Hutson, 1995a,b). However, to formulate and implement a consistent symmetry-adapted theory of nonadditive three-body interactions, the model approach of Jansen had to be abandoned and a rigorous treatment of symmetry and advanced many-electron theory had to be worked out. This was achieved by Jeziorski and his collaborators in the 1970s (Jeziorski, 1974; Jeziorski et al., 1976), who built a framework for the treatment of all nonadditive effects, classic and exchange, within one well-defined and physically sound exchange perturbation theory. Since then the most important many-body contributions have been implemented and evaluated for many-electron systems (Cybulski, 1994). The first-order exchange or Heitler-London (HL) exchange nonadditivity, the second-order induction nonadditivity, and the third order dispersion nonadditivity for monomers described at the Hartree-Fock level, can be routinely calculated using codes developed by Cybulski (1994). Recently, the incomplete third-order induction energy has been added to this list (Cybulski, 1995). However, higher-order exchange corrections (in particular, the exchange–dispersion) have been calculated only for a few model systems (Bulski and Chalasinski, 1987).

The rapid development of SAPT which has occurred in recent years begins to permeate into the theory of nonadditive effects. Two important studies which provide programmable formulae for all first- and second-, and some of the third-order contributions for the many-electron monomer case, have only recently been completed (Moszynski et al., 1995; Lotrich and Szalewicz, 1996). We should keep in mind, however, that SAPT is not free from drawbacks which are mostly related to its convergence properties.

2.2.2 Convergence of SAPT

In clusters composed of many-electron species the viable formulations of SAPT, i.e. with weak symmetry forcing (Jeziorski and Kolos, 1977), are never convergent in the strict sense (Jeziorski et al., 1994; Korona et al., 1996). The 'enfant terrible' is the induction interaction, which reflects the deformation of the wavefunction by the field of a partner molecule. This deformation, if not restricted by the antisymmetry requirement, allows the electrons to flow into forbidden regions and causes the theory to sink into the Pauli-violating continuum of underground unphysical states (Claverie, 1978; Gutowski and Piela, 1988; Frey and Davidson, 1989; Adams, 1992). When the antisymmetrization is forced by the SAPT treatment of induction energy, the following problems arise. The unperturbed wave function Φ_0 should be first antisymmetrized and next perturbed by the field of a partner molecule to yield the induction energy properly restrained by exchange phenomena (Jeziorski and Kolos, 1977). Unfortunately, such a treatment would be untractable for many-electron systems. Instead, Φ_0 is first perturbed by the field and then antisymmetrized, which still leads to a divergent expansion of induction energy. This is the reason why the concept of SCF-deformation energy $\Delta E_{\text{def}}^{\text{SCF}}$ was introduced (Gutowski and Piela, 1988). This energy does not distinguish between induction and exchange–induction effects and treats the wavefunction deformation as resulting from electric polarization which is restrained by the Pauli exclusion principle. The only practical way to calculate SCF-deformation nonadditivity is through the supermolecular SCF interaction energy rather than by means of SAPT.

Yet another difficulty of SAPT arises in the short-range region where the convergence with respect to the interaction operator becomes slow and the application of perturbation theory is not legitimate. In addition, some of the higher-order corrections are derived only through the second or third power of overlap integral (S^2 or S^3) which is not adequate for regions of large overlap.

2.2.3 Intrasystem Correlation Effects

The many-electron theory is conveniently seperated into two levels: the Hartree–Fock (HF) level and the post-HF (correlated) level. The post-HF level may be treated as a sequence of treatments of electron correlation of increasing complexity. If the Møller–Plesset partitioning of the Hamiltonian is used, the perturbation theory can be advanced with two perturbing operators V (interaction) and W (intramonomer correlation) (Szalewicz and Jeziorski, 1979; Williams et al., 1993). Such a partitioning allows us to view each fundamental component of interaction energy as an infinite sum of corrections over the intrasystem correlation operator. So, for example, $\varepsilon_{\text{disp}}^{(2)}$ is a sum of $\varepsilon_{\text{disp}}^{(2k)}$ terms where k denotes the order in W.

2.2.4 Combination of Supermolecular and PT Treatments of Nonadditive Effects

The above-mentioned drawbacks of SAPT may be partially alleviated by applying the supermolecular approach to the calculations of nonadditive interactions. In this approach the total energy of a cluster ABC can be decomposed as (Chalasinski et al., 1987)

$$E_{\text{ABC}}^{(i)} = \sum_{X=\text{A,B,C}} E_X^{(i)} + \sum_{X=\text{A,B,C}} \Delta E_X^{(i)} + \sum_{X>Y=\text{A,B,C}} \Delta E_{XY}^{(i)} + \Delta E_{\text{ABC}}^{(i)} \qquad (2.2)$$

where (i) denotes a particular level of theory, e.g. HF, an order of Møller–Plesset perturbation theory, or any other size-consistent treatment of correlation effects, such as Coupled Cluster (CC) Theory. The second, third and fourth terms describe, respectively, the one- two- and three-body contributions. The one-body term describes the effects of the geometry relaxation of subsystem X in the trimer. A two-body term $\sum \Delta E_{XY}^{(i)}$ describes the pairwise interaction between two monomers, and the $\Delta E_{\text{ABC}}^{(i)}$ term represents the three-body contribution arising between the relaxed-geometry monomers arranged in the same way as they occur in the complex.

A major drawback of the supermolecular approach is that *per se* it does not provide us with a means of partitioning a nonadditivity into physically sound contributions, making the modeling of supermolecular terms particularly difficult. Fortunately, the two approaches may be quite rigorously related to each other (Chałasiński and Szczęśniak, 1988; Chałasiński et al., 1990, 1994). When the supermolecular approach is based upon the Møller–Plesset perturbation theory such a connection can be easily accomplished. The approach that results from it combines the strengths of both methods and allows us to circumvent their respective drawbacks. For example, the problems with the SAPT convergence are avoided; simultaneously the single value of ΔE from the supermolecular approach gains a physical meaning. The relationship between the two theories is shown in Table 2.2.

The nonadditivity of ΔE^{SCF} encompasses nonadditivities of the exchange ε_{exch}^{HL} and deformation ΔE_{def}^{SCF} contributions. The former is always of a short-range character with respect to at least one of three intermonomer distances. (It means that with respect to that distance it decays asymptotically as $e^{-\alpha R}$.) The latter is determined asymptotically by the classic induction effects due to the mutual electrostatic polarization.

The $\Delta E^{(2)}$ nonadditivity may be formally dissected into an exchange part and a deformation part. The second-order deformation-correlation, $\Delta E_{def}^{(2)}$, includes two different effects: the intramonomer correlation correction to the SCF-deformation term (the deformation–correlation which encompasses the induction–correlation energy and allows for exchange effects), and the induction–dispersion effect along with its exchange part. The second-order exchange-correlation term, $\Delta E_{exch}^{(2)}$, includes the exchange counterparts of two different additive effects: second-order electrostatic-correlation, and the second-order dispersion.

The nonadditivity of $\Delta E^{(3)}$ may be thought of as composed of three parts: dispersion, exchange, and deformation. $\varepsilon_{disp}^{(30)}$ is the third-order dispersion nonadditivity of the uncoupled Hartree–Fock (UCHF) type; $\Delta E_{exch}^{(3)}$ is the third-order intracorrelation correction to the exchange effects and also includes the exchange dispersion correction; and $\Delta E_{def}^{(3)}$ is the third-order deformation–correlation effect.

Table 2.2 Summary of nonadditive effects arising in the supermolecular MPPT approach

Order of MPPT	Nonadditive IMPPT (SAPT) term	n-body nonadditivity
ΔE^{SCF}	ε_{exch}^{HL} ($\varepsilon_{exch}^{(10)}$)	3, 4, ..., n-body
	ΔE_{def}^{SCF}	3, 4, ..., n-body
$\Delta E^{(2)}$	$\Delta E_{def}^{(2)}$	3, 4, ..., n-body
	$\Delta E_{exch}^{(2)}$	3, 4, ..., n-body
$\Delta E^{(3)}$	$\varepsilon_{disp}^{(30)}$	3-body
	$\Delta E_{def}^{(3)}$	3, 4, ..., n-body
	$\Delta E_{exch}^{(3)}$	3, 4, ..., n-body
$\Delta E^{(4)}$	$\varepsilon_{disp}^{(40)}$	4-body
	$\varepsilon_{disp}^{(31)}$	3-body
	$\Delta E_{def}^{(4)}$	3, 4, ..., n-body
	$\Delta E_{exch}^{(4)}$	3, 4, ..., n-body

2.3 PHYSICAL INTERPRETATION

2.3.1 Heitler–London Exchange

Exchange effects are perhaps the most difficult to interpret because they result from the quantum character of electrons. Some progress has been achieved lately in the modelng of the Heitler–London exchange nonadditivity.

Let us recall the expression for the three-body HL-exchange term in the trimer composed of the three H atoms (the expression is exact in this case) (Jansen, 1962; Jeziorski *et al.*, 1976):

$$\varepsilon_{\text{exch,ABC}}^{\text{HL}} = N_{abc}\langle ABC|V_{ab} + V_{ac} + V_{bc}|\mathscr{A}_{abc}(ABC)\rangle - \Sigma N_{ab}\langle AB|V_{ab}|\mathscr{A}_{ab}(AB)\rangle \quad (2.3)$$

where N_{abc} and N_{ab} are normalization factors, $\mathscr{A}_{abc}$ and $\mathscr{A}_{ab}$ denote the trimer and dimer antisymmetrizers, respectively, and V denotes the interaction operator. $\mathscr{A}_{abc}$ includes all permutations of electrons among the components of the cluster, i.e. the single exchanges: (12), (13), (23) and triple exchanges (123), (132). Thus $\varepsilon_{\text{exch,ABC}}^{\text{HL}}$ should include the matrix elements of two types:

$$\text{single exchange (SE): e.g. } \langle ABC|P_{12}V_{bc}|ABC\rangle \quad (2.4a)$$

$$\text{triple exchange (TE): e.g. } \langle ABC|P_{123}V_{bc}|ABC\rangle \quad (2.4b)$$

where P denotes the permutation operators related to the above-mentioned exchanges. Upon applying the Mulliken approximation it can be shown that the SE term is proportional to the square of the intersystem overlap integral (S^2) while TE is proportional to S^3 (Kołos, 1974; Kołos and Leś, 1972) From the physical standpoint, the SE term describes the exchange distortion within the charge distribution of a pair of monomers and the energetical effect of this distortion on the interaction with the third monomer. Due to the fact that SE is expressed by the corresponding Coulomb integrals it can be analytically modeled, for selected geometries, using the multipole-expanded electrostatic energy expressions. TE results from the simultaneous exchange distortions of all the three monomers and is expected to decay exponentially with respect to all three distances (Bulski and Chałasiński, 1980). Both terms are thus expected to display a different behavior with respect to the intermonomer degrees of freedom. The results of an exact partitioning into SE and TE are available for only a few complexes and will be discussed in Section 2.4.

2.3.2 Induction Nonadditivity

As mentioned above, the most essential part of induction nonadditivity has a purely classical interpretation. There exists, however, a portion of this effect which originates from a coupling between the induction and exchange effects. The latter part is much more difficult to extract. It is, thus, of fundamental importance (particularly for modeling purposes) to know which systems behave classically and which do not. It should be recognized that the limitations of the classical approach have less to do with the physical properties of the interacting systems and more with the mathematical convergence radius of the SAPT expansion of induction energy.

The classical two-body second-order induction energy, $\varepsilon_{\text{ind,r}}^{(20)}$, describes a favorable deformation of the charge cloud of one of the monomers in the field of the other which results in the energy lowering of this monomer. A description of this energy in terms of coupled Hartree–Fock density matrices and the so-called response properties is preferable (Moszynski et al., 1994), but multipole-expansion-type formulae are used more often. The Cartesian form of the multipole expansion of $\varepsilon_{ind,r}^{(20)}$ for the interaction between two molecules has been given by Buckingham (1978), and the spherical one by Wormer (1975), Stone (1989) and Piecuch (1988). The three-body contribution to $\varepsilon_{ind,r}^{(20)}$ has been analyzed by Stogryn (1971) and, more recently, by Piecuch (1986) who used the irreducible tensor formalism for this purpose. The three-body $\varepsilon_{ind,r}^{(20)}$ effect may be interpreted as an energy lowering due to the polarization of a charge cloud representing one of the monomers, say C, in the combined field of the remaining two molecules, say, A and B. Various simplified models of the second-order induction energy, which account for two- and three-body effects, have been used in molecular dynamics and Monte Carlo simulations. They usually involve electric fields produced by point charges interacting with polarizable atomic centers:

$$E_{\text{ind}} = -1/2 \sum_{i} \mu_i E_i^0 \tag{2.5}$$

where μ_i describe induced dipoles on polarizable centers and E_i^0 the unscreened Coulomb field.

The third-order nonadditivity, $\varepsilon_{ind,r}^{(30)}$, may be interpreted in terms of three possible mechanisms (Piecuch, 1986, 1988). The first is the Coulomb interaction between moments induced on A and B by the field of C. This term is particularly important in the interactions of two closed-shell atoms with a diatomic because it represents the first nonvanishing multipole contribution to three-body induction nonadditivity (Szczęśniak et al., 1993). The second type describes a situation in which the moments on A and B are induced by pairs of A and C or B and C, respectively (in both cases coupling all three monomers). The third type is a third-order analog of $\varepsilon_{ind,r}^{(20)}$. It describes a situation where static fields of A and B create a higher-order contribution to an induced moment on C. The explicit expressions for the multipole expanded nonadditivities of all three types were given by Piecuch (1988), who based his decomposition on the work of Stogryn (1971). Piecuch's definitions of the induction contributions can be easily interpreted using classical electrostatic arguments (see also Stogryn 1972). Interestingly enough, only the first category of the three-body third-order induction interactions survives the orientational averaging (assuming that only neutral species are considered), which underscores their importance in the gas phase interactions (Piecuch, 1988).

The interpretation of the third-order three-body terms becomes complicated by virtue of the fact that the dispersion and induction contributions to the trimer wave function are coupled through the third-order induction-dispersion term (Stogryn, 1971; Piecuch, 1986, 1988). This has recently led Moszynski et al. (1995) to question Piecuch's decomposition of the third-order three-body energy. It should be stressed that Piecuch's decomposition is correct; it agrees with the previous work of Stogryn, and it was recently confirmed by Lotrich and Szalewicz (1996) and by Li and Hunt (1996). To avoid further confusion it should be stressed that their criticism of the numerical values from (Szczęśniak et al., 1993; Cybulski et al., 1994) were also erroneous.

In the above discussion, the multipole expansion of the induction energy is related to multipole properties evaluated at one center at each monomer. A theoretical advantage of such an approach is its uniqueness. However, in practical calculations it is more efficient to use a multi-center expansion in terms of distributed multipole moments (Stone, 1981; Stone and Alderton, 1985) and distributed polarizabilities (Stone, 1989).

As discussed above, the SAPT series, $\Sigma \varepsilon_{ind,r}^{(n0)}$, converges to ΔE_{def}^{SCF} only in the case where no electrons are present on one of the monomers (i.e. no exchange of electrons occurs) (Cybulski, 1992). Otherwise, the induction energy couples very strongly with the exchange effects. The calculation of induction energies which are decoupled from the restraining effects of the Pauli principle may lead to an artificial 'charge transfer' in which the electrons of one monomer try to occupy the already occupied orbitals of the other. In these cases the description of the induction effect in terms of ΔE_{def}^{SCF} (cf. Table 2.2) is superior. An artificial charge transfer has been found in two-body induction energy calculations for Ar—Cl$^-$ (Burcl et al., 1995) In the Ar—Cl$^-$ interaction one could reasonably expect that the energy of Ar in the field of Cl$^-$ will be more significantly lowered than that of Cl$^-$ in the field of Ar. The calculated values of $\varepsilon_{ind,r}^{(20)}$ show that the Ar—> Cl-stabilization is one order of magnitude larger than Cl$^-$—> Ar: a clear manifestation of the incorrect behavior of the induction energy.

2.3.3 Dispersion Nonadditivity

The three-body dispersion energy initially appears in the third order of RSPT as $\varepsilon_{disp}^{(30)}$ and corresponds to the Hartree–Fock description of monomer wavefunctions. Its physical interpretation in terms of the interactions of instantaneous (fluctuating) dipole moments of monomers (ddd) was proposed some 50 years ago by Axilrod and Teller (1943) and independently by Muto (1943):

$$V_{ddd} = 3 C_{ddd}^{(3)} \frac{1 + 3 \cos \theta_1 \cos \theta_2 \cos \theta_3}{R_{12}^3 R_{23}^3 R_{31}^3} \tag{2.6}$$

where R_{ij} describe the sides and θ_i the angles of a triangle formed by three atoms while $C_{ddd}^{(3)}$ stands for a three-body dispersion coefficient. If three monomers are collinear, the arrangement of instantaneous dipoles is favorable and the term is attractive. If, however, the monomers are in the triangular configuration, the instantaneous dipoles induced by monomer A on B and C are in repulsive orientation. In the nonexpanded three-body $\varepsilon_{disp}^{(30)}$ energy the higher multipole terms, such as dipole–dipole–quadrupole, dipole–quadrupole–quadrupole, etc., are also implicitly present along with the charge-overlap effects.

When the correlation effects are included in the description of monomer wavefunctions, the dispersion nonadditivity is said to include the intramonomer correlation effects. These effects are extremely important in trimers of rare gases. For example, it is estimated that they lead to an increase of the three-body dispersion energy by some 36 per cent in the equilibrium He$_3$ trimer, and by nearly 50 per cent in the equilibrium Ne$_3$ (Chałasiński et al., 1994). Calculations of these effects require highly correlated approaches and even the supermolecular calculations are nontrivial. Again, valuable insights have been gained from the relationship between S-MPPT and SAPT (see below).

As indicated in Table 2.2, the four-body dispersion components appear in the fourth order of the perturbation theory as a UCHF $\varepsilon_{disp}^{(40)}$ terms. In the supermolecular MPPT calculations of the four-body effects this term arises as a four-body part of $\Delta E_{\rm D}^{(4)}$ where D denotes double excitations. As a general rule, in order to reproduce an n-body UCHF dispersion term, it is required to carry out at least the nth-order MPPT supermolecular calculations (Chałasiński et al., 1990, 1994) (see Section 2.3.1 for more discussion).

2.4. APPLICATIONS

2.4.1 Rare-gas Trimers

Rare-gas clusters have received a great deal of attention in the literature (Margenau and Kestener, 1967; Jansen, 1962; Klein and Venables, 1976; Meath and Koulis, 1991). One advantage is that their pair potentials have been precisely known for some time. Another attractive feature is the relatively simple composition of interaction energy in these systems. Due to the spherically symmetric charge distribution within the monomers, the long-range parts of the electrostatic and induction effects are absent, and thus the interaction is essentially reduced to the exchange and dispersion components. The atomic character of monomers assures the simplest possible form of PESs. Therefore, the RG clusters are ideally suited for an in-depth study of the many-body effects related to the exchange and dispersion interactions.

The rare-gas clusters are interesting on their own. When studying RG crystalline solids, one is immediately confronted with important and intriguing discrepancies between theoretical model calculations and experiment. Not only has the value of the binding energy of a crystal proved impossible to reproduce without the accurate treatment of pair- and many-body interactions, but the nonadditive effects were also shown to play a crucial role in the resulting fcc (rather than hcp) structure of heavier RG crystals. A reliable calculation and an appropriate analytical modeling of the nonadditivity in rare gases are crucial to solving these and other problems in solid-state simulations.

In this section we will focus on the following problems in this vast area of research: (1) the analysis of the dispersion nonadditivity arising in supermolecular calculations; (2) a case study of the equilateral Ar_3; (3) a discussion of a few of the most reliable results for RG_3. The conclusions for this section are applicable not only to rare gases but also to nonpolar molecular clusters such as $(CH_4)_3$ (Szczęśniak et al., 1990).

2.4.1.1 Dispersion Nonadditivity in Supermolecular Calculations

For the two-body dispersion energy the SAPT formalism leads to the following hierarchy of dispersion-type corrections which can be easily classified according to the degree of the intrasystem correlation effects on dispersion:

$$\text{Two-body: } \varepsilon_{\rm disp}^{(20)}, \varepsilon_{\rm disp}^{(21)}, \varepsilon_{\rm DQ,disp}^{(22)}, \varepsilon_{\rm SDQ,disp}^{(22)}, \varepsilon_{\rm SDQT,disp}^{(22)} \quad (2.7)$$

where the second index in the $\varepsilon_{\rm disp}^{(ij)}$ corrections refers to the intrasystem correlation operator. For the three-body dispersion effects the analogous series can be postulated using diagrammatic arguments:

$$\text{Three-body: } \varepsilon_{\rm disp}^{(30)}, \varepsilon_{\rm disp}^{(31)}, \varepsilon_{\rm DQ,disp}^{(32)}, \varepsilon_{\rm SDQ,disp}^{(32)}, \varepsilon_{\rm SDQT,disp}^{(32)} \quad (2.8)$$

In this series the first term, $\varepsilon_{\text{disp}}^{(30)}$, describes the three-body dispersion nonadditivity arising among the monomers represented at the Hartree–Fock (i.e. uncorrelated) level. The subsequent terms denote the corrections to dispersion resulting from the varying levels of correlation of the monomer wavefunctions. For example, $\varepsilon_{\text{SDQT,disp}}^{(32)}$ denotes the component of the dispersion effect if monomers are correlated with single-, double-, triple- and quadruple-excitations. It is apparent that the three-body effects require one order higher treatments that their two-body analogs. The similarity of the series of equations (2.7) and (2.8) also suggests that the convergence properties of both series should be similar. Thus, if the sum of the two-body series (equation (2.8)) is known (e.g. from *ab initio* calculations or semiempirical data) the bounds to its three-body counterpart can also be provided.

The relationship between the S-MPPT and I-MPPT provides an additional computational advantage in that it allows for the following mapping of the many-body dispersion terms:

```
S-MPPT              I-MPPT order,
order:              dispersion term:
 (i)                     (ij)
  2        20
  3        30 — 21
  4        40 — 31 — 22
  5        50 — 41 — 32 — 23
  6        60 — 51 — 42 — 33 — 24
  •         •     •    •    •    •
  •              6-body 5-body 4-body 3-body 2-body
```

The map describes the dispersion components (right) which are implicitly present in the supermolecular $\Delta E^{(i)}$ terms of the corresponding order (left). For example, if the two-body dispersion effects are reproduced up through the full (22) order (which are implicitly present in MP4 with inclusion of SDT, and Q excitations), the similar balance of intra–inter effects in the three-body case would require the full (32) order calculations (which are implicitly present in MP5 with inclusion of SDT, and Q excitations). Alternatively, we can turn to the Coupled Cluster theory with single-, double-, and noniterative triple excitations (CCSD(T)), since CCSD(T) may be viewed as the infinite-order limit of MPPT with the S, D, Q and T excitations.

$$\Delta E_{\text{ABC}}^{\text{CCSD(T)}} \approx \sum_{i=0}^{\infty} \Delta E_{\text{SDQT,ABC}}^{(i)}$$

where the '≈' sign indicates that the T excitations in CCSD(T) theory are not exact. The supermolecular CCSD(T) calculations of the three-body effect are expected to implicitly include all the three-body dispersion terms present in equation (2.8). We carried out such calculations for He_3, Ne_3, and Ar_3 and provided for these trimers the estimates of dispersion effect which are 5 per cent accurate (He_3, Ar_3) or 10 per cent accurate for Ne_3, which is notorious for its slow convergence. We conclude that the supermolecular CCSD(T) approach, although lacking in insights into the internal structure of dispersion effect, is, so far, the most efficient means of calculating the three-body dispersion term arising among the internally correlated monomers.

2.4.1.2 Analysis of Three-body Effects in the Equilateral Triangle Ar_3

As a case study, in Table 2.3 we show a decomposition of the three-body effects in Ar_3 in the equilateral triangle geometry (Chałasiński et al., 1990). The decomposition and related discussion will be largely representative not only for RG trimers and other cyclic trimers of non-polar species but also for the short-range exchange and long-range dispersion components of any cyclic trimers.

In the short, strongly repulsive region (5 and $6a_0$) the HL-exchange nonadditivity provides a major attractive contribution to the three-body interaction. This attraction is cancelled to a great extent by the repulsive three-body $\Delta E^{(2)}$ and $\Delta E^{(3)}$ corrections. Interestingly, $\Delta E^{(2)}$ which in RG_3 is practically dominated by exchange effects, is larger than $\Delta E^{(3)}$, which includes the dispersion nonadditivity. Indeed, in the repulsive region, all higher correlation terms through the fourth order are expected to be determined by exchange effects, primarily by the exchange–dispersion and the HL-exchange–correlation components.

Around the van der Waals minimum ($7a_0$) the three-body ε_{exch}^{HL} and $\Delta E^{(3)}$ terms represent the two largest contributions which cancel each other to a great extent. The second-order exchange nonadditivity $\Delta E^{(2)}$ is also quite important, as it is more than half as large as $\Delta E^{(3)}$. It is expected to be dominated by the exchange–dispersion component. For example, in He_3 in the equilibrium configuration, the exchange–dispersion and the $\Delta E^{(2)}$ nonadditivities agree quantitatively (Chałasiński et al., 1994).

Finally, at the outset of the long-range distances, $8a_0$, the $\Delta E^{(3)}$ nonadditivity comes to dominate the three-body interaction. Again, this is expected since in this region the $\Delta E^{(3)}$

Table 2.3 Three-body effects (in μE_h; $1\,\mu E_h = 0.219$ cm^{-1}) in the equilateral triangle configuration of Ar_3 (the CP-uncorrected values in parentheses) (from Chałasiński et al., 1990)

	\multicolumn{5}{c}{R, a_0}
	5.0	6.0	\multicolumn{2}{c}{7.0}	8.0
			$[7s4p2d]$	$[7s4p2d1f]$	
ΔE^{SCF}	−2888.0	−230.9	−16.0 (1.3)	−15.8	−1.0
ε_{exch}^{HL}	−2749	−212.9	−14.6		−0.9
SE	2536	138.6	7.0		0.3
ΔE_{def}^{SCF}	−139.0	−18.0	−1.4		−0.1
$\Delta E^{(2)}$	436.0	70.6	9.2 (26.2)	10.8	1.1
$\Delta E^{(3)}$	167.0	57.3	17.7 (17.1)	21.8	5.5
$\varepsilon_{disp}^{(3)}$	381.0	79.1	19.6		5.7
$C_{ddd}^{(3)} R^{9a}$	340.0	65.8	16.4		4.9
$\Delta E^{(4)}$	−66.0	−15.8	−4.9 (−8.8)	−6.3	−1.6
ΔE^{MP4b}	−2351.0	−118.8	6.3 (35.9)	10.5	4.0

$^a C_{ddd}^{(3)} = 663$ au (determined from the calculation of $\varepsilon_{disp}^{(30)}$ at $R = 40\,a_0$).
^{b} Sum of $\Delta E^{SCF} + \Delta E^{(2)} + \Delta E^{(3)} + \Delta E^{(4)}$.

nonadditivity includes the triple dipole ATM term which determines the asymptotic behavior of the total nonadditivity in RG$_3$. The remaining three-body components, $\varepsilon_{\text{exch}}^{\text{HL}}$, $\Delta E^{(2)}$, $\Delta E^{(4)}$, are similar in magnitude and tend to cancel one another. Interestingly, $\Delta E^{(2)}$ is already larger than $\varepsilon_{\text{exch}}^{\text{HL}}$, which is attributed to the fact that at large distances the exchange–dispersion (included in $\Delta E^{(2)}$) becomes the dominant exchange contribution for RG$_3$.

At this point, we cannot help commenting on the importance of using the counterpoise procedure to calculate interaction energy terms. Since the early work of Bulski (1981) and Wells and Wilson (1985) it has been known that the failure to remove BSSE leads to numbers which may be qualitatively wrong. To see this, the entries in parenthesis for $R = 7a_0$ provide an example of the BSSE uncorrected values of the interaction energy. It should be stressed that they appear to be random and elude any rationalization. Yet, time and again, analytical potentials are being fitted to BSSE-uncorrected results: cf. the recent study of helium clusters (Parish and Dykstra, 1993). In the latter study the correlation part of the nonadditivity in the equilibrium is as much as three times too repulsive and predicts the wrong sign of the total effect.

A few general observations should also be made. The ATM dispersion term has so far received the greatest attention from researchers following the widespread belief that it is representative of the total nonadditivity in rare gases. To illustrate how erroneous this belief is, we compare in Table 2.3 the ATM term, $C_{ddd}^{(3)} R^9$, with $\varepsilon_{\text{disp}}^{(30)}$ which includes it as the asymptotic term, with the total nonadditivity evaluated through the full fourth order, ΔE^{MP4}. It is clear that $C_{ddd}^{(3)} R^9$ approximates the total only at long distances. At $R = 6a_0$ and beyond $C_{ddd}^{(3)} R^9$ fails to reproduce both the sign and the magnitude of the total effect. Clearly, the short-range exchange nonadditivity must be included.

The leading contribution to the exchange nonadditivity is $\varepsilon_{\text{exch}}^{\text{HL}}$ which may be further decomposed (see Section 2.3.1) into SE and TE contributions. The SE term (see Table 2.3) is quite large and repulsive, and it reduces the large and attractive effect of the TE term. Thus SE alone cannot be used to model the $\varepsilon_{\text{exch}}^{\text{HL}}$ nonadditivity.

The above analysis sheds some light on the modeling three-body effects in RG$_3$. In the case of Ar$_3$ a very promising model has recently been advanced by Cooper and coworkers (Cooper and Hutson, 1993; Cooper et al., 1993). Following Jansen (1965) they represented the short-range exchange part by a model of a single effective electron in Gaussian orbital on each atom. An important feature of this model is that it correctly predicts the sign and qualitative behavior of the short-range nonadditivity for the full range of RG$_3$ geometries, from triangular to linear. At the same time, the dispersion nonadditivity was represented by the sum of the multipole terms ddd and ddq. The preliminary results are very encouraging and can be perfected in the future if more *ab initio* data are used to adjust the short- and long-range terms. This model is also applicable to mixed trimers, e.g. Ne$_2$–RG and Ar$_2$–RG (Ernesti and Hutson, 1995), although in this case Ernesti and Hutson noted that the two-body potentials were insufficiently accurate. Nevertheless, the work on the mixed species will most likely continue because of the recent microwave measurements (Xu et al., 1994).

One final note: many concepts of Jansen and his collaborators from the 1960s have now proved to be very useful in the modeling of nonadditive terms. Their quantitative aspects were thoroughly criticized (Meath and Aziz, 1984; Bulski and Chałasiński, 1987, and in many references therein), and evidently they cannot compete with the *ab initio* approach. However, as a result of recent work by Hutson and coworkers (Ernesti and

Hutson, 1994, 1995; Cooper *et al.*, 1993), the ideas of effective electron and exchange quadrupole turned out to be valid and insightful.

2.1.4.3 Some Best Results for RG Cyclic Trimers

Our best results for the equilibrium RG trimers (equilateral triangle) are shown in Table 2.4 (Chałasiński *et al.*, 1994). These calculations were carried out with an extended basis set supplemented with bond functions. The entries denoted 'accurate' were obtained by assessing the basis set unsaturation of the dispersion nonadditivity.

He_3: The best estimate for total nonadditivity is equal to -0.33 μE_h at the CCSD(T) level (Table 2.4). Some previous results are listed in Table 2.5. The result of Wells and Wilson (1985) of -0.43 obtained at the MP3 level is more attractive, but agrees perfectly with our value at the same level of theory (-0.44, from Table 2.4). Recently, Tao (1994) reported a value of -0.38 μE_h at the MP4 level, again very close to our MP4 value of -0.37. Tao also noted as 'very surprising' the fact that the two-body energy is much more sensitive to the basis set extensions than the related three-body term. The basis set

Table 2.4 The supermolecular CCSD(T) calculations of the two- and three-body components (in μE_h; 1 $\mu E_h = 0.219$ cm^{-1}) and their decomposition for rare gas trimers in equilateral triangle configuration (Chałasiński *et al.*, 1994)

Energy term[a]	He$_3$ ($R = 5.6$ a_0)		Ne$_3$ ($R = 6.0$ a_0)		Ar$_3$ ($R = 7.0$ a_0)	
	2-body	3-body	2-body	3-body	2-body	3-body
ΔE^{SCF}	29.194	-0.870	61.212	-0.567	516.04	-15.78
$\Delta E^{(2)}$	-49.718	0.155	-134.352	0.204	-895.56	11.39
$\Delta E^{(3)}$	-8.108	0.275	-15.377	1.440	129.51	21.94
$\Delta E^{(4)}_{SDQT}$	-0.260	0.075	-23.136	0.156	-100.46	-6.78
ΔE^{MP4}	-31.867	-0.366	-111.653	1.233	-350.47	10.77
ΔE^{CCSD}	-28.678	-0.374	-96.21	1.37	-245.02	10.43
$\Delta E^{CCSD(T)}$	-33.249	-0.334	-116.02	1.66	-351.04	14.46
Accurate	-34.77[b]	-0.34[c]	-129.3[d]	1.81[c]	-448.5[e]	16.6[c]

[a] $\Delta E^{MP4} = \Delta E^{SCF} + \Delta E^{(2)} + \Delta E^{(3)} + \Delta E^{(4)}_{SDQT}$.
[b] from Vos *et al.* (1991).
[c] Extrapolated in Chalasinski *et al.* (1994).
[d] from Ng *et al.* (1979).
[e] from Tang and Toennies (1984).

Table 2.5 Comparison of literature results (in μE_h; 1 $\mu E_h = 0.219$ cm^{-1}) for the three-body interaction energy in He$_3$ in the equilateral triangle geometry at $R = 5.6$

References	Energy	Method
Wells and Wilson (1985)	-0.43	MP3
Tao (1994)	-0.38	MP4
Mohan and Anderson (1990)	3(32)	Quantum Monte Carlo (sampling of statistical error)
Røeggen (1990)	-0.26	Second-order exchange PT
Røeggen and Almlöf (1995)	-0.16	Extended Group Function Model
Chałasiński *et al.* (1990)	-0.37	MP4
Chałasiński *et al.* (1994)	-0.33	CCSD(T)

dependence of the two-body versus the three-body components has been rationalized by Chałasiński et al. (1989) on the grounds of a different composition of the two- and three-body terms. In He$_2$ at 5.6a_0 the basis set dependence of the two-body term is largely due to a slow saturation of its dominating effects—the second-order dispersion (the exchange repulsion saturates very easily). In the three-body term the second-order dispersion does not appear because it is additive; at the same time, the third-order dispersion plays a much smaller role in the overall three-body effect, and thus its basis set dependence is less pronounced. It should be stressed that the analysis of the effects of the basis set on the individual components is crucial to an understanding of the dependence of the total many-body energies upon the basis set and should always guide the basis set selection.

Other estimates of the three-body effect in He$_3$ have recently been reported. Mohan and Anderson (1990) applied the Quantum Monte Carlo approach, but unfortunately, the effect proved to be too minute to be correctly reproduced in the minimum. However, the method is expected to be much more effective in the large-repulsion region. Røeggen and Almlöf (1995) attempted to calculate the three-body contribution using a variant of Extended Group Function Model (EGFM), (Røeggen et al., 1995). The resulting value is twice as small as our result. Explaining the origin of this discrepancy would require a careful analysis of the EGFM method in terms of the perturbation theory of intermolecular forces.

Ne$_3$: The best estimates of the total three-body contribution for the equilibrium trimers (equilateral triangle) Ne$_3$ and Ar$_3$ amount to 1.81 μE_h and 16.6 μE_h, respectively. It should be emphasized that for multi-shell atoms, such as Ne and Ar, the MP4 and CCSD(T) two-body energies are very close, yet the respective three-body energies are distinctly different. This, of course, reflects the behavior of the dispersion energy rationalized above. As discussed above, the MP4 level provides a reasonably accurate treatment of the two-body dispersion effects in the two-body interactions. To achieve an equal treatment of the three-body dispersion a higher-level approach is necessary, such as CCSD(T). Neither CCSD nor MP4 are sufficient.

It is interesting to compare the role of three-body effects in the three considered trimers. The total nonadditivity is attractive for He$_3$ but repulsive for Ne$_3$ and Ar$_3$, which should lead to a shortening of R_e in He$_3$ and a lengthening of R_e in Ne$_3$ and Ar$_3$. One can also see that the relative magnitude of the total three-body effect versus two-body effect is different in each case: $+1.0$ per cent for He$_3$, -1.5 per cent for Ne$_3$ and -3.7 per cent for Ar$_3$.

2.4.2 RG$_2$-chromophore Clusters

Ar$_2$HCl: In the hope of elucidating the nature of three-body forces in Ar$_3$, researchers chose an isoelectronic model cluster, AR$_2$HCl, in which these forces could be probed by spectroscopy. In addition, the pair Ar–Ar and Ar–HCl interactions were known with high enough accuracy that the small three-body interactions could be determined. The microwave studies found that Ar$_2$HCl, like Ar$_3$, forms a triangular cluster which is nearly equilateral with the H atom pointing toward the center. Subsequently, Hutson et al. (1989) carried out a three-dimensional vibrational analysis of Ar$_2$HCl in the attempt to extract information on three-body potential from the experimental data. They concluded that the microwave spectroscopic information is insufficient to reconstruct the three-body potential uniquely because these spectra do not sample extended regions of the PES. They made the first predictions of the intermolecular bending frequencies of this complex. In a subsequent development, Elrod *et al.* (1991, 1993) observed the three far-IR bending frequencies of Ar_2HCl including the one which samples a secondary Ar_2ClH minimum on the surface. On the basis of these far-IR data Cooper and Hutson (1993) undertook a second attempt to reconstruct the three-body potential of Ar_2HCl by carrying out full-five dimensional vibrational calculations. The potential energy surface for this cluster was constructed by augmenting the known, highly accurate, semiempirical Ar–HCl and Ar–Ar two-body potentials by an *a priori* postulated three-body potential. Their three-body potential contained three terms: exchange, induction, and dispersion nonadditivites. The dispersion nonadditivity was represented by an ATM term, and the induction nonadditivity, as an interaction of multipoles induced by HCl on two Ar atoms up through R^{-9} power. Quite unexpectedly, they found that the sum of these two terms was insufficiently anisotropic. For the exchange nonadditivity they used the model which consisted of an interaction between the exchange-induced quadrupole moment on Ar_2 and the permanent moments on HCl molecule (exchange multipole). Aware that this term did not include the total exchange effect, they supplemented the exchange multipole term by the so-called exchange overlap contribution in which the first-order effective electron model of Jansen (1962) was fit to reproduce *ab initio* results of Chałasiński *et al.* (1991b). Their analysis of the response of the IR frequencies to the inclusion of these terms led them to the conclusion that only the former is anisotropic. Hence in the later calculations on Ar_2DCl and Ar_2HF (Elrod, 1994; Ernesti and Hutson, 1994, 1995a,b) the more cumbersome 'exchange-overlap' term was not included.

The *ab initio* calculations of the three-body potential for this cluster determined the relative importance of each term to the modeling of the total potential. *Ab initio* calculations (Szczęśniak *et al.*, 1993; Cybulski *et al.*, 1994) showed (see Figure 2.1) that the total three-body interaction is very anisotropic with respect to the in- and out-of-plane rotations of HCl. The dispersion component is only moderately anisotropic and the induction component is slightly more anisotropic than dispersion. The most anisotropic is the exchange nonadditivity. Interestingly, two parts of the total exchange nonadditivity, SE and TE, have an opposite behavior. In the multipole approximation the SE term corresponds to the exchange multipole interaction. As seen in Figure 2.1, the SE term is strongly repulsive for the $\Theta = 0°$ geometry, slightly attractive for the $\Theta = 0°$ angle and again becomes slightly attractive for $\Theta = 180°$. The TE contribution is attractive for $\Theta = 0°$ and cancels a large portion of the SE contribution. Thus the approximation of the entire exchange nonadditivity by only its SE part leads to too anisotropic a model.

Ar_2HF: There has been a great deal of interest in this cluster and its larger analogs Ar_nHF. The microwave studies of Gutowsky *et al.* (1987) determined that its structure is similar to that of Ar_2HCl. High-resolution near-IR spectroscopy was used in the study of effects of clustering ($n = 1 - 4$) on the red shift of the HF stretching vibration (McIlroy *et al.*, 1991) and on the DF red-shifts ($n = 1 - 3$; Farrell *et al.* 1995). Using the semiempirical pair potentials for Ar–HF and Ar–Ar, McIlroy and Nesbitt (1992) explored the minimum energy structures for Ar_nHF ($n = 1 - 4$). The effects of zero-point vibrations in these clusters were examined by Lewerenz (1996) using the Diffusion Monte Carlo (DMC) calculations using pair potentials supplemented with the triple-dipole dispersion nonadditivity. Larger clusters leading to the formation of solvation cage were studied by using simulated annealing with pair potentials (Liu *et al.*, 1994c) and by the

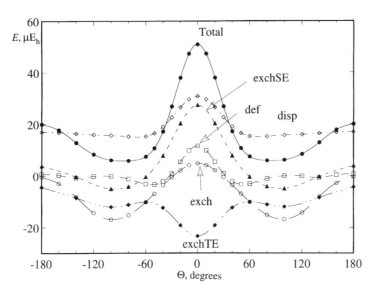

Figure 2.1 The dependence of the three-body components upon the in-plane rotation of HCl in the Ar_2HCl cluster. The following abbreviations have been used: 'exch'—ε_{exch}^{HL}; 'def'—ΔE_{def}^{SCF}; 'disp'—$\varepsilon_{disp}^{(30)}$; 'Total'—$\Delta E^{MP3}$; SE and TE denote single-exchange and triple exchange respectively

five-dimensional quantum bound-state calculations (Liu *et al.*, 1994d). The effect of a solvation cage on the photodissociation of HF was studied by Schröder *et al.* (1995). Although these simulations predominantly used pairwise additive potentials, simultaneous efforts tried to elucidate the three-body potential of Ar_2HF. These efforts proved that there are no universal recipes for doing so, even in the homologous series Ar_2HX.

The *ab initio* calculations of the three-body potential of Ar_2HF by Cybulski *et al.* (1994) reveal considerable differences between this system and Ar_2HCl (see Table 2.6). First, because of the much closer approach of HF to the center of mass of Ar_2, the HL-exchange nonadditivity is larger and much more anisotropic in Ar_2HF. A difference between the maximal and minimal value at a given distance (amplitude) of this interaction amounts to ca 42 μE_h in Ar_2HF versus 22 μE_h in Ar_2HCl. The induction nonadditivity is also much more anisotropic (amplitude of 38 μE_h versus 15 μE_h in Ar_2HCl). Only the dispersion nonadditivity is smaller and slightly less anisotropic, mostly due to the smaller polarizability of HF. The decomposition of the HL-exchange effect into its SE and TE components is seen in Figure 2.2. The TE term is slightly less anisotropic than that in Ar_2HCl; nevertheless, approximating the total term by its SE contribution leads to too high an anisotropy.

The supermolecular three-body $\Delta E^{(2)}$ component also proves very interesting. As stated above, it implicitly contains a number of nonadditive terms of different origins, such as the exchange counterpart of the (additive) electrostatic correlation (i.e. exchange–correlation), exchange–dispersion, deformation–correlation, and induction–dispersion terms. Unfortunately, no rigorous methods of decomposition of $\Delta E^{(2)}$ exist at the moment, and one can only speculate on the relative importance of each term. A comparison of $\Delta E^{(2)}$ nonadditivities in Ar_2HCl and Ar_2HF (Figure 2.3) shows the strikingly

Table 2.6 Comparison of nonadditive effects (in μE_h; $1\ \mu E_h = 0.219\ \text{cm}^{-1}$) the equilibrium Ar_2HCl and Ar_2HF clusters

	Ar_2HCl	Ar_2HF
ΔE^{SCF}	16.3	55.7
$\Delta E^{(2)}$	8.3	−10.0
$\Delta E^{(3)}$	26.3	20.2
ΔE^{MP3}	50.9	65.9
ε^{HL}_{exch}	4.8	24.9
SE	27.2	43.1
TE	−23.5	−20.0
ΔE^{SCF}_{def}	11.3	30.8
$\varepsilon^{(30)}_{ind,r}$	16.5	36.1
$\varepsilon^{(30)}_{ind-r,mult}$ [a]	11.1	27.6
$\varepsilon^{(30)}_{disp}$	30.6	21.5

[a] The third-order induction nonadditivity evaluated using the multipole expansion through the R^{-12} terms.
SE and TE denote the single-exchange and triple-exchange parts of ε^{HL}_{exch} (from Cybulski *et al.*, 1994).

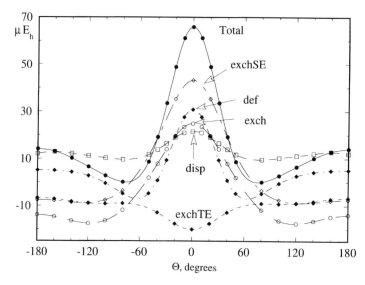

Figure 2.2 The dependence of the three-body components upon the in-plane rotation of HF in the Ar_2HF cluster. (See Figure 2.1 for abbreviations)

different behaviors in both cases. While in Ar_2HCl this term is only repulsive and has a weak maximum for $\Theta = 0°$, in Ar_2HF it varies from attractive, with a deep minimum at $\Theta = 0°$, to repulsive at $\Theta = 180°$. This suggests that the $\Delta E^{(2)}$ nonadditivity has a different composition in both cases. The exchange counterpart (SE) of the dispersion term and the dispersion-induction contribution can be estimated using the so-called pseudodimer approximation (Chałasiński *et al.*, 1996). The results for Ar_2HF displayed in Figure 2.4 indicate that the dispersion-induction has a deep minimum at $\Theta = 0°$, whereas

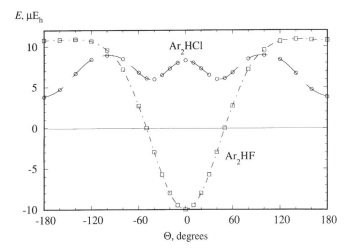

Figure 2.3 Comparison of the angular behavior of the $\Delta E^{(2)}$ nonadditivities in Ar_2HF and Ar_2HCl

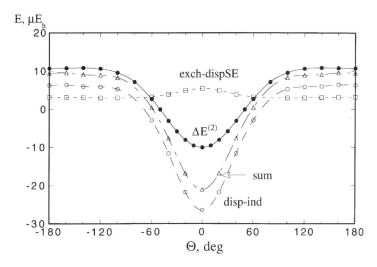

Figure 2.4 Analysis of the contents of $\Delta E^{(2)}$ nonadditivity in Ar_2HF. The SE parts of the exchange–dispersion and the dispersion-induction nonadditivities are estimated using the pseudodimer approach; 'sum' denotes the sum of the exchange–dispersion and dispersion-induction terms

the SE exchange dispersion term is positive for all angles with a maximum for $\Theta = 0°$. On this basis we can infer that the dominating component of the $\Delta E^{(2)}$ effect in Ar_2HF is the dispersion-induction, while the dominating part of the $\Delta E^{(2)}$ effect in Ar_2HCl may be the exchange–dispersion effect.

On the semiempirical front, Ernesti and Hutson (1995) found that the sum of terms modeled in an analogous way to Ar_2HCl was not appropriate for Ar_2HF. They found that the sum of the exchange, induction and dispersion terms was too anisotropic to reproduce observed IR spectra. They considered a number of modifications in the model three-body potential, which were as follows. In addition to the exchange-created quadrupole moment on the Ar_2 part (which was more appropriately represented as distributed dipoles) they considered the fact that the dispersion interaction between Ar atoms also leads to an instantaneous quadrupole moment on Ar_2 of the opposite sign. Since the induction effect also leads to the appearance of dipoles on Ar, there should be some cross-terms which couple the exchange with induction and dispersion effects and the induction with dispersion effects. By adding to it the dispersion nonadditivity, they obtained quite an elaborate three-body potential, much more so than the one for Ar_2HCl, which led to a reasonable description of bending HF frequencies. However, due to the complex form of this potential, with many terms of opposite signs, its unambiguous verification is more difficult.

The coupling terms included in the new three-body potential of Ernesti and Hutson roughly correspond to the contents of the $\Delta E^{(2)}$ nonadditivity discussed above. However, it should be stressed that their model of the exchange–induction and exchange–dispersion effects is capable of describing only some part of the SE contributions to these effects. It is unclear at present how important are the remaining SE terms, as well as the TE contributions. Our *ab initio* calculations confirmed their observations that there should be differences in the modeling of Ar_2HCl and Ar_2HF clusters, and that the differences are due to certain couplings among the basic nonadditive components. The quantitative models, however, must wait until such nonadditive terms as exchange–dispersion or induction–dispersion are rigorously calculated by *ab initio* approaches

RG_2CO_2: The pair potential for the $Ar–CO_2$ dimer has the minimum for the T structure of a C_{2v} symmetry with Ar perpendicular to the CO_2 molecular axis. According to *ab initio* results (Marshall *et al.*, 1996), the collinear position of Ar with respect to the CO_2 axis corresponds to a wide plateau on the PES. Several empirical PESs of $ArCO_2$ have recently been evaluated by Roche *et al.* (1996). The most likely structure of the Ar_2CO_2 trimer is one in which all the three pair interactions would be the least disturbed. Such a structure should involve two Ar atoms arranged in the T position with respect to the CO_2 axis and in the equilibrium distance with one another. In our *ab initio* calculation (Marshall *et al.*, 1996) we also explored a second configuration which is related to the linear structure of the dimer. Three-body effects in both configurations are compared in Table 2.7. In both orientations the three-body term is dominated by the dispersion nonadditivity. They differ markedly in their HL-exchange term. In the T configuration this term is slightly repulsive, while in the L configuration it is larger and attractive. This behavior can be rationalized by considering the SE part of the HL-exchange nonadditivity which can be interpreted in terms of interaction between the exchange quadrupole on Ar_2 and the permanent quadrupole moment of CO_2 (see Table 2.7; entry ($Ar_2 + CO_2$)). If we keep in mind that the quadrupole moment of CO_2 is negative (which means that its ends are charged negatively and its center positively) and the exchange quadrupole of Ar_2 is also negative, the interactions of positively charged centers of both quadrupoles in the T configuration leads to repulsion. In the L configuration, on the other hand, the positively charged center of the Ar_2 quadrupole faces the negatively charged end of the CO_2 molecule, leading to attraction. A similar effect is also observed in He_2CO_2. It should also be mentioned that the HL-exchange nonadditivity in the T structure results from a cancellation of repulsive SE effect and an attractive TE effect of almost equal magnitude,

Table 2.7 Three-body contributions (in μE_h; 1 $\mu E_h =$ 0.219 cm^{-1}) in two configurations of Ar$_2$–CO$_2$

	T	L
ΔE^{SCF}	2.2	−9.6
$\Delta E^{(2)}$	3.1	5.7
$\Delta E^{(3)}$	22.7	15.3
ΔE^{MP3}	27.9	11.5
$\varepsilon^{\text{HL}}_{\text{exch}}$	0.8	−10.7
$\varepsilon^{\text{HL}}_{\text{exch,SE}}(\text{Ar}_2+\text{CO}_2)$	9.1	−6.4
$\varepsilon^{\text{HL}}_{\text{exch,TE}}$	−9.9	−5.0
$\Delta E^{\text{SCF}}_{\text{def}}$	1.4	1.1
$\varepsilon^{(20)}_{\text{ind,r}}$	−3.5	2.8
$\varepsilon^{(30)}_{\text{ind,r}}$	3.7	–
$\varepsilon^{(30)}_{\text{disp}}$	23.3	16.4

The T configuration has two Ar atoms perpendicular to the CO$_2$ axis ($r(\text{C}...\text{AR}) = 7.0a_0$) and in 7.1$a_0$ distance from each other. In the L configuration the c.o.m. of Ar$_2$ is collinear with the CO$_2$ axis with $r(\text{C}...\text{Ar}) = 9.0a_0$ and $r(\text{Ar}...\text{Ar}) = 7.1a_0$ (SE and TE denote, respectively, the single-exchange and triple-exchange parts of $\varepsilon^{\text{HL}}_{\text{exch}}$ (from Marshall et al., 1996)

while in the L configuration SE and TE have the same sign and comparable magnitude (see Table 2.7). Thus, approximating the total HL-exchange effect by its SE component, i.e. by the exchange quadrupole interactions, may lead to large errors.

Ar$_2$CO$_2$ and larger clusters have recently been studied by high-resolution FT-IR spectroscopy by Sperhac et al. (1996). They noticed that because of the unusual structure of these clusters, the measurement of the CO$_2$ ν_3 transition (i.e. an antisymmetric stretch of CO$_2$) could serve as a direct *observable* measure of the many-body effects in these clusters. Their rationale was as follows. Due to the fact that the Ar atoms are in equivalent positions, in the pairwise additive approximation (i.e. when Ar atoms are not allowed to interact with each other), the addition of each Ar atom to CO$_2$ should contribute the same amount to the red-shift of ν_3. Thus one should expect a linear variation of the ν_3 red-shift with the number of Ar atoms. However, the experimentally observed shifts for $n = 0, 1, 2$ showed a slight departure from linearity. This departure indicated that when the Ar atoms are allowed to interact with each other the pair interacts differently with the ground $|000\rangle$ vibrational state of CO$_2$ than with the $|001\rangle$ excited state. Sperhac et al. estimated that the balance of the three interaction energy terms, exchange, induction and dispersion, leads to the net destabilization of the $|001\rangle$ vibrational state of CO$_2$, and thus to the ν_3 red-shift which is no longer linear in n. Although the effect found by Sperhac et al. is very small (0.042 cm^{-1}) this result is exciting in that it provides the first direct observation of a three-body effect.

In summary, the Ar$_2$-chromophore clusters are perhaps of little practical interest compared, for example, to interactions involving all polar species. However, they play a profound role in developing the spectroscopically verifiable models of three-body interactions. The progress achieved in this direction in the last few years has been very impressive due to the synergistic interaction of theory and experiment. A fine-tuning of the analytical models of three-body potentials in these systems requires further developments in computational methodology. It seems that the studies of higher homologues in the Ar_2HX series would also be very interesting.

2.4.3 Polar Trimers

The major difference between these trimers and those discussed in previous sections consists of the overwhelming role of the induction and SCF-deformation nonadditivity which in equilibrium essentially dwarfs all other nonadditive components. Due to its nonspherical charge distribution each monomer serves as a source of the electric field which in turn gives rise to the induction effect. The fact that various sites around monomers are not equivalent in terms of van der Waals interactions leads to an additional anisotropy with respect to monomer orientations in a trimer. The optimal cluster structure is essentially determined by minimizing pair interactions. If the cluster is to be stable, the monomers have to cooperate rather than compete, that is, no pair will assume the optimal isolated pair structure. This may be a large effect, but strictly, it is still a two-body not a three-body interaction. The three-body effects are commonly responsible for small adjustments of geometry. In stronger interactions, such as complexes involving multiple-charged ions, the three-body effects may be large enough to make a qualitative difference. In addition, the internal structure of monomers may also be distorted by the interaction (a one-body effect). This effect may be quite large compared with the interaction, and must be included in a potential function.

2.4.3.1 Water Trimer

The cyclic structure of the water trimer was first postulated by Dyke and Muenter (1972) on the grounds that the complex has no dipole moment. Early *ab initio* calculations of Kistenmacher *et al*. (1974) and Clementi *et al*. (1980) also suggested a cyclic trimer as the equilibrium structure. The gradient optimization study by Honegger and Leutwyler (1988) in a moderate basis set found a nonsymmetric (C_1) cyclic structure with the three sequential O—H...O hydrogen bonds. Two of the three free O—H bonds were found to be above the plane of the three O atoms and one below. Only fairly recently, using far-IR laser spectroscopy, Saykally and collaborators (Pugliano and Saykally, 1992; Liu *et al*., 1994a, b) succeeded in measuring VRT transitions for $(D_2O)_3$ and $(H_2O)_3$, thus providing the first definitive evidence that the water trimer has indeed a C_1 cyclic structure. These studies indicate a complex dynamics of the water trimer involving wide-amplitude torsional or 'flipping' motions of the free O—H bonds with respect to the O_3 plane. More recent *ab initio* studies (Mó *et al*., 1992; Xantheas and Dunning, 1993; Schütz *et al*., 1993; Fowler and Schaefer, 1995) provided a more extensive search of the trimer PES. Many-body effects have been studied by Clementi *et al*. (1980), Chałasiński *et al* (1991a), Xantheas (1994), van Duijneveldt and van Duijneveldt (1993, 1995), Kelterbaum *et al*. (1994), Tachikawa and Iguchi (1994) and Klopper *et al*. (1995). Various aspects of dynamics were investigated by Bürgi *et al*. (1995), Gregory and Clary (1995a,b) and Sabo *et al*. (1995).

The best estimate of the equilibrium interaction energy of -16.32 kcal/mol was obtained by Klopper *et al*. (1995). They found that the equilibrium structure denoted {uud} (for up–up–down) is 0.3 kcal/mol lower than {upd} (up–planar–down) and 0.8 kcal/mol lower than {uuu}. The optimal O...O distance is 2.80 Å and the deviation of hydrogen bonds from linearity amounts to about 20° (van Duijneveldt and van Duijneveldt, 1993; Klopper *et al.*, 1995; Xantheas and Dunning, 1993).

The effect of three-body forces on the optimal structure was studied by van Duijneveldt and van Duijneveldt (1993). Although their optimal O...O distance (2.824 Å) was slightly longer than 2.80 Å, by comparing pairwise approximation and the full calculations, they found that the shortening of O...O due to three-body effects is 0.060 Å (at the MP2 level) and is essentially determined at the SCF level. Interestingly, in the pairwise approximation the correlation effects shorten O...O by even more, 0.095!

Another structural factor in the trimer is the relaxation of the monomer geometries. This effect can be analysed on the basis of the results reported by Xantheas and Dunning (1993) for optimal structures for monomer, dimer and trimer obtained consistently at the same level of theory and with the same basis set. In the equilibrium dimer their aug-cc-pvdz/MP2 calculations predict that the hydrogen donor O—H bond becomes longer by *ca* 0.007 Å, while the free O—H bonds remain the same. In the equilibrium trimer the hydrogen donor O—H bonds become even longer by 0.011–0.012 Å, while the free O—H bonds continue to be the same. However, it is difficult to determine which part of the overall elongation is due to the three-body forces and which to the pair interactions.

The three-body forces also affect the OH stretching frequencies. A careful examination of the frequency shifts due to three-body effects was carried out by van Duijneveldt and van Duijneveldt (1993). They found that the trimer $\Delta\nu_{OH}$ shift is 70 per cent larger than the dimer value, mainly due to the pure nonadditive induction terms. Why is this change so large if the related changes in the intereaction energies do not exceed 15 per cent? This may be rationalized by the fact that the second derivative of the potential energy can be substantially affected by even minute changes in the potential energy. The intermolecular frequencies of $(H_2O)_3$ and $(D_2O)_3$ were calculated by Gregory and Clary (1995b) on PESs with and without the three-body term (induction + dispersion) using DMC. They found that the nonadditivity strongly affects (rigid-body) zero point energies and rotational constants of both trimers, but only slightly the tunneling splittings.

The above discussion pertained to the structural three-body effects. As mentioned above, the noticeable reduction in the O...O...O distances is largely due to electron correlation effects upon the water–water pair potential. The three-body energetics, on the other hand, can be almost completely accounted for at the SCF level of theory. To better appreciate the energetical effects of the structural changes within monomers, the state-of-the-art interaction energy values of the trimer obtained by Klopper *et al.* (1995) are shown in Table 2.8. One can see that the total three-body effect amounts to 2.30 kcal/mole, that is, 14 per cent of the full interaction energy. The electron correlation contribution of 0.02 kcal/mole is negligible. These conlusions agree very well with the previous findings of Chałasiński, Szczęśniak, Cieplak, Scheiner (1991a) and Xantheas (1994) (although in the latter case it was obscured by a rather unfortunate comparison of the SCF and post-SCF energies in their respective optimized geometries rather than at the same geometry). One can also see in Table 2.8 that the overall distortion effect reduces the interaction energy by 0.24 at the MP2 level. Interestingly, SCF provides a reduction of 0.78, which is much too large. Higher orders, the third and the fourth, do not change these results (cf. Table 2.8).

Determination of the origin of the three-body contribution is of utmost importance to

Table 2.8 Interaction energy of the equilibrium water trimer from *ab initio* calculations (in kcal/mol)

	(a)	(b)	(c)	(d)
ΔE^{SCF}(2-body)	−8.61			
ΔE^{SCF}(3-body)	−2.28	−2.45[f]	−2.09	−2.46
ΔE^{SCF}(1-body)[e]	0.78			
ΔE^{SCF}(1 + 2 + 3-body)	−10.11			
$\Delta E^{(2)}$(2-body)	−5.65			
$\Delta E^{(2)}$(3-body)	−0.02	0.00	0.10	0.01
$\Delta E^{(2)}$(1-body)[a]	−0.54			
ΔE^{MP2}(3-body)	−2.30	−2.45	−2.19	−2.45
$\Delta E^{(2)}$(1 + 2 + 3-body)	−6.21			
ΔE^{MP2}(1 + 2 + 3-body)	−16.32	−14.81		
ΔE^{MP3}(3-body)			−2.22	−2.39
ΔE^{MP4}(3-body)		−2.40		

[a]Klopper *et al.* (1995); MP2-R12 method at MP2-optimized {uud} geometry.
[b]Xantheas (1994); aug-cc-p VTZ basis set at MP2-optimized {udd} geometry.
[c]Chałasiński *et al.* (1991a), C_{3h} geometry with $R_{O...O...O} = 2.75$ Å.
[d]Rak (1996); $(5s, 3p, 2d, 1f/3s, 2p, 1d)$ basis set at MP2-optimized {uud} geometry.
[e]Monomer relaxation energy.
[f]Recomputed at the MP2-optimized {uud} geometry by Rak (1996).

the analytical modeling of this effect. Such a task can be accomplished via analysis of the three-body interaction in terms of the intermolecular perturbation theory. Although the inclusion of induction nonadditivity in liquid water simulations dates back to the 1970s (Stillinger and David, 1978; Barnes *et al.*, 1979), the seminal *ab initio* work of Clementi *et al.* (1980) demonstrated that the total nonadditivity in the trimer correlates with the three-body induction terms. A precise dissection of different contributions for the cyclic trimer was performed by Chałasiński, Szczęśniak, Cieplak, Scheiner (1991a). They carried out MP3 calculations combined with I-MPPT for an idealized cyclic C_{3h} structure with water molecules rotating in phase around their O nuclei. The α-dependence of various interaction energy contributions is shown in Figure 2.5.

The three-body ΔE^{SCF} term encompasses the HL-exchange and SCF-deformation contributions, ε_{exch}^{HL} and ΔE_{def}^{SCF}. The SCF-deformation clearly dominates, in particular around the H-bond geometry. The O-to-O and H-to-H configurations also have a significant HL-exchange component. The anisotropy of ΔE_{def}^{SCF} term parallels that of its two-body counterpart (not shown in Figure 2.5). The exchangeless approximation represented by the second-order induction nonadditivity, $\varepsilon_{ind}^{(20)}$, provides a fairly crude approximation to the SCF-deformation term and its usefulness is limited to angles $\alpha > 100°$. The discrepancy is due to higher-order terms and, more importantly, to the exchange effects.

The ε_{exch}^{HL} term behaves in an opposite way. It provides maximal repulsion at the H-to-H geometry; then changes sign around the H-bonded arrangement, and finally gives a stabilizing contribution for the O-to-O form. It acts to damp the SCF-deformation term.

As for the correlation terms, the only component shown is the three-body $\varepsilon_{disp}^{(30)}$ component, which includes the ATM dispersion nonadditivity. The dispersion term is very small and may be essentially neglected together with the other correlation contributions discussed in this work.

Despite their small magnitudes in the equilibrium geometry, the ε_{exch}^{HL} and $\varepsilon_{disp}^{(30)}$ non-

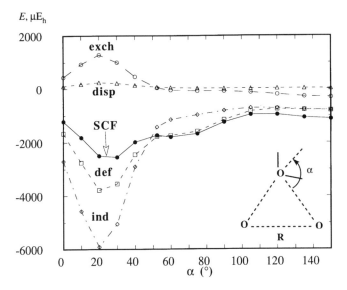

Figure 2.5 Anisotropy of three-body components in $(H_2O)_3$ with respect to concerted α-rotations of all three H_2O molecules at $R = 3.0$ Å (see inset for definition). The following abreviations have been used: 'exch'—ε_{exch}^{HL}; 'SCF'—ΔE^{SCF}; 'def'—ΔE_{def}^{SCF}; 'ind'—$\varepsilon_{ind}^{(20)}$; 'disp'—$\varepsilon_{disp}^{(30)}$

additivities are generally not negligible in other areas of the PES. It is instructive to compare behaviors of the ε_{exch}^{HL} and $\varepsilon_{disp}^{(30)}$ terms to that of the RG$_3$ case discussed in Section 2.4.1. The cyclic water trimer may be roughly related to the RG$_3$ in the equilateral triangle configuration. In both cases the $\varepsilon_{disp}^{(30)}$ term is repulsive, as the triple-dipole ATM term would indicate. The ε_{exch}^{HL} exchange term is attractive for RG$_3$, but for the water trimer, its sign depends on monomer orientation. To explain this high anisotropy, one may further dissect ε_{exch}^{HL} into its SE and TE contributions. The SE contribution was recently evaluated for the water trimer by Tachikawa and Iguchi (1994). As it turns out, the TE term is attractive for all the monomer orientations in the cyclic trimer, which is quite understandable in light of the fact that in both cases, RG$_3$ and $(H_2O)_3$, it is driven essentially by the behavior of overlap integral of monomer wavefunctions. The change of sign of the total ε_{exch}^{HL} nonadditivity is thus due to the SE part. Unlike in RG$_3$, where it was essentially a pure charge overlap effect, in $(H_2O)_3$ the SE term can be described using the electrostatic exchange–multipole–dipole interaction. Such an interaction is, of course, anisotropic and changes its sign with the orientation of monomers. It should be stressed that the SE and TE terms are of similar magnitude.

The existing three-body potentials for simulations of water include the induction term described by equation (2.5) (Lybrand and Kollman, 1985; Rullmann and van Duijnen, 1988; Caldwell et al., 1990; Dang, 1992; Niesar et al., 1989, 1990; Corongiu, 1992, and many others) with parameterizations varying from author to author. The above discussion sheds some light on how this potential could be augmented with the exchange nonadditivity term. A model of the exchange nonadditivity in water should include primarily the SE effect, as indicated by the results of Tachikawa and Iguchi (1994). So far no such attempt has been undertaken.

In the context of the construction of PES for water clusters, the water pair potential should also be mentioned. In the last two decades a number of effective pair potentials for water using polarizable and nonpolarizable models have been proposed (see Wallqvist and Berne, 1994, for a recent review). These potentials are usually rather simple site–site functions which are best suited for the simulations of the liquid properties. For a number of reasons (including lack of the periodic boundry condition) clusters are much more difficult to simulate and require much more sophisticated potential energy functions. Such functions should treat more accurately the directionality of interactions and may include the so-called flexibility, i.e. a dependence upon the intrasystem degrees of freedom. Due to the dimensionality problem, as well as less complete spectroscopic information, a formulation of accurate, spectroscopy-based PES for the water–water interaction is probably impossible at the moment. A fitting of such a (twelve-dimensional) surface to the pure *ab initio* data seems equally improbable. Thus a construction of such a function should rely on the individual modeling of the interaction energy contributions and achieve their correct orientation dependence via the multiple expansion. One such attempt to generate the (intermolecular) PES for water dimer by Millot and Stone (1992) is worth mentioning. Their surface is based on the multi-center multipole expansion of electrostatic and induction effects and also includes the anisotropic exchange sites. Although with the addition of dispersion energy their PES becomes too deep, it is certainly a step in the right direction.

2.4.3.2 HF Trimer

$(HF)_3$ serves as the simplest trimer involving all three molecular subunits. However, the description of the potential energy surface is extremely difficult, inasmuch as it involves twelve dimensions. The major part of the problem is the representation of the two-body potential. Earlier atom–atom-based functions (Hancock *et al.*, 1988) evolved to much more complicated but more physically justified functions, involving representations of individual interaction energy terms (Bunker *et al.*, 1988). A family of pair potentials (1 + 2-body) advanced by Quack and Suhm (1991) on the basis of *ab initio* calculations of Kofranek *et al.* (1988) and including semiempirical adjustments was defined in six dimensions as a sum of electrostatic, induction, damped-dispersion, and short-range atom–atom repulsion terms. The intramolecular degrees of freedom were described by Morse functions. The most reliable one (SQSBDE) has been shown to reproduce reasonably well the intermolecular vibrations (Anderson *et al.*, 1996) and the tunneling splitting (Quack and Suhm, 1995).

The three-body interactions have been examined by Chałasiński *et al.* (1989) in the cyclic $(HF)_3$ trimer. The decomposition of the nonadditive interaction showed that the dominant part of the three-body term is the SCF-deformation term, which originates from the induction effect. The HL-exchange nonadditivity is an order of magnitude smaller, whereas the dispersion effect is negligible. The three-body potential can thus be reliably reproduced even at the SCF level. The orientation dependence of the total three-body is largely determined by the SCF-deformation effect. Only in the H-to-H orientations do the exchange effects become important.

The analytical three-body potential for $(HF)_3$ has been generated by Suhm and coworkers as a twelve-dimensional fit to 1000–3000 *ab initio* points (Quack *et al.*, 1993; Suhm, 1995). The analytical form involved multipole expanded induction terms along with exponential terms. Another formulation of (2 + 3)-body induction potential for HF clusters was done in the framework of the so-called molecular mechanics for clusters (MMC) and was also based on a multipole expansion (Liu *et al.*, 1987; see also Dykstra, 1993).

Apart from its large energetical contribution amounting to *ca* 16 per cent (in the equilibrium C_{3h} cyclic structure) the nonadditive interactions have a major effect on the vibration frequencies and dissociation energies of the HF and DF trimers. Quack *et al.* (1993) carried out DMC calculations on the twelve-dimensional PES of $(HF)_3$ which included three-body interactions. Their calculated $\Delta\nu_{HF}$ frequency shifts were in very good agreement with experimental findings. The three-body effects contributed up to 50 per cent to this shift. The three-body effects were also found to contribute *ca* 25 per cent to the dissociation energies of both $(HF)_3$ and $(DF)_3$ relative to the dimer + monomer dissociation limits.

In conclusion, the theory with the support of the IR spectroscopy predicts that the three-body effects play a major role in the HF trimer. The four-body effects have also been discussed in the context of higher $(HF)_n$, $n = 4 - 7$, oligomers by Suhm (1995). He concluded that these effects should be small. The perturbation interpretation of induction nonadditivity lends some support to this conclusion. For example, in the interactions of polar species the leading three-body nonadditivity appears in the second order of perturbation theory. The four-body effect, on the other hand, should appear in the third order, and is thus expected to be much smaller.

$(HCl)_3$ has received much less attention than its first-row analog $(HF)_3$. The three-body interactions were examined by Chałasiński *et al.* (1989) for the trimer around the equilibrium geometry. Against expectations that the more diffused charge distribution of HCl compared to HF might lead to a more prominent role of the exchange and dispersion effects in the overall nonadditivity, these effects were found to be unimportant and the composition of the three-body effect was quite similar to the one in $(HF)_3$. A more detailed characterization of the three-body potential is needed, including the role of the H–Cl stretches in the nonadditive interactions. The recent successful determination of the intermolecular dimer potential from the VRT spectra by Elrod and Saykally (1995) makes the $(HCl)_3$ cluster an attractive target for attempts to invert the three-body potential as well.

2.4.4 Ion-(molecule)$_2$ Clusters

Clusters involving an ion bound to a number of molecules are of great interest to both theory and experiment since they allow for the investigations of the solvation processes at the microscopic level (see Castleman and Kreesee, 1986; Smith and Pettitt, 1994). The ion–solvent interactions are much stronger than the typical van der Waals interactions or hydrogen bonds, and are believed to be non-pairwise additive to a high degree.

The intrinsic nature of the three-body interactions in these species has not been examined by rigorous *ab initio* approaches as yet, and it has been taken for granted that the induction nonadditivity should be the dominant effect. This assumption has led to the widespread use of the classical polarization of equation (2.5) to describe the overall nonadditivity in the MD simulations of ionic solvation. Such simulations admittedly achieved good agreement with experimentally observed enthalpies of formation even for negative ions (Caldwell et al., 1992; Xantheas and Dang, 1996). However, it is as yet unclear whether the classic model of induction nonadditivity is universally applicable in this case. One could reasonably expect that in the interactions involving especially the negative ions, the quantum exchange effects are quite important. For example, *ab initio* studies of the $X^-(H_2O)_n (X = F, Cl, I)$ clusters (Combariza et al., 1994) suggested that there is a strong delocalization of electrons from X^- to the water molecules. In earlier attempts to fit a three-body potential for $Cl^-(H_2O)_2$ interactions Lybrand and Kollman (1985) observed that the *ab initio* data could not be fitted to the induction term alone unless a short-range exponential term to describe the exchange nonadditivity was also included:

$$E_{exch} = A \exp(-\alpha r_{12}) \exp(-\alpha r_{13}) \exp(-\beta r_{23}) \qquad (2.9)$$

Such a three-body potential has been used in many Monte Carlo and MD simulations of ion-$(H_2O)_n$ clusters (Cieplak et al., 1987; Howard et al. (1988); Dang, 1992; Perera and Berkowitz, 1994).

The simple model in which the nonadditivity was dissected into its components involves the Ar_2Cl^- cluster. As a reasonably strong Lewis base, Ar can serve as a structureless analog of H_2O which, however, has no multipole moments. The three-body potential (see Table 2.9) was found to be composed of three nearly equally important components, induction, exchange, and dispersion (Burcl et al., 1995). Only in the long range of the potential the induction nonadditivity was dominant. Burcl et al. also proposed a three-term analytical three-body potential for this cluster. The induction term was modeled by an electrostatic interaction of the dipoles induced on the Ar atoms by the field of Cl^-. The dispersion effect was described by the ddd and ddq ATM nonadditivities. The exchange nonadditivity was described by the interaction of Cl^- with the exchange-quadrupole on the Ar_2 subunit. The latter term represents the SE part of the

Table 2.9 Decomposition of the three-body interaction energy in the equilateral triangle Ar_2Cl^- cluster at two separations: near the equilibrium ($R = 6.5\ a_0$) and at long range ($R = 10.0 a_0$) (from Burcl, et al., 1995); values in (in μE_h; 1 $\mu E_h = 0.219$ cm^{-1})

Energy term	$R = 6.5$	$R = 10.0$
ΔE^{SCF}	−62.2	11.5
$\Delta E^{(2)}$	116.0	2.3
$\Delta E^{(3)}$	79.9	1.7
ΔE^{MP3}	133.7	15.6
ε_{exch}^{HL}	−285.7	−0.3
$\varepsilon_{exch,SE}^{HL}(Ar_2+Cl^-)$	−143.1	−0.24
$\varepsilon_{exch,TE}^{HL}$	−250.9	−0.06
ΔE_{def}^{SCF}	223.5	11.8
$\varepsilon_{ind,r}^{(30)}$	261.1	12.9
$\varepsilon_{disp}^{(30)}$	102.5	1.8

exchange nonadditivity, which represents a dominant contribution when the charged species are present. The significance of these findings to the potential for X^- $(H_2O)_n$ is that the exponential term of equation (2.9) which was customarily used to model the exchange nonadditivity, in reality, models only its TE part which is expected to be less anisotropic and less important in the case of ionic interactions.

At this point we wish also to clarify one common misnomer in referring to the term in equation (2.9) as 'three-body exchange repulsion'. The exchange nonadditivity can be either repulsive or attractive depending upon a trimer geometry. For example, in the equilateral triangle geometry of Ar_2Cl^- both SE and TE contributions are attractive (see Table 2.9), and the same will most likely be true of $Cl^-(H_2O)_2$ in the triangle geometries.

It seems that as the interacting systems studied will become more complicated, such as transition metal complexes, one may expect the need for the more sophisticated, and more general, many-body potentials than those commonly used. One such a study involving $Pt(NH_3)_3^{2+} + Cl^- + H_2O$ is worth mentioning in this context (Muguruma et al., 1995), since it attempted at designing an analytical three-body potential including exchange, induction, and charge transfer effects (i.e. quantum induction effects). One immediate goal of future studies should be to examine more rigorously the role of exchange effects in the ion–solvent many-body interactions.

In summary, the induction nonadditivity dominates in the polar and ion–polar interactions. It has a major effect on structural properties and on H-bond red-shifts. However, in the construction of many-body potentials the other terms should also be included. The exchange nonadditivity is important at short distances at PESs and in certain non-H-bonded configurations. In trimers of weak proton-donors (such as NH_3) the exchange nonadditivity was shown to determine the angular dependence of the three-body potential (Szcześniak et al., 1991). The dispersion nonadditivity is usually 'clouded' by other larger effects in these clusters; however, it can be easily added to a potential in the form of the ATM term.

2.5. CONCLUSION

In recent years many-body effects have moved to the forefront of studies in the area of intermolecular interactions. Great progress has been achieved in the understanding of these challenging interactions, due in large measure to an unprecedented interplay between theory and experiment. Experimental methods are now able to probe isolated clusters for evidence of a nonpairwise behavior. A great deal of progress has also been made in solving the Schrödinger equation on a PES for bound ro-vibrational states and spectra. PESs can be constructed from *ab initio* two-body and many-body interactions. Theoretical *ab initio* calculations are now capable of evaluating the pair interaction potentials with an accuracy of $\pm 5\%$ and for the three-body interactions with an accuracy which approaches that. In designing the strategies for such reliable calculations, dividing the interaction energy into the SAPT components proved very helpful. The perturbational analysis has been indispensable in analytical representations of PESs, since the individual many-body effects can be more easily cast into analytical formulae. Although a better understanding of a modeling of many-body effects has been achieved in recent years, more work in this direction is necessary.

At the same time, we should not lose sight of the fact that in the future the need for elaborate parametrizations of PESs is likely to diminish. With the advent of 'direct' methods, such as the Carr–Parrinello algorithm (Carr and Parrinello, 1985; Laasonen *et al.*, 1993; see also Gregory and Clary, 1995c), in which intermolecular forces can be directly calculated during molecular dynamics simulations, the need for explicit formulation of PES, and the many-body expansion in particular, could be bypassed. While the analytical formulation of multidimensional PESs in these approaches will not be required, the need for reliable and efficient computations of intermolecular forces will remain. In this task the insights gained from the analysis of individual many-body effects will be crucial.

ACKNOWLEDGEMENTS

This work was supported by the National Science Foundation (CHE-9527099) and by the Polish Committee for Scientific Research (KBN) (3 TO9A 072 09).

REFERENCES

Adams, W. H. (1992) *J. Math. Chem.*, **10**, 1.
Anderson, D. T., Davis, S. and Nesbitt, D. J. (1996) *J. Chem. Phys.*, **104**, 6225.
Arrighini, P. (1981) *Intermolecular Forces and Their Evaluation by Perturbation Theory*, Springer-Verlag, Berlin.
Axilrod, B. M. and Teller, E. (1943) *J. Chem. Phys.*, **11**, 299.
Barker, J. A. (1976) in *Rare Gas Solids*, edited by Klein, M. L. and Venables, J. A., Academic Press, New York.
Barnes, P., Finney, J. L., Nicholas, J. D. and Quinn, J. E. (1979) *Nature*, **282**, 459.
Buckingham, A. D. (1978) in *Intermolecular Interactions: From Diatomics to Biopolymers*, edited by Pullman, B., Wiley, New York.
Bulski, M. (1981) *Chem. Phys. Lett.*, **78**, 361.
Bulski, M. and Chalasinski, G. (1980) *Theor. Chim. Acta*, **56**, 281.
Bulski, M. and Chalasinski, G. (1987) *J. Chem. Phys.*, **86**, 937.
Bunker, P. R., Kofranek, M., Lischka, H. and Karpfen, A. (1988) *J. Chem. Phys.*, **89**, 3002.
Burcl, R., Cybulski, S. M., Szczesniak, M. M. and Chalasinski, G. (1995) *J. Chem. Phys.*, **103**, 299.
Bürgi, T., Graf, S., Leutwyler, S. and Klopper, W. (1995) *J. Chem. Phys.*, **103**, 1077.
Caldwell, J., Dang, L. X. and Kollman, P. A. (1990) *J. Am. Chem. Soc.*, **112**, 9144.
Carr, R. and Parrinello, M. (1985), *Phys. Rev. Lett.*, **55**, 2471.
Castleman, A. W. and Kreesee, R. G. (1986) *Chem. Rev.*, **86**, 589.
Chalasinski, G., Kendall, R. A., and Simons. J. (1987) *J. Chem. Phys.*, **87**, 2965.
Chałasiński, G. and Szczęśniak, M. M. (1988) *Mol. Phys.*, **63**, 205.
Chałasiński, G. and Szczęśniak, M. M. (1994) *Chem. Rev.*, **94**, 1723.
Chałasiński, G., Cybulski, S. M., Szczęśniak, M. M. and Scheiner, S. (1989) *J. Chem. Phys.*, **91**, 7048.
Chałasiński, G., Szczęśniak, M. M. and Cybulski, S. M. (1990) *J. Chem. Phys.*, **92**, 2481.
Chałasiński, G., Szczęśniak, M. M., Cieplak, P. and Scheiner, S. (1991a) *J. Chem. Phys.*, **94**, 2873 (1991).
Chałasiński, G., Szczęśniak, M. M. and Kukawska-Tarnawska, B. (1991b) *J. Chem. Phys.*, **94**, 6677.
Chałasiński, G., Szczęśniak, M. M. and Kendall, R. A. (1994) *J. Chem. Phys.*, **101**, 8860.
Chałasiński, G., Rak, J. A., Cybulski, S. M. and Szczęśniak, M. M. (1996) to be published.
Cieplak, P., Kollman, P. A. and Lybrand, T. P. (1990) *J. Chem. Phys.*, **92**, 6755.
Cieplak, P., Lybrand, T. P. and Kollman, P. A. (1987) *J. Chem. Phys.*, **86**, 6393.

Claverie, P. (1978) in *Intermolecular Interactions: From Diatomics to Biopolymers*, edited by Pullman, B., Wiley, New York.
Clementi, E., Kołos, W., Lie, G. C. and Ranghino, G. (1980) *Int. J. Quantum Chem.*, **17**, 377.
Cohen, R. C. and Saykally, R. J. (1991) *Annu. Rev. Phys. Chem.*, **42**, 369–92.
Cohen, R. C. and Saykally, R. J. (1992) *J. Phys. Chem.*, **96**, 1024.
Combariza, J., Kestner, N. R. and Jortner, J. (1994) *J. Chem. Phys.*, **100**, 2851.
Cooper A. R. and Hutson, J. M. (1993) *J. Chem. Phys.*, **98**, 5337.
Cooper, A. R., Jain, S. and Hutson, J. M. (1993) *J. Chem. Phys.*, **98**, 2160.
Corongiu, G. (1992) *Int. J. Quantum Chem.*, **42**, 1209.
Cybulski, S. M. (1992) *J. Chem. Phys.*, **97**, 7545.
Cybulski, S. M. (1994) Trurl94 package, Rochester, Michigan.
Cybulski, S. M. (1995) *Chem. Phys. Lett.*, **238**, 261.
Cybulski, S. M., Szczęśniak, M. M. and Chałasiński, G. (1994) *J. Chem. Phys.*, **101**, 10708.
Dang, L. X. (1992) *J. Chem. Phys.*, **96**, 6870.
Dotelli, G. and Jansen, L. (1996) to be published.
Dyke, T. R. and Muenter, S. (1972) *J. Chem. Phys.*, **57**, 5011.
Dykstra, C. E. (1993) *Chem. Rev.*, **93**, 2339.
Elrod, M. J. and Saykally, R. J. (1994) *Chem. Rev.*, **94**, 1975.
Elrod, M. J. and Saykally, R. J. (1995) *J. Chem. Phys.*, **103**, 921, 933.
Elrod, M. J., Steyert, D. W. and Saykally, R. J. (1991) *J. Chem. Phys.*, **94**, 58; **95**, 3182.
Elrod, M. J., Loeser, J. G. and Saykally, R. J. (1993) *J. Chem. Phys.*, **98**, 5352.
Elrod, M. J., Saykally, R. J., Cooper, A. M. and Hutson, J. M. (1994) *Mol. Phys.*, **81**, 579.
Ernesti, A. and Hutson, J. M. (1994) *Faraday Discuss.*, **97**, 119.
Ernesti, A. and Hutson, J. M. (1995a) *J. Chem. Phys.*, **103**, 3386.
Ernesti, A. and Hutson, J. M. (1995b) *Phys. Rev. A*, **51**, 239.
Farrell, J. T. Jr, Davis, S. and Nesbitt, D. J. (1995) *J. Chem. Phys.*, **103**, 2395.
Fowler, J. E. and Schaefer, H. F. III (1995) *J. Am. Chem. Soc.*, **117**, 446.
Frey, R. F. and Davidson, E. R. (1989) *J. Chem. Phys.*, **90**, 555.
Gregory, J. K. and Clary, D. C. (1995a) *J. Chem. Phys.*, **102**, 7817.
Gregory, J. K. and Clary, D. C. (1995b) *J. Chem. Phys.*, **103**, 8924.
Gregory, J. K. and Clary, D. C. (1995c) *Chem. Phys. Lett.*, **237**, 39.
Gutowski, M., Piela, L. (1988) *Mol. Phys.*, **64**, 337.
Gutowsky, H. S., Klotz, T. D., Chuang, C., Schmuttenmaer, C. A., Emilsson, T. (1987) *J. Chem. Phys.*, **86**, 569.
Hancock, G. C., Truhlar, D. G. and Dykstra, C. E. (1988) *J. Chem. Phys.*, **88**, 1786.
Hirschfelder, J. O., Curtiss, C. F. and Bird, R. B. (1954) *Molecular Theory of Gases*, Wiley, New York.
Hirschfelder, J. O. and Meath, W. J. (1967) *Adv. Chem. Phys.*, **12**, 3.
Honegger, E. and Leutwyler, S. (1988) *J. Chem. Phys.*, **88**, 2582.
Howard, A. E., Singh, U. C., Billeter, M. and Kollman, P. A. (1988) *J. Am. Chem. Soc.*, **110**, 6984.
Hutson, J. M. (1990) *Annu. Rev. Phys. Chem.*, **41**, 123.
Hutson, J. M. (1991) *Adv. Mol. Vib. Collision Dyn.* **1**, 1.
Hutson, J. M., Beswick, J. A. and Halberstadt, N. (1989) *J. Chem. Phys.*, **90**, 1337.
Jansen, L. (1962) *Phys. Rev.*, **125**, 1798.
Jansen, L. (1965) *Adv. Quantum Chem.*, **2**, 119.
Jeziorski, B. (1974) Dissertation, University of Warsaw, Warsaw.
Jeziorski, B., Bulski, M. and Piela, L. (1976) *Int. J. Quantum Chem.*, **10**, 281.
Jeziorski, B. and Kołos, W. (1977) *Int. J. Quantum Chem. (Suppl.)*, **12**, 91.
Jeziorski, B. and Kołos, W. (1982) in *Molecular Interactions* Vol. 3, edited by Ratajczak, H. and Orville-Thomas, W. J., Wiley, New York.
Jeziorski, B., Moszyński, R. and Szalewicz, K. (1994) *Chem. Rev.*, **94**, 1887.
Jeziorski, B., Moszyński, R., Ratkiewicz, A., Rybak, S., Szalewicz, K. and Williams, H. L. (1993) in *Methods and Techniques in Computational Chemistry, METTEC-94*, edited by Clementi. E., STEF, Cagliari, vol. B, p. 79.
Jorgensen, W. L. (1986) *J. Phys. Chem.*, **90**, 1276.
Jorgensen, W. L. and Tirado-Rives, J. (1988) *J. Am. Chem. Soc.*, **110**, 1657.

Kaplan, I. G. (1986) *Theory of Intermolecular Interactions*, North-Holland, Amsterdam.
Kelterbaum, R., Turki, N., Rahmouni, A. and Kochanski, E. (1994) *J. Mol. Structure (Theochem.)*, **314**, 191.
Kistenmacher, G., Lie, G. C., Popkie, H. and Clementi, E. (1974) *J. Chem. Phys.*, **61**, 546.
Klein, M. L. and Venables, J. A. (eds) (1976) *Rare Gas Solids*, Academic Press, New York.
Klopper, W., Schütz, M., Lüthi, H. P. and Leutwyler, S. (1995) *J. Chem. Phys.*, **103**, 1085.
Kofranek, M., Lischka, H. and Karpfen, A. (1988) *J. Chem. Phys.*, **121**, 137.
Kołos, W. (1974) in *The World of Quantum Chemistry*, edited by Daudel, R. and Pullman, B., Reidel, Dordrecht.
Kołos, W. and Les, A. (1972) *Chem. Phys. Lett.*, **14**, 167.
Korona, T., Moszynski, R. and Jeziorski, B. (1996) *Adv. Quantum Chemistry*, in press.
Laasonen, K., Sprik, M. Parrinello, M. and Car, R. (1993) *J. Chem. Phys.*, **99**, 9080.
Leopold, K. R., Fraser, G. T., Novick, S. E. and Klemperer, W. (1994) *Chem. Rev.*, **94**, 1807.
Lewerenz, M. (1996) *J. Chem. Phys.*, **104**, 1028.
Li, X., Hunt, K. C. L. (1996) *J. Chem. Phys.*, **105**, 4076.
Liu, K., Loeser, J. G., Elrod, M. J., Host, B. C., Rzepiela, J. A., Pugliano, N. and Saykally, R. J. (1994a) *J. Am. Chem. Soc.*, **117**, 3507.
Liu, K., Elrod, M. J., Loeser, J. G., Cruzan, J. D., Pugliano, N., Brown, M. G., Rzepiela, J. A. and Saykally, R. J. (1994b) *Faraday Discuss. Chem. Soc.*, **97**, 35.
Liu, K., Cruzan, J. D., Saykally, R. J. (1996) *Science.*, **271**, 929.
Liu, S., Bačić, Z., Moskowitz, J. W. and Schmidt, K. E. (1994c) *J. Chem. Phys.*, **100**, 7166.
Liu, S., Bačić, Z., Moskowitz, J. W. and Schmidt, K. E. (1994d) *J. Chem. Phys.*, **101**, 10181.
Liu, S. Y., Michael, D. W., Dykstra, C. E. and Lisy, J. M. (1987) *J. Chem. Phys.*, **84**, 5032.
Lotrich, F. V. and Szalewicz, K. (1996) submitted for publication.
Löwdin, P. -O. (1956) *Adv. Phys.*, **5**, 1.
Lybrand, T. P. and Kollman, P. A., (1985) *J. Chem. Phys.*, **83**, 2923.
Margenau, H. and Kestner, N. R. (1967) *Theory of Intermolecular Forces*, Pergamon, Oxford.
Margenau, H. and Stamper, J. (1967) *Adv. Quantum Chem.*, **3**, 129.
Marshall, P. J., Szczęśniak, M. M., Sadlej, J., Chałasiński, G., ter Horst, M. A. and Jameson, C. J. (1996) *J. Chem. Phys.*, **104**, 6569.
McIlroy, A., Lascola, R., Lovejoy, C. M. and Nesbitt, D. J. (1991) *J. Phys. Chem.*, **95**, 2636.
McIlroy, A. and Nesbitt, D. J. (1992) *J. Chem. Phys.*, **97**, 6044.
Meath, W. J. and Aziz, R. A. (1984) *Mol. Phys.*, **52**, 225.
Meath, W. J. and Koulis, M. (1991) *J. Mol. Structure (Theochem.)*, **226**, 1.
Miller, R. E. (1990) *Acc. Chem. Res.*, **23**, 10.
Millot, C. and Stone, A. J. (1992) *Mol. Phys.*, **77**, 439.
Mó, O., Yañez, M. and Elguero, J. (1992) *J. Chem. Phys.*, **97**, 6628.
Mohan, V. and Anderson, J. B. (1990) *J. Chem. Phys.*, **92**, 6971.
Moszyński, R., Cybulski, S. M. and Chałasiński, G. (1994) *J. Chem. Phys.*, **100**, 4998.
Moszyński, R., Wormer, P. E. S., Jeziorski, B. and van der Avoird, A. (1995) *J. Chem. Phys.*, **103**, 8058.
Muguruma, C., Koga, N., Kitaura, K. and Morokuma, K. (1995) *J. Chem. Phys.*, **103**, 9274.
Muto, Y. (1943) *Proc. Phys. Math. Soc. Jpn*, **17**, 629.
Nesbitt, D. J. (1994) *Ann. Rev. Phys. Chem.*, **45**, 367.
Ng, K., Meath, W. J. and Allnatt, A. R. (1979) *Mol. Phys.*, **37**, 237.
Niebel, K. F. and Venables, J. A. (1976) in *Rare Gas Solids*, edited by Klein, M. L. and Venables, J. A., Academic Press, New York.
Niesar, U. Corongiu, G., Huang, M. J., Dupuis, M. and Clementi, E. (1989) *Int. J. Quantum Chem. Sym.* **23**, 421.
Niesar, U., Corongiu, G., Clementi, E., Kneller, G. R. and Bhattacharya, D. K. (1990) *J. Phys. Chem.*, **94**, 7949.
Parish, C. A. and Dykstra, C. E. (1993) *J. Chem. Phys.*, **98**, 437.
Piecuch, P. (1986) *Mol. Phys.*, **59**, 1067, 1085, 1097; *J. Math. Phys.*, **27**, 2165.
Piecuch, P. (1988) in *Molecules in Physics, Chemistry and Biology*, edited by Maruani, J., Kluwer, Dordrecht.
Perera, L. and Berkowitz, M. L. (1991) *J. Chem. Phys.*, **95**, 1954.

Perera, L. and Berkowitz, M. L. (1994) *J. Chem. Phys.*, **100**, 3085.
Pugliano, N. and Saykally, R. J. (1992) *Science*, **257**, 1937.
Quack, M., Stohner, J. and Suhm, M. A. (1993) *J. Mol. Structure*, **294**, 33.
Quack, M. and Suhm, M. A. (1991) *J. Chem. Phys.*, **95**, 28.
Quack, M. and Suhm, M. A., (1995) *Chem. Phys. Lett.*, **234**, 71.
Rak, J. A. (1996) to be published.
Roche, C. F., Ernesti, A., Hutson, J. M. and Dickinson, A. S. (1996) *J. Chem. Phys.*, **104**, 2156.
Røeggen, I. (1990) *Mol. Phys.*, **70**, 353.
Røeggen, I. and Almlöf, J. (1995) *J. Chem. Phys.*, **102**, 7095.
Røeggen, I. Almlöf, J., Ahmadi, R. G. and Wind, P. A. (1995) *J. Chem. Phys.*, **102**, 7088.
Rullmann, J. A. and van Duijnen, P. T. (1988) *Mol. Phys.*, **63**, 451.
Sabo, D., Bačić, Z., Burgi, T. and Leutwyler, S. (1995) *Chem. Phys. Lett.*, **244**, 283.
Saykally, R. J. and Blake, G. A. (1993) *Science*, **259**, 1570.
Schröder, T., Schinke, R., Liu, S., Bačić, Z. and Moskowitz, J. W. (1995) *J. Chem. Phys.*, **103**, 9228.
Schuster, P., Karpfen, A. and Beyer, A. (1980) in *Molecular Interactions*, Vol. 1, edited by Ratajczak, H. and Orville-Thomas, W. J., Wiley, Chichester.
Schütz, M., Bürgi, T., Leutwyler, S. and Bürgi, H. B. (1993), *J. Chem. Phys.*, **99**, 5228; (1994) *ibid.*, **100**, 1780 (errata).
Smith, P. E. and Pettitt, B. M. (1994) *J. Phys. Chem.*, **98**, 9700.
Sperhac, J. M., Weida, M. J. Nesbitt, D. J. (1996) *J. Chem. Phys.*, **104**, 2202.
Stillinger, F. H. and David, C. W. (1978) *J. Chem. Phys.*, **69**, 1473.
Stogryn, D. E. (1971) *Mol. Phys.*, **22**, 81.
Stogryn, D. E. (1972) *Mol. Phys.*, **23**, 897.
Stone, A. J. (1981) *Chem. Phys. Letters*, **83**, 233.
Stone, A. J. and Alderton, M. (1986) *Mol. Phys.*, **56**, 1047.
Stone, A. J. (1989) *Chem. Phys. Letters.*, **155**, 102, 111.
Suhm, M. A. (1995) *Ber. Bunsenges. Phys. Chem.*, **99**, 1159.
Suhm, M. A. and Nesbitt, D. J. (1995) *Chem. Soc. Rev.*, **24**, 45.
Szalewicz, K. and Jeziorski, B. (1979) *Mol. Phys.*, **38**, 191.
Szczęśniak, M. M., Chałasiński, G., Cybulski, S. M. and Scheiner S. (1990) *J. Chem. Phys.*, **93**, 4243.
Szczęśniak, M. M., Kendall, R. A. and Chałasiński, G. (1991) *J. Chem. Phys.*, **95**, 5169.
Szczęśniak M. M. and Chałasiński, G. (1992) *J. Mol. Structure (Theochem.)*, **261**, 37.
Szczęśniak, M. M., Chałasiński, G. and Piecuch, P. (1993) *J. Chem. Phys.*, **99**, 6732.
Tachikawa, M. and Iguchi, K. (1994) *J. Chem. Phys.*, **101**, 3062.
Tang, K. T. and Toennies, J. P. (1984) *J. Chem. Phys.*, **80**, 3726.
Tao, F-M. (1994), *Chem. Phys. Lett.*, **227**, 401.
van der Avoird, A., Wormer, P. E. S. and Moszyński, R. (1994) *Chem. Rev.*, **94**, 1931.
van Duijneveldt-van de Rijdt, J. G. C. M. and van Duijneveldt, F. (1993) *Chem. Phys.*, **175**, 271.
van Duijneveldt-van de Rijdt, J. G. C. M. and van Duijneveldt, F. (1995) *Chem. Phys. Lett.*, **237**, 560.
Vos, R., van Mourik, T., van Lenthe, J. H. and van Duijnveldt, F. B. (1991) unpublished.
Wallqvist, A. and Berne, B. J. (1993) *J. Phys. Chem.*, **97**, 13841.
Wells, B. H. and Wilson, S. (1985) *Mol. Phys.*., **55**, 199.
Williams, H. L., Szalewicz, K., Jeziorski, B., Moszyński, R. and Rybak, S. (1993) *J. Chem. Phys.*, **98**, 1279.
Wormer, P. E. S. (1975) Dissertation, University of Nijmegen, Nijmegen.
Xantheas, S. S. and Dunning, T. H., Jr (1993) *J. Chem. Phys.*, **99**, 8774.
Xantheas, S. S. (1994) *J. Chem. Phys.*, **100**, 7523.
Xantheas, S. S. and Dang, L. X. (1996) *J. Phys. Chem.*, **100**, 3989 (1996).
Xu, Y., Jäger, W. and Gerry, M. C. L. (1994) *J. Chem. Phys.*, **100**, 4171.

3 Basis Set Superposition Error

F. B. VAN DUIJNEVELDT

Utrecht University, The Netherlands

3.1 BSSE: A SHORT HISTORY

When the first computers appeared on the academic scene in the early 1960s, one of the first quantum-chemical projects to be undertaken was a systematic study at the Hartree–Fock (or self-consistent field, SCF) level of the diatomics from H_2 to F_2. Ransil (1961) applied this approach in a calculation on He_2, and somewhat to his surprise he found a potential energy curve with a minimum of roughly the right depth at roughly the right distance. This led him to speculate that the weak attraction in He_2 is already present in an SCF description of the system.

A few years later, Kestner (1968) found that the minimum disappeared in SCF calculations that used larger and more complete basis sets. He noted that this is a reasonable result, since the weak dispersion attraction that stabilizes He_2 cannot be reproduced at the SCF level. The minimum in Ransil's curve was ascribed to the fact that some of Ransil's basis functions could contribute to the dimer energy, but not to that of the free atoms.

Around 1970 it became apparent that Ransil's spurious minimum was but the first example of a fundamental difficulty in what we now call the supermolecular approach. In this approach one calculates the interaction energy, ΔE, of a molecular complex by comparing the total energy $E^{AB}(R)$ for a given geometry (symbolized by R) of the complex to the energy of the complex when the fragments are a large distance apart, or, equivalently (if one employs a size-consistent method like SCF, SCF + MPn, CEPA, CCSD, full-CI), to the sum of the energies of the free monomers, E^A and E^B:

$$\Delta E(R) = E^{AB}(R) - E^A - E^B \tag{3.1}$$

(For simplicity we here assume that the monomers retain their free-monomer geometries when the complex is formed. Monomer deformations caused by the interaction will be discussed in Section 3.5.)

Supermolecular SCF calculations were reported on hydrogen-bonded systems (Clementi, 1967; Morokuma and Pedersen, 1968) which gave unreasonably short equilibrium distances, and much too deep minima, especially when small but flexible basis sets were used. The reason for this behaviour was now apparent. The basis sets one used in calculations on molecular complexes were far from complete, because it would have been too costly to use more complete sets. Now, if one follows the total energy of the complex AB as the distance between the fragments is reduced from infinity to a couple of bohrs, the basis functions of the A fragment will to some extent be used by the B fragment to improve the B energy, and vice versa. This monomer energy lowering lowers the total energy of the complex, but it has nothing to do with the ΔE one is interested in. It is merely an artefact of the calculation caused by the incompleteness of the fragment basis sets. Liu and McLean (1973), noting the problem in correlated calculations on the helium dimer, introduced the term *basis set superposition error* (BSSE) for this type of error in the potential energy curves of molecules and molecular complexes.

BSSE has remained a problem ever since. Present-day program systems such as GAUSSIAN, GAMESS, etc. have gradient options which will automatically search along a supermolecular potential energy surface to find, for example, equilibrium geometries and equilibrium binding energies. The quantity being optimized in this type of work is the total energy of the complex, and since the basis sets used today, while usually better than those of 1970, are still far from complete the binding energies and geometries obtained in this way are subject to BSSE.

Already in 1970 it was clear that BSSE can be of the same order of magnitude as the interaction energy one is trying to calculate, and the first attempts were made to reduce or eliminate the problem. Independently of each other, Jansen and Ros (1969) and Boys and Bernardi (1970) proposed the *counterpoise method* to eliminate BSSE. Recognizing that BSSE arises from the fact that the total A + B basis at a given geometry of the complex is more complete than the A or B basis sets alone, the remedy they proposed is to calculate the total energy of the complex *as well as* the energies of the fragments all in the same basis set, i.e. in the A + B basis. The difference between these energies then reflects the effect of the interaction alone, uncomplicated by spurious improvement of the energies of the fragments present in the complex.

The counterpoise recipe is easy to apply. Having completed a calculation on the full complex, one recalculates the fragment energies at the same geometry in the same basis, i.e. using the same two-electron repulsion integrals, but using a set of one-electron integrals in which the nuclear charges of 'absent' (or 'ghost') fragments has been put to zero. Denoting these fragment energies as E^A (full basis), etc. the counterpoise-corrected interaction energy then becomes

$$\Delta E^{CP}(R) = E^{AB}(R) - E^A \text{ (full basis)} - E^B \text{ (full basis)} \qquad (3.2)$$

(Again, the case that the monomer geometries in the complex differ from those of the free monomers is deferred to Section 3.5.) Several early users of the method were disappointed to find, however, that very little binding energy remained when BSSE was removed in this manner (Johansson *et al.*, 1973; Kocjan *et al.*, 1976). Although we now know that other shortcomings in their calculations were at the root of this problem, these users came to think that there is a basic flaw in the counterpoise recipe. The key reasoning leading to this opinion was the following (Morokuma and Winick, 1970; Johansson *et al.*, 1973; Daudey *et al.*, 1974). BSSE comes about because in the calculation of the total energy of the complex the fragments can use the *unoccupied* part of the basis set of neighbouring fragments to improve their own energy. Occupied orbitals are clearly not available for this purpose. Now, in the counterpoise recipe one compares to fragment energies in which *all* of the ghost basis has been made available. Such energies are lower than the ones one should (according to these critics) have compared to, and so the counterpoise interaction energy is less attractive (or more repulsive) than it should have been. In short: the counterpoise method was felt to overcorrect the BSSE.

Throughout the 1970s and 1980s this 'overcorrection' concept has remained the prevailing opinion on the counterpoise method. The unpleasant implication was that the calculation of accurate interaction energies now seemed an unachievable goal for many years to come. Using very large, nearly complete basis sets (almost) free of BSSE would solve the problem, but even today this is impracticable for systems of an interesting size. The use of BSSE-free methods such as the symmetry-adapted perturbation theory (SAPT) approach in which the interaction energy is obtained directly rather than as a difference (Jeziorski *et al.*, 1994) was not yet a viable option since the theory had not yet been implemented to a sufficiently high level of electron correlation (cf. Chapter 1, this volume). In any case, the supermolecular method, if it could be made to work properly, offers a number of advantages such as the wide availability of highly efficient software for this type of calculation and the applicability to both weak and strong interactions and to complexes comprising any number of fragments.

Work on the supermolecular method then evolved along two lines. A number of methods was investigated in which usage of the partner basis set in calculating the total energy of a complex is prevented from the outset. This type of approach has been termed *a priori* avoidance of BSSE (Mayer and Vibok, 1991). On the other hand, the validity of the original counterpoise proposal has been investigated in detail, leading to the surprising conclusion that it is rigorously correct and that the overcorrection argument is fundamentally wrong (Gutowski *et al.*, 1986, 1987a, 1993; Cybulski and Chalasinski, 1992). A discussion of these developments will be presented in Sections 3.3 and 3.4.

As a result, the present situation is that accurate interaction energies can be obtained more or less routinely for many systems of interest, with basis sets of a manageable size, by simply applying the counterpoise recipe. For a survey of recent results see Chalasinski and Szczesniak (1994). As an example of the accuracy that is attainable Figure 3.1 shows two full-CI curves obtained recently for the ground state of the helium dimer (van Mourik and van Lenthe, 1995). Highly accurate empirical curves have been determined for this system by Aziz *et al.* (1995). Curve (a) follows the total energy, and in spite of the fact that a large basis set was used it seriously suffers from BSSE. On the other hand, curve (b) was obtained (in the same basis set) by applying the counterpoise principle, yielding a curve which is indistinguishable from the best empirical curves available.

The aim of this chapter is not to present an exhaustive review of the vast literature that deals with the BSSE. Such reviews are available elsewhere (van Lenthe *et al.*, 1987; van Duijneveldt *et al.*, 1994). Rather the aim is to discuss the main issues that one should be aware of in dealing with BSSE in practical situations. Section 3.2 gives examples of observable properties of molecular complexes that may be affected by BSSE, and compares the BSSE to other errors that may be present. Section 3.3 reviews the evidence that Boys and Bernardi's counterpoise recipe is rigorously correct. Section 3.4 discusses some alternative ways to avoid BSSE. The handling of BSSE in geometry optimizations is dealt with in Section 3.5, and the final section summarizes the current state of affairs.

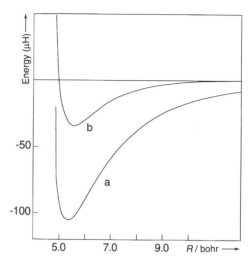

Figure 3.1 Potential energy curves obtained in full-CI calculations on He_2. Curve (a) follows the total energy of the dimer, taking the energy at $R = 100$ bohr as zero. In this approach basis set superposition error (BSSE) remains uncorrected, leading to an equilibrium distance that is too short, and a binding energy that is too large. Curve (b) gives the interaction energy in the dimer, as given by the counterpoise procedure. On the scale of this figure, the latter curve coincides with the most recent empirical He_2 potentials (Aziz *et al.*, 1995). The quality of the agreement confirms that BSSE is completely removed by the counterpoise approach. (Data for basis LARSAT, taken from van Mourik and van Lenthe (1995) and Van Lenthe (private communication). An energy of 100 μH corresponds to 0.063 kcal/mol)

3.2 HOW BSSE AFFECTS THE CALCULATED PROPERTIES OF MOLECULAR COMPLEXES

The aim of this section is to enable the reader to judge the possible adverse effects that BSSE may have on the outcome of a given calculation. Since the counterpoise recipe rigorously removes the BSSE (cf. Section 3.3), we assess the size of BSSE in an uncorrected result by comparing to the counterpoise-corrected value. It must be borne in mind, however, that BSSE is rarely the only error in a given calculation, since the incompleteness of the basis sets used may also lead to an inadequate description of the 'physics' of the interaction. That is, dipoles may be too large, polarizabilities may be too small, and removing the BSSE will not rectify this. Hence to decide whether a removal or reduction of the BSSE would give a marked improvement, some knowledge of the remaining errors must be available as well.

3.2.1 Role of Basis Incompleteness in Deciding the Size of BSSE

By way of example, let us consider how incompleteness of the set of p functions on oxygen affects the SCF + MP2 interaction energy for a near-equilibrium geometry of the water dimer (van Duijneveldt, unpublished). P set incompleteness may be expected to be relatively serious, in view of the importance of these functions in determining the overall shape of the oxygen atom in H_2O. Figure 3.2 shows the counterpoise-corrected (CP) and uncorrected (no CP) results of a series of calculations which were carried out using the EZPPcB basis described in van Duijneveldt-van de Rijdt and van Duijneveldt (1992b). This is a doubly polarized medium-sized [$5s, 3p, 2d|2s, 2p$] basis augmented with a set of (s, p) bondfunctions midway the H$\cdots$O hydrogen bond. When using bondfunctions there is a number of options in doing the uncorrected calculation, since one must decide to which free monomer basis set (if any) the bondfunctions belong. In the results shown here the bondfunctions were taken to form part of the basis of the proton donor molecule. One could also have omitted the bondfunctions from the free monomer calculations altogether. This would have made the uncorrected interaction energies more negative than they are already. No such ambiguity arises in the counterpoise-corrected calculation since here one employs the full dimer basis for the dimer calculation and for both monomer calculations.

In the series shown in Figure 3.2 the oxygen $2p$ parts of the basis were taken from a number of widely used basis sets indicated along the horizontal axis. Three families of sets were used as a source, i.e. the family that ranges from 6-31G to 6-311+G [24], the aug-cc-pVxZ family for $x =$ D and T (Kendall et al., 1992), and a family of basis sets ranging from DZ to L, discussed in van Duijneveldt-van de Rijdt and van Duijneveldt (1992b). Within each family the p sets grow gradually more complete, and so the results within each family should converge to those for calculation L+ (a $9p \rightarrow [6p]$ set which has the L p set augmented by a set of diffuse p functions) in which the p set is nearly saturated.

The CP data in Figure 3.2 are remarkably insensitive to the size of the p set, indicating that even the smallest p sets selected give a decent approximation to a complete p set, as far as ΔE is concerned. Within each family, there is a slight downward trend, but apart

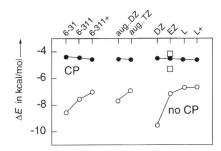

Figure 3.2 Dependence of the water dimer interaction energy on the size of the $2p$ basis set on oxygen. Using the near-equilibrium geometry M2 ($R_{OO} = 2.90$ Å, linear hydrogen bond (van Duijneveldt-van de Rijdt and van Duijneveldt, 1992a)) SCF + MP2 interaction energies (ΔE) have been obtained for a series of basis sets which differ only in the $2p$ basis on the two oxygens. The part of the basis that all calculations have in common is taken from the EZPPcB basis described in (van Duijneveldt-van de Rijdt and van Duijneveldt, 1992b). The $2p$-parts are taken from a number of widely used basis sets indicated along the horizontal axis. The figure shows ΔE-values obtained by the counterpoise procedure (CP), as well as their counterpart in which BSSE has been left uncorrected (no CP).

from sets 6-31G, 6-311G and DZ all basis sets yield a ΔE within 0.04 kcal/mol of the L+ result (-4.56 kcal/mol). The no-CP results are up to twice as attractive due to the presence of BSSE. The trend for these entries is to move up with increasing size of the p set, because (within each family) one gradually removes that part of the BSSE that arises from incompleteness of the p basis. A sizeable discrepancy with the CP results remains even for set L+, because BSSE originating in other parts of the basis (due to incompleteness of $s, d, f, ..$ on O; $s, p, d, ..$ on H; etc.) remains uncorrected even at the p limit. The convergence to the uncorrected L+ value (-6.73 kcal/mol) is quite slow, even the larger sets (6-311+G, aug..TZ and EZ) showing discrepancies of about 0.3 kcal/mol.

The slow convergence of the no-CP results shows how difficult it is to remove BSSE by 'simply' using progressively larger basis sets. The advice to remove BSSE in this way was first given in an often-cited paper by Schwenke and Truhlar (1985) on the basis of their finding that in SCF calculations on (HF)$_2$ in a large series of basis sets, a large scatter remained in ΔE after the BSSE had been removed by the counterpoise procedure. Their advice would have been appropriate if other errors in ΔE will only disappear for very large basis sets with vanishingly small BSSE. But experiments such as the p set saturation test discussed here show that, on the contrary, the 'physics' of the interaction (as represented by the CP ΔE-value) converges quickly and BSSE is then the largest error that remains (if one does not employ CP). Recent experience confirms that to converge the physics embodied in ΔE and simultaneously to suppress BSSE, as one must do in the uncorrected approach, requires significantly larger basis sets than to converge the counterpoise-corrected ΔE (van Duijneveldt-van de Rijdt and van Duijheveldt, 1992b; Chalasinski and Szczesniak, 1994; Novoa et al., 1996). The above results imply that in judging the importance of adding a given basis function for improving the physics in ΔE it is dangerous to rely on the no-CP result, since the effect on ΔE may then be blurred by changes in the BSSE. Thus, diffuse f-functions should be added to the above EZPPcB basis since they change the CP ΔE-value considerably (van Duijneveldt-van de Rijdt and van Duijneveldt, 1992b; Novoa et al., 1996) but their effect on no-CP ΔE results has been found to be small (Del Bene and Shavitt, 1994; Novoa et al., 1996).

Figure 3.2 also shows CP and no-CP energies for a calculation in which the set of (s, p) bondfunctions was removed (cf. the squares for entry EZ). In this more conventional calculation, the size of the BSSE is seen to be much reduced compared to the case with bondfunctions included, but the price one pays is that the CP ΔE-value is now 0.20 kcal/mol less attractive than the result with bondfunctions. To put these numbers in perspective we note that the CP ΔE result for L+ equals -4.56 kcal/mol, and the SCF + MP2 complete basis limit at this water dimer geometry is probably -4.8 kcal/mol. The no-CP calculations overshoot this by a large amount, even the EZ calculation with bondfunctions omitted giving a ΔE as negative as -5.31 kcal/mol.

3.2.2 Equilibrium Distances and Binding Energies

The literature abounds with binding energies (ΔE) and equilibrium distances (R) resulting from automatic minimization of the total (electronic) energy of molecular systems, which is subject to BSSE. We now give a summary of typical BSSE errors in these quantities.

We have already seen (in Figure 3.1) that for very weak complexes, such as those formed by the noble gases, BSSE in the total energy may be larger than ΔE itself. Accordingly, both the uncorrected ΔE-values and the uncorrected R-values must be viewed with suspicion. Eggenberger *et al.* (1991) reported differences as large as 0.58 Å between the counterpoise-corrected and uncorrected R-values of Ne_2, and differences as large as 0.32 Å between the R-values for slipped-parallel geometries of $(CO_2)_2$. The authors noted that in many cases they studied there would be little point in removing the BSSE by applying the counterpoise recipe, since the errors remaining after such correction would still be very large. This is because in these weakly bonded systems the binding energy is dominated by dispersion forces, and unless one carries out well-correlated calculations using well-chosen (and large) basis sets, the precision of the calculation is poor, whether BSSE is removed or not. Thus only the most extensive calculations can be expected to yield useful information on such complexes, and errors of any type, including BSSE, must be avoided. (It seems advisable not to follow up the recent suggestion that MP3 calculations in basis 6-311++G(3d, 3p) uncorrected for BSSE might be of use for such complexes (Yin and MacKerell, 1996).)

Moving now to complexes with a binding energy in the range 2–20 kcal/mol, which includes the case of hydrogen bonding, a rather different situation arises. Examples of R and ΔE values for $(H_2O)_2$ and $(HF)_2$ are given in Table 3.1. Here a large part of the interaction energy stems from electrostatic interactions which are already present at the SCF level of theory. Correlated calculations, for example at the MP2 level, are still necessary to recover the dispersion energy contributions to ΔE (and to adjust overestimation of electrostatics at the SCF level), but since their role is less predominant, accurate binding energies and geometries will still result if only 90 per cent, say, of the correlation interaction energy is reproduced. There may even be cases where the SCF result is already sufficient to rationalize the measured data. Table 3.1 shows that at the counterpoise-corrected SCF level, errors of about 0.1 Å in R and about 1 kcal/mol in ΔE would be made by omitting the MP2 step, whereas the errors made by not correcting for BSSE would be much smaller than this. In other words, BSSE at the SCF level is relatively small. If one decided to use SCF results uncorrected for BSSE there even is some degree of error cancellation, with uncorrected BSSE making up for some missing dispersion attraction. Thus one can understand the success of recent $(H_2O)_n$ (Ojamäe and Hermansson, 1994) calculations, at the uncorrected SCF level, in rationalizing the trends

Table 3.1 Effect of BSSE on the calculated equilibrium distances (R_{OO} or R_{FF}, in Å) as well as on the binding energies (in kcal/mol) of $(H_2O)_2$ and $(HF)_2$

System	Basis set	SCF		SCF + MP2	
		R	$-\Delta E$	R	$-\Delta E$
$(H_2O)_2$	MCY' [a]	3.05 (3.04)	3.8 (4.0)	2.99 (2.90)	4.3 (5.9)
	ESP [b]	3.03 (3.01)	3.89 (4.26)	2.98 (2.84)	4.50 (6.03)
	EZPPPBFD [b]	3.05 (3.03)	3.76 (4.06)	2.94 (2.80)	4.68 (8.21)
$(HF)_2$	aug-cc-pVDZ [c]	2.85 (2.83)	3.67 (3.87)	2.81 (2.75)	4.05 (4.68)
	aug-cc-pVTZ [c]	2.83 (2.83)	3.63 (3.73)	2.77 (2.75)	4.23 (4.71)
	aug-cc-pVQZ [c]	2.82 (2.82)	3.66 (3.71)	2.75 (2.74)	4.38 (4.63)

The values in parentheses were obtained without counterpoise procedure and are subject to BSSE.
References: a) Newton and Kestner (1983); b) van Duijneveldt-van de Rijdt and van Duijneveldt (1992b); Peterson and Dunning (1995).

in the vibrational properties of such clusters, especially if one considers that the trends with increasing n are dominated by non-additive induction effects which are well described at the SCF level (Chalasinski and Szczesniak, 1994) (cf. also Chapter 2, this volume). In a similar vein, for $(HF)_n$ clusters it has been found (Luckhaus et al., 1995) that uncorrected MP2 calculations, in a DZP basis and at the harmonic level, satisfactorily reproduce the experimental trends in the HF frequencies, BSSE effects compensating for the lack of anharmonicity in the theoretical description. On the other hand, in many applications a higher level of accuracy may be needed. Suppose, for example, that the tolerable error in R were 0.01 Å, while that in ΔE were 0.1 kcal/mol. The calculations will then have to be made at the MP2 (or some higher) level. The situation now is much the same as for the noble-gas complexes. Since BSSE at the MP2 level is several times larger than at the SCF level (giving errors of typically 0.05 Å in R and typically 0.5 kcal/mol in ΔE when left uncorrected) an effort to reduce it or to avoid it is called for. Again, correcting for BSSE may not be enough, for even for the largest basis sets shown in Table 3.1 the counterpoise-corrected results barely reach the accuracy aims stated above.

Before leaving Table 3.1 we note the much smaller size of the BSSE errors, especially at the MP2 level, for the aug-cc series of basis sets as compared to the sets shown for $(H_2O)_2$. There are two reasons for this behaviour. First, the ESP and EZPPPBFD sets have been optimized to describe interaction energies well, rather than total energies. As a result they lack high exponent polarization functions. Instead they do have rather diffuse polarization function exponents. The combination of these features leads to a large BSSE. Moreover, basis EZPPPBFD has a set of (s, p) bondfunctions on the H···O hydrogen bond, which improves the dispersion energy (Gutowski et al., 1987b; Burcl et al., 1995) but which also increases the BSSE. The aug-cc sets, on the other hand, have both high- and low-exponent polarization functions, and this tends to reduce the BSSE. The advantage of this is that for these sets it would not be unreasonable to rely on total energies uncorrected for BSSE in the geometry optimization step (i.e. standard gradient optimization routines could be used), since the R-values for the corrected series seem roughly to converge to those for the uncorrected series. To remove the BSSE which contaminates the uncorrected ΔE one may then carry out a single-point counterpoise-corrected calculation at the optimized geometry (cf. Section 3.5). The resulting ΔE^{CP} accurately reflects the physics corresponding to the basis set that was used, and if some of the physics seems missing, one should move on to larger basis sets. The drawback of this approach is that the aug-cc sets are rather large. The aug-cc-pVTZ and QZ sets for $(H_2O)_2$ have 184 and 344 functions, respectively, whereas set EZPPPBFD which has a quality (for ΔE^{CP}) intermediate between these two aug-cc sets has only 140 functions. But EZPPPBFD can clearly only be used at the counterpoise-corrected level.

To conclude this discussion on distances and binding energies we consider the class of (binary) complexes with binding energies in the range 20–50 kcal/mol. This includes solvated ions and weak covalent bonds. The equilibrium distances will be shorter than for the hydrogen-bonded systems discussed above, and BSSE will, accordingly, be a little larger. Compared to ΔE, however, BSSE will be less significant than for the weaker complexes, and we are now in the transition region to the class of ordinary molecules, where BSSE is normally left uncorrected, except when one aims at very high accuracy (Almlöf et al., 1989). The case of solvated ions is special in that electrostatics (and hence the SCF part of ΔE) dominate the interaction even more than for (neutral) hydrogen bonds. The main additional contributions that will be found at the correlated level are a repulsive contribution arising from a reduction of the dipole moment of the solvating molecule and an attractive contribution from dispersion energy. Thus it is not surprising that the overall MP2 contribution is often quite small for these complexes (small in comparison to the total ΔE) and may even be repulsive. Considering that only elaborate calculations will be capable of giving reliable values for such small quantities, the usefulness of performing MP2 calculations in routine-type work on ion-molecule systems may be questioned. An example is provided by the data in Gresh *et al.* (1994), who reported on an extensive series of ion-molecule complexes. They typically found MP2 contributions to ΔE of about -5 kcal/mol at the uncorrected level, and values near zero at the BSSE corrected level. These authors suspected that the counterpoise correction at the MP2 level was too large, and hence chose to omit the correction, but the proper conclusion from their data is that to carry out the MP2 step and then to leave BSSE uncorrected is even more questionable than to omit the MP2 step altogether.

3.2.3 Vibrational Properties

For this topic a distinction may be made between the interfragment vibrations of the complex, i.e. the so-called van der Waals modes, and the modification of intrafragment modes by the complexation. BSSE in the potential energy surface around the equilibrium geometry of the complex clearly translates directly into errors in the frequencies of van der Waals modes. Frequencies too high by a factor of four if BSSE is left uncorrected have been reported for Ne_2 (Eggenberger *et al.*, 1991). For hydrogen-bonded complexes the errors will be less dramatic, especially when basis sets such as the aug-cc sets are used which tend to suppress BSSE. But even here, van der Waals frequencies (especially the stretching frequency) will usually be too high, and this should be taken into account when using such frequencies to determine the zero-point vibrational energy contributions to the enthalpy of formation of the complex.

Shifts of intrafragment modes will also suffer from BSSE. For the XH stretching frequency ν_{XH} of the donor XH bond in XH$\cdots$Y hydrogen bonds it has been found that the error is small at the SCF level (Leclerq *et al.*, 1983; Somasundram *et al.*, 1986; van Duijneveldt-van de Rijdt and van Duijneveldt, 1992a), and no BSSE correction seems necessary. At the correlated level the red shift of ν_{OH}, which is typically -100 cm^{-1} in an isolated OH$\cdots$O contact, can easily be overestimated by some 25 cm^{-1} if BSSE is not corrected for (van Duijneveldt-van de Rijdt and van Duijneveldt, 1992a; Bleiber and Sauer, 1995). Errors of this size would hinder the use of these calculated frequencies in assigning the spectra of such complexes. Doing a more accurate job is not easy, however, since large basis sets and a correlation method beyond MP2 are necessary as well, and it may be necessary to take anharmonicity into account.

3.2.4 Electrical Properties

The changes in overall dipole moment and dipole polarizabilities upon complexation are important parameters in, for example, the calculation of the intensities of infrared and Raman vibrational spectra. To obtain accurate data for these quantities, the counterpoise recipe must now be applied not for the energy but for the property of interest. That is, a dipole moment increment must be evaluated by taking the difference (for each vector component separately) between the dipole moment of the complex and the sum of the dipole moments of the isolated fragments, calculated in the basis set of the whole complex.

An impression of the size of the BSSE in this type of calculation may be gleaned from the following example. The incremental dipole moment for the equilibrium geometry of a system like $(H_2O)_2$ is known to be about 0.1 au (0.25 D). The moment changes in the free fragments by using the basis of the whole complex are typically 0.01 au for a medium sized basis, and around 0.001 au for a large basis, at both the SCF and the correlated levels (Karlström and Sadlej, 1982; Cybulski *et al.*, 1990; van Duijneveldt-van de Rijdt, 1992b). The changes on the donor and acceptor molecules tend to differ in sign, and so the overall error made by not removing the BSSE will (for this type of geometry) be small.

As regards incremental polarizability, in one of the first publications advocating the counterpoise approach, Ostlund and Merrifield (1976) noted that for small basis sets qualitatively different results for this property in He_2 were obtained depending on the removal of BSSE. Relative to monomers in their own basis (i.e. no removal of BSSE) one incorrectly finds an increase of the parallel polarizability. On the other hand, the counterpoise $\Delta\alpha$-calculation in which the monomer calculations employed the full dimer basis correctly gave a polarizability decrease as was also found in much larger sets where BSSE is unimportant.

3.3 REMOVAL OF BSSE: FULL COUNTERPOISE AND VIRTUALS-ONLY COUNTERPOISE

In this section we inspect the evidence that the counterpoise recipe rigorously eliminates the BSSE in a supermolecular calculation. As noted in the introduction, it has long been thought that the recipe overcorrects the BSSE. As a result, a number of incorrect opinions has found its way into the literature, and it may be useful to be aware of some of them. We restrict ourselves to the case of interacting closed-shell fragments.

To put the discussion in the proper perspective let us first examine the orbital spaces available in various counterpoise schemes. Figure 3.3 shows a number of options that exist for performing the monomer and dimer calculations needed in calculating an interaction energy. The horizontal dimension in this figure symbolizes the atomic orbital basis sets of monomers A and B, respectively. The vertical dimension refers to occupied and virtual MOs such as can be obtained in SCF calculations on A or B separately, or on the AB complex. In the latter case, the MOs are assumed to have been localized on A or B by applying a suitable localization algorithm. (For simplicity, we skip discussion of the orthogonalization steps that may be required when working with these orbital sets.)

The arrows point to the orbital spaces that are accessible to the electrons of monomer A in a given calculation. (Similar pictures may, of course, be drawn if one focuses on the description of the B electrons.) In Figure 3.3(f) the entire virtual space of the complex is used in optimizing the A occupieds in an SCF calculation on AB, or in correlating the A electrons in a configuration interaction calculation on AB. Figure 3.3(a)–(d) likewise refer to SCF or CI calculations on free molecule A, in (b)–(d) some or all of the 'ghost' orbitals on B being accessible.

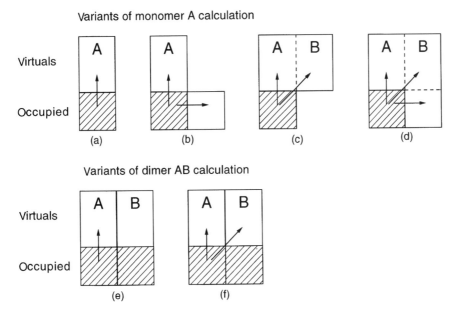

Figure 3.3 Orbital spaces used in various counterpoise schemes. The horizontal dimension in this figure symbolizes the atomic orbital basis sets of monomers A and B, respectively. The vertical dimension refers to occupied and virtual MOs such as can be obtained in SCF calculations on A or B separately, or on the AB complex. In the latter case, the MOs are assumed to have been localized on A or B by applying a suitable localization algorithm. The arrows point to the orbital spaces that are accessible to the electrons of monomer A. Further details are given in the text

A no-counterpoise calculation (no CP) would obtain the binding energy (or any other incremental quantity that arises from the A–B interaction) as the difference between the results of a dimer calculation using orbital spaces for the A electrons as in (f) and monomer calculations as in (a) (and similarly for the B electrons). A standard (also called 'full') counterpoise calculation (CP) would use monomer calculations as in (d) instead.

Note that the 'to be occupied' ghost orbitals on B are made available to A in (d). It is this aspect of the counterpoise recipe that has been criticized most often. The critics of the standard counterpoise scheme (correctly) argued (Johansson *et al.*, 1973; Daudey *et al.*, 1974; Morokuma and Kitaura, 1981) that in the dimer case improvement of the A occupied orbitals by mixing with B occupieds is not possible (cf. the argument given below). Consequently, no arrows have been drawn in (f) from the A occupied to the B occupied space. These critics feel that in the monomers too one should then exclude A improvement by mixing in B occupieds. Only the B *virtual* space, which is accessible in (f), should then be made available in the ghost calculation for monomer A. This leads to the virtuals-only counterpoise (VCP) scheme, where the dimer calculation is as in (f) and the monomers use (c). The debate on the relative merits of CP and VCP has lasted for nearly two decades, but in hindsight it is not difficult to see why the CP recipe is the correct one. (Only an overview of the crucial arguments is given here. More details are given in van Lenthe *et al.*, 1987; Chalasinski and Szczesniak, 1994; van Duijneveldt *et al.*, 1994, and in the original papers quoted therein).

Suppose one plans a dimer calculation of type (f). A trial wavefunction for this calculation may be constructed by first obtaining monomer wavefunctions ψ_0^A and ψ_0^B as in (d) (i.e. employing the full dimer basis) and then taking their antisymmetrized product $A\psi_0^A\psi_0^B$. This 'zeroth order' (because it is of zeroth order in the intermolecular interaction) wavefunction is also called the Heitler–London wavefunction. The antisymmetrization has the consequence that the wavefunction will vanish if an A electron arrives at the same position as a B electron, i.e. the occupied B space has now been made inaccessible to the A electrons and vice versa.

The dimer energy at this stage of the calculation, called the Heitler–London dimer energy, may be found by taking the expectation value of the dimer Hamiltonian, i.e.

$$E_{AB}^{HL} = \frac{\langle A\psi_0^A\psi_0^B | H | A\psi_0^A\psi_0^B \rangle}{\langle A\psi_0^A\psi_0^B | A\psi_0^A\psi_0^B \rangle} \tag{3.3}$$

If one expands this expression one finds, just as in the Heitler–London description of the H_2 molecule, that the antisymmetrization operator gives rise to exchange integrals, and in the present case these are responsible for the short-range repulsion of the two closed-shell fragments A and B. In other words, making the B occupied space unavailable to the A electrons raises the total energy of the dimer. To calculate the corresponding first-order Heitler–London interaction energy ΔE^{HL} we have to subtract monomer energies from E_{AB}^{HL}:

$$\Delta E^{HL} = E_{AB}^{HL} - E^A - E^B \tag{3.4}$$

This is the crucial step, for here we have to decide whether the monomer energies are to be calculated with a wavefunction as in (c), as the VCP recipe would have it, or with (d) monomers, which is the (full) CP recipe. The correct choice is to use (d), for if we are to obtain a proper estimate of the exchange repulsion we have to compare the dimer energy, where the partner space has become inaccessible, to monomer energies in which *no* such restriction has yet been applied (Gutowski et al., 1987a).

While the preceding argument implies that VCP is not a valid recipe, it does not yet prove that the CP scheme gives an interaction energy free from BSSE. However, Chalasinski and Gutowski (1985) have shown that, provided one uses full-CI monomer descriptions ψ_0^A and ψ_0^B which employ the *full* dimer orbital space as in (d), ΔE^{HL} is identical to the BSSE-free perturbational first-order interaction energy of SAPT theory:

$$E^{1,SAPT} = \frac{\langle A\psi_0^A\psi_0^B | V | \psi_0^A\psi_0^B \rangle}{\langle A\psi_0^A\psi_0^B | \psi_0^A\psi_0^B \rangle} \tag{3.5}$$

which implies that ΔE^{HL} too is free from BSSE.

Finally, by extending this argument to higher order it has been possible to prove a Counterpoise Theorem (van Duijneveldt et al., 1994) which states that in finite-basis full-CI calculations on a dimer built from closed-shell fragments the (full) counterpoise recipe will yield a pure interaction energy free from BSSE. In practice, lower-level CI calculations, and even MP2 and SCF calculations will also (almost) be free of BSSE (van Duijneveldt et al., 1994).

A numerical illustration of these statements is given in Figure 3.4, which is based on data in Gutowski et al. (1986, 1993). The idea of these papers was to compare supermolecular interaction energies for two (full-CI) correlated He atoms, in a series of basis sets which markedly differ in the size of the BSSE, to the corresponding BSSE-free

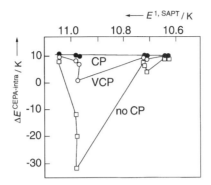

Figure 3.4 Plot comparing the SAPT-theory first-order interaction energy $E^{1,\text{SAPT}}$ of two correlated He atoms with the CEPA-intra supermolecular interaction energy for several atomic basis sets. The entries marked no CP, VCP and CP use monomer orbital spaces as shown in Figures 3.3(a), 3.3(c) and 3.3(d), respectively. Only the CP results closely follow the trend shown by the $E^{1,\text{SAPT}}$ values, confirming that they are free of BSSE. The basis sets used become very large as one moves to the right in the figure. In keeping with this the size of the BSSE (i.e. the difference with the CP curve) becomes small for the rightmost entries. But even here the quality of the calculation would be spoilt by not correcting for BSSE. (Figure based on data in Gutowski *et al.* (1993))

perturbation theory interaction energies employing the same basis set. The simplest such comparison would be between ΔE^{HL} and $E^{1,\text{SAPT}}$, but since programs that can generate ΔE^{HL} for correlated He wavefunctions were unavailable, the closely related CEPA-intra interaction energy was used. Of the various excited configurations that can be present in the CEPA calculation for the dimer, only the intra-correlating single and double excitations from a Hartree–Fock reference configuration are included here (cf. the arrows in Figure 3.3(f)), and those responsible for dispersion attraction are omitted. As a result, $\Delta E^{\text{CEPA-intra}}$, like $E^{1,\text{SAPT}}$, is purely repulsive. Whereas $E^{1,\text{SAPT}}$ is free from BSSE, the $\Delta E^{\text{CEPA-intra}}$ calculation is of the supermolecular type, and so may suffer from BSSE. The amount of BSSE in the final ΔE depends on the way the monomer calculations are carried out. The entries in Figure 3.4 marked no CP, VCP and CP use monomer orbital spaces as shown in Figures 3.3(a), (c) and (d), respectively. The $E^{1,\text{SAPT}}$ results differ only about 0.4 K (i.e. 1.3 μH) as a function of the size of the basis set (the best basis sets being those on the right-hand side of Figure 3.4), but the supermolecular no-CP and VCP results differ up to 50 K and 10 K, respectively, due to the presence of BSSE. On the other hand, the CP results closely follow the trend shown by the $E^{1,\text{SAPT}}$ values, confirming that in accordance with the Counterpoise Theorem they are free of BSSE.

3.3.1 Problems with Virtual Spaces?

A recurring complaint in reports on MP2 (or other correlated) calculations on molecular complexes is that so little of the correlation contribution to ΔE remains after the counterpoise recipe is applied (Gresh *et al.*, 1994; de Oliveira and Dykstra, 1995a; Pudzianowski, 1995). This is often taken as evidence that CP is unreliable when applied at the MP2 level, an opinion which was put forward most forcibly in a series of papers by

Cook *et al.* (1993). Comparing the virtual spaces of a dimer calculation as in Figure 3.3(f) and monomer calculations as in Figure 3.3(d), these authors were concerned about the large differences in orbital energies between these cases, and also about the larger number of virtuals in the case of the monomer calculations. These concerns are reminiscent of the (unfounded) arguments leading to the virtual-only recipe discussed above. If there were any truth in these objections then full-CI calculations, where one employs the same orbital spaces, should give problems as well. But we have seen that CP is rigorously correct in this case. We conclude that a basis for these complaints is lacking, and they certainly do not provide a valid argument for not applying the counterpoise recipe at the correlated level of theory.

3.3.2 Is It Safe to Use CP if BSSE is Large?

Even authors who do not worry about virtual spaces may feel uncomfortable when their BSSE corrections turn out to be large in comparison to the ΔE they are evaluating. It should, however, be recognized that there is no logical connection between the sizes of these two quantities. A proper criterion to decide whether BSSE is large is not how large it is compared to ΔE, but how large it is in comparison to the total energies of the free monomers. Or, if one works at the correlated level (where BSSE tends to be much larger because much more complete basis sets are needed to saturate the correlation energy), one could compare to the total correlation energy of the free monomers. Now correlation energies are typically 1 eV (0.04 H) per electron pair, i.e. about 0.5 H for two molecules of methane or water, whereas the BSSE in the MP2 step of a dimer could be 0.005 H (3 kcal/mol). This is three times larger than the typical MP2 contribution to ΔE in such dimers (about 1 kcal/mol) but it is only 1 per cent of the total correlation energy. When viewed in this perspective there is not much to worry about. (The only problem could be the secondary BSSE, which we will discuss below.)

A case where the large size of the BSSE can be particularly misleading is in calculations which employ sets of bondfunctions in the intermolecular region. For reasons discussed by Burcl *et al.* (1995) such functions are a considerable help in saturating the dispersion energy. The fact that such functions give large BSSE in the no-CP calculation means that the no-CP calculation is less useful than if bondfunctions are not employed, but this carries *no* information on their value in a CP calculation or in a SAPT calculation. Thus, de Oliveira and Dykstra's (1995b) often cited warning against bondfunctions lacks a sound basis (cf. the comments in Williams *et al.*, 1995).

3.3.3 DCBS Versus MCBS, and 'Secondary BSSE'

The perturbation theory approach to molecular interaction energies involves the evaluation of expressions like equation (3.5) in which one needs ground state and excited state wavefunctions for the separate molecules. The user of these theories is free to employ monomer descriptions in which only the monomers own basis set is used (MCBS, monomer certred basis set (Chalasinski and Gutowski, 1988)) or descriptions in which the full basis of the dimer is used (DCBS, dimer centred basis set), or descriptions which employ a suitable selection from the dimer basis set (Williams *et al.*, 1995).

In counterpoise-corrected supermolecular calculations of the type we have discussed so far no freedom of this kind is available: all terms in the interaction energy (when it were analyzed) rely on monomer descriptions in which DCBS, the full dimer set is used. Apart from the lowering of a monomers energy when using DCBS instead of MCBS other properties of a molecule will also change. In Section 3.2 we have noted, for example, that in medium-size-basis calculations on hydrogen-bonded complexes the dipole moments of donor and acceptor may change by about 0.01 au (that of the proton donor will be reduced, and that of the acceptor will be increased because the partner basis extends the charge cloud at the positive and negative ends of the dipole, respectively). Thus it is to be expected that, for example, the dipole–dipole contribution to the total interaction energy will be different when described at the DCBS level or at the MCBS level.

This kind of change is best referred to as 'basis set extension effect' (Chalasinski and Gutowski, 1988). It has also been termed 'secondary BSSE' (Karlström and Sadlej, 1982), but the authors proposing this name already indicated that the change caused by using the full dimer basis may in fact entail an improvement rather than an error. After all, in DCBS one is using a more complete basis set, and moreover it is more complete precisely in the region where the molecules interact and where incompleteness of the basis would be the most serious. Several thorough studies of the relative merits of using DCBS or MCBS are available in the literature (Urban and Hobza, 1975; Chalasinski and Gutowski, 1985, 1988; Cybulski et al., 1990; van Duijneveldt-van de Rijdt and van Duijneveldt, 1992b; Williams et al., 1995). The conclusion of these is that, on the whole, the DCBS description is much superior to the MCBS description. Energy components which are improved are, for example, the exchange repulsion in ΔE^{HL}, the charge-transfer contributions to second-order polarization energies and the London dispersion energies present at the correlated level. Only the electrostatic term in ΔE^{HL} sometimes gives problems. These problems occur in particular when the use of the partner basis causes the appearance of dipole moments which should have been absent, e.g. in a He atom, or perpendicular to the plane of a water molecule. Cases in point are the He–HF interaction (Fowler and Buckingham, 1983) and the He–Li$^+$ interaction (Cybulski and Chalasinski, 1992). Problems of this kind, in principle, occur in other systems too, but the errors resulting from this will often be completely negligible (van Duijneveldt-van de Rijdt and van Duijneveldt, 1992b).

To give an impression of the size of the energy differences caused by using a DCBS than an MCBS monomer description we show in Figure 3.5 the supermolecular first-order (Heitler–London) interaction energies at the SCF level for the $(H_2O)_2$ example mentioned in Section 3.2 and Figure 3.2 (van Duijneveldt, unpublished). That is, a series of calculations was carried out in which the EZPPcB basis was used, but the $2p$ functions on the oxygens were taken from a number of basis sets widely used in the literature. The SCF monomer wavefunctions used in E_{AB}^{HL} (cf. equation (3.3)) were evaluated either in DCBS or in MCBS, and the same then applies for the monomer energies subtracted to obtain ΔE^{HL} (equation (3.4)).

The resulting DCBS ΔE^{HL} values are seen to be significantly more repulsive than the MCBS ΔE^{HL} values. Within each family of basis sets, the MCBS ΔE^{HL} value moves upwards to values near the (more or less constant) DCBS result as the p set grows larger. This suggests that incompleteness of the p set is the primary cause for the MCBS values to differ from the DCBS results, an observation that appears not to have been made previously. Even good sets such as 6-311+G and aug..TZ give poor MCBS ΔE^{HL} values. On the other hand, the fact that the DCBS results do not change much with increasing

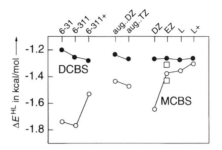

Figure 3.5 Dependence of the water dimer first-order interaction energy on the size of the $2p$ basis set on oxygen. Using the near-equilibrium geometry M2 ($R_{OO} = 2.90$ Å, linear hydrogen bond) SCF Heitler–London first-order interaction energies ΔE^{HL} have been obtained for a series of basis sets which differ only in the $2p$ basis on the two oxygens. (See caption to Figure 3.2 for more details). The monomer wavefunctions needed to compute ΔE^{HL} either used the full basis of the dimer (DCBS) or that of the free monomer (MCBS)

size of the p set (except for sets 6-31G and DZ all sets give an ΔE^{HL} within 0.01 kcal/mol from the L+ value, -1.27 kcal/mol) then implies that the use of the ghost basis in the DCBS case largely compensates for p-shell incompleteness (and presumably for other types of incompleteness as well. Note that the complete basis limit for ΔE^{HL} at this geometry is thought to be -1.22 kcal/mol). Thus the use of the basis functions of B by molecule A (and vice versa) in a supermolecular calculation must not be seen as an error but rather as an improvement of the quality of the calculation. Indeed, users of SAPT theory, where one is free to employ MCBS or DCBS, as seems most appropriate, prefer DCBS descriptions as well (Williams *et al.*, 1995).

3.4 REMOVAL OF BSSE: LOCAL COUNTERPOISE, ALTERNATIVE METHODS

3.4.1 Local Counterpoise

An entirely different class of methods, which may be called *local counterpoise* schemes, has been proposed to deal with the BSSE problem. Here one restricts the dimer calculation in such a way that only the A virtual space is available for improving or correlating the description of the A electrons (Figure 3.3(e)). Use of B basis functions for the A description (which is the root of the BSSE problem) is then excluded from the outset. The corresponding monomer calculation needed to obtain interaction energies is assumed to be that of Figure 3.3(a), which is a calculation on free A in its own basis. (In principle, it would be slightly better to use monomers as in Figure 3.3(b) (van Duijneveldt *et al.*, 1994), but no such calculations have yet been reported.)

Historically, the primary motivation to devise methods of this class has been the opinion that neither the full nor the virtual counterpoise method (cf. Section 3.3) can rigorously solve the BSSE problem. With the advent of the counterpoise theorem, this motivation has become much less relevant. However, local counterpoise schemes are also expected to be more economical than the full counterpoise method, since for each geometry of the complex only a single computation has to be carried out, whereas one needs three (if the complex is a dimer) or more in the full counterpoise approach (not counting the much less costly calculation of monomer deformation energies, cf. Section 3.5).

Local counterpoise schemes have been reported at the SCF level of theory (Mayer and Surjan, 1989; Cullen, 1991; Kapuy and Kozmutza, 1991; Valiron et al., 1993; Muguet and Robinson, 1995) as well as at the correlated level (Meyer and Frommhold, 1986; Kapuy and Kozmutza, 1991; Noga and Vibok, 1991; Røeggen and Skullerud, 1992; Saebø et al., 1993; Wind and Heully, 1994). Within Valence-Bond programs it is usually possible to restrict optimization and excitation to orbitals of a given fragment, and so there also is a series of VB studies on complexes which are claimed to be free of BSSE (Collins and Gallup, 1983; Cooper et al., 1987; Lavendy et al., 1992). A critical survey of some local counterpoise methods has been prepared by Gutowski and Chalasinski (1993). In several local counterpoise studies a comparison was made to standard (i.e. full) counterpoise calculations in the same basis set. It has been found that the ΔE from the local approach is in some cases more attractive but in other cases less attractive than that of the full approach. The proper interpretation of these differences is that, to the extent that the local approach employs an MCBS description of the monomers, this will adversely affect the quality of some terms in the interaction energy. At the SCF level, an MCBS first-order (Heitler–London) interaction energy will be too attractive, as was discussed in the previous section. On the other hand, Cullèn (1991), using his Strictly Monomer MO method, has shown that at the completion of the SCF process the final MCBS ΔE for $(HF)_2$ and $(H_2O)_2$ is much too repulsive, compared to the full counterpoise result. He ascribed this to the fact that most of the charge-transfer stabilization of the hydrogen bond in these systems will be absent in the MCBS description, even when the MCBS basis is very large. At the MP2 level, the use of an MCBS description for the intramolecular correlation effects implies that the (repulsive) correlation contribution to the exchange repulsion energy will be underestimated. An MCBS description of the dispersion energy would lead to a lack of dispersion attraction, but some of the local approaches avoid this problem by using the full virtual space for the dispersion term. Unfortunately, only a few of the authors (Cullen, 1991; Wind and Heully, 1994) who proposed local methods have explicitly considered the problems an MCBS description may give, and so the judgment that was given on the quality of the local results was frequently, shall we say, incomplete.

Special mention should be made here of the Chemical Hamiltonian Approach (CHA) developed by Mayer and coworkers (Mayer and Surjan, 1989; Noga and Vibok, 1991). In this approach, the monomers in the dimer are not allowed to improve their own energy by using the partner basis, but mixing of A and B orbitals induced by the intermolecular terms in the dimer Hamiltonian does take place, and so charge transfer effects are accounted for. For a thorough analysis see Gutowski and Chalasinski (1993). Different implementations of the method at the SCF level have been reported, the most accurate one (CHA/CE) (Mayer and Surjan, 1989) employing the energy expectation value over the CHA wavefunction to obtain the interaction energy. Correlated versions of the approach are available (Noga and Vibok, 1991) but these have not yet given convincing results (Gutowski and Chalasinski, 1993) (cf. also Mayer et al., 1994).

CHA calculations in several basis sets have been carried out at the SCF level, for a large number of hydrogen-bonded complexes (Valiron et al., 1990; Pye et al., 1994) and it was found that the CHA energies are usually more attractive (by amounts up to 2 kcal/mol) than full counterpoise interaction energies. However, the differences become small when large basis sets are used, and interestingly, they even become small compared to the BSSE in these large sets. The authors ascribed the differences to overcorrection of BSSE by the full counterpoise approach, an interpretation which is not in accord with the counterpoise theorem, and which, moreover, does not explain why the differences become small while BSSE is still sizeable. The proper explanation of the differences is probably that the first-order (Heitler–London) interaction in CHA is obtained at the MCBS level (Gutowski and Chalasinski, 1993). In the previous section we have seen that this gives rise to a significant lack of repulsion, except when very large basis sets are used. This is precisely the pattern exhibited by the CHA interaction energies in Valiron *et al.* (1993) and Pye *et al.* (1994).

In summary, local counterpoise schemes do avoid the occurrence of BSSE, and have the advantage that for each geometry of the complex only a single calculation has to be done. However, the problems generated by the lower quality of an MCBS monomer description appear to have been underestimated. The size of these errors is such that in many cases one might just as well perform an (equally economical) no-CP calculation in a basis with small BSSE.

3.4.2 Alternative Methods: Larger Basis Sets, R12-methods, Geminals

BSSE can, in principle, be eliminated by employing monomer descriptions giving energies so close to the basis set limit that improvements by partner basis functions are negligibly small. Using conventional basis-set oriented methods, this strategy readily succeeds at the SCF level, where BSSE is small, but at the correlated level special precautions are necessary. Thus one should employ basis sets such as the aug-cc sets (Kendall *et al.*, 1992) which have large sets of polarization functions with both high and low exponents, and, moreover, one should not employ bondfunctions. In this way BSSE can be suppressed to the level where it is comparable in size to the other errors in ΔE. However, if the final results are to be highly accurate it turns out that very large sets of this kind must be used (cf. the discussion in Section 3.2).

It has recently been shown (Klopper, 1995) that it is possible to reach total energies very close to the basis set limit for correlated methods such as MP2, CCSDT, etc. by including terms in the correlating part of the wavefunction which linearly depend on the interelectronic distances. These methods, called MP2-R12, CCSDT-R12, etc., have been applied to systems like He_2 (Klopper and Noga, 1995) and $(H_2O)_2$ (Klopper *et al.*, 1995), and the BSSE was found to be small. Moreover, these methods seem to give a good description of the subtle correlation effects in the intermolecular region which are partly responsible for the very slow convergence of conventional ΔE calculations to the basis set limit (unless bondfunctions are used). The only drawback of the R12-methods is that their implementation relies on a resolution of the identity operator in terms of the conventional basis set in a given calculation, and this results in exaggerated binding energies unless the conventional basis is chosen to be larger than usual. Indeed the BSSE of conventional MP2 or CCSDT calculations in these sets was also small. Therefore the main advantage of the R12 calculations lies in the better convergence of the counterpoise-corrected ΔE to its basis set limit.

Yet another method to reach low total energies for the fragments in a complex is the use of a basis set of explicitly correlated Gaussian-type geminals. Such methods have very recently been applied to the He$_2$ system and very accurate results were obtained at the MP2 level (Bukowski *et al.*, 1996) and at the full-CI level (Komasa and Rychlewski, 1996). BSSE is small in these calculations, but it is a non-trivial problem to remove this small BSSE (Bukowski *et al.*, 1996).

3.5 GEOMETRY OPTIMIZATION. NON-UNIQUENESS OF CALCULATED INTERACTION ENERGIES

3.5.1 Algorithm for Geometry Optimization

One of the useful features of current quantum-chemical program packages is the option of automated optimization, using analytical derivative techniques, of the interfragment geometrical parameters (R) of a complex as well as of the intrafragment geometrical parameters (r^A, r^B). Such methods operate on the uncorrected total energy $E^{AB}(R, r^A, r^B)$ of the complex, and so the results are contaminated with BSSE. Denoting the equilibrium geometry of the free monomers by (r^{Ae}, r^{Be}), the uncorrected binding energy $U(R, r^A, r^B)$ of the complex at a chosen geometry (R, r^A, r^B), relative to the energy of the free fragments, is

$$U(R, r^A, r^B) = E^{AB}(R, r^A, r^B) - E^A(r^{Ae}) - E^B(r^{Be}) \tag{3.6}$$

This binding energy may be partitioned into the deformation energy $E_{\text{def}}(r^A, r^B)$, required to deform the fragments from their unperturbed geometries (r^{Ae}, r^{Be}) to the desired geometry (r^A, r^B), and the 'vertical' interaction energy $\Delta E(R, r^A, r^B)$ resulting when we bring the deformed fragments (without further geometry change) together in the complex:

$$E_{\text{def}}(r^A, r^B) = E^A(r^A) - E^A(r^{Ae}) + E^B(r^B) - E^B(r^{Be}) \tag{3.7}$$

$$\Delta E(R, r^A, r^B) = E^{AB}(R, r^A, r^B) - E^A(r^A) - E^B(r^B) \tag{3.8}$$

$$U(R, r^A, r^B) = \Delta E(R, r^A, r^B) + E_{\text{def}}(r^A, r^B) \tag{3.9}$$

Note that in this formulation the interaction energy ΔE provides the driving force causing the monomers to deform from their equilibrium geometries (r^{Ae}, r^{Be}). Although U is usually formulated as in equation (3.6) and not as in equation (3.9), the latter formulation makes it easier to locate the BSSE which enters in the process of evaluating U. This happens in the step leading to $\Delta E(R, r^A, r^B)$, if in equation (3.8) one employs energies for the deformed fragments which do not use the full basis set used in $E^{AB}(R, r^A, r^B)$. Thus the BSSE-free method of calculating a binding energy U^{CP} of the complex with monomers deformed from their equilibrium geometries is to use, instead of equations (3.8) and (3.9) (Emsley *et al.*, 1978; Smit *et al.*, 1978; Maggiora and Williams, 1982),

$$\Delta E^{CP}(R, r^A, r^B) = E^{AB}(R, r^A, r^B) - E^A(R, r^A, r^B, \text{ full basis}) - E^B(R, r^A, r^B, \text{ full basis}) \tag{3.10}$$

$$U^{CP}(R, r^A, r^B) = \Delta E^{CP}(R, r^A, r^B) + E_{\text{def}}(r^A, r^B) \tag{3.11}$$

while $E_{\text{def}}(r^A, r^B)$ remains formally the same as in equation (3.7).

Finding the (uncorrected) equilibrium geometry and equilibrium binding energy of the complex amounts to the minimization of $U(R, r^A, r^B)$. This is entirely equivalent to minimizing $E^{AB}(R, r^A, r^B)$, as done in gradient optimizations. After completion of the optimization it is common practice to eliminate BSSE from the final U-value by doing a counterpoise calculation of ΔE^{CP} (equation (3.10)) at the final geometry. However, as recently emphasized by Xantheas (1996), if one looks for a BSSE-free quantity that is strictly comparable to the uncorrected U of equations (3.6) and (3.9) it is more appropriate to evaluate U^{CP}, i.e. to allow for E_{def} as well, unless the monomer deformations are so small that E_{def} is negligible. (The nomenclature used in the Xantheas paper is as follows: his $\Delta E(BSSE)$ is what here is called the counterpoise-corrected binding energy U^{CP}, and the deformation energy E_{def} is labelled 'relaxation' energy in his paper.)

Thus, whereas in gradient optimization using equation (3.6) the evaluation of E_{def} remains hidden, the evaluation of E_{def} becomes an explicit step when using equation (3.11). The evaluation of E_{def} can be done in calculations entirely separate from those of ΔE^{CP}, using basis sets and computational schemes that are optimal for this purpose. (This is actually an advantage compared to the case where one optimizes E^{AB} since in the latter case the description of E^{AB} (basis set and theoretical model) must simultaneously be adequate for intramolecular distortions and for intermolecular interactions.)

Since analytical derivatives for all energies entering equation (3.11) have already been implemented in current programs, there is no fundamental problem in designing automated optimization algorithms based on equation (3.11), but no such algorithm has yet been implemented. Geometry optimizations directly using equation (3.11) therefore remain rather laborious. Nevertheless a fair number of structures in which both inter- and intramolecular coordinates were optimized, using equation (3.11), have already appeared in the literature (e.g. see Xantheas, 1996, and the references quoted therein).

For completeness we note that U^{CP} of equation (3.11) is also the expression that must be used if one wants to study the effect of complexation on intramolecular vibration frequencies, free from BSSE. The derivatives at the optimum of equation (3.11) then define the required force field. More generally, U^{CP} is the potential energy surface that must be used in the nuclear Schrödinger equation governing the nuclear motions in the complex.

3.5.2 Ambiguities in Defining the Fragments

The simplest way of forming a trimer is by the process $A + B + C \rightarrow ABC$, and the CP approach to calculating the ΔE for this process would be to use the trimer basis for all fragments. The formation of a trimer ABC can also be done in steps such as $A + B \rightarrow AB$, followed by $AB + C \rightarrow ABC$. If one adopts the CP method in calculating the binding energies for each individual step, then in the second step one would employ the trimer basis {ABC} for all fragments involved. But in the first step one has two choices, i.e. to calculate all fragment energies in the {AB} set or to calculate all fragment energies in the {ABC} set. Both choices adhere to the CP principle, but it has been found (Wells and Wilson, 1983; Hermansson, 1988; Turi and Dannenberg, 1993) that the two choices for the first step give different ΔE's and hence different overall ΔE's as well. If one considers the alternative path $A + C \rightarrow AC$ followed by $AC + B \rightarrow ABC$ then again two choices can be made. From the point of view of thermodynamics, the overall binding energy should be independent of the chosen path, but only the choice to calculate all individual energies in basis {ABC} will yield the same overall ΔE as that for the first path (Turi and Dannenberg, 1993). Thus the use of the {ABC} set for all fragments was recommended for the sake of consistency. Likewise, in larger clusters, ΔE^{CP} should preferably be calculated by using monomer energies evaluated in the full cluster basis (although sets on distant ghosts may make a very small contribution in practice (Hermansson, 1988; Turi and Dannenburg, 1993).

One may wonder about the physical significance of the various overall ΔE-values that could have been obtained by making different choices. Since the CP method is followed in all variants, all ΔE's can rightfully be considered as 'true interaction energies' for system ABC. The reason that the results can be different must therefore be that the different basis sets, such as {AB} or {ABC} when studying the step A + B → AB, are of different quality in describing the physics of this step. Thus the differences are a symptom of lingering basis set incompleteness, and they are an indication of the error margin in the final ΔE, as compared to the result in a complete basis.

Note that the ambiguities are not inherent to the use of the CP approach, as they would also arise when applying a BSSE-free method such as SAPT. Indeed, the ambiguities result from the fact that in these approaches one partitions the total complex in a number of polyatomic fragments. Already a system like $(HF)_2$, normally treated by starting from two HF fragments, could also be studied by starting from F^- interacting with HFH^+. A full geometry optimization should lead to the same final geometry, but in practice small differences will no doubt occur, for the reasons discussed above. Problems of this kind have been encountered also in studies of proton-transfer reactions, where the more natural choice of fragments will switch as the proton approaches the partner molecule (Latajka *et al.*, 1992).

Problems of a different kind arise for interactions where no simple fragmentation in closed shell fragments is possible, such as in calculations on intramolecular hydrogen bonds. Perhaps one could tear the molecule apart in open-shell fragments in order for a SAPT or counterpoise calculation to become possible. But the prospects of this approach leading to better accuracy than large-basis calculations based on the uncorrected total energy of such systems are not promising.

3.6 CONCLUSION

- The question of how to deal with the BSSE in supermolecular calculations on molecular complexes has long been a controversial issue. However, it has now been established for closed-shell interactions that BSSE can rigorously be avoided by applying the Boys and Bernardi counterpoise recipe. The interaction energy ΔE^{CP} obtained in this way, as well as all properties derived from it, such as the geometry of a complex, are free from BSSE and correspond precisely to the physics embodied in the computational model employed (basis set, level of correlation treatment). This allows the user to focus on improving the physics until some desired accuracy is reached, without having to worry about the size of the BSSE.
- In the light of these developments, phrases such as 'the counterpoise method is known to overcorrect the BSSE' and 'we estimated the BSSE by applying the counterpoise

recipe' should be avoided since they incorrectly imply that the CP method is only approximate.
- The size of the BSSE in a given calculation carries *no* information on the quality of ΔE^{CP}, and should not be used as an argument for or against certain aspects of the computational method, such as the use of bondfunctions.
- These remarks are not meant to imply that all calculations on complexes should henceforth be done using the counterpoise method. On the contrary, there is no point in applying it in cases where BSSE is very small or where many other sources of error would remain uncorrected anyway. Furthermore, given the advantages of derivative techniques based on direct use of the total energy of a system, it may be more convenient not having to correct for BSSE. In such cases one may try to keep BSSE small by using basis sets with added diffuse functions (such as 6-311+G and aug-cc) and by not using bondfunctions. Very large basis sets of this kind are needed, however, if one looks for high accuracy, and it is likely that counterpoise-corrected calculations using basis sets designed to converge ΔE^{CP} to the basis set limit provide a more efficient route to highly accurate results.

ACKNOWLEDGEMENTS

I am indebted to Grzegorz Chalasinski, Jeanne van Duijneveldt-van de Rijdt, Maciej Gutowski and Joop van Lenthe for many enlightening discussions on the subject of BSSE. I thank Jeanne also for critically reading the manuscript and for suggesting several improvements, and Joop van Lenthe for providing the He_2 BSSE-uncorrected full-CI data used in Figure 3.1.

REFERENCES

Almlöf, J., de Leeuw, B. J., Taylor, P. R., Bauschlicher, C. W. and Siegbahn, P. (1989) *Int. J. Quant. Chem. Symp.*, **23**, 345.
Aziz, R. A., Janzen, R. A. and Moldover, M. R. (1995) *Phys. Rev. Lett.*, **74**, 1589.
Bleiber, A. and Sauer, J. (1995) *Chem. Phys. Lett.*, **238**, 243.
Boys, S. F. and Bernardi, F. (1970) *Mol. Phys.*, **19**, 553.
Bukowski, R., Jeziorski, B. and Szalewicz, K. (1996) *J. Chem. Phys.*, **104**, 3306.
Burcl, R., Chalasinski, G., Bukowski, R. and Szczesniak, M. M. (1995) *J. Chem. Phys.*, **103**, 1498.
Chalasinski, G. and Gutowski, M. (1985) *Mol. Phys.*, **54**, 1173.
Chalasinski, G. and Gutowski, M. (1988) *Chem. Rev.*, **88**, 943.
Chalasinski, G. and Szczesniak, M. M. (1994) *Chem. Rev.*, **94**, 1723.
Clementi, E. (1967) *J. Chem. Phys.*, **46**, 3851.
Collins, J. R. and Gallup, G. A. (1983) *Mol. Phys.*, **49**, 871.
Cook, D. B., Sordo, J. A. and Sordo, T. L. (1993) *Int. J. Quant. Chem.*, **48**, 375.
Cooper, D. L., Gerratt, J. and Raimondi, M. (1987) *Adv. Chem. Phys.*, **69**, 319.
Cullen, J. M. (1991) *Int. J. Quant. Chem. Symp.*, **25**, 193.
Cybulski, S. M. and Chalasinski, G. (1992) *Chem. Phys. Lett.*, **197**, 591.
Cybulski, S. M., Chalasinski, G. and Moszynski, R. (1990) *J. Chem. Phys.*, **92**, 4357.
Daudey, J. P., Claverie, P. and Malrieu, J. P. (1974) *Int. J. Quant. Chem.*, **8**, 1.
de Oliveira, G. and Dykstra, C. E. (1995a) *Chem. Phys. Lett.*, **243**, 158.
de Oliveira, G. and Dykstra, C. E. (1995b) *J. Mol. Struct. (Theochem.)*, **337**, 1.
Del Bene, J. E. and Shavitt, I. (1994) *J. Mol. Struct.*, **307**, 27.

Eggenberger, R., Gerber, S., Huber, H. and Searles, D. (1991) *Chem. Phys. Lett.*, **183**, 223.
Emsley, J., Hoyte, O. P. A. and Overill, R. E. (1978) *J. Am. Chem. Soc.*, **100**, 3303.
Fowler, P. W. and Buckingham, A. D. (1983) *Mol. Phys.*, **50**, 1349.
Gresh, N., Leboeuf, M. and Salahub, D. (1994) *ACS Symp. Series*, **569**, 82.
Gutowski, M. and Chalasinski, G. (1993) *J. Chem. Phys.*, **98**, 5540.
Gutowski, M., van Duijneveldt, F. B., Chalasinski, G. and Piela, L. (1987a) *Mol. Phys.*, **61**, 233.
Gutowski, M., van Duijneveldt-van de Rijdt, J. G. C. M., van Lenthe, J. H. and van Duijneveldt, F. B. (1993) *J. Chem. Phys.*, **98**, 4728.
Gutowski, M., van Lenthe, J. H., Verbeek, J., van Duijneveldt, F. B. and Chalasinski, G. (1986) *Chem. Phys. Lett.*, **124**, 370.
Gutowski, M., Verbeek, J., van Lenthe, J. H. and Chalasinski, G. (1987b) *Chem. Phys.*, **111**, 271.
Hermansson, K. (1988) *J. Chem. Phys.*, **89**, 2149.
Jansen, H. B. and Ros, P. (1969) *Chem. Phys. Lett.*, **3**, 140.
Jeziorski, B., Moszynski, R. and Szalewicz, K. (1994) *Chem. Rev.*, **94**, 1887.
Johansson, A., Kollman, P. and Rothenberg, S. (1973) *Theoret. Chim. Acta*, **29**, 167.
Kapuy, E. and Kozmutza, C. (1991) *J. Chem. Phys.*, **94**, 5565.
Karlström, G. and Sadlej, A. (1982) *Theoret. Chim. Acta*, **61**, 1.
Kendall, R. A., Dunning, T. H. and Harrison, R. J. (1992) *J. Chem. Phys.*, **96**, 6796.
Kestner, N. R. (1968) *J. Chem. Phys.*, **48**, 252.
Klopper, W. (1995) *J. Chem. Phys.*, **102**, 6168.
Klopper, W. and Noga, J. (1995) *J. Chem. Phys.*, **103**, 6127.
Klopper, W., Schütz, M., Lüthi, H. P. and Leutwyler, S. (1995) *J. Chem. Phys.*, **103**, 1085.
Kocjan, D., Koller, J. and Azman, A. (1976) *J. Mol. Struct.*, **34**, 145.
Komasa, J. and Rychlewski, J. (1996) *Chem. Phys. Lett.*, **249**, 253.
Latajka, Z., Scheiner, S. and Chalasinski, G. (1992) *Chem. Phys. Lett.*, **196**, 384.
Lavendy, H., Robbe, J. M. and Flament, J. P. (1992) *Chem. Phys. Lett.*, **196**, 377.
Leclerq, J. M., Allavena, M. and Bouteiller, Y. (1983) *J. Chem. Phys.*, **78**, 4606.
Liu, B. and McLean, A. D. (1973) *J. Chem. Phys.*, **59**, 4557.
Luckhaus, D., Quack, M., Schmitt, U. and Suhm, M. (1995) *Ber. Bunsenges. Phys. Chem.*, **99**, 275.
Maggiora, G. M. and Williams, I. H. (1982) *J. Mol. Struct.*, **88**, 23.
Mayer, I. and Surjan, P. R. (1989) *Int. J. Quant. Chem.*, **36**, 225.
Mayer, I. and Vibok, A. (1991) *Int. J. Quant. Chem.*, **40**, 139.
Mayer, I., Vibók, A. and Valiron, P. (1994) *Chem. Phys. Lett.*, **224**, 166.
Meyer, W. and Frommhold, L. (1986) *Phys. Rev.*, **A33**, 1791.
Morokuma, K. and Kitaura, K. (1981), in *Chemical Application of Atomic and Molecular Electrostatic Potentials*, edited by Politzer, P. and Truhlar, D. G., Plenum, New York, p. 215.
Morokuma, K. and Pedersen, L. (1968) *J. Chem. Phys.*, **48**, 3275.
Morokuma, K. and Winick, J. R. (1970) *J. Chem. Phys.*, **52**, 1301.
Muguet, F. F. and Robinson, G. W. (1995) *J. Chem. Phys.*, **102**, 3648.
Newton, M. D. and Kestner, N. R. (1983) *Chem. Phys. Lett.*, **94**, 198.
Noga, J. and Vibok, A. (1991) *Chem. Phys. Lett.*, **180**, 114.
Novoa, J. J., Planas, M. and Rovira, M. C. (1996) *Chem. Phys. Lett.*, **251**, 33.
Ojamäe, L. and Hermansson, K. (1994) *J. Phys. Chem.*, **98**, 4271.
Ostlund, N. S. and Merrifield, D. L. (1976) *Chem. Phys. Lett.*, **39**, 612.
Peterson, K. A. and Dunning, T. H. (1995) *J. Chem. Phys.*, **102**, 2032.
Pudzianowski, A. (1995) *J. Chem. Phys.*, **102**, 8029.
Pye, C. C., Poirier, R. A., Yu, D. and Surján, P. R. (1994) *J. Mol. Struct. (Theochem.)*, **307**, 239.
Ransil, B. J. (1961) *J. Chem. Phys.*, **34**, 2109.
Røeggen, I. and Skullerud, H. J. (1992) *J. Phys. B.*, **25**, 1795.
Saebø, S., Tong, W. and Pulay, P. (1993) *J. Chem. Phys.*, **98**, 2170.
Schwenke, D. W. and Truhlar, D. G. (1985) *J. Chem. Phys.*, **82**, 2418.
Smit, P. H., Derissen, J. L. and van Duijneveldt, F. B. (1978) *J. Chem. Phys.*, **69**, 4241.
Somasundram, K., Amos, R. D. and Handy, N. C. (1986) *Theoret. Chim. Acta*, **69**, 491.
Turi, L. and Dannenberg, J. J. (1993) *J. Phys. Chem.*, **97**, 2488.
Urban, M. and Hobza, P. (1975) *Theoret. Chim. Acta*, **36**, 215.
Valiron, P., Vibok, A. and Mayer, I. (1993) *J. Comp. Chem.*, **14**, 401.

van Duijneveldt, F. B., unpublished results.
van Duijneveldt, F. B., van Duijneveldt-van de Rijdt, J. G. C. M. and van Lenthe, J. H. (1994) *Chem. Rev.*, **94**, 1873.
van Duijneveldt-van de Rijdt, J. G. C. M. and van Duijneveldt, F. B. (1992a) *J. Comp. Chem.*, **13**, 399.
van Duijneveldt-van de Rijdt, J. G. C. M. and van Duijneveldt, F. B. (1992b) *J. Chem. Phys.*, **97**, 5019.
van Lenthe, J. H., private communication.
van Lenthe, J. H., van Duijneveldt-van de Rijdt, J. G. C. M. and van Duijneveldt, F. B. (1987) *Adv. Chem. Phys.*, **69**, 521.
van Mourik, T. and van Lenthe, J. H. (1995) *J. Chem. Phys.*, **102**, 7479.
Wells, B. H. and Wilson, S. (1983) *Chem. Phys. Lett.*, **101**, 429.
Williams, H., Mas, E. M., Szalewicz, K. and Jeziorski, B. (1995) *J. Chem. Phys.*, **103**, 7374.
Wind, P. and Heully, J.-L. (1994) *Chem. Phys. Lett.*, **230**, 35.
Xantheas, S. S. (1996) *J. Chem. Phys.*, **104**, 8821.
Yin, D. and MacKerell, D. (1996) *J. Phys. Chem.*, **100**, 2588.

4 Theory and Computation of Vibration, Rotation and Tunneling Motions of Van der Waals Complexes and their Spectra

A. VAN DER AVOIRD[a], P. E. S. WORMER[a], and R. MOSZYNSKI[b]

[a]*University of Nijmegen, The Netherlands*
[b]*University of Warsaw, Poland*

4.1 INTRODUCTION

The experimental spectrum of a van der Waals dimer is a rich mine of information on its intermolecular potential, but one that is not easily exploited. One usually starts by obtaining a first approximation of the potential energy surface (PES) from an *ab initio* calculation or from an empirical force field, and then calculates the spectrum from this potential. Subsequent comparison of the computed and experimental spectrum leads to refinement of the potential, often in an iterative fashion.

In this chapter we will explain and illustrate the theoretical and computational methods for obtaining the spectrum from a given PES. We concentrate on the solution of the nuclear motion problem—the second step in the Born–Oppenheimer approximation—and the ensuing computation of vibration–rotation–tunneling (VRT) spectra. Such a calculation consists of two steps: first, one calculates the bound states and sometimes also the resonances of the van der Waals complex, and next, one computes the intensities of the transitions between these states. To that end, one needs the dipole operator as a function of the nuclear coordinates for emission or absorption spectra, while the polarizability function is required for Raman scattering.

A hierarchy of nuclear motions appear in van der Waals complexes. Within the chemically bound monomers which constitute the complex, the nuclei vibrate with high frequency. The motions of the monomers as a whole are much slower, as they move in the field of the weak van der Waals forces holding them together. This allows an adiabatic separation between the intramolecular vibrations and the intermolecular motions. The latter have usually large amplitudes and, since there are often multiple minima in the potential surface with only low barriers between them, the intermolecular 'vibrations' may look more like hindered rotations or tunneling motions.

Clearly, the standard methods, based on the harmonic oscillator/rigid rotor model with perturbation corrections, traditionally used to study the rovibrational spectra of nearly rigid molecules (Wilson *et al.*, 1955; Califano, 1976; Papoušek and Aliev, 1982), are not applicable to the intermolecular modes in van der Waals complexes. New methods for the computation of VRT states of van der Waals molecules were developed over the past two decades and are still being improved.

This chapter is organized as follows. First, we discuss the different choices of coordinate systems used in the study of van der Waals molecules and then give the corresponding kinetic energy expressions. The derivations of these expressions are outlined in a recent review (van der Avoird *et al.*, 1994a). This is followed by a discussion of the analytic forms of the intermolecular potential energy surfaces. Given the Hamiltonians, we go on to discuss how to obtain their eigenstates. From the eigenstates we may obtain dimer properties and transition intensities. Permutation-inversion symmetry plays an important role in van der Waals molecules, which are highly non-rigid, and therefore we touch briefly upon this subject.

The last section is devoted to concrete results of calculations and their comparison with experiment. As examples we discuss He–HF, He–CO, the ammonia dimer and the CO molecule enclosed in buckminsterfullerene C_{60}.

4.2 CHOICE OF COORDINATES

It is customary to use normal coordinates for the fast vibrations of the nearly rigid nomomers. Their coefficients in terms of the nuclear displacements might be determined by the standard Wilson GF-matrix method (Wilson *et al.*, 1955; Califano, 1976; Papoušek and Aliev, 1982). On the other hand, the optimum strategy for the calculation of the bound states of a van der Waals molecule requires a careful choice of coordinates. By the very nature of the motions in such a complex, it is rare that Cartesian coordinates can be used; almost always one is forced to employ curvilinear coordinates. If a monomer in a weakly bound complex is considered to be rigid, one needs angular coordinates to describe its rotational motion. If this motion is not infinitesimal, as is the case in most van der Waals complexes, then a linearization of the angular coordinates will not do, and the full range of these curvilinear coordinates must be considered.

For instance, Euler angles $\zeta \equiv (\phi, \theta, \psi)$, which define the orientation of a system of axes on a monomer, are a natural choice of van der Waals coordinates for a non-spherical monomer in a dimer. These Euler angles may be defined with respect either to a laboratory (or space-fixed) frame or to a frame that is somehow embedded in the dimer. Their definition is as follows. One starts in a situation where the monomer axes are parallel to the reference frame, which is the space-fixed or the dimer-embedded frame. Then one rotates the molecule and its frame about the z-axis over the angle ϕ. This sends the molecular y-axis to the y'-axis and (formally) the z-axis to the z'-axis. Second, one rotates the molecule and its frame over θ about the y'-axis. The z'-axis is rotated to the z''-axis. The final rotation is over ψ around the z''-axis. See Talman (1968) for a proof that any rotation can be described in this manner.

It goes without saying that the distance R between the centers of mass of the monomers A and B is usually included among the coordinates, although sometimes R is kept fixed in model calculations.

The angular momentum associated with the overall rotation of a dimer is a conserved quantity, provided the dimer rotates in a field-free space. In most computational methods one takes advantage of this fact. The use of a dimer embedded frame enables a separation of this overall rotation from the 'vibrations' of the dimer, i.e. from its internal (van der Waals) motions. Unfortunately, however, an exact separation of vibrations and rotations is not possible, even in the case of nearly rigid molecules, and for highly non-rigid van der Waals complexes there will be strong vibration–rotation coupling. Nevertheless, it may be advantageous to start with assuming separability, and to account for the coupling terms in a second step of the calculations. We will return to this point later, when we discuss the dimer Hamiltonians.

With regard to the choice of a dimer versus a space-fixed frame, we will see that this is in fact determined by some characteristic properties of the system at hand, such as the rotational constants of the monomers and the dimer and the anisotropy of the intermolecular potential. These properties are partly responsible for the nature of the VRT motions in the complex and for the existence of approximate constants of the motion, and they determine which coordinates achieve optimum separability.

It may occur that the rotations of the monomers are strongly hindered in some directions and less in others, so that the van der Waals motions follow certain pathways. In such a case one can introduce special curvilinear coordinates that describe the motions along these pathways and the motions orthogonal to them. Examples are given by the semi-rigid bender coordinates in the HF-dimer (Bunker *et al.*, 1990) and the specific tunneling pathways in the H_2O-dimer (Coudert and Hougen, 1988; Fraser, 1991) and H_2O-trimer (Van der Avoird *et al.*, 1996; Olthof *et al.*, 1996).

4.3 KINETIC ENERGY EXPRESSIONS

In principle, the nuclear motion Hamiltonian of a van der Waals complex depends on intra- as well as on intermolecular coordinates. In this section we will not explicitly write its dependence on the intramolecular (normal) coordinates of the nearly rigid monomers, though, and follow the simplest road to avoiding the dependence on the intramonomer coordinates: We make the assumption that the monomers are rigid. This seemingly crude model works well in many cases, especially if one adopts vibrationally averaged geometries of the nearly rigid monomers, instead of their equilibrium geometries.

The kinetic energy T for a set of general curvilinear coordinates q_i was given by Podolsky (1928) (see also Essén, 1978):

$$2T = -\hbar^2 g^{-1/2} \sum_{ij} \frac{\partial}{\partial q_i} g^{1/2} (\mathbf{G}^{-1})_{ij} \frac{\partial}{\partial q_j} \quad (4.1)$$

Here **G** is the metric tensor and g is the determinant of **G**; see Appendix A of van der Avoird *et al.* (1994). In that Appendix the Podolsky expression was explicitly worked out for different sets of coordinates useful in the study of van der Waals dimers. Observe that the inverse of the metric tensor appears in this expression. This has the consequence that constraints are often difficult to introduce directly into a Hamiltonian. A valid procedure of deriving a constrained Hamiltonian is by first writing down the kinetic energy part of the corresponding constrained Lagrangian expressed in generalized coordinates, which defines **G**. One then inverts this tensor and substitutes the inverse into equation (4.1). Since very often generalized coordinates are non-orthogonal, i.e. **G** contains non-zero off-diagonal elements, this procedure gives a Hamiltonian which differs from a Hamiltonian obtained by simply—and erroneously—omitting constrained terms. Only if **G** is (block) diagonal the omission of terms in a Hamiltonian is allowed, because a blocked matrix may be inverted block by block.

The simplest Hamiltonian for a dimer consisting of two general, non-linear, monomers is obtained by defining the Euler-angles $\zeta_A = (\phi_A, \theta_A, \psi_A)$ and $\zeta_B = (\phi_B, \theta_B, \psi_B)$ and the polar angles $\hat{\mathbf{R}} = (\beta, \alpha)$ of the vector **R**, pointing from A to B with respect to a space-fixed (SF) frame. The kinetic energy expression in this Hamiltonian simply reads

$$T = T_A + T_B + T_{AB} \tag{4.2}$$

with T_X, $X = $ A or B, given by the standard rigid rotor expression

$$T_X = A_X (j_{Xx}^{BF})^2 + B_X (j_{Xy}^{BF})^2 + C_X (j_{Xz}^{BF})^2 \tag{4.3}$$

and

$$T_{AB} = \frac{1}{2\mu_{AB} R^2} \left[-\hbar^2 \frac{\partial}{\partial R} R^2 \frac{\partial}{\partial R} + (l^{SF})^2 \right] \tag{4.4}$$

Here A_X, B_X and C_X are the rotational constants of monomer X, the components of $\mathbf{j}_X^{BF}$ are the usual monomer angular momentum operators given by

$$\begin{cases} j_{Xx}^{BF} = -i\hbar \left(\cos \psi_X \cot \theta_X \frac{\partial}{\partial \psi_X} + \sin \psi_X \frac{\partial}{\partial \theta_X} - \frac{\cos \psi_X}{\sin \theta_X} \frac{\partial}{\partial \phi_X} \right) \\ j_{Xy}^{BF} = -i\hbar \left(-\sin \psi_X \cot \theta_X \frac{\partial}{\partial \psi_X} + \cos \psi_X \frac{\partial}{\partial \theta_X} + \frac{\sin \psi_X}{\sin \theta_X} \frac{\partial}{\partial \phi_X} \right) \\ j_{Xz}^{BF} = -i\hbar \frac{\partial}{\partial \psi_X} \end{cases} \tag{4.5}$$

The components of $\mathbf{j}_X$ are with respect to the body-fixed (BF) principal axes frame of monomer X. The dimer reduced mass is designated by $\mu_{AB} \equiv m_A m_B / (m_A + m_B)$, with m_X the total mass of X, and $\mathbf{l}^{SF} \equiv -i\hbar \mathbf{R} \times \nabla_R$ is the end-over-end angular momentum operator.

Although this choice of SF coordinates leads to the simplest kinetic energy expression, it poses the problem that the intermolecular potential is not easily expressed in these coordinates. Instead, the potential is naturally dependent on the internal angles of the complex, e.g. the Euler angles of the monomers with respect to the dimer-embedded frame. Still, it may be advantageous, when the strength of the anisotropy in the intermolecular potential is small in comparison with the end-over-end rotational constant

$\hbar^2/(2\mu_{AB}R^2)$ of the dimer, to use the space-fixed coordinates that lead to equations (4.2)–(4.4). This situation is called coupling case (*a*) in the early paper on van der Waals molecules by Bratoz and Martin (1965) and case 1 in a review by Hutson (1991). It corresponds to the nearly free rotation of the monomers in the complex. In other words, the monomer rotational quantum numbers (e.g. j_A and j_B in the case of linear molecules) and the end-over-end angular momentum l are nearly good quantum numbers of the dimer states. In practice this coupling case arises only for van der Waals complexes that contain very light and/or weakly anisotropic monomers, such as He, H_2, HD, D_2, or NH.

In most other cases it is convenient to use a dimer-embedded frame. A 'two-angle embedded' frame is characterized by the fact that the intermolecular vector **R** lies along its *z*-axis. The expression of the potential causes no problems, and the kinetic energy operator reads

$$T = T_A + T_B + \frac{1}{2\mu_{AB}R^2}\left[-\hbar^2 \frac{\partial}{\partial R} R^2 \frac{\partial}{\partial R} + (\mathbf{J}^{SF})^2 + (\mathbf{j}_A + \mathbf{j}_B)^2 - 2(\mathbf{j}_A + \mathbf{j}_B)\cdot\mathbf{J}\right] \quad (4.6)$$

Here $\mathbf{J}^{SF}$ is given by the well-known expressions

$$\begin{cases} J_x^{SF} = -i\hbar\left(-\cos\alpha \cot\beta \frac{\partial}{\partial \alpha} - \sin\alpha \frac{\partial}{\partial \beta} + \frac{\cos\alpha}{\sin\beta} \frac{\partial}{\partial \gamma}\right) \\ J_y^{SF} = -i\hbar\left(-\sin\alpha \cot\beta \frac{\partial}{\partial \alpha} + \cos\alpha \frac{\partial}{\partial \alpha} + \cos\alpha \frac{\partial}{\partial \beta} + \frac{\sin\alpha}{\sin\beta} \frac{\partial}{\partial \gamma}\right) \\ J_z^{SF} = -i\hbar \frac{\partial}{\partial \alpha} \end{cases} \quad (4.7)$$

The Euler angles α and β are the polar angles of **R** with respect to a space-fixed frame and $\gamma \equiv (\phi_A + \phi_B)/2$. The operator $\mathbf{j}_X$, $X = A, B$, has the SF form of equation (4.7), but with α, β, and γ replaced by ϕ_X, θ_X, and ψ_X, respectively. The operator **J** is a special form of the total angular momentum of the dimer,

$$\begin{cases} J_x = -i\hbar\left(-\frac{1}{\sin\beta} \frac{\partial}{\partial \alpha} + \cot\beta \frac{\partial}{\partial \gamma}\right) \\ J_y = -i\hbar \frac{\partial}{\partial \beta} \\ J_z = -i\hbar \frac{\partial}{\partial \gamma} \end{cases} \quad (4.8)$$

The monomer kinetic energies T_A and T_B in equation (4.6), finally, are the same rigid rotor energies as given in equation (4.3). Equation (4.6) was derived from equations (4.2)–(4.4) by Brocks *et al.* (1983) with the use of chain rules. An alternative derivation was given in van der Avoird *et al.* (1994a). At first sight it seems that one may simply obtain equation (4.6) by introducing the dimer frame and replacing $(J^{SF})^2$ by $(\mathbf{J} - \mathbf{j}_A - \mathbf{j}_B)\cdot(\mathbf{J} - \mathbf{j}_A - \mathbf{j}_B)$ in equation (4.4). This is assumed in many expositions of the present theory for instance, Le Roy and Carley, 1980; Pack, 1974; Cooper and Hutson, 1993; Brookes *et al.*, 1996). However, this ignores the fact that **J** and $\mathbf{j}_X$ do not commute. Moreover, this procedure does not yield the explicit expressions of equation (4.8) for the components of **J**, which are quite unusual and which do not obey the standard, or even the so-called anomalous (Biedenharn and Louck, 1981) commutation relations. It implies that one must accept without proof that **J** acts in the usual manner on rotation functions, but, as was shown by Brocks *et al.* (1983) and van der Avoird *et al.* (1994), this only holds for a specific choice of basis and is not true in general.

For dimers consisting of a rather large non-linear molecule A and an atom or linear molecule B, such as benzene-Ar (van der Avoird, 1993; Riedle and van der Avoird, 1996) or CO@C_{60} (Olthof *et al.*, 1996), it may be advantageous to use the principal axes frame of molecule A as the dimer BF frame. The corresponding van der Waals coordinates are $\zeta_A = (\phi_A, \theta_A, \psi_A)$, the Euler angles that describe the orientation of the molecule-embedded frame with respect to a space-fixed frame, the Cartesian or polar components of the vector **R** pointing from the mass center of A to that of B, and of **r**, describing the orientation and bondlength of B. The components of the latter two vectors are with respect to the molecular frame attached to A.

The kinetic energy expression

$$T = \frac{1}{2}(\mathbf{J} - \mathbf{L} - \mathbf{j})^T \mathbf{I}^{-1}(\mathbf{J} - \mathbf{L} - \mathbf{j}) - \frac{\hbar^2}{2\mu_{AB}} \nabla_R^2 - \frac{\hbar^2}{2\mu_B} \nabla_r^2 \qquad (4.9)$$

has been given in Olthof *et al.* (1996). Here **I** is the inertia tensor of the molecule A in the principal axes frame of A and hence has a diagonal form. The operator **J** is defined in equation (4.5). The operators $\mathbf{L} \equiv -i\hbar \mathbf{R} \times \nabla_R$ and $\mathbf{j} \equiv -i\hbar \mathbf{r} \times \nabla_r$ are expressed with respect to the frame on monomer A. Although, as we pointed out above, one must be careful in introducing constraints into a Hamiltonian, we can in this case specialize directly the kinetic energy of equation (4.9) to a molecule-atom system by virtue of the fact that the coordinates **R** and **r** are orthogonal. Thus, by removal of **j** and its last term from it, equation (4.9) is transformed into the kinetic energy of a molecule-atom system.

If one prefers to use still other coordinates to describe the motions in van der Waals complexes, one has to derive the metric tensor that corresponds to these coordinates and to substitute this tensor into the Podolsky formula for the kinetic energy. The same prescription can be followed if one wishes to consider van der Waals complexes consisting of more than two monomers (van der Avoird *et al.*, 1996; Xantheas and Sutcliffe, 1995).

4.4 (*AB INITIO*) INTERMOLECULAR POTENTIALS; REPRESENTATIONS

Since there are several chapters in this book which deal with the *ab initio* calculation of intermolecular potentials, we will not discuss this problem. Let us just mention that these computations can be divided into two categories: the supermolecule and symmetry-adapted perturbation theory (SAPT) approach. Both methods have shown to be able to yield pointwise accurate potentials for (small) van der Waals dimers. Later in this section we will make some comments on the various possible ways to fit these potentials—given on discrete grids—to analytic functions. The choice of fit function depends on the strategy chosen to calculate the VRT states.

Let us start by generalizing van der Avoird *et al.* (1980), the well-known Legendre expansion (Hutson, 1991) for atom-diatom systems to arbitrarily shaped molecules. In its most general form the spherical expansion is expressed in the Euler angles ζ_A^{SF} and ζ_B^{SF} of the monomers and the polar angles $\hat{\mathbf{R}} = (\beta, \alpha)$ of **R** with respect to a space-fixed frame

$$V(R, \zeta_A, \zeta_B, \hat{\mathbf{R}}) = \sum_{\{\Lambda\}} v_{\{\Lambda\}}(R) A_{\{\Lambda\}}(\zeta_A, \zeta_B, \hat{\mathbf{R}}) \qquad (4.10)$$

The orthogonal set of angular functions, labeled by $\{\Lambda\} \equiv \{L_A, K_A, L_B, K_B, L\}$, is given by

$$A_{\{\Lambda\}}(\zeta_A^{SF}, \zeta_B^{SF}, \hat{\mathbf{R}}) = (-1)^{L_A+L_B+L} \sum_{M_A M_B M} \begin{pmatrix} L_A & L_B & L \\ M_A & M_B & M \end{pmatrix} D_{M_A K_A}^{(L_A)}(\zeta_A^{SF})^* D_{M_B K_B}^{(L_B)}(\zeta_B^{SF})^* C_M^L(\hat{\mathbf{R}}) \qquad (4.11)$$

where the functions $D_{mk}^{(l)}$ are Wigner rotation functions, C_m^l are spherical harmonics in the Racah normalization, and the expression in large brackets is a 3-j symbol (Brink and Satchler, 1975). Since the functions $A_{\{\Lambda\}}$ form a complete set, the expansion equation (4.10) is exact. In practice, one may truncate the expansion when the coefficients $v_{\{\Lambda\}}(R)$ have become sufficiently small. These coefficients depend only on the distance R; if we include the dependence of the intermolecular potential on the molecular geometries they depend on the intramolecular (normal) coordinates too. One advantage of the spherical expansion is that it shows explicitly the anisotropy of the potential; the term with $\{L_A, K_A, L_B, K_B, L\} = \{0, 0, 0, 0, 0\}$ is the isotropic potential. Another advantage is that it can immediately be written in terms of coordinates expressed with respect to a dimer frame. One just has to realize that, by construction, the angular functions $A_{\{\Lambda\}}$ are invariant with respect to any frame rotation and to use the property that the polar angles (β, α) are $(0, 0)$ with respect to the dimer-embedded frame. Substitution of $C_M^L(0, 0) = \delta_{M0}$ yields, then,

$$A_{\{\Lambda\}}(\zeta_A^{BF}, \zeta_B^{BF}) = (-1)^{L_A+L_B+L} \sum_{\substack{M_A \\ (M_B=-M_A)}} \begin{pmatrix} L_A & L_B & L \\ M_A & M_B & 0 \end{pmatrix} D_{M_A K_A}^{(L_A)}(\zeta_A^{BF})^* D_{M_B K_B}^{(L_B)}(\zeta_B^{BF})^* \qquad (4.12)$$

For atom–molecule dimers $L_A = K_A = M_A = 0$. Using the properties of Wigner D-functions and spherical harmonics given in, for example, van der Avoird et al. (1994a), we find that the angular expansion functions become

$$A_{L_B K_B}(\zeta_B^{BF}) = (-1)^{L_B-K_B}(2L_B + 1)^{-1/2} C_{K_B}^{L_B}(\theta_B^{BF}, \psi_B^{BF}). \qquad (4.13)$$

The well-known Legendre expansion for atom–diatom systems, where $K_B = 0$, is obtained by the simple substitution of $C_0^L(\theta, \psi) = P_L(\cos\theta)$.

In *ab initio* calculations of the potential one always chooses a BF frame. The expansion coefficients can be written as

$$v_{\{\Lambda\}}(R) = \frac{(2L_A + 1)(2L_B + 1)(2L + 1)}{64\pi^4} \int d\zeta_A^{BF} \int d\zeta_B^{BF} A_{\{\Lambda\}}(\zeta_A^{BF}, \zeta_B^{BF})^* V(R, \zeta_A^{BF}, \zeta_B^{BF}) \qquad (4.14)$$

with $d\zeta_X = \sin\theta_X d\phi_X d\theta_X d\psi_X$. After calculation of the potential V on a grid of angles ζ_A^{BF} and ζ_B^{BF}, the integration in equation (4.14) can be performed by numerical quadrature (Abramowitz and Stegun, 1964), for each distance R. Actually, one may choose the BF frame such that one of the Euler angles, either ϕ_A^{BF} or ϕ_B^{BF}, is equal to zero and can be omitted from the integration. If we deal with simpler dimers, e.g. if A or B is an atom or a linear molecule, this procedure can be further simplified. The expansion coefficients $v_{\{\Lambda\}}(R)$ which are thus obtained define the potential, both with respect to the BF frame, via equations (4.10) and (4.12), and with respect to the SF frame, via equations (4.10) and (4.11).

In practical calculations of the VRT states of a van der Waals dimer the spherical expansion of the potential is very convenient. If the angular basis in such calculations is chosen as Clebsch–Gordan coupled products of monomer and overall rotor functions, all the angular integrals in the matrix elements of the potential are just 6- and 9j-symbols. For the same reason the spherical expansion is used in most scattering calculations. Only when the potential is too strongly anisotropic does this procedure become inefficient, since one needs too many terms in the spherical expansion and too large a basis for the diagonalization of the Hamiltonian.

We can explain now why in most cases the use of a dimer frame is the most convenient. As follows from the relation $M_B = -M_A$ in equation (4.12) and from the definition of the Wigner D-functions (Brink and Satchler, 1975), the intermolecular potential depends only on the difference angle $\phi_B - \phi_A$, not on ϕ_A or ϕ_B themselves. Hence, in a BF angular basis used to describe the bound states of the dimer, functions with different K are not mixed by the potential. Off-diagonal matrix elements between such functions are given only by the Coriolis terms $(\mathbf{j}_A + \mathbf{j}_B)\cdot \mathbf{J}/(\mu_{AB}R^2)$ in the BF kinetic energy operator (equation (4.6)). In almost all cases (except for very light dimers or very high values of J), these terms are much smaller than the anisotropy of the potential. This anisotropy is dominated by the leading terms $v_{\{\Lambda\}}(R)$ with $\{\Lambda\} \neq \{0,0,0,0,0\}$. In all these cases K, which is the eigenvalue of both J_z and $j_{Az} + j_{Bz}$, is a nearly good quantum number. These cases are treated as coupling case (b) by Bratoz and Martin (1965) and as cases 2 and 3 by Hutson (1991). Even when the complex becomes nearly rigid, K is still a good quantum number in many van der Waals dimers, because such dimers are often prolate near-symmetric tops due to the relatively large van der Waals bonding distance R.

We end this section by making some remarks regarding other, analytic or discrete, representations of the potential. These are closely connected to the method chosen to calculate the VRT states, so it is not so easy to keep the discussion general. If one chooses a discrete variable representation (DVR) (Light et al., 1985) of the VRT states, for example, then the potential must be known only on a grid of quadrature points. But even then, the DVR method may require too dense a grid to evaluate the potential in all points directly by *ab initio* methods. Analytic fitting (globally or by splines) or some other interpolation technique may reduce the number of required *ab initio* calculations.

The atom–atom model (Kitaigorodski, 1973), $V_{AB} = \Sigma_{i\in A}\Sigma_{j\in B} v_{ij}(r_{ij})$, with a Lennard–Jones $v_{ij}(r_{ij}) = A_{ij}r_{ij}^{-12} - B_{ij}r_{ij}^{-6}$ or exp-6 potential $v_{ij}(r_{ij}) = A_{ij}\exp(-B_{ij}r_{ij}) - C_{ij}r_{ij}^{-6}$, is a widely used global fitting model. An advantage of atom–atom potentials is that they simultaneously model the dependence of the potential on the intra- and intermolecular degrees of freedom. Usually it is assumed that the atom–atom potentials are isotropic, which seriously limits their accuracy. In a few cases, however, anisotropic atom–atom potentials were introduced (Stone and Price, 1988).

Another way of representing the anisotropy of the intermolecular potential is by the use of a parameterized R-dependent form with parameters that depend on the orientations of the molecules. An example is the Lennard–Jones potential

$$V(R,\zeta_A,\zeta_B) = \epsilon(\zeta_A,\zeta_B)\left[\left(\frac{R_m(\zeta_A,\zeta_B)}{R}\right)^{12} - 2\left(\frac{R_m(\zeta_A,\zeta_B)}{R}\right)^6\right] \quad (4.15)$$

with the parameters ϵ and R_m depending on the Euler angles ζ_A and ζ_B (Pack, 1978). In potentials that are used especially to fit the spectra of van der Waals molecules (Hutson, 1992; Cohen and Saykally, 1990, 1993), the short-range repulsion is modeled by

$$A(\zeta_A, \zeta_B) \exp\left[-\beta(\zeta_A, \zeta_B)R\right] \qquad (4.16)$$

and the long-range electrostatic, induction and dispersion terms by

$$-\sum_n C_n(\zeta_A, \zeta_B) D_n(R, \zeta_A, \zeta_B) R^{-n} \qquad (4.17)$$

The damping functions $D_n(R, \zeta_A, \zeta_B)$ correct the long-range contributions for overlap effects (Ahlrichs *et al.*, 1977; Tang and Toennies, 1984). Usually the parameters in equation (4.16) and (4.17) are not directly optimized. Instead, one adopts some reasonable (*ab initio*) values for all but the highest long-range coefficients C_n and then writes the highest $C_n(\zeta_A, \zeta_B)$ and the short-range coefficient $A(\zeta_A, \zeta_B)$ as functions of the (angular dependent) well depth $\epsilon(\zeta_A, \zeta_B)$ and position of the minimum $R_m(\zeta_A, \zeta_B)$ in the potential. The latter quantities and the exponent $\beta(\zeta_A, \zeta_B)$ are written as truncated expansions in the angular functions $A_{\{\Lambda\}}(\zeta_A, \zeta_B)$ of equation (4.12). The actual fitting parameters are the coefficients in these expansions. Therefore these occur in the potential in a highly non-linear way. This procedure is chosen to reduce the number of fitting parameters and to avoid a high correlation between them.

4.5 METHODS FOR THE CALCULATION OF VRT STATES

Methods for the computation of VRT states in van der Waals molecules can be classified as variational and non-variational. In the linear variational methods one chooses a basis, the functional dependence of which depends on the choice of the (intermolecular) coordinates (see Section 4.2). Using space-fixed coordinates, for instance, we may choose the following basis for a dimer consisting of two arbitrary non-linear molecules:

$$|n, j_A, k_A, j_B, k_B, j_{AB}, l; J, M\rangle = \Phi_n(R) \left[\frac{(2j_A + 1)(2j_B + 1)(2l + 1)}{256\pi^5}\right]^{1/2}$$

$$\times \sum_{m_A m_B} \sum_{K_m} D_{m_A k_A}^{(j_A)}(\zeta_A^{SF})^* D_{m_B k_B}^{(j_B)}(\zeta_B^{SF})^* \langle j_A m_A; j_B m_B | j_{AB} K\rangle C_m^l(\hat{\mathbf{R}}) \langle j_{AB} K; lm | JM\rangle \qquad (4.18)$$

The angular momentum coupling in this basis, by means of the Clebsch–Gordan coefficients $\langle j_1 m_1; j_2 m_2 | jm\rangle$, takes into account that the total angular momentum J and its z-component M are exact quantum numbers, provided the complex moves in an isotropic space. One may use analytic functions for the radial basis $\Phi_n(R)$, such as the associated Laguerre functions (Tennyson and Sutcliffe, 1982, 1983), which resemble the eigenfunctions of a Morse oscillator, or distributed Gaussians (Bačić and Light, 1986; Clary and Nesbitt, 1989), or numerical functions defined on a grid of R points. If the intermolecular potential is just weakly anisotropic, a convenient numerical basis may be obtained by solving the one-dimensional Schrödinger equation with the isotropic potential and the radial terms in the kinetic energy. If, on the other hand, the dimer potential has a deep well at a certain orientation of the monomers, one may solve the one-dimensional equation with the R-dependent potential at fixed angles corresponding to this well. The eigenvectors from the secular equation for the one-dimensional problem can be used as contraction coefficients for the radial basis functions in the full problem.

In the case of a somewhat stronger anisotropy it is more natural (because of the nearly conserved quantum number K, see Section 4.4), and also more convenient, to use the BF basis

$$|n, j_A, k_A, j_B, k_B, j_{AB}, K; J, M\rangle = \Phi_n(R) \left[\frac{(2j_A+1)(2j_B+1)(2J+1)}{256\pi^5}\right]^{1/2}$$

$$\times \sum_{m_A m_B} D^{(j_A)}_{m_A k_A}(\zeta^{BF}_A)^* D^{(j_B)}_{m_B k_B}(\zeta^{BF}_B)^* \langle j_A m_A; j_B m_B | j_{AB} K\rangle D^{(J)}_{MK}(\alpha, \beta, 0)^* \quad (4.19)$$

For fixed $j_A, k_A, j_B, k_B, j_{AB}, J, M$ and K running from $-\min(J, j_{AB})$ to $+\min(J, j_{AB})$ the BF basis in equation (4.19) spans the same space as the SF basis in equation (4.18) with l running from $|J - j_{AB}|$ to $J + j_{AB}$. So the final VRT states will be the same in both bases. In the SF coordinate system one has to use the kinetic energy operator of equations (4.2)–(4.4). The monomer terms, T_A and T_B in equation (4.3), act on the functions $D^{(j_A)}_{m_A k_A}(\zeta^{SF}_A)^*$ and $D^{(j_B)}_{m_B k_B}(\zeta^{SF}_B)^*$ and they yield the standard rigid rotor expressions (Papoušek and Aliev, 1982; Biedenharn and Louck, 1981). For example, for symmetric tops with $A_X = B_X$ the operator T_X is diagonal, with eigenvalues $A_X j_X (j_X + 1) + (C_X - A_X) k_X^2$. Since the basis functions in equation (4.18) are eigenfunctions of $(l^{SF})^2$, with eigenvalue $l(l+1)$, the dimer term T_{AB} in equation (4.4) is diagonal in the angular basis. If the potential is expanded as in equations (4.10) and (4.11), its matrix elements are

$$\langle n', j'_A, k'_A, j'_B, k'_B, j'_{AB}, l'; JM | V | n, j_A, k_A, j_B, k_B, j_{AB}, l; J, M\rangle =$$

$$\sum_{\{\Lambda\}} \langle \Phi_{n'}(R) | v_{\{\Lambda\}}(R) | \Phi_n(R)\rangle (-1)^{j_A + j_B + j'_{AB} + l + J - k'_A - k'_B} \quad (4.20)$$

$$\times [(2j'_A + 1)(2j'_B + 1)(2j_A + 1)(2j_B + 1)(2j'_{AB} + 1)(2j_{AB} + 1)(2l' + 1)(2l + 1)]^{1/2}$$

$$\times \begin{pmatrix} j'_A & L_A & j_A \\ -k'_A & K_A & k_A \end{pmatrix} \begin{pmatrix} j'_B & L_B & j_B \\ -k'_B & K_B & k_B \end{pmatrix} \begin{pmatrix} l' & L & l \\ 0 & 0 & 0 \end{pmatrix} \begin{Bmatrix} l & l' & L \\ j'_{AB} & j_{AB} & J \end{Bmatrix} \begin{Bmatrix} j'_A & L_A & j_A \\ j'_B & L_B & j_B \\ j'_{AB} & L & j_{AB} \end{Bmatrix}$$

where the expressions in large braces are 6-j and 9-j symbols (Brink and Satchler, 1975), respectively.

In the 'two-angle embedded' BF coordinate system one must use the kinetic energy operator of equation (4.6). The monomer terms T_A and T_B yield the same standard rigid rotor expressions as in the SF case. The dimer term T_{AB} is different, however. In Appendix B of van der Avoird et al. (1994a) it is explicitly shown how it acts on the angular BF basis of equation (4.19). Most of its terms are diagonal in this basis, but the terms $[(j_{Ax} + j_{Bx})J_x + (j_{Ay} + j_{By})J_y]/(\mu_{AB}R^2)$ connect the basis functions K with functions $K' = K \pm 1$. After expanding the potential as in equations (4.10) and (4.12), the potential matrix elements over the BF basis in equation (4.19) are

$$\langle n', j'_A, k'_A, j'_B, k'_B, j'_{AB}, K'; JM | V | n, j_A, k_A, j_B, k_B, j_{AB}, K; J, M\rangle =$$

$$\delta_{K'K} \sum_{\{\Lambda\}} \langle \Phi_{n'}(R) | v_{\{\Lambda\}}(R) | \Phi_n(R)\rangle (-1)^{j_A + j_B + j_{AB} + L - k'_A - k'_B - K}$$

$$\times \left[(2j'_A + 1)(2j'_B + 1)(2j_A + 1)(2j_B + 1)(2j'_{AB} + 1)(2j_{AB} + 1)\right]^{1/2}$$

$$\times \begin{pmatrix} j'_A & L_A & j_A \\ -k'_A & K_A & k_A \end{pmatrix} \begin{pmatrix} j'_B & L_B & j_B \\ -k'_B & K_B & k_B \end{pmatrix} \begin{pmatrix} j'_{AB} & L & j_{AB} \\ -K & 0 & K \end{pmatrix} \begin{Bmatrix} j'_A & L_A & j_A \\ j'_B & L_B & j_B \\ j'_{AB} & L & j_{AB} \end{Bmatrix} \quad (4.21)$$

Additional advantages of the BF basis now become directly apparent. The potential matrix elements are simpler than in the SF basis. This is especially advantageous for atom–molecule systems, where $j'_A = j_A = L_A = 0$, and the 9-j symbol in equations (4.20) and (4.21) becomes simply $\delta_{j'_B j'_{AB}} \delta_{j_B j_{AB}} \delta_{L_B L} [(2j'_B + 1)(2j_B + 1)(2L_B + 1)]^{-1/2}$. The remaining angular factors in equation (4.20) contain the 6-j symbol—in contrast to the corresponding expression in the BF coordinates, (equation (4.21)). Moreover, it is obvious from equation (4.21) that the potential does not couple BF basis functions with different K. Although, as we discussed above, such functions are coupled by a Coriolis term in the kinetic energy, the corresponding off-diagonal matrix elements are small and they occur only for $K' = K \pm 1$. As a first approximation one may neglect these couplings, reducing the size of the Hamiltonian matrix by a factor $(2J + 1)$. If one wishes to go beyond this 'helicity decoupling' approximation, one may solve in a second step a secular problem in a truncated basis of eigenstates of the simpler Hamiltonian while reintroducing the off-diagonal Coriolis terms, or one may take the latter into account by perturbation theory.

The use of free rotor functions in the basis has the additional advantage that it does not introduce any bias for specific orientations of the monomers in the complex. These are free to find their most favorable orientational wavefunctions, depending on the barriers in the potential surface. Note that the average orientation of the monomers often depends strongly on the VRT state of the dimer.

When the monomers in a van der Waals complex are strongly aspherical, they may be larger in some directions than the van der Waals bonding distance R. Then the potential becomes too strongly anisotropic and the use of the free rotor basis is no longer appropriate. A border case is benzene–Ar (Brocks and Huygen, 1986), where the spherical expansion of the potential needs terms up to $L_B = 36$ and the convergence of the VRT states requires angular basis functions as high as $j_B = 27$. Other types of basis functions have to be applied in such cases, and it may be better to use other coordinates too.

Sometimes, for instance in the case of benzene–Ar (van der Avoird, 1993), a product basis of harmonic oscillator functions $H_k(x)H_l(y)H_m(z)$, centered at the equilibrium position $\mathbf{R}_e = (x_e, y_e, z_e)$ can be applied usefully. Here $\mathbf{R} = (x, y, x)$ is the position vector of the atom expressed with respect to the frame of the molecule. If the kinetic energy of equation (4.9), restricted to the molecule–atom case, and the coordinate and momentum operators are expressed in the ladder operators of the harmonic oscillator (Messiah, 1969; Huber, 1985), it is easy to evaluate all the kinetic energy matrix elements analytically. The matrix elements of the potential $V(x, y, z)$ over the harmonic oscillator basis can be evaluated by Gauss–Hermite quadrature (Abramowitz and Stegun, 1964) with the same center and scaling as the basis functions $H_k(x)$, $H_l(y)$ and $H_m(z)$. This procedure works well, even if the molecule becomes as large as fluorene (Brocks and van Koeven, 1988).

Let us now discuss some non-variational methods. The traditional non-variational method to obtain the bound states of van der Waals dimers is the close-coupling method, as implemented for scattering calculations (Arthurs and Dalgarno, 1960; Child, 1974). The angular basis functions used in such calculations are the same as in equation (4.18), for SF coordinates, and equation (4.19), for BF coordinates. The angular matrix elements are the same as in equation (4.20) and (4.21), respectively. The radial functions are not expanded in a basis, however, but they are written as the R-dependent 'coefficients' in the expansion of the exact wavefunction in the complete set of angular (channel) functions. When this expansion is substituted into the Schrödinger equation one obtains a set of coupled differential equations for the radial functions of the different channels (Hutson, 1991). In practice, this set is truncated, of course. The coupled differential equations are solved by the numerical propagator methods (Johnson, 1973; Manolopoulos, 1986) developed for scattering calculations. For bound states, it is not possible to choose the energy, however. One has to find, by iteration, those energies that produce the radial wavefunctions which vanish at $R \to \infty$ and remain finite at $R = 0$ (Dunker and Gordon, 1976; Johnson, 1978; Shapiro and Balint-Kurti, 1975; Danby, 1983). Since this may be a rather time-consuming process, special methods for bound state calculations have been devised. In the SEPT (secular equation perturbation theory) method (Hutson and Le Roy, 1985) one first calculates a (small) set of uncoupled channel functions, then solves a secular problem with these functions as a basis and, next, includes more channels by perturbation theory. The (first-order) perturbation equations are again a set of coupled differential equations in the radial coordinate, but these do not contain the unknown energy. Recent improvements of this method, such as the ISE (iterative secular equation) method (Slee and Le Roy, 1993) include the perturbed wavefunctions as additional basis functions in a (larger) secular problem. An advantage of these methods is that they are directly applicable to the resonances, vibrational and rotational predissociation states, which are often found in van der Waals complexes.

Non-variational approaches which are based on discrete representatives of the wavefunction are the discrete variable representation (DVR) (Light et al., 1985; Choi and Light, 1990; Mandziuk and Bačić, 1991, 1993) and the collocation method (Cohen and Saykally, 1990, Peet and Yang, 1989a,b; Yang et al., 1989; Block et al., 1993). Let us briefly describe the DVR method, starting in one dimension. We consider functions on the interval $[a, b]$, $a < b$, of the real axis and a positive definite weight function (volume element) $w(x)$. A discrete variable representation consists of a discrete grid $a < x_1 < x_2 < \cdots < x_N < b$, together with a set of orthonormal functions $\phi_k(x)$, $k = 1, \ldots, N$, satisfying (in Dirac notation)

$$\langle x_i | k \rangle \equiv \phi_k(x_i) = \delta_{ki} w_k^{-1/2} \qquad (4.22)$$

where the $\{w_k\}$ are positive definite weights. This set of functions defines an approximate resolution of the identity in the x-representation

$$\hat{1} = \int_a^b |x\rangle w(x) \langle x| \, dx \approx \sum_{i=1}^N |\phi_i\rangle\langle\phi_i| = \sum_{i=1}^N |x_i\rangle w_i \langle x_i| \qquad (4.23)$$

with $|x_i\rangle \equiv w_i^{-1/2}|\phi_i\rangle$.

A local (multiplicative) operator $\hat{V}$ has a very simple diagonal matrix in a DVR basis. Recalling that $\langle x|\hat{V}|x'\rangle = \langle x|x'\rangle V(x) = \delta(x - x') w(x)^{-1} V(x)$, we find

$$\langle k|\hat{V}|l\rangle = \int_a^b \langle k|x\rangle V(x)\langle x|l\rangle w(x)\,dx \approx \sum_{i=1}^N \langle k|x_i\rangle V(x_i)\langle x_i|l\rangle w_i$$

$$= \sum_{i=1}^N \phi_k(x_i)^* V(x_i)\phi_l(x_i) w_i = \delta_{kl} V(x_k) \tag{4.24}$$

where we used equation (4.22) and its complex conjugate. Of course, this equation is in fact a numerical quadrature of the matrix element.

A first example of a DVR basis is the 'classical' DVR (Light et al., 1985) based on the theory of orthogonal polynomials $p_k(x)$, $k = 0, \ldots, N-1$. The overlap between these polynomials can be computed exactly by Gaussian quadrature (Stoer and Bulirsch, 1980)

$$\int_a^b p_k(x) p_l(x) w(x)\,dx = \sum_{i=1}^N w_i p_k(x_i) p_l(x_i) \tag{4.25}$$

On the other hand, we have

$$\int_a^b p_k(x) p_l(x) w(x)\,dx = \delta_{kl}\langle p_k|p_k\rangle \equiv \delta_{kl}\lambda_k \quad \text{hence } \lambda_k > 0 \tag{4.26}$$

Introducing the matrix $T_{ki} \equiv \lambda_k^{-1/2} p_k(x_i) w_i^{1/2}$, we find from equations (4.25) and (4.26) that

$$\mathbf{T}\mathbf{T}^T = \mathbf{1} \text{ and, since } T \text{ is non-singular, also that } \mathbf{T}^T\mathbf{T} = \mathbf{1} \tag{4.27}$$

The functions $\lambda_k^{-1/2} p_k(x)$ form an orthonormal set for $k = 0, \ldots, N-1$, and hence the functions obtained from them by the orthogonal transformation

$$\psi_k(x) \equiv \sum_{l=0}^{N-1} \lambda_l^{-1/2} p_l(x) T_{lk} \tag{4.28}$$

are also orthonormal. These functions satisfy the DVR condition (equation (4.22)), which follows by the definition of **T**, that is,

$$\psi_k(x_i) = \sum_{l=0}^{N-1} \lambda_l^{-1/2} p_l(x_i) T_{lk} = \sum_{l=0}^{N-1} w_i^{-1/2} T_{li} T_{lk} = w_i^{-1/2} \delta_{ki} \tag{4.29}$$

where we used equation (4.27). Thus, the functions (4.28), together with a Gaussian quadrature grid determined by $[a, b]$ and $w(x)$, allow a DVR.

Another example used in practice is the periodic Fourier DVR (Muckerman, 1990; Sabo et al., 1995). Define $\omega \equiv \exp(2\pi i/N)$. With the use of the summation formula for a geometric series, it is easily shown that

$$\sum_{j=0}^{N-1} (\omega^{jm})^* \omega^{jn} = \delta_{mn} N \tag{4.30}$$

Note that this is the orthogonality relation for characters of the cyclic group of order N. The functions $\exp(li\alpha)/\sqrt{2\pi}$, $l = 0, \ldots, N-1$, are orthonormal on $[0, 2\pi]$ with weight $w(\alpha) = 1$, and so are the following functions:

$$\psi_k(\alpha) \equiv \sqrt{\frac{1}{2\pi N}} \sum_{l=0}^{N-1} \exp(il\alpha)(\omega^{kl})^* \qquad (4.31)$$

Substitution of $\alpha = 2\pi j/N$ together with the orthogonality relations (4.30) shows that

$$\psi_k(2\pi j/N) = \sqrt{\frac{N}{2\pi}} \delta_{kj} \qquad (4.32)$$

Hence the functions (4.31) together with the grid $\alpha_j = 2\pi j/N$, $j = 1, \ldots, N$, enable a DVR on the interval $[0, 2\pi]$ with weight $w_j = 2\pi/N$.

A final example is formed by the sinc functions (Colbert and Miller, 1992; Groenenboom and Colbert, 1993), which are defined on the whole real axis $(-\infty, \infty)$ with weight $w(x) = 1$,

$$\psi_k(x) = \Delta^{-1/2} \operatorname{sinc}\left[\pi\left(\frac{x}{\Delta} - k\right)\right] \equiv \Delta^{-1/2} \frac{\sin \pi\left(\frac{x}{\Delta} - k\right)}{\pi\left(\frac{x}{\Delta} - k\right)} \qquad (4.33)$$

By substitution of the Fourier transform

$$\Delta^{-1/2} \operatorname{sinc}\left[\pi\left(\frac{x}{\Delta} - k\right)\right] = \frac{1}{\sqrt{2\pi}} \int_{-\pi/\Delta}^{\pi/\Delta} \exp(-ipx)\tilde{\psi}_k(p)\,\mathrm{d}p$$

$$\text{with } \tilde{\psi}_k(p) = \sqrt{\frac{\Delta}{2\pi}} \exp(ikp\Delta) \qquad (4.34)$$

one proves easily

$$\int_{-\infty}^{\infty} \psi_k(x)\psi_l(x)\,\mathrm{d}x = \frac{\sin \pi(k-l)}{\pi(k-l)} = \delta_{kl} \qquad (4.35)$$

and hence the set $\psi_k(x)$ is orthonormal. Further it is clear that the DVR condition (equation 4.22) is satisfied, i.e. $\psi_k(j\Delta) = \Delta^{-1/2}\delta_{kj}$.

The kinetic energy matrix may be calculated in a finite basis of polynomials or Fourier functions and then transformed to the DVR basis $\psi_k(x)$ by equations (4.28) and (4.31), respectively. For the sinc function DVR it is most convenient to calculate the kinetic energy integrals in the p representation and then to transform them to the sinc basis.

In applying DVR to multi-dimensional systems, it is most common to use a direct product basis, and a direct product of quadrature grids in the individual coordinates. Improvements of this scheme have been proposed recently (Friesner et al., 1993). Alternatively, one may use discrete representations in some coordinates and analytic bases in others. For instance, in atom–diatom systems the DVR in the angular coordinate—with Legendre functions $P_l(\cos\theta)$ as orthogonal polynomials and Gauss–Legendre quadrature for $\cos\theta$—has been combined (Mladenović and Bačić, 1991; Henderson and Tennyson, 1993) with a basis of distributed Gaussians for the radial coordinate R. Although DVR has in common with the strictly variational methods that one has to diagonalize a symmetric Hamiltonian matrix, the lowest DVR eigenvalue is not necessarily an upper bound to the exact ground state energy, because of the approximation of the potential matrix elements by the quadrature formula.

A closely related non-variational method is the collocation method (Peet and Yang, 1989a,b; Block et al., 1993). In this method the exact wavefunction is expanded in a finite (analytic) N-dimensional basis: $\Psi(x) = \Sigma_k \phi_k(x) c_k$. The Schrödinger equation is required to be satisfied exactly, for N points x_i in the coordinate space. The resulting equation, in Dirac notation,

$$\sum_{k=1}^{N} [\langle x_i | H | \phi_k \rangle - E \langle x_i | \phi_k \rangle] c_k = 0 \qquad \text{for } i = 1, \ldots, N \qquad (4.36)$$

is an $N \times N$ eigenvalue equation for a non-symmetric matrix $H_{ik} = \langle x_i | H | \phi_k \rangle$ and 'overlap' matrix $S_{ik} = \langle x_i | \phi_k \rangle = \phi_k(x_i)$. The advantage of this method is that it is easy to program; one has just to compute at the grid points x_i the values of the potential, of the basis functions ϕ_k and of their second derivatives (which occur in $H\phi_k$). If the basis ϕ_k consists of orthogonal polynomials, it is advised to choose the associated quadrature points x_i. Formally, this method can be justified in the limit of a complete basis ϕ_k. Alternatively, it can be derived variationally, by searching for a stationary point of the asymmetric functional $\langle \Psi' | H | \Psi \rangle$, while expanding Ψ in the analytic basis ϕ_k and Ψ' in a basis localized at the grid points x_i (Yang et al., 1989). This stationary point is not required to be a minimum, however, and the collocation method is not variational in the sense that it gives an upper bound to the exact ground state energy. The eigenvalues of the non-symmetric matrix H_{ik} may even become complex. In practice, the collocation method seems to work well (Cohen and Saykally, 1990; Peet and Yang, 1989a,b; Yang et al., 1989; Block et al., 1993). However, in spite of its simplicity, it may not be the most efficient method from a computational point of view. If the diagonalization of the H-matrix is the most time-consuming step (as it is in most practical calculations), the time gained by the easier construction of this matrix is more than lost by the slower diagonalization of a non-sparse, non-symmetric matrix.

A common property of all basis set and discrete representation methods is that in the end one has to solve the (symmetric or non-symmetric) matrix eigenvalue problem. Standard library routines are available for this purpose. If the basis becomes too large to store the Hamiltonian matrix in computer memory, one may also use a different type of iterative procedure, such as the Lanczos (Parlett, 1980; Cullum and Willoughby, 1985) or Davidson (1975) algorithm. If the system has many degrees of freedom, or if the construction and diagonalization of the H-matrix has to be repeated many times in the process of improving the potential by fitting the experimental spectrum, it is desirable to reduce the size of the basis. Early work in this direction (Holmgren et al., 1977) used BOARS: the Born–Oppenheimer (or adiabatic) separation of the angular and radial motion. More recently, it has become common practice to use (sequential) adiabatic reduction methods (Choi and Light, 1990; Mandziuk and Bačić, 1993): one or more coordinates are clamped and the eigenvalue problem is solved for the remaining degrees of freedom. The eigenvalues, for different values of the clamped coordinates, form the effective potentials for the second step in the calculation. Adiabatic (or quasi-adiabatic) reduction implies that in this second step, which yields the final wavefunction, one uses a truncated set of eigenfunctions from the first step. In multi-dimensional systems this procedure may be followed sequentially, in the different coordinates. It is easily implemented in DVR methods, which already use a finite grid representation for some of the coordinates. But, as we have seen in the treatment of the off-diagonal Coriolis coupling (equation (4.21)) similar simplifications can be achieved in other methods.

We end this discussion of methods for the calculation of the VRT states of van der Waals molecules by briefly mentioning the quantum Monte Carlo method. The variational Monte Carlo procedure (Rick et al., 1991; Bačić et al., 1992) is, in essence, a method for the numerical computation of the multidimensional integrals of the Hamiltonian over a trial wavefunction. So the accuracy of this method is limited by the trial function chosen. The Green's function or diffusional Monte Carlo method (Bačić et al., 1992; Quack and Suhm, 1991) is very powerful, however, since it converges to the exact ground state of the system. It has been applied to several van der Waals complexes including even multiple monomers (Bačić et al., 1992; Quack et al., 1993; Gregory and Clary, 1995; Lewerenz, 1996). For a more detailed description of these Monte Carlo methods we refer to the papers mentioned in this paragraph.

4.6 PROPERTIES, TRANSITIONS, INTENSITIES

Most of the methods used for the calculation of the VRT states yield explicitly the wavefunctions of these states. It is then relatively easy to compute the different measurable properties and to evaluate the intensities of the transitions observed in spectra. The (infrared) absorption coefficient for the transition between two thermally populated VRT levels (i, J) and (i', J') is given by (McQuarrie, 1976)

$$\frac{\pi N_A g_i}{3\hbar^2 \varepsilon_0 c Z}(E_{i',J'} - E_{i,J})[\exp(-E_{i,J}/kT) - \exp(-E_{i',J'}/kT)]S(i, J \to i', J') \quad (4.37)$$

where $E_{i,J}$ is the energy of the VRT state (i, J) and Z is the partition function:

$$Z = \sum_{i,J} g_i(2J+1)\exp(-E_{i,J}/kT) \quad (4.38)$$

It is assumed here that the distribution over the VRT levels is a Boltzmann distribution with temperature T; g_i is the nuclear spin statistical weight of the level i, N_A is Avogadro's number and the other constants are well-known fundamental constants. The (calculated) wavefunctions of the VRT states are the kets $|iJM\rangle$; in the absence of external fields these are degenerate for $M = -J, -J+1, \ldots, J$. The line strengths in equation (4.37) are defined as

$$S(i, J \to i', J') = \sum_{MM'm} |\langle i'J'M'|\mu_m^{SF}|iJM\rangle|^2 \quad (4.39)$$

If the wavefunctions $|iJM\rangle$ of the VRT states have been calculated in terms of the SF basis in equation (4.18), it is convenient to express the space-fixed spherical components μ_m^{SF} of the dipole moment operator in the same basis:

$$\mu_m^{SF}(R, \zeta_A^{SF}, \zeta_B^{SF}, \hat{\mathbf{R}}) = \sum_{\{\Lambda\}\lambda} d_{\{\Lambda\}\lambda}^{SF}(R) B_{\{\Lambda\}\lambda m}^{SF}(\zeta_A^{SF}, \zeta_B^{SF}, \hat{\mathbf{R}}) \quad (4.40)$$

The angular functions $B_{\{\Lambda\}\lambda m}$ must transform as a vector quantity. In terms of the space-fixed orientation angles ζ_A^{SF}, ζ_B^{SF} and end-over-end angles $\hat{\mathbf{R}}$ they read

$$B^{\text{SF}}_{\{\Lambda\}\lambda m}(\zeta^{\text{SF}}_A, \zeta^{\text{SF}}_B, \hat{\mathbf{R}}) = \sum_{M_A M_B} \sum_{M_{AB} M} D^{(L_A)}_{M_A K_A}(\zeta^{\text{SF}}_A)^* D^{(L_B)}_{M_B K_B}(\zeta^{\text{SF}}_B)^* C^{\lambda}_M(\hat{\mathbf{R}})$$

$$\times \langle L_A M_A; L_B M_B | L M_{AB} \rangle \langle L M_{AB}; \lambda M | 1 m \rangle \quad (4.41)$$

with the composite index $\{\Lambda\} \equiv \{L_A, K_A, L_B, K_B, L\}$. This might be compared with the functions in equation (4.11) for the angular expansion of the potential.

If the dipole moment given by equation (4.40) and the wavefunctions $|iJM\rangle$ in terms of the basis in equation (4.18) are substituted into equation (4.39) for the line strength, all the occurring angular matrix elements are just $3n$-j symbols. The result is similar to but slightly more complicated than the corresponding result for the potential, (equation (4.20)).

If the wavefunctions $|iJM\rangle$ have been calculated in terms of the BF basis (equation (4.19)) it is preferable to express the dipole moment operator on that basis too. The dipole components relative to the BF frame are given by

$$\mu^{\text{BF}}_k(R, \zeta^{\text{BF}}_A, \zeta^{\text{BF}}_B) = \sum_{\{\Lambda\}} d^{\text{BF}}_{\{\Lambda\}k}(R) B^{\text{BF}}_{\{\Lambda\}k}(\zeta^{\text{BF}}_A, \zeta^{\text{BF}}_B) \quad (4.42)$$

and the angular functions are

$$B^{\text{BF}}_{\{\Lambda\}k}(\zeta^{\text{BF}}_A, \zeta^{\text{BF}}_B) = \sum_{M_A M_B} D^{(L_A)}_{M_A K_A}(\zeta^{\text{BF}}_A)^* D^{(L_B)}_{M_B K_B}(\zeta^{\text{BF}}_B)^* \langle L_A M_A; L_B M_B | L k \rangle \quad (4.43)$$

The relation between the SF and BF expansion coefficients is given (van der Avoird *et al.*, 1994) by

$$d^{\text{BF}}_{\{\Lambda\}k}(R) = \sum_{\lambda} d^{\text{SF}}_{\{\Lambda\}\lambda}(R) \langle Lk; \lambda 0 | 1 k \rangle \quad (4.44)$$

The SF and BF dipole components are related as

$$\mu^{\text{SF}}_m(R, \zeta^{\text{SF}}_A, \zeta^{\text{SF}}_B, \hat{\mathbf{R}}) = \sum_k \mu^{\text{BF}}_k(R, \zeta^{\text{BF}}_A, \zeta^{\text{BF}}_B) D^{(1)}_{mk}(\alpha, \beta, 0)^* \quad (4.45)$$

and the BF label k indicates whether a given transition has a parallel ($k = 0$) or a perpendicular ($k = \pm 1$) component.

If the monomers have large permanent dipole moments, the infrared transitions are strongly determined by these, and the corresponding coefficients $d^{\text{SF}}_{\{\Lambda\}\lambda}(R)$ in equation (4.40) are simple (R-independent) functions of these dipoles. If we wish to include also the dipole moment induced on monomer A by the permanent multipole moments $Q^{L_B}_{K_B}$ on monomer B, we must add terms depending on the (mixed) dipole – 2^{l_A}-pole polarizability tensor of monomer A, which for $l_A = 1$ is the normal dipole polarizability. For the most common cases of a dipole moment induced by a monopole (charge), dipole or quadrupole through the normal dipole polarizability the coefficients have been listed in Table 2 of van der Avoird *et al.* (1994). Formulas have been derived for special cases, such as atom–polyatom (van Bladel *et al.*, 1991), atom–diatom (Dunker and Gordon, 1978; Brocks *et al.*, 1984) and diatom–diatom complexes (Poll and Hunt, 1976; Brocks and van der Avoird, 1985). General formulas are given in a recent paper (Heijmen *et al.*, 1996).

The line strengths (equation (4.39)), are calculated in the BF basis of equation (4.19) by the use of equations (4.42) and (4.43) for the dipole moment and an appropriate model for $d^{\text{BF}}_{\{\Lambda\}k}(R)$. Again, the result is similar to the expression in equation (4.21) for the potential matrix elements, but slightly more complicated. For other (tensorial) properties, such as the polarizability function needed for the calculation of Raman intensities, it is easy to write similar expressions (see Heijmen *et al.*, 1996).

Van der Waals complexes are formed in relatively high concentrations during the expansion of a supersonic nozzle beam; the use of such beams for spectroscopy has two other important advantages. First, when the spectra are taken somewhat downstream from the expansion, they are practically free of collision- and Doppler broadening. The spectral resolution can be considerably increased, so that the individual rotational $J \rightarrow J'$ transitions are resolved, even for rather large complexes. This yields a wealth of detailed and accurate information (Fraser, 1991; Nelson *et al.*, 1987a; Novick *et al.*, 1990; Klemperer, 1992; Saykally, 1989; Cohen and Saykally, 1992; Saykally and Blake, 1993; Meerts *et al.*, 1984; Champagne *et al.*, 1981; Nesbitt, 1988; Miller, 1988). Second, the molecules have become very cold, typically a few degrees kelvin. Only some J levels of the ground state are populated, which leads to simple spectra that can be (relatively) easily interpreted. Also the calculation of such spectra from the VRT states presents no special problems, once the wavefunctions of these states are known (Hutson, 1991; Clary and Nesbitt, 1989; Van Bladel *et al.*, 1991). In gas-phase spectra higher states are populated too, which causes a multitude of hot bands. In combination with the lower resolution this leads to very complex spectra, with composite, overlapping bands (Henderson and Ewing, 1973, 1974; Long *et al.*, 1973; Long and Ewing, 1973; McKellar, 1988). Also the computation of such spectra from the VRT states becomes a major task (Brocks and van der Avoird, 1985; Brocks, 1988; Garcia Ayllon *et al.*, 1990; Wang *et al.*, 1996). As the van der Waals or hydrogen bonds are weak, the transitions between different VRT levels in a van der Waals complex are observed in the far-infrared, typically below 300 cm^{-1}. They may also be seen in the mid- or near-infrared, however, or even in visible or UV spectra, if they occur simultaneously with vibrational or electronic transitions in the monomers.

4.7 SYMMETRY ASPECTS

In 'normal' nearly rigid molecules it is customary to use the point group of the equilibrium structure to classify the vibrations and the electronic states. This is just an approximate symmetry, however. In van der Waals molecules with multiple minima in the potential surface and large-amplitude vibrations it is no longer valid. The symmetry group of such molecules contains (1) permutations of identical nuclei, (2) space-inversion, and (3) products of (1) and (2). Usually not all permutation-inversions (PIs) are physically meaningful in the sense that they give rise to observable splittings; one only has to consider the so-called (Hougen, 1962; Longuet-Higgins, 1963) feasible PIs. There are two kinds of these: the first kind is equivalent to a rotation of the (rigid) complex in isotropic space. In this case no energy barrier has to be surmounted. The second kind of feasible PI requires the tunneling through some barrier, deforming the complex to another equivalent structure that is distinguished from the earlier structure by the change in one or more internal coordinates. It is very hard to predict *a priori* if an operation of the second kind is feasible. Detailed experiments or elaborate calculations are required to do so. Furthermore, whether or not an operation is considered to be feasible depends on the resolution of the measuring device.

The application of the molecular symmetry group, i.e. the group of all feasible PIs, is treated in several textbooks (Bunker, 1979; Ezra, 1982). It is known (van der Avoird *et al.*, 1994a) for the various coordinate systems commonly used in van der Waals molecules, how the PIs act on the coordinates. The action on the different basis functions then follows rather easily from the analytic properties of these functions. In SF coordinates this derivation is rather trivial, and not very useful, since it is not possible in these coordinates to separate the overall rotations of the complex from its internal motions. With the use of BF coordinates such a separation is possible, although approximate. The action of the PIs becomes more complicated: each PI corresponds with an 'equivalent rotation' of the BF frame (Bunker, 1979) and a transformation of the internal coordinates of the complex. If the complex is nearly rigid and has a single equilibrium structure, the PI group contains just the operations of the first kind and it is isomorphic to the point group of the equilibrium structure. The action of the PIs on the internal coordinates is equivalent to that of the point group operations on the small vibrational displacements. It is the additional PIs, of the second kind, which make the VRT states of van der Waals molecules so interesting, however.

The PI group symmetry can be used for several purposes. In the calculation of the VRT states, the adaptation of the basis to the irreducible representations (irreps) of the PI group leads to a separation of the Hamiltonian matrix into smaller blocks. In some examples, such as $(NH_3)_2$ (van Bladel *et al.*, 1992a; Olthof *et al.*, 1994a), this simplification was essential to make the calculations possible. Also the VRT states are symmetry adapted and, since the dipole operator is invariant under all permutations of identical nuclei and antisymmetric under space inversion E^*, this causes the (exact) selection rules. Further, approximate selection rules may be derived as well, by considering the separate PI group adaptation of the overall rotation functions and of the internal VRT wavefunctions. For this purpose, the components of the dipole operator should be expressed with respect to the BF frame, as in equation (4.42). The PI group symmetry of the 'parallel' and 'perpendicular' dipole components follows easily from the transformation properties of the coordinates.

Finally, we note that also the nuclear spin functions must be adapted to the permutations of (all) identical nuclei. The spin functions are invariant under space inversion. Since the nuclei are bosons (for integer I) or fermions (for half-integer I), it follows from the Pauli principle that the spatial wavefunctions of the VRT states are explicitly related, through their permutation symmetry, to the occurrence of specific nuclear spin quantum numbers. It is this relation that determines the nuclear spin statistical weight (Bunker *et al.*, 1979) of each VRT level. For nearly rigid molecules this is only relevant for the rotational structure in the spectra, but for floppy van der Waals molecules the permutation symmetry will strongly affect the internal states. Therefore, the nature of the internal VRT states and the spectra that pertain to the different nuclear spin species will be rather different. As in almost all practical cases the various nuclear spin species occur simultaneously, the measured spectra in fact consist of a set of overlapping spectra for all the species. In high-resolution spectra it is no problem to separate the individual species and to relate their spectra to the spectra calculated for the corresponding species.

4.8 COMPARISON WITH EXPERIMENTAL SPECTRA, TEST AND IMPROVEMENT OF AB INITIO POTENTIALS, SEMI-EMPIRICAL POTENTIAL FITS

We will give a few examples of cases where an initial potential is used to compute spectra, followed by a refinement of the potential by comparison with the corresponding experimental high-resolution spectra. First, we describe two atom–diatom complexes, He–HF and He–CO, where recently calculated *ab initio* potentials give already very precise predictions of the spectra, but the experimental information can still aid in improving on some subtle details of the surface. Next, we discuss two dimers, $(NH_3)_2$ and CO in CO_{60}, where existing force fields gave a first guess of the potential. In the former case the potential was improved by comparison with the very complete spectral work of the Berkeley group (Loeser *et al.*, 1992), while in the latter case further refinement awaits the synthesis and measurement of the dimer (Meijer, private communication).

These examples, which are drawn from our own research, are by no means complete. For instance, we forgo a discussion of the simplest of all van der Waals molecules: the rare gas (Rg) pairs. The spectroscopy of those dimers is discussed in Maitland *et al.* (1981), Grabow *et al.* (1995) and Xu *et al.* (1995) and their potentials are described by Aziz (1984) and Janzen and Aziz (1995). Let us further mention the work on the Rg–H_2 systems (Le Roy and Hutson, 1987), which were the first van der Waals molecules studied (Le Roy and Van Kranendork, 1974) in the manner outlined in this chapter. Recent work on H_2–Ar which involves the test of an accurate *ab initio* potential can be found in Williams *et al.* (1993a), Moszynski *et al.* (1994a) and Bisonette *et al.* (1996). Extensive work, which we will not explicitly review either, has been performed for Rg–HX dimers, with X = F, Cl or Br. This work has yielded accurate semi-empirical potentials and is summarized in (Hutson, 1990). The high-resolution spectra of van der Waals molecules have also yielded semi-empirical potentials for somewhat more complex systems already, such as Ar–H_2O (Cohen and Saykally, 1993), Ar–NH_3 (Schmuttenmaer *et al.*, 1994) and $(HCl)_2$ (Elrod and Saykally, 1995). Finally, we mention the experimental and theoretical studies on the $(HF)_2$ dimer (Bunker *et al.*, 1990; Quack and Suhm, 1991; Dyke *et al.*, 1972; Howard *et al.*, 1984; Gutowsky *et al.*, 1985; Pine and Howard, 1986; Kofranck *et al.*, 1988; Bunker *et al.*, 1988; Marshall *et al.*, 1991; Quack and Suhm, 1995; Collins *et al.*, 1995; Wu *et al.*, 1995; Zang *et al.*, 1995; Anderson *et al.*, 1996) and the $(H_2O)_2$ dimer (Fraser, 1991; Gregory and Clary, 1995; Dyke and Muenter, 1974; Kistenmacher *et al.*, 1974; Jeziorski and van Hemert, 1976; Coker and Watts, 1987; Szalowicz *et al.*, 1988; Zwart *et al.*, 1991; Rybak *et al.*, 1992; Millot and Stone, 1992; van Duijneveldt-van de Rijdt and van Duijneveldt, 1992; Pugliano *et al.*, 1993; Scheiner, 1994; Karyakin *et al.*, 1995; Althorpe and Clary, 1995); this work includes extensive *ab initio* computations on the potential surface, several calculations of the vibration–rotation–tunneling states and detailed spectroscopic studies of the splittings associated with these motions.

4.8.1 He–HF

The He–HF complex is very weakly bound and for some time it was investigated only by scattering techniques (Frick, 1984; Boughton *et al.*, 1986). In 1990 Lovejoy and Nesbitt reported the first study of the high-resolution near-infrared vibration-rotation spectra, corresponding to the simultaneous excitation of the vibration and rotation of HF within the He–HF complex. They also considered line broadening due to rotational predissociation.

The few dynamical calculations for this complex (Lovejoy and Nesbitt, 1990; Tennyson and Sutcliffe, 1983; Gianturco et al., 1984) were based on the *ab initio* potential of Rodwell et al. (1981). The most advanced of these studies was reported by Lovejoy and Nesbitt (1990). Their calculations of the (quasi)-bound rovibrational levels and of the linewidths revealed that the *ab initio* potential (Rodwell et al., 1981) does not reproduce correctly the near-infrared spectrum of the complex. Comparison of the experimental results with the *ab initio* predictions suggested that the van der Waals well in this potential is 11 per cent too shallow and that its anisotropic terms are 30 per cent too large in the repulsive region. This is not entirely surprising since the potential developed by Rodwell et al. follows the 'Hartree–Fock plus dispersion' model (Ahlrichs et al., 1977), and neglects important intramonomer correlation effects. However, by a simple scaling of the long-range dispersion coefficients in this potential Lovejoy and Nesbitt were able to obtain an anisotropic potential surface which reproduced all spectroscopic data available for He–HF.

Recently, Moszynski et al. (1994b) reported a SAPT calculation of the three-dimensional potential energy surface for the He–HF complex. This potential was represented by an expansion in Legendre polynomials $P_L(\cos\theta_B)$, where θ_B is the angle between the diatomic bond axis and the intermolecular vector **R** connecting the mass centers. The expansion coefficients $v_L(R, r)$, calculated for different values of the HF bond length r, were fitted by analytic functions of R which represent the various long-range R^{-n} and short-range (exponential) contributions. Further improvement of the important dispersion term was achieved by the computation of high-quality long-range dispersion coefficients at the same level of electron correlation (Rijks and Wormer, 1988; Wormer and Hettema, 1992a,b) in a large *spdfg* basis set. The SAPT potential surface is in very good agreement with the semi-empirical potential of Lovejoy and Nesbitt (1990) (see Figure 4.1). In Moszynski et al. (1994c) the accuracy of the SAPT potential was checked by direct comparison with experiment (Lovejoy and Nesbitt, 1990) after computation of the near-infrared spectrum and linewidths. Here, we present a brief summary of the VRT states and spectrum of He–HF, as calculated with the SAPT (Moszynski et al., 1994a) and semi-empirical (Lovejoy and Nesbitt, 1990) interaction potentials.

Although the He–HF interaction potential in the region of the van der Waals minimum is only weakly anisotropic, BF coordinates were used, with the kinetic energy operator given by equation (4.6) with $T_A = 0$ and $T_B = C(r)j_B^2$. The Legendre expansion of the potential can be directly used in these coordinates and the basis of equation (4.19) with $j_A = k_A = k_B = 0$ and after parity adaptation, is given by

$$|n, j, K; \sigma, J, M\rangle = \Phi_n(R)\left[\frac{2J+1}{4\pi}\right]^{1/2}[Y_K^j(\hat{\mathbf{r}})D_{M,K}^{(J)}(\alpha,\beta,0)^* + \sigma Y_{-K}^j(\hat{\mathbf{r}})D_{M,-K}^{(J)}(\alpha,\beta,0)^*]$$

(4.46)

where $Y_K^j(\hat{\mathbf{r}}) = Y_K^j(\theta_B, \phi_B)$ are spherical harmonics in the polar coordinates of the diatom bond axis relative to the dimer BF frame, and σ is the spectroscopic parity. This parity is related to the conventional parity p under space inversion E^* as $\sigma = p(-1)^J$ (Brown et

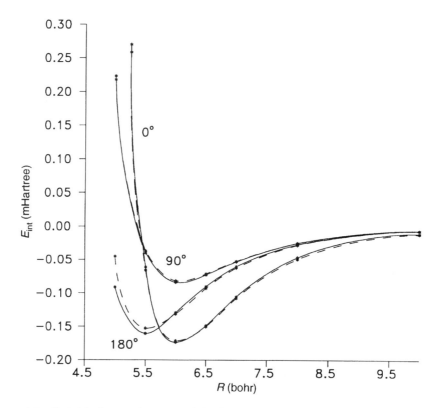

Figure 4.1 *Ab initio* interaction potential of He–HF computed by SAPT (Moszynski *et al.*, 1994a) (solid line) and semi-empirical potential (Lovejoy and Nesbitt, 1990) (dashed line), for $\theta = 0°, 90°, 180°$ and $r = 1.7328$ bohr

al., 1975). The potential matrix elements are given by equation (4.21), with the 9-j symbol substituted as indicated below this equation and the labels k'_A, K_A, k_A and k'_B, K_B, k_B in the 3-j symbols equal zero. The only rigorous quantum numbers are the total angular momentum J, M and the parity σ. The HF rotational quantum number j, and the projection K of **J** (and **j**) onto the body-fixed intermolecular axis, are nearly conserved (coupling case b of Bratoz and Martin (1965)). Functions with different j are mixed by the anisotropic potential, functions with different K only by the off-diagonal Coriolis interactions. States with (approximately) $K = 0, \pm 1$, etc. are denoted as Σ, Π, etc. Levels with $\sigma = +1$ and $\sigma = -1$ are designated by the superscripts e and f, respectively. For $K = 0$ only e parity states exist. The splitting of the states with $|K| > 0$ (the so-called l-doubling) into states with e and f parity is caused by the Coriolis interactions.

The allowed dipole transitions between the VRT states of the complex can be deduced from equation (4.39). They lead to the following rigorous selection rules for the transition $(v'', J'', \sigma'') \to (v', J', \sigma')$:

$$J'' = J' \quad \sigma'' = -\sigma' \quad \text{or } J'' = J' \pm 1, \sigma'' = \sigma' \quad (4.47)$$

Since the quantum number K is nearly conserved, an additional selection rule

$$K'' - K' = 0, \pm 1 \quad (4.48)$$

holds to a good approximation. Thus, the observed bands in the cold He–HF near-infrared spectrum correspond to the transitions from the bound Σ states of He–HF ($v = 0$) to Σ^e, Π^e, and Π^f states of He–HF ($v = 1$). In view of equation (4.47), two branches (P and R) corresponding to $J' = J'' + 1$ and $J' = J'' - 1$, respectively, are observable for $\Sigma \to \Sigma^e$ and $\Sigma \to \Pi^e$ bands, for the $\Sigma \to \Pi^f$ transitions one should see only one (Q) branch. A schematic diagram of the energy levels and observed near-infrared transitions is depicted in Figure 4.2.

The only truly bound states in He–HF are those e levels which lie below the $j = 0$

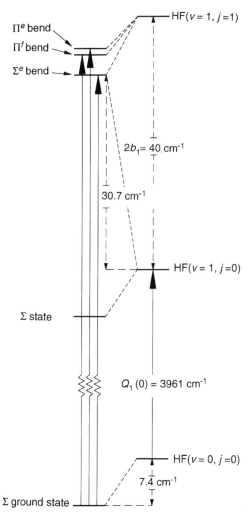

Figure 4.2 Schematic diagram of the rovibrational levels and near-infrared transitions in He–HF, according to Moszynski *et al.* (1994b). $Q_1(0) = 3961.4229$ cm^{-1} corresponds to the stretch fundamental of HF. The Π^f state is located 33.9 cm^{-1} above the He–HF ($v = 1$, $j = 0$) threshold, while the l-doubling (the splitting of the Π^e and Π^f energy levels) is 0.4 cm^{-1}

states of the free HF and, since the parity must be conserved upon dissociation, the f levels which lie below the $j = 1$ state of HF. Other states are either so-called 'shape' or 'orbiting' resonances which dissociate directly by tunneling through the centrifugal barrier, or Feshbach or 'compound' resonances (Child, 1974) which decay via rotational predissociation. The latter mechanism implies that the energy of the rotational excitation to $j = 1$ is converted into translation energy of the dissociating fragments. Of course, all states of He–HF ($v = 1$) may undergo vibrational predissociation, which utilizes the HF vibrational energy to dissociate the He–HF ($v = 1$) complex into He and HF ($v = 0$) fragments. This process was found to be extremely slow (Lovejoy and Nesbitt, 1990), however, and was ignored in Moszynski et al. (1994b). The lifetimes of metastable states of He–HF are not negligible, and the corresponding line broadenings have been measured (Lovejoy and Nesbitt, 1990). In Moszynski et al. (1994b) the positions and widths of these upper states have been obtained from close-coupling scattering calculations. The resonance parameters have been extracted from the behavior of the S-matrix as function of the energy (Ashton et al., 1983).

In Table 4.1 we report the results of bound state and close-coupling calculations (Moszynski et al., 1994b) of the energy levels in He–HF obtained from the *ab initio* potential energy surface (Moszynski et al., 1994a). As expected, the He–HF complex is very weakly bound. The potential energy surface for $v = 0$ supports only five bound states: the ground rovibrational state ($J = 0$) and four rotationally excited levels. The

Table 4.1 Calculated energy levels (in cm^{-1}) of the He–HF ($v = 0$) and He–HF ($v = 1$) complexes

	J	E_J (Moszynski et al., 1994b)[a]	E_J (Lovejoy and Nesbitt, 1990)[b]
Ground state[c]	0	−7.380	−7.347
	1	−6.608	−6.572
	2	−5.085	−5.043
	3	−2.861	−2.812
	4	−0.040	+0.011
Σ^e bend[d,e]	0	+30.684	+30.725
	1	+31.056	+31.124
	2	+32.013	+32.109
	3	+33.656	+33.776
	4	+35.982	+36.124
	5	+38.915	+39.071
Π^e bend[d,e]	1	+34.298	+34.508
	2	+36.371	+36.565
	3	+39.120	+39.294
Π^f bend[d,f]	1	+33.885	+34.112
	2	+35.364	+35.592
	3	+37.512	+37.736
	4	+40.206[g]	+40.371

[a] Energy levels computed using the *ab initio* SAPT potential (Moszynski et al., 1994a).
[b] Energy levels computed using the empirical potential (Lovejoy and Nesbitt, 1990).
[c] Energies relative to HF($v = 0, j = 0$).
[d] Energies relative to HF($v = 1, j = 0$).
[e] Resonance states determined from close-coupling scattering calculations.
[f] Bound states relative to HF($v = 1, j = 1$).
[g] Obtained from a variational calculation without the $j = 1$ function in the basis.

energy levels computed (Lovejoy and Nesbitt, 1990) from the semi-empirical potential are also included in Table 4.1. The agreement is very good: the energies of bound states agree within 0.05 cm^{-1} or better, the positions of the Σ^e resonances within 0.1 cm^{-1}, and the positions of the Π^e resonances and the energies of the Π^f states within 0.2 cm^{-1}. The theoretical dissociation energy, $D_0 = 7.38$ cm^{-1}, compares very well with the result obtained from the semi-empirical potential, $D_0 = 7.35$ cm^{-1}.

The computed transition frequencies corresponding to the experimentally observed $\Sigma \to \Sigma^e$, $\Sigma \to \Pi^e$, and $\Sigma \to \Pi^f$ bands are presented in Table 4.2. The SAPT potential surface predicts all infrared transitions with errors smaller than 0.1 cm^{-1}. For comparison we also report in Table 4.2 the transition frequencies computed from the semi-empirical potential (Lovejoy and Nesbitt, 1990). In general, the *ab initio* SAPT potential reproduces the experimental data with similar accuracy as the semi-empirical potential, which is fitted to these data.

It is also worth noting that the *ab initio* SAPT potential (Moszynski *et al.*, 1994a) accurately reproduces the experimental vibrational frequency shift of the HF molecule embedded in helium clusters. Recent diffusion quantum Monte Carlo calculations (Blume *et al.*, 1996) with this potential predict a red shift of 2.7 ± 0.1 cm^{-1}, which compares very favorably with the experimental value of 2.65 ± 0.15 cm^{-1}.

For dimers consisting of linear molecules, and in particular for an atom–diatom system (with $L_A = j_A = j'_A = 0$) as we have here, the labels K_A, K_B in the angular expansion of the potential and the labels k_A, k_B in the basis functions are zero. From equation (4.21) it

Table 4.2 Near-infrared transitions in He–HF (in cm^{-1}) accompanying the fundamental band of HF. The frequency corresponding to the HF stretch fundamental is $Q_1(0) = 3961.4229$ cm^{-1}

Transition	J''	J'	$\Delta E(J'' \to J')$, Computed (Moszynski et al., 1994b)[a]	$\Delta E(J'' \to J')$, Observed (Lovejoy and Nesbitt, 1990)[b]	$\Delta E(J'' \to J')$, Computed (Lovejoy and Nesbitt, 1990)[c]	Δ^d	Δ^e
$\Sigma \to \Sigma^e$	0	1	3999.860	3999.953	3999.894	−0.094	−0.059
	1	2	4000.044	4000.137	4000.104	−0.093	−0.033
	2	3	4000.164	4000.251	4000.242	−0.087	−0.009
	3	4	4000.266	4000.345	4000.359	−0.079	+0.014
	4	5	4000.378	4000.449	4000.483	−0.071	+0.034
$\Sigma \to \Pi^e$	0	1	4003.102	4003.161	4003.278	−0.059	+0.117
	1	2	4004.402	4004.418	4004.560	−0.016	+0.142
	2	1	4000.806	4000.904	4000.974	−0.098	+0.070
	3	2	4000.655	4000.735	4000.800	−0.080	+0.065
	4	3	4000.583	4000.639	4000.706	−0.056	+0.067
$\Sigma \to \Pi^f$	1	1	4001.916	4002.005	4002.107	−0.089	+0.102
	2	2	4001.871	4001.952	4002.058	−0.081	+0.106
	3	3	4001.796	4001.859	4001.971	−0.063	+0.112
	4	4	4001.669	4001.680	4001.783	−0.011	+0.103

[a] Computed transitions from the *ab initio* SAPT potential (Moszynski *et al.*, 1994a).
[b] Measured transitions (Lovejoy and Nesbitt, 1990).
[c] Computed transitions from the empirical potential (Lovejoy and Nesbitt, 1990).
[d] Absolute error of the transition frequency computed from the *ab initio* potential.
[e] Absolute error of the transition frequency computed from the empirical potential.

follows then that the diagonal potential matrix elements vanish for odd values of $L = L_B$. This suggests that the energy levels and transition frequencies are mainly sensitive to the terms with even L in the Legendre expansion of the intermolecular potential. The results reported in Table 4.2 confirm that these terms in the *ab initio* potential (Moszynski *et al.*, 1994a) are indeed very accurate. The correctness of the (much smaller) terms with odd L can be checked by computing the widths of resonances which decay via rotational predissociation. Rotational predissociation lifetimes can be defined via the Fermi golden rule expression (Tucker and Truhlar, 1986) which, indeed, mixes states with $j = 0$ and $j = 1$ via the $L = 1$ term in the Legendre expansion of the potential. Since Lovejoy and Nesbitt (1990) measured the linewidths of all $\Sigma \to \Sigma^e$ and $\Sigma \to \Pi^e$ transitions in He–HF, a direct comparison of the computed and measured widths serves as a further test of the accuracy of the *ab initio* potential.

In Table 4.3 we report the widths of the Σ^e and Π^e resonances computed on the *ab initio* potential. The agreement with experiment is less satisfactory here: all computed widths are too large by a factor of two. This suggests that the small $L = 1$ anisotropy in the *ab initio* potential is not correct. To check this assumption, the short-range contribution to the $L = 1$ angular component of the potential was scaled by a factor of 0.95 and the widths of the Σ^e and Π^e resonances were recomputed. The results are given in parentheses in Table 4.3. The agreement with the measured linewidths (Lovejoy and Nesbitt, 1990) is very good now: almost all widths computed from the scaled potential agree with the experimental data within the error bars. Also the agreement with the widths computed from the semi-empirical potential (Lovejoy and Nesbitt, 1990) is very satisfactory. It should be noted that this scaling introduces a very small change in the interaction potential (see Figure 4.3). In the repulsive region the scaled potential is ≈ 4 per cent smaller than the original *ab initio* potential (Moszynski *et al.*, 1994a), while the depth of the van der Waals well is only 2 per cent lower. These results clearly show that the widths of resonances are extremely sensitive to the accuracy of the small odd terms. The quantitative prediction of the rotational predissociation lifetimes is a challenge for *ab initio* calculations. At the same time, we see that spectral information aids in refinement of the intermolecular potential.

Table 4.3 Calculated widths (MHz) of the He–HF resonance states. The widths corresponding to the scaled potential are given in parenthesis

	J	Γ_J Computed (Moszynski *et al.*, 1994b)[a]	Γ_J Observed (Lovejoy and Nesbitt, 1990)[b]	Γ_J Computed (Lovejoy and Nesbitt, 1990)[c]
Σ^e bend	1	7203 (3452)	3020 ± 500	3550
	2	5731 (2673)	2830 ± 200	2730
	3	4453 (2001)	1640 ± 150	1999
	4	3280 (1397)	1260 ± 100	1349
	5	2158 (848)	770 ± 100	780
Π^e bend	1	1080 (575)	530 ± 100	532
	2	1773 (928)	890 ± 150	900
	3	1930 (993)	1000 ± 400	990

[a] Linewidths computed using the *ab initio* SAPT potential (Moszynski *et al.*, 1994a).
[b] Measured linewidths (Lovejoy and Nesbitt, 1990).
[c] Linewidths computed using the empirical potential (Lovejoy and Nesbitt, 1990).

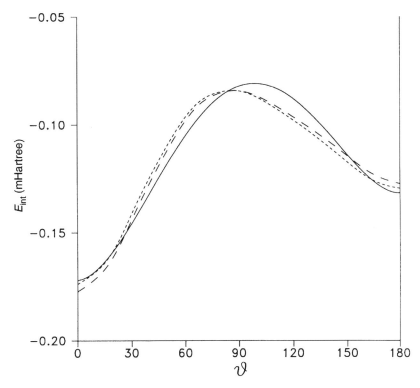

Figure 4.3 Anisotropies of the *ab initio* (– – –), scaled (———) (Moszynski *et al.*, 1994b), and semi-empirical (solid line) (Lovejoy and Nesbitt, 1990) potentials of He–HF in the region of the van der Waals minimum ($R = 6$ bohr and $r = 1.7328$ bohr)

4.8.2 He–CO

Because of its astrophysical interest and its relatively small number of electrons, the He–CO dimer has been the subject of many *ab initio* studies (Green and Thaddeus, 1976; Thomas *et al.*, 1980; Schinke and Diercksen, 1985; Parish *et al.*, 1992; Zolothoukina and Kotake, 1993; Kukawska–Tarnawska *et al.*, 1994; Tao *et al.*, 1994; Moszynski *et al.*, 1995). A recent infrared spectrum, recorded in a low temperature cell, was fitted by Chuaqui *et al.* (1994). They assigned the experimental spectrum by means of a two-dimensional potential energy surface that was adapted to give the correct line positions and intensities. Recently Le Roy *et al.* (1994) fitted a new potential to reproduce the experimental spectrum. The starting point for the new fit was the exchange-coulomb (XC) model potential (Meath and Koulis, 1991), partly based on *ab initio* information on the exchange-repulsion energy and long-range induction and dispersion coefficients. In this section we summarize the recent work of Moszynski *et al.* (1995a) on the SAPT calculations of the PES of He–CO and the nuclear motion calculations based on this SAPT potential. The latter give the positions and intensities of lines in the infrared spectrum of the complex in the region of the fundamental band of CO.

Just as in the case of He–HF, the He–CO PES was described in terms of the Jacobi coordinates (R, θ_B), where R is the distance from the center of mass of CO to the He atom, and θ_B is the angle between the vector pointing from the center of mass of CO to He and the vector pointing from the carbon to the oxygen atom. The CO bondlength was kept fixed at the experimental value 2.132 bohr. Separate fits of the sum of short-range contributions, of the induction energy and of the dispersion energy were made in the manner of Moszynski *et al.* (1995b). All the different contributions were expanded as in equation (4.10) and we recall again that the angular function in equation (4.12) reduces to a Legendre polynomial $P_L(\cos \theta_B)$ for the present atom–diatom case. The fits were constrained to coincide with asymptotic long-range results obtained in the same atomic orbital basis and at the same correlation level.

In Figure 4.4 we find an angular scan of the various PESs in the region of the van der Waals minimum ($R = 6.85$ bohr). An inspection of this figure shows that the potential is rather flat as function of the angle for $\theta_B \leqslant 120°$, and shows a strong barrier to internal rotation of the CO subunit around $\theta_B = 180°$. This barrier originates from the more diffuse character of the CO charge distribution near the carbon atom (see Kukawska–Tarnawska *et al.*, 1994). Since for larger R the height of this barrier is considerably less, we can expect the CO monomer in the complex to behave as a slightly hindered rotor. The flatness of the potential for $\theta_B \leqslant 120°$ explains why the two empirical potentials

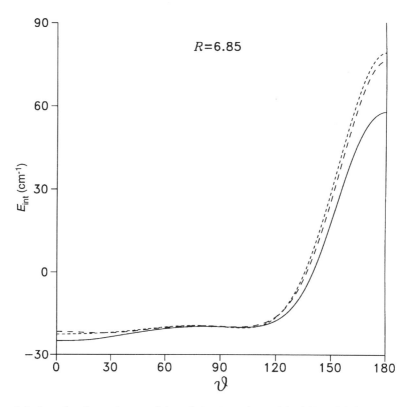

Figure 4.4 Angular dependence of the *ab initio* and empirical He–CO interaction potentials in the region of the van der Waals minimum ($R = 6.85$ bohr). Solid line represents the *ab initio* potential, while lines with small and large dashes show the empirical potentials from Le Roy *et al.* (1994) and Chuaqui *et al.* (1994), respectively

fitted to reproduce the spectrum (Chuaqui et al., 1994; Le Roy et al., 1994) predict different positions for the minima, and shows that the spectroscopic data are not sensitive enough to locate the minimum. This, in turn, suggests that the He atom in the complex will be strongly delocalized. The angular dependence of the empirical potentials (Chuaqui et al., 1994; Le Roy et al., 1994) agrees fairly well with the potential of Moszynski et al. (1995a). The agreement deteriorates somewhat at large angles, but in this region the interaction energy is repulsive, and the spectroscopic data are probably less sensitive to the shape of the repulsive wall. The anisotropy of the potential in the repulsive region, which is dominated by the exchange repulsion, is much more pronounced.

We discussed in Section 4.2 that nuclear motions in van der Waals complexes can be described in a set of coordinates related to either a space-fixed or a body-fixed frame. Although the anisotropy of the He–CO potential in the region of the van der Waals minimum is not particularly weak, Moszynski et al. (1995a) followed Chuaqui et al. (1994) and assumed that the space-fixed description is appropriate. If this is the case, the energy levels and infrared transitions in He–CO can be approximately classified by case a coupling (Bratoz and Martin, 1965). It was also assumed in Moszynski et al. (1995a) that the intramolecular vibration can, to a good approximation, be decoupled from the intermolecular modes due to its high frequency (2143 cm^{-1}), and vibrationally averaged rotational constants C_v of CO can be used instead. Thus, the Hamiltonian of equations (14.2)–(14.4) describes the nuclear motion. The rigid rotor Hamiltonian (4.3) takes the form $C_v j^2$ for a diatomic.

In the limit of vanishing anisotropy the following are good quantum numbers: j, which corresponds to the rotation of CO in the dimer, and $l \equiv l^{SF}$, which corresponds to the end-over-end rotation of the vector $\mathbf{R}$. The total angular momentum $\mathbf{J} = \mathbf{j} + \mathbf{l}$ is always conserved, due to the isotropy of space, but j and l are broken by the anisotropy in the potential. A degenerate (j, l)-level splits into sublevels $J = |j - l|, \ldots, j + l$ under the influence of the anisotropy. If these splittings are small, as in Ar–NH (Jansen et al., 1993), He–C$_2$H$_2$ (Moszynski et al., 1995b), or Ar–H$_2$ (Moszynski et al., 1994), the states can still be labeled to a good approximation by j and l.

The wavefunction $\Psi^{JM}(R, \hat{\mathbf{R}}, \hat{\mathbf{r}})$ was expanded in a basis of products of radial functions $\Phi_n(R)$ (consisting of Morse-type oscillator functions (Tennyson and Sutcliffe, 1983) and angular functions (cf. equation (4.18)), which are in this case Clebsch–Gordan coupled spherical harmonics $[Y^j(\hat{\mathbf{r}}) \otimes Y^l(\hat{\mathbf{R}})]^J_M$. The angular basis functions have a well-defined parity $p = (-1)^{j+l}$, so the full Hamiltonian is blocked in both p and J. Within each block states of various j and l values are mixed through the potential.

In the calculation of the intensities of the intermolecular transitions, accompanied by the monomer transition $v'' = 0 \to v' = 1$, a Boltzmann average was performed over the $v = 0$ van der Waals states (with a temperature of $T = 50$ K). The contribution to the infrared absorption coefficient from the transition $(0, i'', J'') \to (v', i', J')$ is proportional to the quantity in equation (4.37). Assuming that the vibrational transition dipole moment of CO, μ^{0v}, is not affected by the intermolecular interaction, we can use in equation (4.39) the simple dipole model

$$\mu_m^{SF} \to \mu_m^{0v} = \mu^{0v} C_m^1(\hat{\mathbf{r}}) \tag{4.49}$$

where $C_m^l(\hat{\mathbf{r}})$ are spherical harmonics in the Racah normalization (Brink and Satchler, 1975). The explicit formula for the matrix element of the transition dipole in equation

(4.49) can be found in Moszynski *et al.* (1995b). The infrared transitions in He–CO accompanying the fundamental band of CO will obey the following selection rules: $|\Delta p| = 1$, and $|\Delta J| = 1$ or 0. Additionally, if the energy levels can be labeled with the quantum numbers j and l, the selection rules $\Delta l = 0$ and $|\Delta j| = 1$ hold approximately.

In Table 4.4 we summarize the results of the bound state calculations. The SAPT potential energy surface supports 13 bound states. (Note that the states of positive energy with parity $p = (-1)^{J+1}$ that lie below the $j = 1$ state of CO are truly bound, because they cannot mix with the $j = 0$ continuum due to the parity constraint.) As expected, He–CO is a very weakly bound complex. The ground state of ^{4}He–CO is bound only by 6.962 cm^{-1}. All other states correspond to angularly excited energy levels of the complex. No states excited in the van der Waals stretch are supported by the SAPT potential. As discussed above, the anisotropy of the He–CO potential in the region of the van der Waals minimum is relatively strong, so it is of interest to see if the energy levels of He–CO can still be approximately labeled with j and l quantum numbers. The labeling of the states with these approximate quantum numbers is listed in Table 4.4. Also shown in this table are contributions of the dominant (j, l) angular function to the wave function of the ith state for a given J, defined as

$$\%(j, l) = 100 \sum_n |C^J_{i,j,l,n}|^2 \qquad (4.50)$$

where $C^J_{i,j,l,n}$ is the coefficient of the basis function $\Phi_n(R)[Y^j(\hat{\mathbf{r}}) \otimes Y^l(\hat{\mathbf{R}})]^J_M$ in the eigenvector of the ith state with J, p, and v fixed. For most of the eigenstates, the dominant (j, l) contribution to the wave function is of the order of 90 per cent. This result suggests that the CO molecule in the dimer behaves like a slightly hindered rotor, and that the van der Waals states of the He–CO complex will be strongly delocalized, both in R and θ_B. The anisotropy of the present *ab initio* potential is quite similar to that of the empirical potentials (Chuaqui *et al.*, 1994; Le Roy *et al.*, 1994) fitted to reproduce the infrared spectrum, so the pattern of the levels is quantitatively the same.

The highly delocalized character of the He–CO van der Waals states suggests that the average position of He may be very different from the minimum-energy geometry. Indeed, the expectation values of the van der Waals stretch coordinate $\langle R \rangle$, the values of $R_0 = \langle R^{-2} \rangle^{-1/2}$, and the values of the angle $\theta_0 = \arccos((2\langle P_2(\cos \theta_B) \rangle + 1)/3)^{1/2}$ are 7.87 bohr, 7.64 bohr, and 53.82°, and 8.59 bohr, 8.25 bohr, and 42.08° for the $(j = 0, J = 0)$ and $(j = 1, J = 0)$ states, respectively. Both $(\langle R \rangle, \theta_0)$ and (R_0, θ_0) are rather different from the position corresponding to the minimum of the interaction potential ($R_m = 6.85$ bohr, $\theta_m = 0°$).

Table 4.4 Energy levels (in cm^{-1}) of the ^{4}He–CO ($v = 0$) complex (Moszynski *et al.*, 1995a)

	$j = 0$	$j = 1$		
l	$J = l$	$J = l - 1$	$J = l$	$J = l + 1$
0	-6.962 (91.61%)			-2.869 (90.02%)
1	-6.389 (91.26%)	-1.974 (81.95%)	-2.485 (98.07%)	-2.221 (88.34%)
2	-5.255 (90.62%)	-1.100 (83.08%)	-1.323 (98.19%)	-1.074 (86.66%)
3	-3.584 (89.83%)		$+0.384$ (98.39%)	
4	-1.420 (89.45%)			

EXPERIMENTAL SPECTRA, TEST AND IMPROVEMENT   135

The infrared transition frequencies and intensities for the ^{4}He–CO complex are illustrated in Figure 4.5. Since the CO vibration was assumed to be decoupled from the intermolecular modes, the transition frequencies could be computed from the formula

$$\Delta E(J'' \to J') = E^{J'}_{i',1} - E^{J''}_{i'',0} + Q_1(0) \qquad (4.51)$$

where $Q_1(0) = 2143.2712$ cm^{-1} (Varberg and Evenson, 1992) is the frequency of the CO stretching fundamental. First, we note that unlike in the He–C$_2$H$_2$ case (Moszynski et al., 1995b) most transitions do not obey selection rules corresponding to the case (a) coupling of Bratoz and Martin (1965), although the most intense lines do correspond to the free internal rotor limit. This again confirms that CO behaves in the complex like a slightly hindered rotor. An inspection of Figure 4.5 shows that the agreement of theoretical transition frequencies and intensities with the results of high-resolution measurements (Chuaqui et al., 1994) is satisfactory. Most of the line positions agree within 0.1–0.2 cm^{-1} or better. The intensities are also accurately predicted by the ab initio potential. However, four transition frequencies are in error by ≈ 0.5 cm^{-1}, suggesting that the anisotropy of the ab initio potential is not entirely correct. These transitions involve angularly excited states (bending states) with $(j, l) = (1, 1)$. Perturbation theory analysis reported by Chuaqui et al. (1994) shows that the energies of these states are mainly sensitive to even terms in the Legendre expansion of the potential. The agreement with experiment being not perfect, the bound state calculations were repeated with a potential in which the contribution of the short-range energy to the $V_2(R)$ component was increased by 4.5 per cent. In general, the agreement with the measured line positions is much better

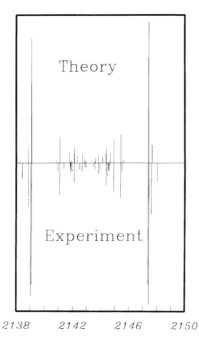

Figure 4.5 Comparison of theoretical and experimental infrared spectra of the ^{4}He–CO complex accompanying the fundamental band of CO. The temperature is 50 K

(≈ 0.1 cm^{-1}), although some lines are reproduced with a somewhat lower accuracy. It should be stressed, however, that no attempt was made to fit the potential to the spectrum.

Finally, in Figure 4.6 we illustrate the transition frequencies and intensities for the ^{3}He–CO complex (generated from the *ab initio* potential). Here, the agreement is even better: all line positions are reproduced within 0.2 cm^{-1} or better. Also the infrared intensities are correctly predicted. This confirms that the error in the anisotropy of the present potential is not very large.

In summary, the predicted positions and intensities of lines in the infrared spectrum are in good agreement with the experimental spectrum (Chuaqui *et al.*, 1994). Still, a minor improvement in the agreement between theory and experiment could be achieved by empirically adjusting the anisotropy of the exchange-repulsion energy

4.8.3 NH$_3$–NH$_3$

It is a fact, well established theoretically (Bunker *et al.*, 1990; Quack and Suhm, 1991; Scheiner, 1994) and experimentally (Fraser, 1991; Dyke *et al.*, 1972; Anderson *et al.*, 1996; Dyke and Muenter, 1974), that the dimers (HF)$_2$ and (H$_2$O)$_2$ have a hydrogen-bonded structure. Until 1985 it was generally believed that the ammonia dimer, too, had a 'classical' hydrogen-bonded structure with a proton of one monomer pointing to the nitrogen lone pair of the other. In that year Nelson *et al.* (1985) interpreted their microwave spectra by assuming that this dimer has a nearly cyclic structure in which the two umbrellas are almost antiparallel. This finding was surprising in view of the fact that

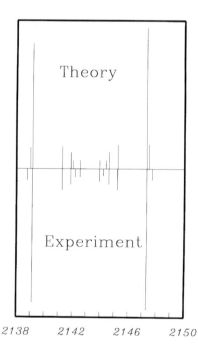

Figure 4.6 Comparison of theoretical and experimental infrared spectra of the ^{3}He–CO complex accompanying the fundamental band of CO. The temperature is 50 K

most *ab initio* calculations (Liu *et al.*, 1986; Frisch *et al.*, 1986) predicted the classical, nearly linear, hydrogen bonded structure. Two recent calculations differ in the prediction of the equilibrium structure: Hassett *et al.*, (1991) found a hydrogen bonded structure, whereas Tao and Klemperer (1993) found a cyclic structure thanks to the addition of bondfunctions.

An obvious explanation of the discrepancy between the outcome of most calculations and the microwave data might be found in the effect of vibrational averaging: whereas the electronic structure calculations focus mainly on finding the minimum of the intermolecular potential, the experiment gives a vibrationally averaged structure. This question was addressed experimentally by Nelson *et al.* (1987a,b) by means of various isotope substitutions. From the fact that the relevant intermolecular bond angles hardly change with isotope substitution they concluded that $(NH_3)_2$ is fairly rigid and that also its equilibrium structure must be (nearly) cyclic. They supported this latter conclusion by the observation that the dipole moment of $(ND_3)_2$—in which the vibrational averaging effects are expected to be less than in $(NH_3)_2$—is 0.17 D smaller than the value of 0.74 D found for the $(NH_3)_2$ dimer. Nelson and co-workers took this as an indication that, indeed, the equilibrium structure is nearly cyclic. Note, parenthetically, that the dipole of the free ammonia monomer is 1.47 D, which means that the sum of the components of the permanent dipoles along the dimer axis in the linear hydrogen bonded structure is about 2.0 D.

The effects of vibrational averaging have been assessed theoretically by Van Bladel *et al.* (1992). Using the model potential of Sagarik *et al.* (1986), which was the only full potential surface available from *ab initio* calculations, they solved the six-dimensional Schrödinger equation for the intermolecular motions in a basis of coupled internal rotor functions and Morse-type stretch functions (see equation (4.19)). Although it was found that the vibrationally averaged structure shifts from the equilibrium hydrogen bonded structure toward the cyclic geometry, the work did not produce complete reconciliation with the microwave geometry. Further Van Bladel *et al.* obtained indirect evidence that the umbrella inversion of the two monomers is not completely quenched, as was assumed by Nelson *et al.* (1985).

The latter conclusion was also reached by Loeser *et al.* (1992), who reported an extensive set of new far-infrared and microwave measurements and gave a very detailed analysis of these—as well as previous (Havenith *et al.*, 1991)—experimental data. They concluded that the group of feasible operations (permutations, inversion and their products) is of the order 144, which implies that they observed the tunneling splittings associated with the two umbrella inversions, as well as the interchange tunneling in which the role of the two monomers is reversed. The same conclusion was reached by the Nijmegen/Bonn group (Havenith *et al.*, 1992), on the basis of infrared/far-infrared double-resonance experiments. The latter authors also measured the dipole moments in various $|K| = 1$ states of *G*-symmetry (Linnartz *et al.*, 1993; Cotti *et al.*, 1996). Thus, the various experimental approaches present evidence that seems conflicting regarding the rigidity of $(NH_3)_2$ and its equilibrium structure. Also the different *ab initio* calculations lead to different pictures. Multiple discussions (Nelson *et al.*, 1987a; Saykally and Blake, 1993; Baum, 1992) have been devoted to this problem.

Recently Olthof *et al.* (1994a–c and van der Avoird *et al.* (1994b) presented a more complete theoretical approach. They constructed a family of model potentials which contain the electrostatic interactions between the dipoles, quadrupoles and octupoles of the ammonia monomers and an exp-6 site–site representation (with sites on the nuclei, as well as on the lone pairs) of the exchange repulsion and dispersion interactions. By changing the strength parameters in the exchange repulsion they varied the height of the barriers in the interchange motion and in the hindered rotations of the two NH_3 monomers around their C_3 axes. For each of these potentials they calculated the six-dimensional vibration–rotation–tunneling (VRT) states and the various transition frequencies that have been observed. For various states they computed the expectation values of the dipole moment and the nuclear quadrupole splittings, which are indicative of the orientations of the NH_3 monomers in the complex. By improving the parameters they arrived at a model potential that was able to reproduce all the observed splittings with deviations of about 0.2 cm^{-1}. Also the dipole and the nuclear quadrupole splittings were in good agreement with the observed values, both for $(NH_3)_2$ and $(ND_3)_2$. The minimum in this potential does not correspond to a cyclic equilibrium structure with two equivalent monomers, but the barrier between equivalent global minima with bent hydrogen bonded structures in which the donor/acceptor roles of the monomers are interchanged is only 7 cm^{-1}.

Before use in the calculation of the VRT states the potentials were expanded in the complete set of angular functions of equation (4.12). The R-dependent coefficients were computed by numerical quadrature (cf. equation (4.14)). The kinetic energy part of the Hamiltonian, which has to be diagonalized in order to obtain the VRT states, is given by equation (4.6). The body-fixed basis and the calculation of the matrix elements are described in Section 4.5.

Because of the size of the basis, the full symmetry of the system had to be taken into account. The molecular symmetry group is of order 36, provided the umbrella inversions are frozen. Otherwise it is of order 144. These groups are denoted by G_{36} and G_{144}, respectively. Olthof *et al.* mainly focused on G_{36}, which has four one-dimensional irreducible representations (irreps), designated A_i, $i = 1,...,4$, four two-dimensional irreps (E_i, $i = 1,...,4$) and one four-dimensional irrep G. Recall that the three proton spins of NH_3 can couple either to a quartet: *ortho* ammonia, or to a doublet: *para* ammonia. The kets of A_i symmetry belong to two *ortho* monomers, those of E_i symmetry belong to two *para* monomers and G kets describe a mixed *ortho–para* dimer. For more details on symmetry adaptation we refer to Appendix C of van der Avoird *et al.* (1994b).

The results from the calculations on $(NH_3)_2$ are summarized in Table 4.5 for $K = 0$ and $|K| = 1$. Note that K, which is the projection of the total angular momentum J on the dimer bond axis, is not an exact quantum number. Since the off-diagonal Coriolis coupling is small, the observed states can be well characterized by K and therefore the Coriolis coupling was neglected. The energy differences $E_{A_4} - E_{A_1}$ and $E_{E_2} - E_{E_1}$ in Table 4.5 are due to the interchange tunneling. Note that these differences are large, in the order of 20 cm^{-1}, which confirms that the interchange between the donor and the acceptor molecule in the hydrogen bond takes place rapidly. Also the splitting $E'_G - E_G$ between the lowest G states is partly due to this interchange tunneling and partly to the difference between the *ortho* and the *para* monomers that form these G states. We present values of $180° - \theta_B$ in Table 4.5, rather than of θ_B, because whenever $\theta_A \approx 180° - \theta_B$, we have a cyclic structure. The observed and calculated energy levels are visualized in Figure 4.7, which clearly shows their surprisingly good agreement.

Owing to the fact that the G states belong to two non-identical molecules, i.e. *ortho* and *para*, they are localized to some extent on one side of the interchange barrier. This is in contrast to the A_i and E_i states, which are either symmetric or antisymmetric with

Table 4.5 Comparison of calculated and measured properties of $(NH_3)_2$. All values pertain to $K = 0$ states, unless indicated otherwise

Property	Calculation (Olthof et al., 1994a)	Experiment
Equilibrium dipole	1.08 D	–
Equilibrium θ_A	40.47°	–
Equilibrium $180° - \theta_B$	84.49°	–
Dipole G	0.66 D	0.74 D (Nelson et al., 1987a)
Dipole G ($|K| = 1$)	0.19 D	0.10 D (Linnartz et al., 1993)
$\theta_A{}^a$	48.2°	48.6° (Nelson et al., 1987a)
$180° - \theta_B{}^b$	65.1°	64.5° (Nelson et al., 1987a)
$E_{A_4} - E_{A_1}$	15.85 cm^{-1}	16.12 cm^{-1} (Loeser et al., 1992)
$E_{E_2} - E_{E_1}$	19.14 cm^{-1}	19.36 cm^{-1} (Loeser et al., 1992)
$E'_G - E_G$	20.25 cm^{-1}	20.50 cm^{-1} (Loeser et al., 1992)
$E_{G_2^-} - E_{G_2^+}$	2.05 GHz	3.31 GHz (Loeser et al., 1992)
$E'_{G_2^-} - E'_{G_2^+}$	1.24 GHz	2.39 GHz (Loeser et al., 1992)

^{a} From $\langle P_2(\cos\theta_A)\rangle$.
^{b} From $\langle P_2(\cos\theta_B)\rangle$.

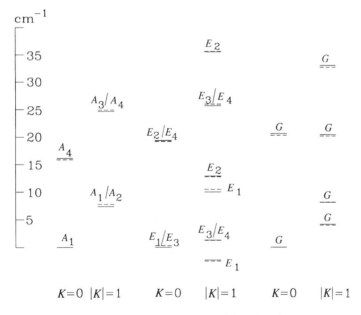

Figure 4.7 Comparison of computed and observed levels of NH_3-NH_3. Dashed lines: computed; full lines: observed (Loeser et al., 1992). The $K = 0$ ground level of each species A, E and G is adjusted. In several cases the difference between computed and observed values is within the width of the lines

respect to interchange. See Figure 4.8 for contour plots of the G-symmetry wavefunctions. These plots show clearly that the dimer is highly non-rigid, a fact which is confirmed by the difference between the equilibrium dipole and the G-state dipoles (see Table 4.5). Another important observation is that the partial localization, which manifests

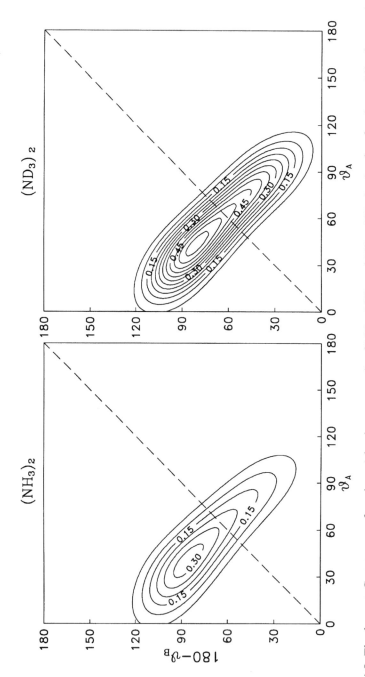

Figure 4.8 The lowest G-state wavefunctions (absolute squared) of NH_3-NH_3 and ND_3-ND_3 in the $\theta_A-\theta_B$ plane. All other angles are fixed at their equilibrium values and $R = 3.373$ Å. Note that the wavefunctions are not symmetric with respect to reflection in the diagonal, unlike the potential from which they are obtained

itself in the G state expectation values of the dipole moment, depends also on the barriers to rotation around the symmetry axes over ψ_A and ψ_B. It was found that addition of octupole interactions to the potential gave a substantial increase in the dipole moment; note that the first ψ_X-dependent electrostatic term is due to the octupole on monomer X.

The final two splittings in Table 4.5 are due to monomer umbrella inversion. If the monomer retains its threefold symmetry during its 'umbrella in the storm' inversion, one coordinate ρ, the angle of the NH bonds with the threefold symmetry axis of the monomer, suffices to describe this motion. Under this assumption an exact calculation requires the solution of an eight-dimensional dynamics problem: six intermolecular coordinates plus the two umbrella angles ρ_A and ρ_B. The group of this system is G_{144} and the labels $G_2^\pm$ refer to irreps of this group. These irreps correlate with the irrep G of $G_{36} \subset G_{144}$. A dynamics problem of this size cannot be handled at present, so that Olthof *et al.* (1994c) employed an approximate model, which is an extension of a model proposed and tested earlier for Ar–NH$_3$ (Van Bladel, 1991, 1992b).

Let us end this section by discussing the isotope substitution effects. It was observed that the G-state dipole moment decreases when going from $(NH_3)_2$ to $(ND_3)_2$. Since the value of the dipole calculated at the equilibrium geometry is 1.08 D, much larger than the average value of 0.66 D, and since one would expect $(ND_3)_2$ to stay closer to equilibrium than $(NH_3)_2$, it is not *a priori* clear that this decrease will also be found in calculations. However, the rovibrationally averaged computed dipole moment *does* decrease, from 0.66 D for $(NH_3)_2$ to 0.38 D for $(ND_3)_2$. This decrease follows nicely the experimentally observed (Nelson *et al.*, 1987b) decrease from 0.74 D for $(NH_3)_2$ to 0.57 D for $(ND_3)_2$. Also the accompanying changes in the angles θ_A and θ_B obtained from the expectation values $\langle P_2(\cos\theta_A) \rangle$ and $\langle P_2(\cos\theta_B) \rangle$ agree well with the changes observed by measuring the nuclear quadrupole splittings in $(NH_3)_2$ and $(ND_3)_2$: $(\theta_A, 180° - \theta_B)$ change from $(48.2°, 65.1°)$ to $(51.2°, 61.7°)$, experimentally they change from $(48.6°, 64.5°)$ to $(49.6°, 62.6°)$. So it appears that $(ND_3)_2$ is more nearly cyclic than $(NH_3)_2$. In Van der Avoird *et al.* (1994b) this rather unexpected observation is explained by analysis of the wavefunctions (see Figure 4.8). One has to remember that, in spite of the equivalence of the two minima in the potential, the G-state wavefunctions are mainly localized on one side because of the *ortho–para* differences. This difference in the behavior of *ortho* and *para* monomers will be less for ND$_3$ than for NH$_3$, because its rotational constants are smaller by about a factor of 2. Consequently, the asymmetry in the G-state wavefunctions which is caused by these *ortho–para* differences will be smaller in $(ND_3)_2$. In other words, $(ND_3)_2$ is more nearly cyclic (in its G state) because of the smaller *ortho–para* differences. This explains the observed decrease of the average dipole moment upon deuteration.

Other results from the calculations on $(ND_3)_2$ which are most relevant for comparison with the observed quantities (Nelson *et al.*, 1987b; Loeser, to be published) are collected in Table 4.6. The interchange tunneling frequencies predicted by Olthof *et al.* (1994a) are about 30 per cent smaller than in $(NH_3)_2$, but in recent measurements (Loeser, to be published) it was found that the calculated values are still (systematically) too high by about 20 per cent. For the small umbrella inversion splittings in $(ND_3)_2$ the predictions of Olthof *et al.* (1994c) proved to be quite good. For $(NH_3)_2$ there have been new measurements of the nuclear quadrupole splittings in $|K| = 1$ substates of E_3 and E_4 symmetry (i.e. the *para–para* species) which probed, for the first time, the dependence of the corresponding VRT states on the dihedral angle between the C_3 axes of the

Table 4.6 Comparison of calculated and measured properties of $(ND_3)_2$. All values pertain to $K = 0$

Property	Calculation (Olthof et al., 1994a)	Experiment
Dipole G	0.38 D	0.57 D (Nelson et al., 1987a)
$\theta_A{}^a$	51.2°	49.6° (Nelson et al., 1987a)
$180° - \theta_B{}^b$	61.7°	62.6° (Nelson et al., 1987a)
$E_{A_4} - E_{A_1}$	11.03 cm^{-1}	8.82 cm^{-1} (Loeser et al., to be published)
$E_{E_2} - E_{E_1}$	13.78 cm^{-1}	11.21 cm^{-1} (Loeser et al., to be published)
$E'_G - E_G$	13.06 cm^{-1}	10.81 cm^{-1} (Loeser et al., to be published)

^{a} From $\langle P_2(\cos\theta_A)\rangle$.
^{b} From $\langle P_2(\cos\theta_B)\rangle$.

monomers (Heineking et al., 1995). In order to get more precise quantitative agreement with all the measurements, the model potential of Olthof et al. will have to be refined.

We may conclude that, within one consistent computational model and by the use of a single parameterized potential, Olthof et al. were able to reproduce the observed level splittings, the observed dimer geometries and properties in various VRT states, and the fact that the deuterated dimer has a smaller dipole than the protonated one. In the explanation of these features there was no need to invoke near-rigidity or a nearly cyclic equilibrium structure. On the contrary, the ammonia dimer is so highly non-rigid that its average structure differs strongly for the different VRT states, and is influenced also by the *ortho/para* nature of the monomers.

4.8.4 CO in C_{60}

It has been found (Meijer, private communication; Weiske et al., 1991; Wan et al., 1992; Wang et al., 1993; Heath et al., 1985; Bethune et al., 1993; Saunders et al., 1994; Holleman et al., 1994) that atoms (He, Ne, Ar, Ca, and La) can be trapped inside the fullerene C_{60} and form so-called endohedral complexes denoted as X@C_{60}. In some theoretical studies (Williams et al., 1993b; Pang and Brisse, 1993; Cioslowski, 1991) it was predicted that also small molecules such as CO and HF can form stable endohedral complexes with C_{60}. There is at least one ongoing experimental project aimed at the production of CO@C_{60} (Meijer, private communication). A recent theoretical study (Olthof et al., 1996) was intended to describe the hindered rotations and vibrations of CO inside C_{60} and to predict its infrared spectrum. Here we summarize this study.

The C_{60} molecule has some rather soft modes, starting at 273 cm^{-1} (Dong et al., 1993), which might couple with the van der Waals modes of the complex, but in view of the exploratory nature of the work, it was decided to neglect this coupling and to consider C_{60} as rigid. Also the CO molecule was assumed to be rigid, since its stretch frequency is high, 2143 cm^{-1}. In principle, CO@C_{60} can be studied in two forms: as a freely rotating complex in the gas phase, or dissolved in solid C_{60}. In the latter case the C_{60} cage is fixed and the motion of the CO molecule is determined by an external potential that is provided mainly by this cage. In the former case the kinematic coupling of the motions of the cage and molecule must be taken into account. Since the experimental (Meijer, private communication) attempt to produce CO@C_{60} and to measure its (far-)infrared spectrum in the solid, we will concentrate in this review on non-rotating CO@C_{60}. Note, however, that Olthof *et al.* (1996) treats both cases. Similar calculations have been performed for rare gas atoms (Joslin *et al.*, 1993; Mandziuk and Bačić, 1994a, b) and alkali ions (Li^+ and Na^+) (Hernández-Rojas *et al.*, 1995, 1996a) inside fullerenes; another recent study of CO, LiH and LiF inside C_{60} (Hernández-Rojas *et al.* 1996b) concentrates on the hindered rotations.

In Olthof *et al.* (1996) an orthogonal BF frame was attached to C_{60} with its axes coinciding with three of the 15 twofold rotation axes. The vector **R** points from the origin (the center of the cage) to the mass center of CO. The vector **r** is along CO, pointing towards the C-atom. The pertinent kinetic energy operator is given in equation (4.9). If the C_{60} cage is fixed, the first term in this operator can be omitted and the reduced mass μ_{AB} must be replaced by the mass of the CO molecule. The intermolecular potential was modeled as a sum of exp-6 atom–atom potentials (cf. Section 4.4) with the parameters from Mirsky (1978).

The potential has a minimum of depth -1073 cm^{-1} at $R_{eq} = 0.192$ Å, with **R** pointing toward the midpoint of a six-ring, i.e. the mass center of CO is shifted into the [111] direction, and the CO axis **r** is antiparallel to **R**. Note that this is just one of the 20 symmetry-related minima of CO@C_{60}. The position of this minimum can be rationalized by a simple (and somewhat crude) hard sphere model. We define the 'geometric center' of CO as the center of the sphere which circumscribes the van der Waals spheres of the C and O atom. This center does not coincide with the nuclear mass center of CO, but is shifted by about 0.20 Å towards the carbon atom. According to this simple model, the steric hindrance is minimal if the 'geometric center' of CO coincides with the midpoint of C_{60}. Indeed, $R_{eq} = 0.192$ Å is not far from the point of minimum repulsion. An *ab initio* calculation of the steric repulsion of CO in C_{60} by the Hartree–Fock method (Cioslowski, 1991) yields $R_{eq} = 0.175$ Å. We will see below that the rotational structure in the infrared spectrum of CO@C_{60} can be reasonably well understood if we assume that the molecule rotates around the 'geometric center', rather than around its mass center.

Fixing the angles at their equilibrium values, one obtains from the model potential a 'fixed angle' radial potential $V_{rad}(R)$. This radial potential is nearly harmonic around R_{eq}. Moving (with constant $R = R_{eq}$) from one minimum to a neighbouring equivalent minimum, a barrier of only 1.2 cm^{-1} has to be surmounted, which shows that the potential is rather flat for motions in which **r** and **R** remain antiparallel. The small barrier is caused by the corrugation of the inner wall of C_{60}. However, if one moves away from the antiparallel orientation the potential rises steeply.

The bound states were calculated according to the linear variation procedure, and prior to that, the model potential was expanded in terms of coupled spherical harmonics. As we discussed in Section 4.4, this has the advantage that all angular matrix elements can be calculated analytically, provided that the wave function is also expanded in terms of such functions. Observe that for non-rotating C_{60} the configuration space does not contain the external Euler angles describing the rotation of C_{60} and hence the basis does not include the Wigner D-functions depending on these Euler angles. So there are only four angular quantum numbers: L, j, λ, and μ, with the corresponding coupled basis

$$|L_j\lambda\mu\rangle \equiv [Y^L(\hat{\mathbf{R}}) \otimes Y^j(\hat{\mathbf{r}})]^\lambda_\mu \qquad (4.52)$$

The bound state problem was solved in two steps. In the first step radial basis functions

$R^{-1}\chi_n(R)$ were determined by diagonalizing the radial Hamiltonian, containing the fixed angle radial potential $V_{\text{rad}}(R)$. The eigenfunctions $\chi_n(R)$ with energies ϵ_n of this radial Hamiltonian were obtained by the discrete variable representation (DVR) method of Colbert and Miller (1992) and Groenenboom and Colbert (1993) based on sinc functions (see equation (4.33)). In the second step, the computation of the five-dimensional bound state wave functions, the lowest three radial eigenfunctions were retained and multiplied by the angular basis of equation (4.52). It was checked that a larger radial basis did not change the results significantly.

The icosahedral buckyball cage provides a potential for CO that is invariant under the point group I_h. By taking linear combinations of angular expansion functions $A_{L_1 L_2 \Lambda M_\Lambda}$ (cf. equation (4.12)), which in this case are coupled spherical harmonics similar to the basis in equation (4.52), it is possible to reduce the number of expansion coefficients in the potential considerably. It can be shown that only $\Lambda = 0, 6, 10, 12, 15, \ldots$, functions afford the totally symmetric irrep A_g. The terms with $\Lambda = 0$ represent the potential of a spherical cage and the terms with $\Lambda = 6$ are already sufficient to describe the small (≈ 0.5 per cent) corrugation effects.

For the far-infrared transitions that correspond to the vibrations and (hindered) rotations of rigid CO in C_{60} the dipole operator is modeled simply by the permanent dipole μ^{CO} of the CO molecule. Thus one neglects all the terms due to the interaction of the buckyball and the diatom. The dipole operator expressed with respect to the frame fixed on C_{60} is

$$\mu_\nu^{BF} = \mu^{CO} C_\nu^{(1)}(\hat{\mathbf{r}}) \qquad \text{with} \qquad \nu = 0, \pm 1 \qquad (4.53)$$

If one wants to study the van der Waals side bands of the fundamental stretch of CO in the infrared region, then (neglecting the coupling of the intra-monomer and inter-monomer vibrations) one has to use the monomer vibrational transition dipole $\mu^{01} = \langle 0|\mu(r)|1\rangle$ instead of the permanent dipole μ^{CO}. However, since the two dipoles have the same angular dependence, the theory for the line intensities of the far- and mid-infrared part of the spectrum is the same. The dipole operator is an irreducible tensor operator transforming as T_{1u} under I_h. This leads to the selection rules $A \leftrightarrow T_1$, $T_2 \leftrightarrow G$, $T_2 \leftrightarrow H$, $T_1 \leftrightarrow H$, $G \leftrightarrow H$, $H \leftrightarrow H$, $T_1 \leftrightarrow T_1$, and $G \leftrightarrow G$, in combination with an obligatory change of parity: $g \leftrightarrow u$. The line strengths of the allowed transitions (cf. equation (4.39)), have been calculated from the five-dimensional bound state wave functions. The energy level differences, together with these line strengths, provide the predicted (far-)infrared spectra in Figures 4.9 and 4.10, computed with the aid of equation (4.37) at $T = 77$ K.

The energy levels and spectra show some very characteristic features. Apart from the 'crystal field' splittings due to the icosahedral field of C_{60}, they look like the levels and spectra of a linear triatomic molecule. The rotational bands do not correspond to a simple rotation of the CO monomer (with quantum number j), but rather to a simultaneous rotation of $\mathbf{r}$ and $\mathbf{R}$ (remaining antiparallel) with quantum number λ. The effective rotational constant $B = 1.73$ cm^{-1} obtained from a fit of the computed energy levels is substantially smaller than that of free CO ($B = 1.92$ cm^{-1} (Mantz et al., 1975). This can be understood by viewing the rotation of CO in C_{60} as a rotation around its 'geometric center', defined above. Indeed, the moment of inertia for rotation of CO around this center gives a rotational constant close to 1.73 cm^{-1}. Also the rovibrational transitions are very similar to those of a linear triatom (Herzberg, 1991). The 'stretch', i.e. radial, fundamental vibration band has its origin at 209 cm^{-1} and corresponds to a $\Sigma \leftarrow \Sigma$ transition, the twofold degenerate 'bending' (or librational) mode corresponds to a $\Pi \leftarrow \Sigma$ transition, with l-type doubling, and an allowed Q band in addition to the P and R bands (see Figure 4.9). The origin of this 'bending' transition is at 162 cm^{-1} and even its overtones, a $\Sigma \leftarrow \Sigma$ transition at 327 cm^{-1} and a $\Delta \leftarrow \Sigma$ transition at 325 cm^{-1}, could be identified in the calculations (see Table 4.7). It is evident from these frequencies that the 'bending' mode is nearly harmonic; from various other observations it follows that this holds also for the 'stretch' mode at 209 cm^{-1}. This is related to the steepness of the potential of the inner wall of C_{60}; 'bending' implies in fact that **r** and **R** move away from their antiparallel equilibrium configuration.

In the far-infrared spectrum of Figure 4.9 it can be seen also that the 'bending' and 'stretch' fundamentals have sufficient intensity to be observable. Note the $P(5)/P(6)$ band head in the latter band, which is due to the large increase of the effective rotational constant B upon stretch excitation (see Table 4.7). This increase is actually caused by the strong Coriolis coupling of the stretch excited Σ state with one of the components of the l-type doubled Π bending state. Another effect of Coriolis coupling (of the Π bending state to the ground Σ state) is the unusually high intensity in the P and R branches of the bending band for higher values of λ. In the mid-infrared spectrum of Figure 4.10 which accompanies the $v = 1 \leftarrow 0$ transition in the CO monomer, only the rotational P and R lines have relatively high intensity. The Q band is due to hot-band transitions between bending excited levels in the upper and lower states. Also some of the weaker lines in the P and R branches of both spectra are caused by hot-band transitions.

We mentioned above that the rovibrational lines are split by the icosahedral field of C_{60}. From group theory it follows easily that this splitting occurs only for the levels with $\lambda \geqslant 3$ and, hence, for the $P(3), Q(3), R(2)$ and higher transitions. Although the $(\Lambda = 6)$ corrugation terms in the potential are relatively small, the resulting splittings are of the order of a few cm^{-1} and clearly observable in the P and R branches in the spectra of Figures 4.9 and 4.10.

All these observations are so characteristic that they can be considered as a signature of CO@C_{60}. They are mainly determined by geometric effects, such as the head-tail asymmetry of the CO molecule, and the nearly perfect spherical shape of C_{60} together with the icosahedral symmetry of the corrugation which breaks this perfect shape, so that they are rather independent of the (approximate) potential used in the calculations. When they will be observed, one can be sure that indeed CO molecules have been trapped inside

Table 4.7 Spectral properties of CO@C_{60}: band origins ΔE, effective rotational constants B, and l-type doubling constants q. Also given are the average position $\langle R \rangle$ of the CO mass center and the vibrational amplitude ΔR of each state for the lowest values of λ

States	ΔE (cm^{-1})	$B(q)$ (cm^{-1})	$\langle R \rangle$ (Å)	ΔR (Å)
$\Sigma, \sigma = +$	0.0	1.73	0.211	0.052
$\Pi, \sigma = \pm$[a]	162.2	1.73 (−0.18)	0.228	0.051
$\Sigma, \sigma = +$[b]	209.3	1.98	0.219	0.088
$\Sigma, \sigma = +$[c]	326.6	1.64	0.242	0.055
$\Delta, \sigma = \pm$[c]	324.8	1.64 (−0.03)	0.242	0.050

[a] Libration fundamental.
[b] Radial stretch fundamental.
[c] Libration overtone.

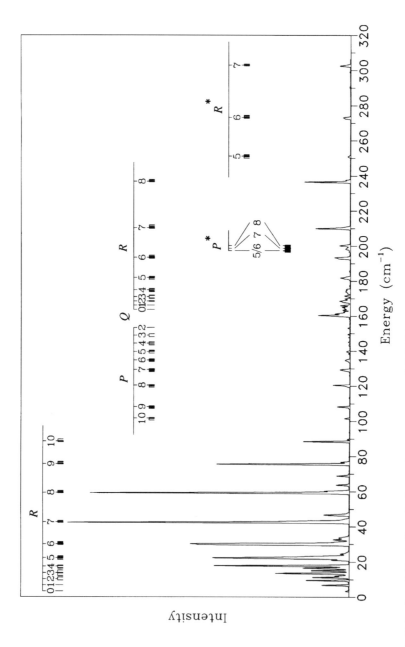

Figure 4.9 Calculated far-infrared absorption spectrum of CO@C_{60} at 77 K. A Gaussian line shape is assumed, full width at half maximum 0.5 cm^{-1}. The P and R branches marked with an asterisk (*) belong to the radial excitation band with origin 209.3 cm^{-1}. The vertical bars that contain the labels of the lines mark the frequencies from the calculations with the $\Lambda = 0$ potential. The icosahedral field splittings are indicated by the vertical bars below the labels (transitions with intensities less than 10 per cent of the strongest ones are omitted)

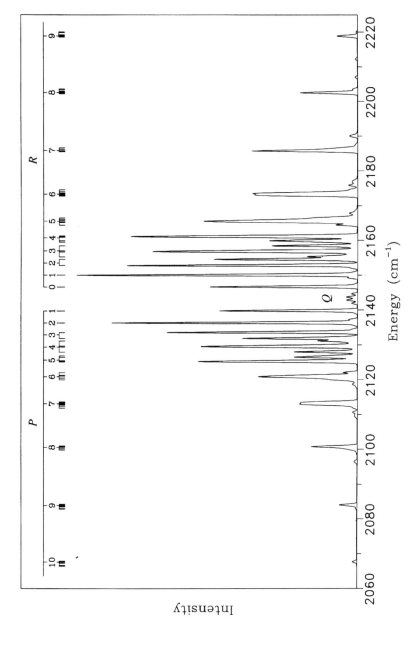

Figure 4.10 Calculated mid-infrared absorption spectrum of CO@C$_{60}$ at 77 K; the CO fundamental stretch frequency is 2143.27 cm^{-1}. For details, see the caption for Figure 4.9

the C_{60} cages. However, the band origins in particular will probably differ from the predicted frequencies. These (and other) observed values can then be used to improve the atom–atom model potential.

4.9 CONCLUSION

The aim of this chapter was to show the reader that much can be learned about intermolecular potentials from the spectroscopic study of van der Waals and hydrogen-bonded complexes. We paid special attention to the theory and computational methods required to calculate the bound states and spectrum of a van der Waals dimer from a given anisotropic pair potential. The examples given in the latter part of this chapter are meant to illustrate how the experimentalist and theoretician working in tandem can gather a large amount of useful quantitative information about the interaction between molecules. We hope that we succeeded in convincing the reader that the spectroscopy of van der Waals molecules is a branch of science well worth pursuing, also from the theoretical side.

ACKNOWLEDGEMENTS

We thank Dr G. C. Groenenboom for valuable discussions. This work was supported in part by the Netherlands Foundation for Chemical Research (SON) with financial aid from the Netherlands Organization for Scientific Research (NWO). One of us (R.M.) also thanks the Polish Scientific Research Council (grant KBN 3 T09A 072 09).

DEDICATION

This chapter is dedicated to the memory of Professor Włodzimierz Kołos.

REFERENCES

Abramowitz, M. and Stegun, I. A. (eds) (1964) *Handbook of Mathematical Functions*, Natl. Bur. Standards, Washington, DC.
Ahlrichs, R., Penco, R. and Scoles, G. (1977) *Chem. Phys.*, **19**, 119.
Althorpe, S. C and Clary, D. C. (1995) *J. Chem. Phys.*, **102**, 4390.
Anderson, D. T., Davis, S. and Nesbitt, D. J. (1996) *J. Chem. Phys.*, **104**, 6225.
Arthurs, A. M. and Dalgarno, A. (1960) *Proc. Royal Soc. (London) A*, **256**, 540.
Ashton, C. J., Child, M. S. and Hutson, J. M. (1983) *J. Chem. Phys.*, **78**, 4025.
Aziz, R. A. (1984) in *Inert Gases*, edited by Klein, M. L., Springer-Verlag, Berlin.
Bačić, Z., Kennedy-Mandziuk, M. and Moskowitz, J. W. (1992) *J. Chem. Phys.*, **97**, 6472.
Bačić, Z. and Light, J. C. (1986) *J. Chem. Phys.*, **85**, 4594.
Baum, R. M. (1992) *Chem. Eng. News*, 19 October, p. 20.
Bethune, D. S., De Vries, M. S. Johnson, R. D., Salem, J. R. and Yannoni, C. S. (1993) *Nature*, **366**, 123.
Biedenharn, L. C. and Louck, J. D. (1981) *Angular Momentum in Quantum Physics*, Addison-Wesley, London.

References

Bisonette, C., Chuaqui, C. E., Crowell, K. G. and Le Roy, R. J. (1996) *J. Chem. Phys.*, accepted.
Block, P. A., Pedersen, L. G. and Miller, R. E. (1993) *J. Chem. Phys.*, **98**, 3754.
Blume, D., Lewerenz, M., Huisken, F. and Kaloudis, M. (1996) *J. Chem. Phys.*, submitted.
Boughton, C. V., Miller, R. E., Vohralik, P. F. and Watts, R. O. (1986) *Mol. Phys.*, **58**, 827.
Bratoz, S. and Martin, M. L. (1965) *J. Chem. Phys.*, **42**, 1051.
Brink, D. M. and Satchler, G. R. (1975) *Angular Momentum*, Clarendon Press, Oxford.
Brocks, G. (1988) *J. Chem. Phys.*, **88**, 578.
Brocks, G. and Huygen, T. (1986) *J. Chem. Phys.*, **85**, 3411.
Brocks, G., Tennyson, J. and Van der Avoird, A. (1984) *J. Chem. Phys.*, **80**, 3223.
Brocks, G., Van der Avoird, A., Sutcliffe, B. T. and Tennyson, J. (1983) *Mol. Phys.*, **50**, 1025.
Brocks, G. and Van der Avoird, A. (1985) *Mol. Phys.*, **55**, 11 (Note that the coefficient $M_{2,2,3,1}$ in this paper is too large by a factor of $\sqrt{2}$.)
Brocks, G. and Van Koeven, D. (1988) *Mol. Phys.*, **63**, 999.
Brookes, M. D., Hughes, D. J. and Howard, B. J. (1996) *J. Chem. Phys.*, **104**, 5391.
Brown, J. M., Hougen, J. T., Huber, K.-P., Johns, J. W. C., Kopp, I., Lefebvre-Brion, H., Merer, A. J., Ramsay, D. A., Rostas, J. and Zare, R. N. (1975) *J. Mol. Spectrosc.*, **55**, 500.
Bunker, P. R. (1979) *Molecular Symmetry and Spectroscopy*, Academic Press, New York.
Bunker, P. R., Jensen, P., Karpfen, A., Kofranek, M. and Lischka, H. (1990) *J. Chem. Phys.*, **92**, 7432.
Bunker, P. R., Kofranek, M., Lischka, H. and Karpfen, A. (1988) *J. Chem. Phys.*, **89**, 3002.
Califano, S. (1976) *Vibrational States*, Wiley, London.
Champagne, B. B., Plusquellic, D. F., Pfanstiel, J. F., Pratt, D. W., Van Herpen, W. M. and Meerts, W. L. (1991) *Chem. Phys.*, **156**, 251.
Child, M. S. (1974) *Molecular Collision Theory*, Academic Press, New York.
Chuaqui, C. E., Le Roy, R. J. and McKellar, A. R. W. (1994) *J. Chem. Phys.*, **101**, 39.
Choi, S. E. and Light, J. C. (1990) *J. Chem. Phys.*, **92**, 2129.
Cioslowski, J. (1991) *J. Am. Chem. Soc.*, **113**, 4139.
Clary, D. C. and Nesbitt, D. J. (1989) *J. Chem. Phys.*, **90**, 7000.
Cohen, R. C. and Saykally, R. J. (1990) *J. Phys. Chem.*, **94**, 7991.
Cohen, R. C. and Saykally, R. J. (1992) *J. Phys. Chem.*, **96**, 1024.
Cohen, R. C. and Saykally, R. J. (1993) *J. Chem. Phys.*, **98**, 6007.
Coker, D. F. and Watts, R. O. (1987) *J. Phys. Chem.*, **91**, 2513.
Colbert, D. T. and Miller, W. H. (1992) *J. Chem. Phys.*, **96**, 1982.
Collins, C. L., Morihashi, K., Yamaguchi, Y. and Schaefer, H. F. (1995) *J. Chem. Phys.*, **103**, 6051.
Cooper, A. R. and Hutson, J. M. (1993) *J. Chem. Phys.*, **98**, 5337.
Cotti, G., Linnartz, H., Meerts, W. L., Van der Avoird, A. and Olthof, E. H. T. (1996) *J. Chem. Phys.*, **104**, 3898.
Coudert, L. H. and Hougen, J. T. (1988) *J. Mol. Spectrosc.*, **130**, 86.
Cullum, J. K. and Willoughby, R. A. (1985) *Lanczos Algorithms for Large Symmetric Eigenvalue Computations*, Birkhaüser, Boston.
Danby, G. (1983) *J. Phys. B*, **16**, 3393.
Davidson, E. R. (1975) *J. Comput. Phys.*, **17**, 87.
Dong, Z.-H., Zhou, P., Holden, J. M., Eklund, P. C., Dresselhaus, M. S. and Dresselhaus, G. (1993) *Phys. Rev. B.*, **48**, 2862.
Dunker, A. M. and Gordon, R. G. (1976) *J. Chem. Phys.*, **64**, 4984.
Dunker, A. M. and Gordon, R. G. (1978) *J. Chem. Phys.*, **68**, 700.
Dyke, Th. R., Howard, B. J. and Klemperer, W. (1972) *J. Chem. Phys.*, **56**, 2442.
Dyke, Th. R. and Muenter, J. S. (1974) *J. Chem. Phys.*, **60**, 2929.
Elrod, M. J. and Saykally, R. J. (1995) *J. Chem. Phys.*, **103**, 921, 933.
Essén, H. (1978) *Am. J. Phys.*, **46**, 983.
Ezra, G. S. (1982) *Symmetry Properties of Molecules*, Springer-Verlag, Berlin.
Fraser, G. T. (1991) *Intern. Rev. Phys. Chem.*, **10**, 189.
Frick, J. (1984) Thesis, MPI Strömungsforschung, Bericht **9**, Göttingen.
Friesner, R. A., Bentley, J. A. Menou, M. and Leforestier, C. (1993) *J. Chem. Phys.*, **99**, 324.
Frisch, M. J., Del Bene, J. E., Binkley, J. S. and Schaefer, H. F. (1986) *J. Chem. Phys.*, **84**, 2279.
Garcia Ayllon, A., Santamaria, J., Miller, S. and Tennyson, J. (1990) *Mol. Phys.*, **71**, 1043.

Gianturco, F. A., Palma, A., Villarreal, P. and Delgado-Barrio, G. (1984) *Chem. Phys. Lett.*, **111**, 399.
Grabow, J.-U., Pine, A. S., Fraser, G. T., Lovas, F. J., Suenram, R. D., Emilsson, T., Arunan, E. and Gutowski, H. S. (1995) *J. Chem. Phys.*, **102**, 1181.
Green, S. and Thaddeus, P. (1976) *Astrophys. J.*, **205**, 766.
Gregory, J. K. and Clary, D. C. (1995) *J. Chem. Phys.*, **102**, 7817.
Groenenboom, G. C. and Colbert, D. T. (1993) *J. Chem. Phys.*, **99**, 9681.
Gutowsky, H. S., Chuang, C., Keen, J. D., Klots, T. D. and Emilsson, T. (1985) *J. Chem. Phys.*, **83**, 2070.
Hassett, D. M., Marsden, C. J. and Smith, B. J. (1991) *Chem. Phys. Lett.*, **183**, 449.
Havenith, M., Cohen, R. C., Busarow, K. L., Gwo, D.-H., Lee, Y. T. and Saykally, R. J. (1991) *J. Chem. Phys.*, **94**, 4776.
Havenith, M., Linnartz, H., Zwart, E., Kips, A., Ter Meulen, J. J. and Meerts, W. L. (1992) *Chem. Phys. Lett.*, **193**, 261.
Heath, J. R., O'Brien, S. C., Zhang, Q., Liu, Y., Curl, R. F., Kroto, H. W., Tittel, F. K. and Smalley, R. E. (1985) *J. Am. Chem. Soc.*, **107**, 7779.
Heijmen, T. G. A., Moszynski, R., Wormer, P. E. S. and Van der Avoird, A. (1996) *Mol. Phys.*, **89**, 81.
Heineking, N., Stahl, W., Olthof, E. H. T., Van der Avoird, A., Wormer, P. E. S. and Havenith, M. (1995) *J. Chem. Phys.*, **102**, 8693.
Henderson, G. and Ewing, G. E. (1973) *J. Chem. Phys.*, **59**, 2280.
Henderson, G. and Ewing, G. E. (1974) *Mol. Phys.*, **27**, 903.
Henderson, J. R. and Tennyson, J. (1993) *Comput. Phys. Commun.* **75**, 365.
Hernández-Rojas, J., Bretón, J. and Gomez-Llorente, J. M. (1995) *Chem. Phys. Lett.*, **237**, 115.
Hernández-Rojas, J., Bretón, J. and Gomez-Llorente, J. M. (1996a) *J. Chem. Phys.*, **104**, 1179.
Hernández-Rojas, J., Bretón, J. and Gomez-Llorente, J. M. (1996b) *J. Chem. Phys.*, **104**, 5754.
Herzberg, G. (1991) *Molecular Spectra and Molecular Structure, II. Infrared and Raman Spectra of Polyatomic Molecules*, Krieger, Malabar.
Holleman, I., Boogaarts, M. G. H. and Meijer, G. (1994) *Recl. Trav. Chim. Pays-Bas*, **113**, 543.
Holmgren, S. L., Waldman, M. and Klemperer, W. (1977) *J. Chem. Phys.*, **67**, 4414.
Hougen, J. T. (1962) *J. Chem. Phys.*, **37**, 1433.
Howard, B. J., Dyke, Th. R. and Klemperer, W. (1984) *J. Chem. Phys*, **81**, 5417.
Huber, D. (1985) *Int. J. Quant. Chem.*, **28**, 245.
Hutson, J. M. (1990) *Annu. Rev. Phys. Chem.*, **41**, 123.
Hutson, J. M. (1991) *Adv. Mol. Vib. Collision Dyn.*, **1**, 1.
Hutson, J. M. (1992) *J. Chem. Phys.*, **96**, 6752.
Hutson, J. M. and Le Roy, R. J. (1985) *J. Chem. Phys.*, **83**, 1197.
Jansen, G., Hess, B. A and Wormer, P. E. S. (1993) *Chem. Phys. Lett.*, **214**, 103.
Janzen, A. R. and Aziz, R. A. (1995) *J. Chem. Phys.*, **103**, 9626.
Jeziorski, B. and Van Hemert, M. C. (1976) *Mol. Phys.*, **31**, 713.
Johnson, B. R. (1973) *J. Comput. Phys.*, **13**, 445.
Johnson, B. R. (1978) *J. Chem. Phys.*, **69**, 4678.
Joslin, C. G., Gray, C. G., Goddard, J. D., Goldman, S., Yang, J. and Poll, J. D. (1993) *Chem. Phys. Lett.*, **213**, 377.
Karyakin, E. N., Fraser, G. T., Lovas, F. J., Suenram, R. D. and Fujitake, M. (1995) *J. Chem. Phys.*, **102**, 1114.
Kitaigorodski, A. I. (1973) *Molecular Crystals and Molecules*, Academic Press, New York.
Kistenmacher, H., Popkie, H., Clementi, E. and Watts, R. O. (1974) *J. Chem. Phys.*, **60**, 4455.
Klemperer, W. (1992) *Science*, **257**, 887.
Kofranek, M., Lischka, H. and Karpfen, A. (1988) *Chem. Phys.*, **121**, 137.
Kukawska-Tarnawska, B., Chalasinski, G. and Olszewski, K. A. (1994) *J. Chem. Phys.*, **101**, 4964.
Lewerenz, M. (1996) *J. Chem. Phys.*, **104**, 1028.
Le Roy, R. J., Bissonnette, C., Wu, T. H., Dham, A. K. and W. J. Meath, W. J. (1994) *Faraday Disc. Chem. Soc.*, **97**, 81.
Le Roy, R. J. and Carley, J. S. (1980) *Adv. Chem. Phys.*, **42**, 353.
Le Roy, R. J. and Hutson, J. M. (1987) *J. Chem. Phys.*, **86**, 837.

Le Roy, R. J. and Van Kranendonk, J. (1974) *J. Chem. Phys.*, **61**, 4750.
Light, J. C., Hamilton, I. P. and Lill, J. V. (1985) *J. Chem. Phys.*, **82**, 1400.
Linnartz, H., Kips, A., Meerts, W. L. and Havenith, M. (1993) *J. Chem. Phys.*, **99**, 2449.
Liu, S., Dykstra, C. E., Kolenbrander, K. and Lisy, J. M. (1986) *J. Chem. Phys.*, **85**, 2077.
Loeser, J. G., Fraser, G. T. and Saykally, R. J. to be published.
Loeser, J. G., Schmuttenmaer, C. A., Cohen, R. C. Elrod, M. J. Steyert, D. W., Saykally, R. J., Bumgarner, R. E. and Blake, G. A. (1992) *J. Chem. Phys.*, **97**, 4727.
Long, C. A., Henderson, G. and Ewing, G. E. (1973) *Chem. Phys.*, **2**, 485.
Long, C. A. and Ewing, G. E. (1973) *J. Chem. Phys.*, **58**, 4824.
Longuet-Higgins, H. C. (1963) *Mol. Phys.*, **6**, 445.
Lovejoy, C. M. and Nesbitt, D. J. (1990) *J. Chem. Phys.*, **93**, 5387.
Maitland, G. C., Rigby, M., Smith, E. B. and Wakeham, W. A. (1981) *Intermolecular Forces*, Oxford University Press, Oxford.
Mandziuk, M. and Bačić, Z. (1993) *J. Chem. Phys.*, **98**, 7165.
Mandziuk, M. and Bačić, Z. (1994a) *J. Chem. Phys.*, **101**, 2126.
Mandziuk, M. and Bačić, Z. (1994b) *Faraday Disc. Chem. Soc.*, **97**, 265.
Manolopoulos, D. E. (1986) *J. Chem. Phys.*, **85**, 6425.
Mantz, A. W., Maillard, J.-P., Roh, W. B. and Rao, K. N. (1975) *J. Mol. Spectrosc.*, **57**, 155.
Marshall, M. D. Jensen, P. and Bunker, P. R. (1991) *Chem. Phys. Lett.*, **176**, 255.
McKellar, A. R. W. (1988) *J. Chem. Phys.*, **88**, 4190.
McQuarrie, D. (1976) *Statistical Mechanics*, Harper & Row, New York.
Meath, W. J. and Koulis, J. (1991) *J. Mol. Struct. (Theochem)*, **226**, 1.
Meerts, W. L., Majewski, W. A. and Van Herpen, W. M. (1984) *Can. J. Phys.*, **62**, 1293.
Meijer, G., private communication.
Messiah, A. (1969) *Quantum Mechanics*, North-Holland, Amsterdam.
Miller, R. E. (1988) *Science*, **240**, 447.
Millot, C. and Stone, A. J. (1992) *Mol. Phys.*, **77**, 439.
Mirsky, K. (1978) *Computing in Cristallography*, Chapter 5, page 167, Delft University Press.
Mladenović, M. and Bačić, Z. (1991) *J. Chem. Phys.*, **94**, 4988.
Moszynski, R., Jeziorski, B., Wormer, P. E. S. and Van der Avoird, A. (1994a) *Chem. Phys. Lett.*, **221**, 161.
Moszynski, R., Wormer, P. E. S., Jeziorski, B. and Van der Avoird, A. (1994b) *J. Chem. Phys.*, **101**, 2811.
Moszynski, R., Jeziorski, B., Van der Avoird, A. and Wormer, P. E. S. (1994c) *J. Chem. Phys.*, **101**, 2825.
Moszynski, R., Korona, T., Wormer, P. E. S. and Van der Avoird, A. (1995a) *J. Chem. Phys.*, **103**, 321.
Moszynski, R., Wormer, P. E. S. and Van der Avoird, A. (1995b) *J. Chem. Phys.*, **102**, 8385.
Muckerman, J. T. (1990) *Chem. Phys. Lett.*, **173**, 200.
Nelson, D. D., Fraser, G. T and Klemperer, W. (1985) *J. Chem. Phys.*, **83**, 6201.
Nelson, D. D., Fraser, G. T. and Klemperer, W. (1987a) *Science*, **238**, 1670.
Nelson, D. D., Klemperer, W., Fraser, G. T., Lovas, F. J. and Suenram, R. D. (1987b) *J. Chem. Phys.*, **87**, 6364.
Nesbitt, D. J. (1988) *Chem. Rev.*, **88**, 843.
Novick, S. E., Leopold, K. R. and Klemperer, W. (1990) in *Atomic and Molecular Clusters*, edited by Bernstein, E. R., Elsevier, Amsterdam.
Olthof, E. H. T., Van der Avoird, A. and Wormer, P. E. S. (1994a) *J. Chem. Phys.*, **101**, 8430.
Olthof, E. H. T., Van der Avoird, A. and Wormer, P. E. S. (1994b) *J. Mol. Struct. (Theochem.)*, **307**, 201.
Olthof, E. H. T., Van der Avoird, A., Wormer, P. E. S., Loeser, J. and Saykally, R. J. (1994c) *J. Chem. Phys.*, **101**, 8443.
Olthof, E. H. T., Van der Avoird, A., Wormer, P. E. S., Liu, K. and Saykally, R. J. (1996a) *J. Chem. Phys.*, **105**, 8051.
Olthof, E. H. T., Van der Avoird, A. and Wormer, P. E. S. (1996b) *J. Chem. Phys.*, **104**, 832.
Pack, R. T. (1974) *J. Chem. Phys.*, **60**, 633.
Pack, R. T. (1978) *Chem. Phys. Lett.* **55**, 197.

Pang, L. and Brisse, F. (1993) *J. Phys. Chem.*, **97**, 8562.
Papoušek, D. and Aliev, M. W. (1982) *Molecular Vibration-Rotational Spectra*, Elsevier, Amsterdam.
Parish, C. A., Augspurger, J. D. and Dykstra, C. E. (1992) *J. Phys. Chem.*, **96**, 2069.
Parlett, B. N. (1980) *The Symmetric Eigenvalue Problem*, Prentice Hall, Englewood Cliffs, NJ.
Peet, A. C. and Yang, W. (1989a) *J. Chem. Phys.*, **91**, 6598.
Peet, A. C. and Yang, W. (1989b) *J. Chem. Phys.*, **90**, 1746.
Pine, A. S. and Howard, B. J. (1986) *J. Chem. Phys.*, **84**, 590.
Podolsky, B. (1928) *Phys. Rev.*, **32**, 812.
Poll, J. D. and Hunt, J. L. (1976) *Can. J. Phys.*, **54**, 461.
Pugliano, N., Cruzan, J. D., Loeser, J. G. and Saykally, R. J. (1993) *J. Chem. Phys.*, **98**, 6600.
Quack, M. and Suhm, M. A. (1991) *J. Chem. Phys.*, **95**, 28.
Quack, M., Stohner, J. and Suhm, M. A. (1993) *J. Mol. Struct.*, **294**, 33.
Quack, M. and Suhm, M. A. (1995) *Chem. Phys. Lett.*, **234**, 71.
Riedle, E. and Van der Avoird, A. (1996) *J. Chem. Phys.*, **104**, 882.
Rick, S. W., Lynch, D. L. and Doll, J. D. (1991) *J. Chem. Phys.*, **95**, 3505.
Rijks, W. and Wormer, P. E. S. (1988) *J. Chem. Phys.*, **88**, 5704.
Rodwell, W. R., Sim Fai Lam, L. T. and Watts, R. O. (1981) *Mol. Phys.*, **44**, 225.
Rybak, S., Jeziorski, B. and Szalewicz, K. (1992) *J. Chem. Phys.*, **95**, 6576.
Sabo, D., Bačić, Z., Bürgi, T. and Leutwyler, S. (1995) *Chem. Phys. Lett.*, **244**, 283.
Sagarik, K. P., Ahlrichs, R. and Brode, S. (1986) *Mol. Phys.*, **57**, 1247.
Saunders, M., Jimenez-Vazquez, H. A., Cross, R. J., Mroczkowski, S., Freedberg, D. I. and Anet, F. A. L. (1994) *Nature*, **367**, 256.
Saykally, R. J. (1989) *Acc. Chem. Res.*, **22**, 295.
Saykally, R. J. and Blake, G. A. (1993) *Science*, **259**, 1570.
Scheiner, S. (1994) *Annu. Rev. Phys. Chem.*, **45**, 23.
Schinke, R. and Diercksen, G. H. F. (1985) *J. Chem. Phys.*, **83**, 4516.
Schmuttenmaer, C. A., Cohen, R. C. and Saykally, R. J. (1994) *J. Chem. Phys.*, **101**, 146.
Shapiro, M. and Balint-Kurti, G. G. (1979) *J. Chem. Phys.*, **71**, 1461.
Slee, T. and Le Roy, R. J. (1993) *J. Chem. Phys.*, **99**, 360.
Stoer, J. and Bulirsch, R. (1980) *Introduction to Numerical Analysis*, Springer-Verlag, New York.
Stone, A. J. and Price, S. L. (1988) *J. Phys. Chem.*, **92**, 3325.
Szalewicz, K., Cole, S. J., Kolos, W. and Bartlett, R. J. (1988) *J. Chem. Phys.*, **89**, 3662.
Talman, J. D. (1968) *Special Functions*, Benjamin, New York.
Tang, K. T. and Toennies, J. P. (1984) *J. Chem. Phys.*, **80**, 3726.
Tao, F.-M. and Klemperer, W. (1993) *J. Chem. Phys.*, **99**, 5976.
Tao, F.-M., Drucker, S., Cohen, R. C. and Klemperer, W. (1994) *J. Chem. Phys.*, **101**, 8680.
Tennyson, J. and Sutcliffe, B. T. (1982) *J. Chem. Phys.*, **77**, 4061; (1983) *J. Chem. Phys.*, **79**, 42.
Tennyson, J. and Sutcliffe, B. T. (1983) *J. Chem. Phys.*, **79**, 43.
Thomas, L. D., Kraemer, W. P. and Diercksen, G. H. F. (1980) *Chem. Phys*, **51**, 131.
Tucker, S. C. and Truhlar, D. G. (1986) *J. Chem. Phys.*, **86**, 6251.
Van Bladel, J. W. I., Van der Avoird, A. and Wormer, P. E. S. (1991) *J. Phys. Chem.*, **95**, 5414.
Van Bladel, J. W. I., Van der Avoird, A., Wormer, P. E. S. and Saykally, R. J. (1992a) *J. Chem. Phys.*, **97**, 4750.
Van Bladel, J. W. I., Van der Avoird, A., and Wormer, P. E. S. (1992b) *Chem. Phys.*, **165**, 47.
Van der Avoird, A. (1993) *J. Chem. Phys.*, **98**, 5327.
Van der Avoird, A., Wormer, P. E. S., Mulder, F. and Berns, R. M. (1980) *Topics in Current Chemistry*, **93**, 1.
Van der Avoird, A., Wormer, P. E. S. and Moszynski, R. (1994a) *Chem. Rev.*, **94**, 1931.
Van der Avoird, A., Olthof, E. H. T. and Wormer, P. E. S. (1994b) *Faraday Disc. Chem. Soc.*, **97**, 43.
Van der Avoird, A., Olthof, E. H. T. and Wormer, P. E. S. (1996) *J. Chem. Phys.*, **105**, 8034.
Van Duijneveldt-Van de Rijdt, J. G. C. M. and Van Duijneveldt, F. B. (1992) *J. Chem. Phys.*, **97**, 5019.
Varberg, T. D. and Evenson, K. M. (1992) *Astrophys. J.*, **385**, 763.
Wan, Z., Christian, J. F. and Anderson, S. L. (1992) *J. Chem. Phys.*, **96**, 3344.
Wang, L. S., Alford, J. M., Chai, Y., Diener, M., Zhang, J., McClure, S. M., Guo, T., Scuseria, G. E. and Smalley, R. E. (1993) *Chem. Phys. Lett.*, **207**, 354.

Wang, F., McCourt, F. R. W. and Le Roy, R. J. (1996) *Mol. Phys.*, in press.
Weiske, T., Hrušác, J., Böhme, D. K. and Schwartz, H. (1991) *Chem. Phys. Lett.*, **186**, 459.
Williams, H. L., Szalewicz, K., Jeziorski, B., Moszynski, R. and Rybak, S. (1993a) *J. Chem. Phys.*, **98**, 1279.
Williams, C. I., Whitehead, M. A. and Pang, L. (1993b) *J. Phys. Chem.*, **97**, 11652.
Wilson, E. B., Decius, J. C. and Cross, P. C. (1955) *Molecular Vibrations*, McGraw-Hill, New York.
Wormer, P. E. S. and Hettema, H. (1992a) *J. Chem. Phys.*, **97**, 5592.
Wormer, P. E. S. and Hettema, H. (1992b) *Polcor Package*, Nijmegen.
Wu, Q., Zhang, D. H. and Zang, J. Z. H. (1995) *J. Chem. Phys.*, **103**, 2548.
Xantheas, S. S. and Sutcliffe, B. T. (1995) *J. Chem. Phys.*, **103**, 8022.
Xu Y., Jäger, W., Djauhari, J. and Gerry, C. L. (1995) *J. Chem. Phys.*, **103**, 2827.
Yang, W., Peet, A. C. and Miller, W. H. (1989) *J. Chem. Phys.*, **91**, 7537.
Zang, D. H., Wu, Q., Zang, J. Z. H., Von Dirke, M. and Bačić, Z. (1995) *J. Chem. Phys.*, **102**, 2315.
Zwart, E., Ter Meulen, J. J. and Meerts, W. L. (1991) *J. Mol. Spectrosc.*, **147**, 27.
Zolothoukina, T. N. and Kotake, S. (1993) *J. Chem. Phys.*, **99**, 2855.

Section II

APPLICATIONS

5 The Quest for Reliability in Calculated Properties of Hydrogen-bonded Complexes

J. E. DEL BENE

Youngstown State University, USA

I. SHAVITT

The Ohio State University, USA

5.1 INTRODUCTION

It has long been recognized that an atom of hydrogen which is covalently bonded to atom A in one molecule can interact with a region of high electron density in another molecule. If the electron density in the second molecule is associated with atom B, then atoms A and B will approach closer than expected from the sum of their van der Waals radii as the A—H...B hydrogen bond is formed. Although the hydrogen bond is the strongest type of interaction between a pair of neutral covalent molecules, in an absolute sense it is relatively weak, ranging from about 2 to 10 kcal/mol, and can be easily formed or broken. The importance of hydrogen bonds cannot be overestimated, since they are ubiquitous in nature, being found in systems ranging from ice to the molecules of life itself. They are responsible for the unusual properties of water, are mediators of chemical reactions, provide for drug–molecule interactions in the body, and are important in the structure of DNA. Since the publication of the book *The Hydrogen Bond* by Pimentel and McClellan (1960), there have appeared in the literature myriad papers on various aspects of hydrogen bonding. These papers have addressed hydrogen bonding both experimentally and theoretically, and have shaped the present-day view of this important intermolecular interaction (Murthy and Rao, 1970; Schadd, 1974; Kollman and Allen, 1972; Ratner and Sabin, 1973; Schuster *et al.*, 1974; Allen, 1975; Kollman, 1977; Ratajczak and Orville-Thomas, 1980; Huyskens *et al.*, 1991; Jeffrey and Saenger, 1991; Scheiner, 1994; *Molecular Interactions*, 1994; *Horizons in Hydrogen Bonding*, 1994; *Hydrogen Bonding*, 1994). Understanding in detail the nature and consequences of hydrogen bonding is a continuing challenge to both chemical theorists and experimentalists.

Ab initio quantum-chemical calculations on a pair of hydrogen-bonded molecules date back about 30 years. These early studies, which were carried out at the single-determinant Hartree–Fock level with small basis sets, demonstrated that *ab initio* quantum-chemical calculations can provide important information about and insight into the nature of the hydrogen bond (Kollman and Allen, 1972). Although the level of theory used for these studies would be judged to be totally inadequate today, they did provide a first-principles basis for a description of the hydrogen bond. For example, studies of the water dimer found that this complex has an open structure of C_s symmetry with an essentially linear O—H...O hydrogen bond, in agreement with experimental data (Dyke and Muenter, 1973; Dyke *et al.*, 1977; Odutola *et al.*, 1979). However, as the number and sophistication of theoretical studies increased, it became evident that the detailed description of the structure and binding energy of this dimer is very dependent on the theoretical method used for the calculation. This is apparent from Table 5.1, which reports the computed

Table 5.1 Water dimer intermolecular distance R (Å) and electronic binding energy ΔE_e (kcal/mol) at various levels of theory[a]

Method	Basis set	R	ΔE_e
HF	STO–3G	2.740	−5.9
	3–21G	2.797	−11.0
	6–31G(d)	2.971	−5.6
	6–31G(d,p)	2.980	−5.5
	DZP	2.986	−5.0
	6–31+G(d)	2.964	−5.4
	6–31++G(d)	2.959	−5.4
	6–31+G(d,p)	2.988	−5.0
	6–31++G(d,p)	2.987	−5.0
	6–311+G(d,p)	2.999	−4.8
	6–311++G(2d,2p)	3.035	−4.1
	11+7+(2d,2p)	3.033	−3.9
	6–311++G(3df,3pd)	3.026	−4.0
MP2	3–21G	2.802	−12.7
	6–31G(d)	2.913	−7.4
	6–31G(d,p)	2.910	−7.1
	DZP	2.909	−6.3
	6–31+G(d)	2.901	−7.1
	6–31++G(d)	2.895	−7.2
	6–31+G(d, p)	2.914	−6.5
	6–311+G(d,p)	2.908	−6.1
	6–311++G(d,p)	2.910	−6.1
	6–311++G(2d,2p)	2.911	−5.4
CISD	6–31G(d)	2.937	
	6–31+G(d)	2.920	

[a] Data taken from Frisch *et al.* (1986). Binding energies were computed at the same level of theory as used for structure optimization.

intermolecular O—O distance and the electronic binding energy of the water dimer at various levels of theory (Frisch *et al.*, 1986). Computed intermolecular distances were found to be sometimes too short and sometimes too long, and binding energies could be significantly in error.

Over the last decade dramatic progress in *ab initio* computational quantum chemistry has been made, owing to the development of new theoretical models, improvements in algorithms for carrying out *ab initio* calculations, and the advent of larger and faster computing machines. These improvements have allowed high-level *ab initio* calculations to be performed on relatively small hydrogen-bonded complexes so that questions concerning the methodological dependence of computed results can be addressed. Such studies provide the data needed to answer the following very important questions:

1. Has the value of a computed property of a hydrogen-bonded complex converged with respect to further extension of the methodology?
2. At what level of theory does convergence occur, that is, at what level is reliability obtained at minimal computational expense?
3. How do properties computed at this level compare with reliable experimental data?

Because of the complexity and magnitude of the task, in practice it may not yet be possible to demonstrate unequivocally convergence of all computed properties of hydrogen bonded complexes as a function of methodology. However, progress has been made, and studies aimed at accomplishing this goal are ongoing. The importance of this task cannot be overemphasized, since the results of such studies define the minimum level of theory required for reliability, and provide benchmark data for comparisons with calculations on larger systems carried out at lower levels of theory.

In evaluating the reliability of theoretical models, we have chosen to concentrate on three key properties of hydrogen-bonded complexes. In our view, these must be well described by theory if the results of the calculations are to be used with confidence to interpret experimental data, and if the computed results are to have predictive value. These are also properties which can be measured experimentally, so that comparisons between computed and experimental data may be made. These properties are:

1. The A—B intermolecular distance (hydrogen bond distance) between a pair of molecules containing an A—H...B hydrogen bond
2. The binding enthalpy of the hydrogen-bonded complex
3. The frequency shift of the A—H stretching vibration in the infrared (IR) spectrum of a complex which is a consequence of hydrogen bonding

5.2 METHODOLOGICAL CONSIDERATIONS

5.2.1 Basis sets

It has been well documented that there is a significant basis set and correlation energy dependence of computed properties of hydrogen-bonded complexes (Frisch *et al.*, 1986; Kollman and Rothenberg, 1977; Latajka and Scheiner, 1984, 1987; Newton and Kestner, 1983; Kestner *et al.*, 1983; Schwenke and Truhlar, 1985; Gaw *et al.*, 1984; Del Bene, 1987, 1988, 1989; Backskay, 1992; Del Bene and Shavitt, 1989, 1994a; Chakravorty and

Davidson, 1993; Nowek and Leszczynski, 1996). Minimal basis sets, split-valence basis sets, split-valence sets augmented with polarization and diffuse functions have all been used for the calculation of structures, vibrational spectra, and binding energies. The weight of the evidence suggests that basis sets which are not at least augmented split-valence basis sets are inadequate. In this chapter we will limit our critique of basis sets to polarization-augmented and diffuse-augmented split-valence basis sets. We will examine the results of calculations performed using the split-valence basis sets developed by Pople and coworkers in the 1970s (Harriharan and Pople, 1973; Dill and Pople, 1975; Spitznagel et al., 1982; Clark et al., 1983; Krishnan et al., 1980) and the more recent correlation-consistent polarized split-valence basis sets developed by Dunning and coworkers (Dunning, 1989; Kendall et al., 1992; Woon and Dunning, 1993). The Pople basis sets (Harriharan and Pople, 1973; Dill and Pople, 1975; Spitznagel et al., 1982; Clark et al., 1983; Krishnan et al., 1980) are perhaps the most widely used basis sets for computing the structures and vibrational spectra of molecules. Such calculations are often done at the single-determinant Hartree–Fock (HF) level of theory, with the split-valence plus polarization (on nonhydrogen atoms) 6–31G(d) basis set. The many successes enjoyed at the HF/6–31G(d) level for monomers, and the ease and relatively low cost of optimizing hydrogen-bonded structures at this level of theory using the Gaussian system of programs (Frisch et al., 1995), have made HF/6–31G(d) a popular choice. The valence double-split 6–31G and triple-split 6–311G basis sets were developed at a time when integral evaluation was the bottleneck of quantum chemical calculations. Hence, in these basis sets a given set of s and p orbitals share the same exponent. These basis sets have been augmented with diffuse functions and polarization functions, although there is no set formula for how such augmentation should be done. Among the more widely used augmented 6–31G basis sets are 6–31G(d) [previously designated 6–31G*] with d polarization functions on nonhydrogen atoms, 6–31G(d, p) [6–31G**] with first polarization functions on all atoms, 6–31+G(d, p) with diffuse s and p functions added to nonhydrogen atoms, 6–31++G(d, p) with a diffuse s function also added to hydrogen atoms, and 6–31G($2df, 2pd$) with two sets of first polarization functions and one set of second polarization functions on all atoms. Various other combinations of diffuse and polarization functions are possible. The same types of extensions made for the double-split 6–31G basis set can be made for the valence triple-split 6–311G basis set.

Recently, Dunning and coworkers (Dunning, 1989; Kendall et al., 1992; Woon and Dunning, 1993) have published a series of basis sets which were optimized for calculations carried out with electron correlation effects included. In this series of basis sets both the valence space and the polarization space of the atoms are systematically expanded as the basis set size increases. These basis sets are known as correlation-consistent polarized split-valence basis sets (cc-pVXZ), where X = D for double-, T for triple-, Q for quadruple-, and 5 for quintuple-split. The smallest of these sets, cc-pVDZ, is a valence double-split basis set with first-polarization d functions on nonhydrogen atoms and p functions on hydrogen. The next set, cc-pVTZ, is a valence triple-split basis with two sets of d functions and a set of f functions on nonhydrogen atoms, and two sets of p functions and a set of d functions on hydrogen. The pattern continues as the valence space is further split and functions with higher angular momentum quantum numbers are added to the polarization space. These basis sets have also been augmented with diffuse functions (aug-cc-pVXZ), with one set of diffuse functions added for each value of the quantum number l which appears in the cc-pVXZ basis. For example, aug-cc-pVTZ has diffuse s,

p, d, and f functions on nonhydrogen atoms, and diffuse s, p, and d functions on hydrogens. These basis sets are particularly well-suited for convergence studies because of the very systematic way in which they have been constructed.

5.2.2 Wavefunction Model

The first *ab initio* calculations on small hydrogen-bonded complexes and many present-day calculations on large systems have been carried out at the single-determinant HF level of theory. With a sufficiently large basis set, it is possible to approach the Hartree–Fock limit for the binding energy of a hydrogen-bonded complex. At the Hartree–Fock level dispersion effects are neglected, and independent of the basis set used, quantitatively accurate binding energies cannot be obtained. In those cases in which computed Hartree–Fock binding energies are in agreement with experimental binding energies, a fortuitous cancellation of errors has occurred due to limitations in the basis set and wave function model. To account for dispersion effects, it is necessary to go beyond Hartree–Fock to a level of theory with includes electron correlation.

Describing electron correlation effects is one of the challenges of *ab initio* computational quantum chemistry today. In principle, the problem can be solved by doing a full configuration interaction (CI) treatment (Shavitt, 1977) with a large basis set. Unfortunately, such calculations are not feasible for any but very small molecular systems. However, there are different electron correlation methods which aim to approximate the full CI results, with the following among the most widely used:

1. Configuration interaction with all single and double excitations (CISD) (Shavitt, 1977)
2. Perturbation theory, such as many-body Møller–Plesset perturbation theory (MBPT or MP) truncated at some order (Pople *et al.*, 1976; Krishnan and Pople, 1978; Bartlett and Silver, 1975, 1976; Bartlett and Purvis, 1978)
3. Coupled-cluster (CC) methods (Purvis and Bartlett, 1982; Urban *et al.*, 1985; Bartlett *et al.*, 1990; Raghavachari *et al.*, 1989)

The principal deficiency of the CISD model is its lack of size consistency. As defined by Pople *et al.* (1976), a computational model is size consistent if the energy computed by that model for a system made up of widely separated (i.e. noninteracting) subsystems is equal to the sum of the energies computed separately by the same model for the subsystems. Lack of size consistency is a particularly serious problem in the calculation of small interaction energies, such as hydrogen bond energies. The Møller–Plesset model is size consistent at each order, has been highly successful, and is the most widely used method. However, it is not variational, and in certain cases may not be sufficiently converged even at fourth order. The infinite-order coupled-cluster method with noniterative inclusion of triple excitations [CCSD(T)] may be the method of choice, but it is the most computationally expensive of those methods which are feasible for hydrogen-bonded systems. The performance of these methods will be examined below.

5.3 STRUCTURES AND VIBRATIONAL SPECTRA OF HYDROGEN-BONDED COMPLEXES

Early studies of hydrogen bonding employed experimental geometries. This, of course, is a severe limitation, since such geometries are not available for most complexes, particularly in the gas phase. The advent of gradient optimization computational techniques and the implementation of analytical first and second derivatives in quantum chemistry software codes have had a dramatic impact on this situation. Fully optimized geometries can be computed routinely for hydrogen-bonded complexes. The quality of these structures can be judged internally in terms of convergence with respect to further extension of the methodology, and externally by comparison with reliable experimental gas-phase structures when these are available. Once reliability has been established, *ab initio* calculations can be used with confidence to predict the structures of hydrogen-bonded complexes which are not known experimentally.

After an optimized structure has been obtained at a particular level of theory, it should be characterized as an equilibrium structure (a local minimum) on the potential energy surface with no imaginary frequencies, or a saddle point with one or more imaginary frequencies. This is done by computing the vibrational frequencies at the same level of theory, usually within the harmonic approximation. The computed frequencies and IR intensities produce a computed vibrational spectrum which can be compared (within the limitations of the harmonic approximation) with the experimental spectrum, and also provide the data needed to evaluate the zero-point and thermal vibrational contributions to the binding enthalpy of the complex. The shift to lower frequency of the A—H stretching band and the increase in intensity of this band upon complex formation are well-known experimentally as the IR spectroscopic signatures of hydrogen bonding (Pimentel and McClellan, 1960). The ability of theory to predict the intermolecular distance and these vibrational spectral properties will now be addressed.

5.3.1 Intermolecular Distances

Perhaps the most important structural feature of a hydrogen-bonded complex is the intermolecular, or hydrogen bond A—B distance. This distance is a structural feature which is amenable to experimental measurement. Table 5.1 shows significant variations among computed O—O distances in the water dimer, ranging from 2.740 Å at HF/STO-3G to 3.035 Å at HF/6−311++G(2d,2p). The experimental O—O distance, R_0, in the water dimer is 2.976 Å. Corrected for anharmonicity, the intermolecular distance, R_e, was estimated to be 2.946 Å (Dyke and Muenter, 1973; Dyke *et al.*, 1977; Odutola *et al.*, 1979). Closer analysis of the data of Table 5.1 shows that water dimer structures optimized at the Hartree–Fock level with all the augmented split-valence basis sets (i.e. all basis sets except STO-3G and 3-21G) have intermolecular distances which are overestimated by 0.013–0.089 Å. At the Hartree–Fock level the best agreement between theory and experiment is obtained at HF/6−31++G(d) with $R = 2.959$ Å. Unfortunately, increasing the basis set size in the 6−31G and 6−311G series of basis sets does not necessarily lead to better agreement between theory and experiment. For example, at HF/6−311++G(2d,2p) R_e is 3.035 Å. This, of course, is an unsatisfactory situation.

Including electron correlation effects in *ab initio* calculations of the structure of the water dimer reduces the intermolecular O—O distance. It is interesting to note that the range of MP2 distances computed with augmented split-valence basis sets (all except 3-21G) is smaller than the range at Hartree–Fock (0.076 Å at HF versus 0.028 Å at MP2). However, at MP2 the intermolecular O—O distance appears to be underestimated by 0.023–0.051 Å relative to experiment.

Recently Xantheas (private communication) has undertaken a very systematic study of the structure of the water dimer using various wavefunction models and large augmented split-valence basis sets. This study is the most comprehensive of its kind, and may be viewed as providing benchmark results for the water dimer. Table 5.2 reports values of the computed intermolecular O—O distance (Xantheas, private communication; Feller *et al.*, 1994). At the Hartree–Fock level the O—O distance in the water dimer is predicted to be about 3.03 Å, which is significantly longer than the experimental distance. Calculations which include electron correlation effects yield significantly shorter O—O distances relative to Hartree–Fock. At correlated levels of theory, increasing the basis set size from aug-cc-pVDZ to aug-cc-pVTZ leads to a small decrease in the computed O—O distance of about 0.01 Å at MP2(fc), MP4(fc), CCSD and CCSD(T), and a larger decrease of about 0.02 Å at MP2(fu) and MP4(fu). (The notation (fc) indicates that electrons below the valence shell have been frozen in the Hartree–Fock MOs for the correlation calculations, while (fu) indicates that all electrons have been correlated.) Distance changes upon further extension of the basis set are negligible. The most extensive electron correlation treatment (CCSD(T) with the aug-cc-pVQZ′ basis set, where aug-cc-pVQZ′ is the aug-cc-pVQZ basis without *f* functions on hydrogens) leads to a value of the intermolecular distance of 2.913 Å. Thus, the computed distances appear to converge to about 2.91 Å, with all the calculated distances being within 0.02 Å of this value. The computed results suggest that the often-quoted experimental intermolecular O—O distance of 2.946 Å in the water dimer may be too long, and should be reinvestigated.

The levels of structure optimization carried out by Xantheas for the water dimer are quite computationally demanding, and at this time are not feasible for routine studies of larger hydrogen-bonded complexes. Is there a lower level of theory which can consistently predict reasonable intermolecular distances and vibrational frequency shifts, and which can be applied routinely to larger complexes? An examination of Table 5.1 indicates that for the water dimer, MP2 calculations with several of the augmented basis sets derived from 6-31G (the smallest being 6-31G(d)) yield intermolecular distances in the range 2.90–2.91 Å. Will calculations at the MP2 level on other hydrogen-bonded complexes exhibit such a small dependence on basis set selection?

Table 5.2 The intermolecular O—O distance (Å) in the water dimer obtained from recent high-level *ab initio* calculations[a]

Basis set	HF	MP2(fc)	MP2(fu)	CCSD	CCSD(T)	MP4(fc)	MP4(fu)
aug-cc-pVDZ	3.030	2.916	2.911	2.944	2.925	2.921	2.917
aug-cc-pVTZ	3.039	2.907	2.889	2.933	2.914	2.912	2.892
aug-cc-pVQZ′	3.037	2.907		2.932	2.913	2.912	
aug-cc-pVQZ	3.036	2.903					
aug-cc-pV5Z		2.905					

[a] Some HF, MP2(fc) and MP4(fc) data are taken from Feller *et al.* (1994). The remainder are from Xantheas (private communication).

To answer this question, calculations have been carried out on a set of nine small hydrogen-bonded complexes which contain A—H...B hydrogen bonds, and for which some experimental data are available (Legon and Millen, 1986, 1988, 1992; Wofford et al., 1986, 1987; Legon et al., 1981, 1987; Barnes, 1983; Ballard and Henderson, 1991; Kyro et al., 1983; Jucks and Miller, 1987; Dyke et al., 1969; Quack and Suhm, 1991; Pine and Lafferty, 1983; Engdahl and Nelander, 1989). Table 5.3 presents computed intermolecular distances and A—H vibrational frequency shifts (Del Bene et al., 1995a) for these complexes computed at Hartree–Fock and at MP2 using two different basis sets: 6–31G(d) and 6–31+G(d,p). Several observations about trends in computed intermolecular distances can be made from these data:

1. As found for the water dimer, intermolecular distances computed at the Hartree–Fock level of theory can be significantly different from experimental distances. Hartree–Fock distances are usually too long, often by more than 0.1 Å. Overestimation of the intermolecular distance in hydrogen-bonded complexes has severe consequences relative to the nature of the intermolecular hydrogen-bonding surface. This will be discussed further below.
2. The equilibrium HF/6–31G(d) and MP2/6–31G(d) structures of (HF)$_2$ are predicted to be cyclic, in contrast to the open structure found experimentally and computed at higher levels of theory.
3. In contrast to the water dimer, in which distances computed at MP2 with various augmented split-valence basis sets are similar, intermolecular distances computed at MP2/6–31G(d) and MP2/6–31+G(d,p) may be different. When significant differences are found, the MP2/6–31+G(d,p) distances are in better agreement with experimental data.

It is appropriate at this time to elaborate on the consequences of the overestimation of intermolecular distances at the Hartree–Fock level of theory, as noted in point 1 above. At these longer distances, Hartree–Fock potential curves for proton transfer often exhibit two distinct minima, corresponding to traditional A—H...B hydrogen bonds and to hydrogen-bonded A$^-$...$^+$H—B ion pairs. Because early studies which examined proton transfer in hydrogen-bonded complexes between elements of the second period were carried out at the Hartree–Fock level of theory, models for proton transfer were developed which include double minima along the proton transfer coordinate (Murthy and Rao, 1970; Schadd, 1974; Kollman and Allen, 1972; Ratner and Sabin, 1973; Schuster et al., 1974; Allen, 1975; Kollman, 1977; Ratajczak and Orville-Thomas, 1980; Huyskens et al., 1991; Jeffrey and Saenger, 1991; Scheiner, 1994; *Molecular Interactions*, 1994; *Horizons in Hydrogen Bonding*, 1994; *Hydrogen Bonding* 1994; Somerjai and Hornig, 1962; Ceulemans, 1991). These models have been used in discussions of proton-transfer in various chemical and biological systems (Cleland and Kreevoy, 1994). However, more recent calculations which include electron correlation effects at the MP2 level indicate that only a single minimum exists along the proton transfer coordinate (Latajka et al., 1984, 1987, 1992; Del Bene et al., 1995b, 1996). These same MP2 calculations can also reproduce the complex IR spectral patterns found experimentally in some hydrogen-bonded complexes, without invoking splittings due to the existence of double minima (Del Bene et al., 1996; Person et al., in preparation). It would appear therefore, that overestimation of the intermolecular distance at Hartree–Fock leads to distortion of the potential energy surface along the proton transfer coordinate. Models for proton transfer based on Hartree–Fock potentials need to be reexamined.

The weight of the evidence suggests that Hartree–Fock calculations should not be used if reliable structures and potential energy surfaces for hydrogen-bonded systems are needed. The inclusion of electron correlation effects, at least at the level of MP2, is absolutely essential. It also appears that diffuse functions must be included in the basis set, so that MP2/6–31+G(d,p) would appear to be the minimum level of theory required for reliability. The data of Table 5.3 suggest that, at this level, computed intermolecular distances agree with experimental distances to about 0.03 Å. Whether this is acceptable will depend upon the particular application.

5.3.2 A—H Vibrational Frequency Shifts in Complexes with A—H...B Hydrogen Bonds

It is well known that intermolecular vibrational frequencies are usually only slightly perturbed by hydrogen bonding. The exception, of course, is the shift of the A—H stretching band to lower frequency upon complex formation. This shift is so dramatic that it is a spectroscopic signature of hydrogen bonding. If theory is to be used to understand and characterize the spectra of hydrogen-bonded complexes, it is important that theory be able to reproduce this shift. Table 5.3 provides computed and experimental shifts of the A—H band in some hydrogen-bonded complexes. Once again, the Hartree–Fock level of theory is found to be unreliable, often significantly underestimating this shift. This occurs because Hartree–Fock bond-stretching potentials are too steep, leading to an underestimation of the lengthening of the A—H bond and of the shift of the A—H band in the complex. Potentials are softer at MP2, with the result that the computed A—H shift at MP2 with a given basis set is greater than the shift at HF. When significant differences occur between computed MP2/6–31G(d) and MP2/6–31+G(d,p) frequency shifts, it is the MP2/6–31+G(d,p) shift which is in better agreement with experimental data. The calculation of the vibrational frequencies for hydrogen-bonded complexes, even within the harmonic approximation (see discussion below), is quite demanding of computer resources, and there is a scarcity of results obtained at higher levels of theory for comparison. Additional studies are needed in this area. However, it appears that MP2/6–31+G(d,p) intermolecular distances and A—H frequency shifts are reasonable. MP2/6–31+G(d,p) is therefore recommended as the minimum level of theory required for reliable calculation of the structures of hydrogen-bonded complexes and the shift of the A—H stretching band in complexes with A—H...B hydrogen bonds.

5.4 BINDING ENERGIES AND ENTHALPIES

The binding enthalpy (ΔH^T) of a hydrogen-bonded complex is the enthalpy of the reaction for the formation of the complex from the isolated monomers:

$$\Delta H^T = \Delta E_e^0 + \Delta E_v^0 + \Delta(\Delta E_v)^T + \Delta E_r^T + \Delta E_t^T + \Delta nRT$$

The term ΔE_e^0 is the electronic energy difference between the complex and the isolated

Table 5.3 Intermolecular distances R (Å) and A—H vibrational frequency shifts $\delta\nu$ (cm^{-1}) in hydrogen-bonded complexes[a]

Complex	HF/6−31G(d)	HF/6−31+G(d,p)	MP2/6−31G(d)	MP2/6−31+G(d,p)	Expt.[b]
FH...NCH					
R	2.923	2.900	2.875	2.808	2.805[c]
$\delta\nu$	106	176	119	262	(v)245[d]
FH...NCCH$_3$					
R	2.878	2.850	2.835	2.758	
$\delta\nu$	146	234	162	343	(v)334[d,e]
ClH...NCH					
R	3.498	3.522	3.380	3.376	3.402[f]
$\delta\nu$	73	68	126	113	(v)79[g]
ClH...NCCH$_3$					
R	3.434	3.462	3.324	3.316	3.291[f]
$\delta\nu$	110	103	173	167	(v)155[h]
FH...CO					
R	3.145	3.198	3.080	3.062	3.047[i]
$\delta\nu$	37	75	55	134	(v)117[j,k]
ClH...OH$_2$					
R	3.242	3.288	3.152	3.185	3.215[l]
$\delta\nu$	163	121	276	177	(Ar)207[g]

ClH...ClH					
R	4.132	4.139	3.913	3.868	(Ar)53[g]
$\delta\nu$	13	14	28	32	
FH...FH					
R	2.596[m]	2.811	2.535[m]	2.777	2.79[n]
					2.72 ± 0.03[o]
$\delta\nu$		89		116	(v)93[p]
HOH...OH$_2$					
R	2.972	2.989	2.914	2.914	2.946[q]
$\delta\nu$	42	50	67	79	(Ar)64[r]

[a] Data for complexes of HF and HCl with HCN and CH$_3$CN taken from Del Bene (1992). MP2/6−31+G(d,p) data for other complexes from Del Bene et al. (1995a).
[b] Experimental vibrational data are shifts to lower frequencies from either vapor (v) or argon matrix (Ar) data.
[c] Legon and Millen (1986).
[d] Wooford et al. (1986, 1987).
[e] Legon et al. (1987).
[f] Legon and Millen (1988).
[g] Barnes (1983).
[h] Ballard and Henderson (1991).
[i] Legon et al. (1981).
[j] Kyro et al. (1983).
[k] Jucks and Miller (1987).
[l] Legon and Millen (1992).
[m] The computed structure is cyclic with a short F—F distance.
[n] Dyke et al. (1969).
[o] Quack and Suhm (1991).
[p] Pine and Lafferty (1983).
[q] Dyke and Muenter (1973); Dyke et al. (1977); Odutola et al. (1979).
[r] Engdahl and Nelander (1989).

monomers, evaluated at a particular level of theory. Computed electronic binding energies show a dependence of the level of theory employed for the calculations, that is, they depend on both basis set and wavefunction choices, as evident from Table 5.1 for the water dimer.

The second term in the enthalpy expression is the zero-point vibrational energy contribution to the binding enthalpy. In the formation of a complex from two nonlinear molecules, three degrees of rotational and three degrees of translational freedom are lost, and these appear as low-frequency vibrational modes in the complex, often referred to as 'dimer modes'. The hydrogen-bonded complex may have as many as six more vibrational modes than the sum of the vibrational modes of the isolated monomers, and experience shows that ΔE_v^0 has a destabilizing effect on the complex. Usually, the vibrational modes of the monomers are changed little by hydrogen bond formation between neutral molecules, except for a decrease in the stretching frequency of the A—H bond. The remaining vibrational term $\Delta(\Delta E_v)^T$ is the change in the vibrational energies in going from 0 K to some higher temperature. This term also has a destabilizing effect since the low-frequency dimer modes are populated at higher temperatures. The vibrational modes of monomers and complexes are usually computed within the harmonic approximation, and this is another source of error in the evaluation of the enthalpy of formation of the hydrogen-bonded complex. The remaining terms which contribute to ΔH^T are thermal terms which account for the loss of rotational and translational degrees of freedom, and the change in the number of moles of gas (ΔE_r, ΔE_t, and ΔnRT, respectively). These are usually evaluated classically.

Because the electronic binding energy ΔE_e^0 is usually the dominant term in the enthalpy expression, it is imperative that this energy be computed at a sufficiently high level of theory to yield enthalpies which are consistent and in reasonable agreement with available experimental values. As noted above, many studies have shown the dependence of calculated binding energies of complexes on the basis set and the wavefunction model used for the calculations (Frisch et al., 1986; Kollman and Rothenberg, 1977; Latajka and Scheiner, 1984, 1987; Newton and Kestner, 1983; Kestner et al., 1983; Schwenke and Truhlar, 1985; Gaw et al., 1984; Del Bene, 1987, 1988, 1989; Backskay, 1992; Del Bene and Shavitt, 1989, 1994a; Chakravorty and Davidson, 1993; Nowek and Leszczynski, 1996). In the following sections, these dependencies will be examined. The aim is to identify that level of theory which yields reliable energies at minimal computational expense.

5.4.1 Basis Set Effects

Binding enthalpies of hydrogen-bonded complexes calculated from binding energies ($D_e = \Delta E_e^0$) which have been obtained with correlated methods and a polarized split-valence basis set of moderate size [i.e. 6−31G(d,p) and 6−311G(d,p)] are generally too large. This has been attributed to two factors:

1. These basis sets are too contracted around the nucleus, and therefore do not describe adequately the spatial extent of lone pairs of electrons at proton-acceptor sites.
2. The use of an incomplete basis set leads to increased stabilization of one monomer due to the presence of basis functions on the other. This nonphysical stabilization is known as the basis set superposition error (BSSE), and a so-called counterpoise

correction has been proposed to compensate for this error (Xanatheas, 1996; Boys and Bernardi, 1970; Newton and Kestner, 1983; Kestner et al., 1983; Schwenke and Truhlar, 1985; Somasundram et al., 1986; Latajka and Scheiner, 1989; van Lenthe et al., 1987; Bouteiller and Behrouz, 1992; Saebo et al., 1993; Davidson and Chakravorty, 1994; van Duijneveldt et al., 1994).

Studies which have addressed the question of the basis-set dependence of computed binding energies have shown that the presence of diffuse functions in the basis set significantly reduces computed binding energies (Del Bene, 1987, 1988, 1989). This is illustrated by the MP4 energies presented in Table 5.4 for $(NH_3)_2$, $(H_2O)_2$, and $(HF)_2$ using augmented 6–31G and 6–311G basis sets. For example, the presence of diffuse functions can lower the binding energy of $(HF)_2$ by as much as 3 kcal/mol out of a total binding energy of about 7.5 kcal/mol. Moreover, the inclusion of diffuse functions in these basis sets appears to increase in importance as the electronegativity of the proton acceptor atom of the dimer increases.

Figure 5.1 shows plots of counterpoise-corrected and uncorrected MP4 binding energies for $(HF)_2$, $(H_2O)_2$, and FH...NCH (Del Bene, 1992) computed with the Dunning correlation-consistent basis sets. These calculations were done before diffuse functions had been published for these sets, so diffuse s and p functions only were added to nonhydrogen atoms, with exponents taken from the 6−31+G(d,p) basis set. These modified sets are denoted cc−pVXZ+. (It had previously been demonstrated that diffuse functions on hydrogen atoms do not have a significant effect on binding energies (Del Bene, 1987, 1988, 1989).) Once again, these data illustrate that diffuse functions lower computed binding energies, and also significantly reduce the basis set superposition error, as estimated by the counterpoise correction (Boys and Bernardi, 1970). For example, the counterpoise corrections for the FH...NCH complex are 1.8, 1.1, and 0.5 kcal/mol at MP4/cc-pVDZ, cc-pVTZ, and cc-pVQZ′, respectively. (cc-pVQZ′ is cc-pVQZ without g functions on nonhydrogen atoms and f functions on hydrogen.) These corrections are reduced to 1.1, 0.6, and 0.3 kcal/mol at cc-pVDZ+, cc-pVTZ+, and cc-pVQZ′+, respectively. Figure 5.1 also suggests that computed binding energies appear to converge with increasing basis set size, with satisfactory convergence occurring at cc-pVTZ+. Moreover, uncorrected cc-pVTZ+ binding energies lie closer to the interval between corrected and uncorrected cc-pVQZ′+ binding energies than the counterpoise-corrected

Table 5.4 MP2 electronic binding energies (kcal/mol) computed from polarization-augmented 6-31G and 6-311G basis sets with and without diffuse functions[a,b]

Basis set	$(HF)_2$	$(H_2O)_2$	$(NH_3)_2$
6−31G(d,p)	−7.45/−4.69	−6.88/−6.16	−4.12/−3.76
6−31G($2d,2p$)	−7.71/−4.61	−7.10/−5.25	−4.10/−3.12
6−311G(d,p)	−6.22/−4.40	−6.97/−5.80	−4.23/−3.50
6−311G($2d,2p$)	−6.56/−4.66	−7.12/−5.33	−4.51/−3.11

[a] Data from Del Bene (1987, 1988, 1989). These energies were obtained from single-point calculations carried out at optimized HF/6−31G(d) geometries.
[b] Energies are given as X/Y:
X = the binding energy with the listed basis set
Y = the binding energy with the same basis set augmented with diffuse s and p functions on nonhydrogen atoms.

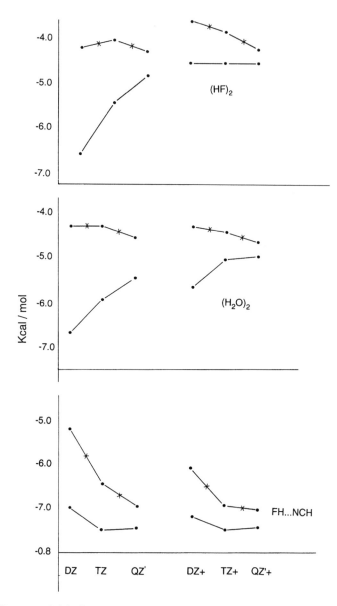

Figure 5.1 Computed binding energies and counterpoise-corrected binding energies for $(HF)_2$, $(H_2O)_2$, and FH...NCH, from Del Bene (1992). (– – – – –) MP4 energies, (– –X– –) counterpoise-corrected MP4 energies

cc-pVTZ+ energies. These observations led to the recommendation that the cc-pVTZ+ basis set without the counterpoise correction should be used to calculate reliable hydrogen bond energies at minimal computational expense. Subsequent studies with the fully augmented basis sets (aug-cc-pVXZ) and with these basis sets with diffuse functions on nonhydrogen atoms only (aug'-cc-pVXZ) confirmed these earlier observations. (Del Bene and Shavitt, 1989, 1994a; Feller, 1992; Xantheas and Dunning, 1993; Feyereisen et al., 1996).

Feyereisen et al. (1996) have recently reported an exhaustive study of the MP2 binding energy of the water dimer using the Dunning basis sets, with the aim of establishing the basis set limit. Some of their results are given in Table 5.5. They found that the computed MP2 binding energy of $(H_2O)_2$ appears to converge to -4.9 ± 0.1 kcal/mol. (Recently, Xantheas, 1996, has reported a complete basis set (CBS) limit of the MP2 binding energy of -4.94 kcal/mol.) The data of Table 5.5 show that the counterpoise-corrected binding energies approach this limit from below, that is, they underestimate the binding, while the uncorrected energies approach from above. It is significant that the aug-cc-pVTZ binding energy without the counterpoise correction is closer to this limit than the counterpoise-corrected aug-cc-pVTZ energy, and also closer than the corrected and uncorrected cc-pVQZ energies, despite the fact that cc-pVQZ is a larger basis set. The aug-cc-pVXZ binding energies without the counterpoise correction show the smallest variation with basis set size, with the MP2/aug-cc-pVDZ energy being within 0.2 kcal/mol of the basis set limit.

Single-point calculations on hydrogen-bonded complexes of moderate size at correlated levels of theory may still be computationally quite challenging with the aug'-cc-pVTZ basis set. There are two alternatives which appear to give reasonable hydrogen bond energies for neutral complexes, and which are more feasible (Del Bene and Shavitt, 1989, 1994a). The first is to remove the diffuse d and f functions from the aug'-cc-pVTZ basis set. Full MP4 calculations on a set of neutral hydrogen-bonded complexes including $(NH_3)_2$, $(H_2O)_2$, $(HF)_2$, and FH...NCH show that omitting diffuse f functions reduces computed binding energies by no more than 0.1 kcal/mol, while omitting both diffuse d and f functions changes binding energies by no more than 0.2 kcal/mol. However, without the diffuse d and f functions, the computed binding energies may be larger or smaller than the aug'-cc-pVTZ energies. The second option is to use the aug'-cc-pVDZ basis set for studies of neutral hydrogen-bonded complexes. For the same set of four dimers, MP4/aug'-cc-pVDZ binding energies differ from MP4/aug'-cc-pVTZ energies by no more than 0.2 kcal/mol, but again the aug'-cc-pVDZ binding energies may be larger or smaller than the aug'-cc-pVTZ energies. Both these options appear viable, but both should be further investigated in other hydrogen-bonded complexes.

Table 5.5 MP2 binding energies (kcal/mol) for the water dimer

Basis set	ΔE^a	Counterpoise-corrected ΔE^a	ΔE^b
cc-pVDZ	-7.29	-3.94	
cc-pVTZ	-6.03	-4.40	
cc-pVQZ	-5.45	-4.67	
aug-cc-pVDZ	-5.16	-4.32	-4.43
aug-cc-pVTZ	-5.10	-4.64	-4.71
aug-cc-pVQZ	-5.03	-4.79	-4.84
aug-cc-pV5Z			-4.88

[a]Feyereisen et al. (1996).
[b]Counterpoise-corrected binding energies including fragment relaxation, from Xantheas (1996).

5.4.2 Electron Correlation Effects

As noted above, the calculation of quantitatively accurate hydrogen bond energies requires the inclusion of the effects of electron correlation. While there are many approaches for treating electron correlation, three will be discussed here: configuration interaction with all single and double excitations (CISD); many-body Møller–Plesset perturbation theory truncated at some order (MPx); and coupled-cluster theory with singles and doubles and noniterative inclusion of triple excitations [CCSD(T)].

1. *CISD.* Although configuration interaction is a popular electron correlation method, its lack of size-consistency is a serious drawback. (Chakravorty and Davidson, 1993; Gaw *et al.*, 1984; Alrichs *et al.*, 1969). Size consistency of the computational method is always important in the calculation of binding energies, but it is of paramount importance for weakly bound hydrogen-bonded complexes because small binding energies can be overshadowed by errors due to the lack of size-consistency. For example, neutral hydrogen-bonded complexes such as $(H_2O)_2$, $(HF)_2$, and $(HCl)_2$ are predicted to be unbound at CISD relative to the isolated monomers (Del Bene and Shavitt, 1989, 1994a). The size-consistency error arises because wave function terms in which both monomers are doubly excited are not included in the CISD calculation on the complex. Such terms, referred to as disconnected cluster contributions, can only be accounted for in CI by including higher-order (i.e. quadruple) excitations in the CI expansion. Corrections for the size-consistency error have been developed (Langhoff and Davidson, 1974; Bartlett and Shavitt, 1977, 1978; Siegbahn, 1978; Pople *et al.*, 1977). Application of modified Davidson corrections (Del Bene and Shavitt, 1989, 1994a; Langhoff and Davidson, 1974; Bartlett and Shavitt, 1977, 1978; Siegbahn, 1978) or of the Pople correction (Pople *et al.*, 1977) can lead to binding of the above dimers, but the binding energies are still underestimated relative to size-consistent results.

   Another means of reducing the size-consistency error is to evaluate the CISD binding energy in the supermolecule approach. (Del Bene and Shavitt, 1989, 1994a). In this approach the binding energy is obtained as the difference between the total energy of the complex minus the energy of a supermolecule composed of the two optimized monomer units separated by a very large distance (e.g. 500 Å). The supermolecule approach works quite well for the above complexes, and in combination with the Davidson or Pople corrections, is particularly effective for overcoming size-consistency errors associated with truncation of the CI expansion. However, there is a computational price, that of an additional calculation on the complex at the supermolecule distance. While this does not double the computational effort, it is still more expensive than other correlation methods which produce essentially the same binding energies. It is not the best method for routine calculations of hydrogen bond energies.

2. *MPx* Møller–Plesset many-body perturbation theory is a size-consistent method, although it is not variational. It is the most widely used method (particularly at the MP2 level) for computing correlated binding energies of hydrogen-bonded complexes. Its popularity is due to several factors:

   - It is size-consistent.
   - It leads to binding energies for neutral complexes which are usually within a few

tenths of a kcal/mol of binding energies computed at higher levels of theory (i.e. MP4 and CCSD(T)).
- Its implementation using the DIRECT algorithm in the Gaussian suite of programs (Frisch *et al.*, 1995) makes it generally available and the least expensive method computationally.

Investigations of the correlation energy dependence of binding energies of complexes of first- and second-row hydrides AH_n (A = N, O, F, P, S, and Cl) with H_2O and NH_3 using the 6−31+G(d,p), 6−31++G(d,p) and 6−31+G($2d,2p$) basis sets have been carried out (Del Bene, 1987, 1988, 1989). These studies show that the second-order contribution to the correlation energy increases the stabilities of all complexes, and is significantly larger than any other correlation term in the MP expansion. Small correlation energy contributions which are often of opposite sign arise at third and fourth order for these complexes. As a result, MP2 and MP4 binding energies differ by no more than 0.2 kcal/mol, except for ClH...NH_3, where the difference is 0.5 kcal/mol. These studies also demonstrate that unless the full MP4 binding energy, including triples (MP4SDTQ), is computed, going beyond MP2 to MP3 or MP4SDQ is of little value, since the binding energies at these levels are further from MP4 binding energies than are the MP2 energies.

Xantheas (private communication) has examined the binding energy of the water dimer using the aug-cc-pVDZ, aug-cc-pVTZ, and aug-cc-pVQZ basis sets at MP2(fc), MP4(fc), and CCSD(T) levels of theory. MP4 and CCSD(T) binding energies computed with a particular basis set were found to agree with each other to 0.01 kcal/mol, while MP2 energies differed from MP4 and CCSD(T) energies obtained with the same basis set by no more than 0.04 kcal/mol. In another study, MP2/cc-pVTZ+ binding energies for three complexes of HF with CH_3Cl were found to agree with MP4/cc-pVTZ+ binding energies to 0.1 kcal/mol. (Del Bene and Shavitt, 1994b). Finally, differences of 0.1 and 0.2 kcal/mol were found between the MP2/cc-pVTZ+ binding energy of FH...NCH and MP4/cc-pVTZ+ and QCISD(T)/cc-pVTZ+ energies, respectively (Del Bene, 1992). QCISD(T) is an approximate coupled-cluster correlation method including the effects of triple excitations (Pople *et al.*, 1987). Thus, second-order many-body Møller–Plesset perturbation theory appears to be the method of choice for neutral hydrogen-bonded complexes, giving reasonable binding energies at minimal computational expense.

It is important to note, however, that for certain molecular complexes the MP expansion may not be sufficiently converged at either MP2 or MP4. This was observed for complexes with N_2O as the proton-acceptor molecule, with hydrogen bonding occurring at either N or O (Del Bene and Frisch, 1989; Del Bene *et al.*, 1990). Because of the severe oscillation of the dipole moment of N_2O in the perturbation series, Møller–Plesset theory fails to provide a reliable description of the association of this molecule with HF and various other acids, including H^+ and Li^+. The preferred site of hydrogen bonding at MP2 and MP4 for the HF:N_2O complex is in disagreement with all other correlated wavefunction models, including the approximate coupled-cluster QCISD(T) method. (MP2 and MP4 also predict the wrong protonation site when compared with infinite-order correlation methods, and yield proton affinities which have not converged and do not agree with experimental data.) Thus, while MP2 may be the wavefunction model of choice for the routine calculation of the binding energies of neutral hydrogen-bonded complexes, care must be taken when applying this method in unusual cases, as exemplified by the complexes of N_2O with HF.

3. *Coupled-cluster methods.* The methods of coupled clusters with single- and double-excitation terms (CCSD), CCSD with noniterative triples [CCSD(T)] (Purvis and Bartlett, 1982; Urban *et al.*, 1985; Bartlett *et al.*, 1990; Raghavachari *et al.*, 1989) and the closely related approximate method of quadratic CI with noniterative triples (QCISD(T)) (Pople *et al.*, 1987) have been used to compute the hydrogen bond energies of a few complexes. These are size-consistent infinite-order methods and, as such, are deemed to be the most reliable. The inclusion of triple excitations at MP4 (full MP4SDTQ) was found to be important previously (Del Bene, 1987, 1988, 1989). Triples are also important in infinite order methods, as evident from the data of Xantheas (private communication) on the water dimer. Computed CCSD binding energies with the aug-cc-pVDZ, aug-cc-pVTZ, and aug-cc-pVQZ basis sets were found to differ from the corresponding CCSD(T) binding energies by about 0.3 kcal/mol. CCSD(T)/aug'-cc-pVTZ binding energies were also found to yield binding enthalpies in good agreement with experimental data not only for neutral complexes but also for positive- and negative-ion hydrogen bonded complexes (Del Bene and Shavitt, 1989, 1994a). It would appear that CCSD(T) is the correlation method of choice for consistency and reliability. Unfortunately, it is computationally the most expensive method, and, as such, is not feasible for routine use on larger systems. However, it is the correlation method which should be used for benchmarking purposes in small systems, and the method of choice for difficult cases.

5.4.3 Comparisons with Experimental Data

The first requirement for reliability of a theoretical model for describing a particular property of a hydrogen-bonded complex is internal convergence, that is, convergence of the computed property with respect to further extension of the methodology. Once a sufficienctly converged level of theory has been identified, then the model can be tested externally by comparison with reliable experimental data. Table 5.6 presents experimental (Curtiss *et al.*, 1979; Bohac *et al.*, 1992; Quack and Suhm, 1995; Wooford *et al.*, 1987; Legon *et al.*, 1987; Mettee, private communication) and computed binding energies and enthalpies, reported as D_e, D_0 or ΔH^T. D_0 is the binding enthalpy at 0 K which is the sum of the electronic binding energy (D_e) and the zero-point vibrational contribution. ΔH^T is the binding enthalpy at some temperature T.

The most extensively studied neutral hydrogen-bonded complexes are $(H_2O)_2$ and $(HF)_2$, and high-level theoretical studies of these complexes have appeared recently (Xantheas, private communication; Feller *et al.*, 1994; Feller, 1992, 1993; Xantheas and Dunning, 1993; Feyereisen *et al.*, 1996; van Duijneveldt-van de Rijdt and van Duijneveldt, 1992; Kim *et al.*, 1992; Saebo *et al.*, 1993; Chakravorty and Davidson, 1993; Racine and Davidson, 1993; Kim and Jordan, 1994; Collins *et al.*, 1995; Peterson and Dunning, 1995). Feyereisen *et al.* (1996) obtained the binding energy of $(H_2O)_2$ at correlated levels of theory by extrapolating to the complete basis set limit using a sequence of aug-cc-pVXZ basis sets. They estimated a binding enthalpy at 375 K of -3.2 ± 0.1 kcal/mol, which is at the lower end of the range of the experimental binding enthalpy of -3.6 ± 0.5 kcal/mol (Curtiss *et al.*, 1979). Peterson and Dunning (1995)

Table 5.6 Computed and experimental binding enthalpies (kcal/mol) of selected neutral hydrogen-bonded complexes

Complex	Energy reported	Method	Computed energy[a]	Experimental
$(H_2O)_2$	ΔH^{375}	MP2/Extrapolated	-3.2 ± 0.1[b]	-3.6 ± 0.5[c]
		MP2/aug'-cc-pVTZ	-3.3	
$(HF)_2$	D_0	CCSD(T)/TZ2P(f, d)	-2.94[d]	-3.04[e,f]
	D_e	CCSD(T)/TZ2P(f, d)	-4.73[d]	-4.47[f]
	D_e	CCSD(T)/CBS limit	-4.60[g]	
		MP2/aug'-cc-pVTZ	-4.5	
FH...NCH	D_0	MP2/aug'-cc-pVTZ	-5.7[h]	-5.0[i]
FH...NCCH$_3$	D_0	MP2/aug'-cc-pVTZ	-7.4[h]	-6.2[j]
ClH...NCH	ΔH^{298}	MP2/aug'-cc-pVTZ	-4.1[h]	-4.4[k]
ClH...NCCH$_3$	ΔH^{298}	MP2/aug'-cc-pVTZ	-5.3[h]	-5.7[k]

[a] MP2/aug'-cc-pVTZ//MP2/6−31+G(d,p).
[b] Feyereisen *et al.* (1996).
[c] Curtiss *et al.* (1979).
[d] Collins *et al.* (1995).
[e] Bohac *et al.* (1992).
[f] Quack and Suhm (1995).
[g] Peterson and Dunning (1995).
[h] Del Bene (1992).
[i] Wooford *et al.* (1987).
[j] Legon *et al.* (1987).
[k] Mettee (private communication).

have investigated $(HF)_2$ using different correlation treatments and aug-cc-pVXZ basis sets. Their best estimate of the CCSD(T) complete basis set limit binding energy (D_e) is -4.60 kcal/mol, in good agreement with the experimental estimate of -4.47 kcal/mol (Quack and Suhm, 1995). Collins *et al.* (1995) have also studied $(HF)_2$ and obtained a D_0 value of -2.94 kcal/mol based on CCSD(T)/TZ2P(d, f) electronic energies. This value is in agreement with the experimentally determined D_0 of -3.04 kcal/mol (Bohac *et al.*, 1992; Quack and Suhm, 1995). Their computed value for D_e of -4.73 kcal/mol is slightly greater than that of Peterson and Dunning (1995), but still in agreement with experiment (Quack and Suhm, 1995). Since low-temperature experiments measure D_0 directly and estimate D_e, while theoretical studies compute D_e and estimate D_0 from harmonic frequencies, exact agreement between theoretical and experimental D_e and D_0 values should not be expected. However, the high-level theoretical results quoted for $(H_2O)_2$ and $(HF)_2$ demonstrate that even within these limitations, theory can predict reliable binding energies. Such theoretical results also serve as benchmarks with which calculations carried out at lower levels of theory should be compared. Since experimental binding enthalpies are also difficult to measure, high-level theoretical enthalpies can be used to resolve discrepancies between conflicting experimental data, and to challenge experimental results in those cases for which discrepancies exist between theory and experiment.

For comparative purposes, Table 5.6 provides binding enthalpies for $(H_2O)_2$ and $(HF)_2$ computed at MP2/aug'-cc-pVTZ, the minimum level of theory recommended for the calculation of binding energies. The MP2/aug'-cc-pVTZ binding energies for $(H_2O)_2$ and $(HF)_2$ are in good agreement with the benchmark results. The data of Table 5.6 also show good agreement between computed and experimental binding enthalpies (ΔH^{298}) for complexes of HCl with HCN and CH$_3$CN. Experimental D_0 values are available for FH...NCH and FH...NCCH$_3$. Differences between MP2/aug'-cc-pVTZ and experimental D_0 values of 0.7 and 1.2 kcal/mol are found for these two complexes, respectively. The large difference of 1.2 kcal/mol for the binding enthalpy of FH...NCCH$_3$ suggest that the experimental value should be reevaluated.

5.5 CONCLUSIONS

Over the past thirty years dramatic progress has been made in *ab initio* studies of hydrogen-bonded complexes. Through the dedicated efforts of many research groups, a prescription has been written for obtaining from first principles the structures and binding energies of hydrogen-bonded complexes with an accuracy comparable to that which can be measured experimentally. Indeed, theory is now in a position to challenge experimental results when there is disagreement between high-level *ab initio* theoretical results and either gas-phase or low-temperature matrix-isolation experimental data. The major problem which remains is a practical one, namely, the application of these methods to larger systems, including those of biological interest. Here, the recommended levels of theory may not be feasible today, but the definition of feasibility continuously changes as new computer architectures and new algorithms are developed. Judging from the past, the future appears bright indeed.

A problem which still remains in the quest for a full first-principles description of hydrogen-bonded complexes is related to the inability to compute anharmonic frequencies routinely. Almost all the studies of hydrogen bonding in the literature report spectral data and zero-point and thermal vibrational energy contributions to binding enthalpies based on harmonic frequencies. While it is often stated that anharmonicity, particularly of the dimer modes and the proton-stretching mode, is important, a satisfactory treatment of these modes which has general applicability has not been forthcoming. Despite this, comparisons of computed binding enthalpies based on high-level electronic binding energies and harmonic vibrational energies obtained at correlated levels of theory with appropriate basis sets are often in good agreement with experimental data. Moreover, computed vibrational spectra based on harmonic frequencies continue to be useful for assigning bands in the experimental vibrational spectra of hydrogen-bonded complexes. There have been some attempts to examine anharmonicity effects (Botschwina, 1987, 1988; Botschwina *et al.*, 1988; Bouteiller *et al.*, 1991; Bacskay and Craw, 1994; Ojamäe *et al.*, 1995) and coupling between vibrational modes in hydrogen-bonded complexes (Del Bene *et al.*, 1995b, 1996; Person *et al.*, in preparation; Ojamäe *et al.*, 1995) but considerable work remains to be done. It is not unrealistic to expect that in the not too distant future dramatic progress will be made in this area as well.

ACKNOWLEDGMENTS

The research reported in this chapter was supported by grants of computer time from the Ohio Supercomputer Center and by a grant to JEDB from the National Science Foundation (CHE-9505888). The authors would also like to thank Dr Sotiris Xantheas for providing his data on the water dimer prior to publication.

REFERENCES

Allen, L. C. (1975) *J. Am. Chem. Soc.*, **97**, 6921.
Alrichs, R., Scharf, P. and Ehrhardt, C. (1969) *J. Chem. Phys.*, **14**, 35.
Bacskay, G. B. (1992) *Mol. Phys.*, **77**, 61.
Bacskay, G. B. and Craw, J. S. (1994) *Chem. Phys. Lett.*, **221**, 167.
Ballard, L. and Henderson, G. (1991) *J. Phys. Chem.*, **95**, 660.
Barnes, A. J. (1983) *J. Mol. Struct.*, **100**, 259.
Bartlett, R. J. and Purvis, G. D. (1978) *Int. J. Quantum Chem.*, **14**, 561.
Bartlett, R. J. and Shavitt, I. (1978) *Int. J. Quantum Chem., Quantum Chem. Symp.*, **12**, 543.
Bartlett, R. J. and Silver, D. M. (1975) *J. Chem. Phys.*, **62**, 3258.
Bartlett, R. J. and Silver, D. M. (1976) *J. Chem. Phys.*, **64**, 4578.
Bartlett, R. J. and Shavitt, I. (1977) *Int. J. Quantum Chem., Quantum Chem. Symp.*, **11**, 165.
Bartlett, R. J., Watts, J. D., Kucharski, S. A. and Noga, J. (1990) *Chem. Phys. Lett.*, **165**, 513.
Bohac, E. J., Marshall, M. D. and Miller, R. E. (1992) *J. Chem. Phys.*, **96**, 6681.
Botschwina, P. (1987) in *Structure and Dynamics of Weakly Bound Molecular Complexes*, edited by Weber, D., Reidel, Dordrecht.
Botschwina, P. (1988) *J. Chem. Soc. Faraday Trans.*, **2**, 84, 1263.
Botschwina, P., Sebald, P. and Burmeister, R. (1988) *J. Chem. Phys.*, **88**, 5246.
Bouteiller, Y. and Behrouz, H. (1992) *J. Chem. Phys.*, **96**, 6033.
Bouteiller, Y., Latajka, Z., Ratajczak, H. and Scheiner, S. (1991) *J. Chem. Phys.*, **94**, 2956.
Boys, S. F. and Bernardi, F. (1970) *Mol. Phys.*, **19**, 553.
Ceulemans, A. (1991) in *Intermolecular Forces. An Introduction to Modern Methods and Results*, edited by Huyskens, D. P. L., Luck, W. A. and Zeegers-Huyskens, T., Springer-Verlag, Berlin.
Chakravorty, S. J. and Davidson, E. R. (1993) *J. Phys. Chem.*, **97**, 6373.
Clark, T., Chandrasekhar, J., Spitznagel, G. W. and Schleyer, P. v. R. (1983) *J. Comp. Chem.*, **4**, 294.
Cleland, W. W. and Kreevoy, M. M. (1994) *Science*, **264**, 1887.
Collins, C. L., Morihashi, K., Yamaguchi, Y. and Schaefer, H. F. III (1995) *J. Chem. Phys.*, **103**, 6051.
Curtiss, L. A., Frurip, D. J. and Blander, M. (1979) *J. Chem. Phys.*, **71**, 2703.
Davidson, E. R. and Chakravorty, S. J. (1994) *Chem. Phys. Lett.*, **217**, 48.
Del Bene, J. E. (1987) *J. Chem. Phys.*, **86**, 2110.
Del Bene, J. E. (1988) *J. Phys. Chem.*, **92**, 2874.
Del Bene, J. E. (1989) *J. Comp. Chem.*, **10**, 603.
Del Bene, J. E. (1992) *Int. J. Quantum Chem., Quantum Chem. Symp.*, **26**, 527.
Del Bene, J. E. and Frisch, M. J. (1989) *Int. J. Quantum Chem., Quantum Chem. Symp.*, **23**, 371.
Del Bene, J. E. and Shavitt, I. (1989) *Int. J. Quantum Chem., Quantum Chem. Symp.*, **23**, 445.
Del Bene, J. E. and Shavitt, I. (1994a) *J. Mol. Struct. (Theochem.)*, **307**, 27.
Del Bene, J. E. and Shavitt, I. (1994b) *J. Mol. Struct. (Theochem.)*, **314**, 9.
Del Bene, J. E., Stahlberg, E. and Shavitt, I. (1990) *Int. J. Quantum Chem., Quantum Chem. Symp.*, **24**, 445.
Del Bene, J. E., Person, W. B. and Szczepaniak, K. (1995a) *J. Phys. Chem.*, **99**, 10705.
Del Bene, J. E., Person, W. B. and Szczepaniak, K. (1995b) *Chem. Phys. Lett.*, **247**, 89.
Del Bene, J. E., Person, W. B. and Szczepaniak, K. (1996) *Mol. Phys.*, **89**, 47.
Dill, J. D. and Pople, J. A. (1975) *J. Chem. Phys.*, **62**, 2921.
Dunning, T. H. Jr (1989) *J. Chem. Phys.*, **90**, 1007.
Dyke, T. and Muenter, J. S. (1973) *J. Chem. Phys.*, **59**, 3125.
Dyke, T. R., Howard, B. J. and Klemperer, W. (1969) *J. Chem. Phys.*, **56**, 2442.
Dyke, T. R., Mack, K. M. and Muenter, J. S. (1977) *J. Chem. Phys.*, **66**, 498.
Engdahl, A. and Nelander, B. (1989) *J. Mol. Struct.*, **193**, 101.
Feller, D. (1992) *J. Chem. Phys.*, **96**, 6104.
Feller, D. (1993) *J. Chem. Phys.*, **98**, 7059.
Feller, D., Glendening, E. D., Kendall, R. A. and Peterson, K. A. (1994) *J. Chem. Phys.*, **100**, 4981.
Feyereisen, M. W., Feller, D. and Dixon, D. A. (1996) *J. Phys. Chem.*, **100**, 2993.
Frisch, M. J., Del Bene, J. E., Binkley, J. S. and Schaefer, H. F. III (1986) *J. Chem. Phys.*, **84**, 2279.

Frisch, M. J., Trucks, G. W., Schlegel, H. B., Gill, P. M. W., Johnson, B. G., Robb, M. A., Cheeseman, J. R., Keith, T. A., Petersson, G. A., Montgomery, J. A., Raghavachari, K., Al-Laham, M. A., Zakrzewski, V. G., Ortiz, J. V., Foresman, J. B., Cioslowski, J., Stevfanov, B. B., Nanayakkara, A., Challacombe, M., Peng, C. Y., Ayala, P. Y., Chen, W., Wong, M. W., Andres, J. L., Replogle, E. S., Gomperts, R., Martin, R. L., Fox, D. J., Binkley, J. S., DeFrees, D. J., Baker, J., Stewart, J. P., Head-Gordon, M., Gonzalez, C. and Pople, J. A. (1995) GAUSSIAN 94, Gaussian, Inc., Pittsburgh, PA.

Gaw, J. F., Yamaguchi, Y., Vincent, M. A. and Schaeffer, H. F. III (1984) *J. Am. Chem. Soc.*, **106**, 3133.

Harriharan, P. C. and Pople, J. A. (1973) *Theor. Chim. Acta.*, **28**, 213.

Horizons in Hydrogen Bonding (1994) *J. Mol. Struct. (Theochem.)*, **322**.

Huyskens, P. L., Luck, W. A. P. and Zeegers-Huyskens, T. (eds). *Intermolecular Forces. An Introduction to Modern Methods and Results.* (1991) Springer-Verlag, Berlin.

Hydrogen Bonding (1994) *J. Mol. Struct. (Theochem.)*, **314**.

Jeffrey, G. A. and Saenger, W. (1991) *Hydrogen Bonding in Biological Structures*, Springer-Verlag, Berlin.

Jucks, K. W. and Miller, R. E. (1987) *J. Chem. Phys.*, **86**, 6637.

Kendall, R. A., Dunning, T. H. Jr and Harrison, R. J. (1992) *J. Chem. Phys.*, **96**, 6796.

Kestner, N. R., Newton, M. D. and Mathers, T. L. (1983) *Int. J. Quantum Chem., Quantum Chem. Symp.*, **17**, 431.

Kim, K. and Jordan, K. D. (1994) *J. Phys. Chem.*, **98**, 10089.

Kim, K. S., Mhin, B. J., Choi, U.-C. and Lee, K. (1992) *J. Chem. Phys.*, **97**, 6649.

Kollman, P. A. (1977) in *Applications of Electronic Structure Theory*, edited by Schaefer, H. F. III, Plenum, New York.

Kollman, P. A. and Allen, L. C. (1972) *Chem. Rev.*, **72**, 283.

Kollman, P. and Rothenberg, S. (1977) *J. Am. Chem. Soc.*, **99**, 1333.

Krishnan, R. and Pople, J. A. (1978) *Int. J. Quantum Chem.*, **14**, 91.

Krishnan, R., Binkley, J. S., Seeger, R. and Pople, J. A. (1980) *J. Chem. Phys.*, **72**, 650.

Kyro, E. K., Shoja-Chaghervand, P., McMillan, K., Eliades, M., Danzeiser, D. and Bevan, J. W. (1983) *J. Chem. Phys.*, **79**, 28.

Langhoff, S. R. and Davidson, E. R. (1974) *Int. J. Quantum Chem.*, **8**, 61.

Latajka, Z. and Scheiner, S. (1984) *J. Chem. Phys.*, **81**, 407.

Latajka, Z. and Scheiner, S. (1987) *J. Comp. Chem.*, **8**, 663.

Latajka, Z. and Scheiner, S. (1989) *Theochem.*, **58**, 9.

Latajka, Z., Scheiner, S. and Ratajczak, H. (1987) *Chem. Phys. Lett.*, **135**, 367.

Latajka, Z., Scheiner, S. and Ratajczak, H. (1992) *Chem. Phys.*, **166**, 85.

Latajka, Z., Sukai, S., Morokuma, K. and Ratajczak, H. (1984) *Chem. Phys. Lett.*, **100**, 464.

Legon, A. C. and Millen, D. J. (1986) *Chem. Rev.*, **86**, 635.

Legon, A. C. and Millen, D. J. (1988) *Proc. Roy. Soc.*, **A417**, 21.

Legon, A. C. and Millen, D. J. (1992) *Chem. Soc. Rev.*, **21**, 71.

Legon, A. C., Soper, P. D. and Flygare, W. H. (1981) *J. Chem. Phys.*, **74**, 4944.

Legon, A. C., Millen, D. J. and North, H. M. (1987) *J. Chem. Phys.*, **86**, 2530.

Mettee, H. D. private communication.

Molecular Interactions (1994) *J. Mol. Struct. (Theochem.)*, **307**.

Murthy, A. S. N. and Rao, C. N. R. (1970) *J. Mol. Struct.*, **6**, 253.

Newton, M. D. and Kestner, N. R. (1983) *Chem. Phys. Lett.*, **94**, 198.

Nowek, A. and Leszczynski, J. (1996) *J. Chem. Phys.*, **104**, 1441.

Odutola, J. A., Viswanathan, R. and Dyke, T. R. (1979) *J. Am. Chem. Soc.*, **101**, 4787.

Ojamäe, L., Shavitt, I. and Singer, S. H. (1995) *Int. J. Quantum Chem., Quantum Chem. Symp.*, **29**, 657.

Person, W. B., Szczepaniak, K. and Del Bene, J. E. (in preparation).

Peterson, K. A. and Dunning, T. H. Jr (1995) *J. Chem. Phys.*, **102**, 2032.

Pimentel, G. C. and McClellan, A. L. (1960) *The Hydrogen Bond*, W. H. Freeman, San Francisco.

Pine, A. S. and Lafferty, W. J. (1983) *J. Chem. Phys.*, **78**, 2154.

Pople, J. A., Binkley, J. S. and Seeger, R. (1976) *Int. J. Quantum Chem., Quantum Chem. Symp.*, **10**, 1.

Pople, J. A., Seeger, R. and Krishnan, R. (1977) *Int. J. Quantum Chem., Quantum Chem. Symp.*, **11**, 149.
Pople, J. A., Head-Gordon, M. and Raghavachari, K. (1987) *J. Chem. Phys.*, **87**, 5968.
Purvis, G. D. III and Bartlett, R. J. (1982) *J. Chem. Phys.*, **76**, 1910.
Quack, M. and Suhm, M. A. (1991) *J. Chem. Phys.*, **95**, 28.
Quack, M. and Suhm, M. A. (1995) *Chem. Phys. Lett.*, **234**, 71.
Racine, S. C. and Davidson, E. R. (1993) *J. Phys. Chem.*, **97**, 6367.
Raghavachari, K., Trucks, G. W., Pople, J. A. and Head-Gordon, M. (1989) *Chem. Phys. Lett.*, **157**, 479.
Ratajczak, H. and Orville-Thomas, W. J. (eds) (1980) *Molecular Interactions*. Vols 1–3, John Wiley, Chichester.
Ratner, M. A. and Sabin, J. R. (1973) in *Wave Mechanics—The First Fifty Years*, edited by Chissick, S. S., Price, W. C. and Ravendale, T., Butterworths, London.
Saebo, S., Tong, W. and Pulay, P. (1993) *J. Chem. Phys.*, **98**, 2170.
Schaad, L. J. (1974) in *Hydrogen Bonding*, edited by Joesten, M. D. and Schaad, L. J. Marcel Dekker, New York.
Scheiner, S. (1994) *Annu. Rev. Phys. Chem.*, **45**, 23.
Schuster, P., Zundel, G. and Sandorfy, C. (eds) (1974) *The Hydrogen Bond. Recent Developments in Theory and Experiment*, Vols 1–3, North-Holland, Amsterdam.
Schwenke, D. W. and Truhlar, D. G. (1985) *J. Chem. Phys.*, **82**, 2418.
Shavitt, I. (1977) in *Methods of Electronic Structure Theory*, edited by Schaefer, H. F., Plenum, New York.
Siegbahn, P. E. M. (1978) *Chem. Phys. Lett.*, **55**, 386.
Somasundram, K., Amos, R. D. and Handy, N. C. (1986) *Theor. Chim. Acta.*, **69**, 491.
Somerjai, R. L. and Hornig, D. J. (1962) *J. Chem. Phys.*, **36**, 1980.
Spitznagel, G. W., Clark, T., Chandrasekhar, J. and Schleyer, P. v. R. (1982) *J. Comp. Chem.*, **3**, 363.
Urban, M., Noga, J., Cole, S. J. and Bartlett, R. J. (1985) *J. Chem. Phys.*, **83**, 4041.
van Lenthe, J. H., van Duijneveldt-van Rijdt, J. G. C. M. and van-Duijneveldt, F. B. (1987) in *Ab Initio Methods in Quantum Chemistry*, Part 2, edited by Lawley, K. P., John Wiley, New York.
van Duijneveldt-van de Rijdt, J. G. C. M. and van-Duijneveldt, F. B. (1992) in *J. Chem. Phys.*, **97**, 5019.
van Duijneveldt, F. B., van Duijneveldt-van de Rijdt, J. G. C. M. and van Lenthe, J. H. (1994) *Chem. Rev.*, **94**, 1873.
Wooford, B. A., Lieb, S. G. and Bevan, W. J. (1987) *J. Chem. Phys.*, **87**, 4478.
Wooford, B. A., Bevan, J. W., Olson, W. B. and Lafferty, W. J. (1986) *Chem. Phys. Lett.*, **124**, 579.
Wooford, B. A., Eliades, M. E., Lieb, S. G. and Bevan, J. W. (1987) *J. Chem. Phys.*, **87**, 5674.
Woon, D. E. and Dunning, T. H. Jr (1993) *J. Chem. Phys.*, **98**, 1358.
Xantheas, S. S. *J. Chem. Phys.* (1996) *J. Chem. Phys.*, **104**, 8821.
Xantheas, S. S. (private communication; to be submitted for publication).
Xantheas, S. S. and Dunning, T. H. Jr (1993) *J. Chem. Phys.*, **99**, 8774.

6 *Ab Initio* Predictions of the Vibrational Spectra of Some Molecular Complexes: Comparison with Experiment

T. A. FORD

University of Natal, Durban, South Africa

6.1 INTRODUCTION

The introduction of the normal mode analysis package in the GAUSSIAN family of computer programs, first implemented in GAUSSIAN-80 (Binkley *et al.*, 1980), represented the first opportunity for carrying out automatic *ab initio* calculations of the vibrational properties of polyatomic molecules. Prior to that time, spectroscopists interested in computing vibrational spectra were obliged to go through a laborious process of determining the energy and the dipole moment components of a molecule in its equilibrium structure and in a number of distorted configurations simulating the normal modes of vibration. These energy and dipole moment data were then manipulated to yield the Cartesian force constants, by finite differences, and the infrared band intensities, through the atomic polar tensors (Person *et al.*, 1981; Steele *et al.* 1981; Chin and Person, 1984). The availability of the normal mode facility in GAUSSIAN-80, and later versions, has led to the accumulation of an extensive library of computed vibrational spectroscopic data, generated using a number of combinations of level of theory and basis set (Besler *et al.*, 1988; Thomas *et al.*, 1993). These options are stored internally in the later versions of GAUSSIAN, making vibrational property calculations a relatively routine procedure.

Having recognized the ease with which the vibrational spectra of simple monomeric species could now be predicted theoretically, spectroscopists became interested in the possibility of exploring the ways in which the spectra were perturbed as a result of the interaction of a pair of molecules in a 1:1 molecular complex. In particular, the effects on the spectra of a proton donor (e.g. the red shift of the AH stretching mode and the increase in its integrated intensity) due to the formation of a hydrogen-bonded complex, AH...B, as described by Pimentel and McClellan (1960), have received much attention, in attempts to reproduce the essential features of the spectra of complexes observed in the gas phase (Miller, 1988, 1990; Saykally, 1989; Howard and Brown, 1992), in cryogenic matrices (Andrews, 1988, 1990) and in liquefied noble gases (Tokhadze and Tkhorzhevskaya, 1992).

In this chapter we explore the results of some *ab initio* studies of the infrared and Raman spectra of a variety of complexes, some hydrogen bonded and some of the electron donor–acceptor (EDA) type. The differences in the modes of binding in these complexes are manifested in the different types of behaviour of the computed spectroscopic properties. These observations lead to the conclusion that such calculations may be of great value in analysing the spectra of associated species, and to the suggestion that, with the ready availability of *ab initio* computer packages, vibrational spectroscopists should nowadays consider performing *ab initio* calculations of the structures and spectra of their target molecules as a matter of routine.

6.2 COMPUTATIONAL DETAILS

All calculations described in this review were carried out using the GAUSSIAN-86 program (Frisch *et al.*, 1984) and later versions (Frisch *et al.*, 1988, 1990, 1992, 1995). Most computations were performed at the second order Møller-Plesset perturbation theory (MP2) level (Møller and Plesset, 1934), except those for the larger complexes, which were done at the restricted Hartree–Fock (RHF) level (Hehre *et al.*, 1986). The 6−31G** split valence polarized Gaussian basis set (Francl *et al.*, 1982) was used exclusively. Specific details of individual calculations are described in the cited references. In general, full geometry optimizations were carried out at the tightest level of convergence (VERYTIGHT, or TIGHT in some more troublesome cases), subject to the constraints of the point group symmetry of the particular complex or monomer. Subsequent relaxation of the symmetry constraints led in some cases (e.g. $NH_2OH.OH_2$, $CH_4.HF$ and $SiH_4.HF$) to a lower symmetry structure with improved energy, but in most cases left the original symmetry unaltered. All normal mode calculations were carried out on the appropriate converged geometries at the same level of theory and using the same basis set.

The GAUSSIAN programs perform the vibrational analysis within the standard Wilson FG matrix formalism, using the equation (Wilson *et al.*, 1955)

$$GF = E\lambda$$

where G is the inverse kinetic energy matrix, calculated from the equilibrium structural parameters and the atomic masses, F is the potential energy (force constant) matrix, E is the identity matrix and λ is the diagonal matrix whose elements are

$$\lambda_i = 4\pi^2 c^2 \tilde{v}_i^2,$$

where c is the speed of light and $\tilde{v}_i$ is the ith normal mode wavenumber.

The algorithm involves computing the forces (first derivatives of the molecular energy with respect to the nuclear coordinates), differentiating the forces to derive the Hessian matrix (the matrix of second derivatives of the energy with respect to nuclear coordinates), and then transforming to a mass-weighted force constant matrix. The atomic masses are stored internally for all isotopic species. By default the program uses that for the most abundant isotope of each element, but the atomic mass may be selected by the user for a calculation on a specific isotopomer. The first derivatives were initially carried out analytically and the second by numerical differentiation of the gradient vector (Komornicki and Jaffe, 1979). In later versions, computation of the second derivatives is available analytically for self-consistent field (SCF) (Pople et al., 1979), configuration interaction (CI) (Krishnan et al., 1980) and MP2 (Schlegel et al., 1984) wavefunctions. The wavenumbers and normal modes are then determined by diagonalization of the mass-weighted force constant matrix.

Similarly, the intensities of the infrared and Raman bands are computed by differentiation of the molecular dipole moment, μ (Yamaguchi et al., 1986), or polarizability, α (Frisch et al., 1986), with respect to the nuclear coordinates to obtain the dipole moment and polarizability derivatives, from which the infrared intensity is derived using the equation

$$A_i = \frac{N\pi}{3c^2}\left(\frac{\partial \mu}{\partial Q_i}\right)^2$$

where N is Avogadro's constant, c is the speed of light and Q_i is the normal coordinate of the ith mode. The Raman intensity is obtained from the equation

$$I_i = 45\alpha_i'^2 + 7\gamma_i'^2$$

where α' and γ' are derivatives of the trace and anisotropy of the polarizability of the ith normal mode, respectively.

In this chapter the assignments of the normal modes were made with the aid of the Cartesian atomic displacement coordinates, and are described in the following tables using the standard spectroscopic notation for group vibrations; ν (stretching), δ (in-plane bending), γ (out-of-plane bending), ρ (rocking), ω (wagging), τ (twisting, or torsion) and l (libration). The subscripts s and a stand for symmetric and antisymmetric coupled vibrations. The mode numbering for each molecular species follows the standard Mulliken spectroscopic convention, the modes being numbered in decreasing wavenumber order, grouped within each symmetry species (Mulliken, 1955). In many cases, more detailed assignments are given in the original publications, including the percentage contributions of individual group modes. Those given here are intended to emphasize the correspondence between the forms of the normal modes of the complexes and those of the parent monomer vibrations.

6.3 COMPUTED VIBRATIONAL SPECTRA

6.3.1 The Water–Ammonia–Hydroxylamine System

The binary complexes formed from water, ammonia and hydroxylamine create a set of related adducts illustrating a number of different intermolecular bonding situations. The HOH.NH$_3$ complex may be confidently predicted to adopt a classical hydrogen-bonded structure, with the water molecule donating one of its protons to the ammonia nitrogen lone pair. This is indeed the structure observed in the gas phase by means of infrared (Millen and Mines, 1977) and molecular beam electric resonance (MBER) spectroscopy (Herbine and Dyke, 1985), and the structure optimized by Yeo and Ford (1991a) is consistent with the experimental observations. This structure possesses C_s symmetry. The results of this theoretical study are in agreement with those of some earlier *ab initio* computations by Diercksen *et al.* (1972), Kerns and Allen (1978), Del Bene (1988) and Latajka and Scheiner (1990). The preferred structure is illustrated in Figure 1(a).

The $NH_2OH.NH_3$ complex is expected to adopt a similar classically hydrogen-bonded structure, with the hydroxylamine molecule acting as the proton donor and ammonia the proton acceptor. However, Yeo and Ford (1991b) found that, on optimization, a second, weaker hydrogen bond interaction is present, involving one of the ammonia hydrogen

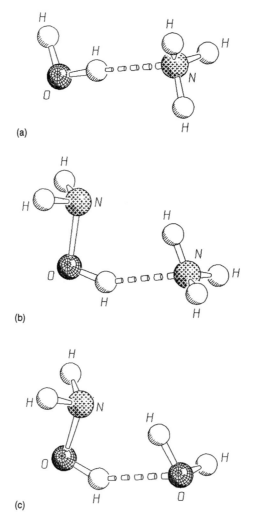

Figure 6.1 Structures of the (a) water–ammonia, (b) hydroxylamine–ammonia and (c) hydroxylamine–water complexes

atoms and the hydroxylamine nitrogen lone pair, in a five-membered cyclic structure, also of symmetry C_s. The presence of a secondary interaction in the calculated structures of weak complexes of this type is not uncommon, and is also observed experimentally in the gas phase (Legon, 1995), where the absence of the effects of the environment characteristic of condensed phases allows for some unusual cooperative bonding. The structure of $NH_2OH.NH_3$ has apparently not been determined in the gas phase; the most probable structure derived by Yeo and Ford (1991b) is shown in Figure 6.1(b).

Likewise, although the gas-phase structure of the 1:1 $NH_2OH.OH_2$ complex is not known, Yeo and Ford (1991c) showed that it, too, contains a secondary hydrogen bond interaction. The hydroxyl proton of hydroxylamine is bonded to the water oxygen atom, while one of the water protons interacts with the nitrogen atom of hydroxylamine. This five-membered cyclic complex has C_1 symmetry, and its structure is shown in Figure 6.1(c).

The predicted wavenumbers of these three complexes are listed in Table 6.1 (Yeo and Ford 1991a–c). The extents to which vibrational wavenumbers are overestimated using

Table 6.1 Computed wavenumbers (MP2/6–31G**) of the 1:1 water–ammonia, hydroxylamine–ammonia and hydroxylamine–water complexes

Complex	Symmetry species	Mode	Wavenumber (cm^{-1})	Approximate description
HOH.NH$_3$[a]	a'	ν_1	3989	ν(OH) (free)
		ν_2	3728	ν(OH) (bonded)
		ν_3	3721	ν_a(NH$_3$)
		ν_4	3567	ν_s(NH$_3$)
		ν_5	1743	δ(HOH)
		ν_6	1715	δ_a(NH$_3$)
		ν_7	1158	δ_s(NH$_3$)
		ν_8	460	δ(OH...N)(IM)
		ν_9	202	ν(H...N)(IM)
		ν_{10}	172	δ(HN...H)(IM)
	a"	ν_{11}	3723	ν_a(NH$_3$)
		ν_{12}	1724	δ_a(NH$_3$)
		ν_{13}	712	γ(OH...N)(IM)
		ν_{14}	161	l(NH$_3$)(IM)
		ν_{15}	30	γ(HN...H)(IM)
H$_2$NOH.NH$_3$[b]	a'	ν_1	3708	ν_a(NH$_3$)
		ν_2	3614	ν(OH)
		ν_3	3553	ν_s(NH$_3$)
		ν_4	3518	ν_s(NH$_2$)
		ν_5	1719	δ_a(NH$_3$)
		ν_6	1702	δ(NH$_2$)
		ν_7	1574	δ(NOH)
		ν_8	1220	ω(NH$_2$)
		ν_9	1141	δ_s(NH$_3$)
		ν_{10}	963	ν(NO)
		ν_{11}	433	l(NH$_3$)(IM)
		ν_{12}	234	ν(OH...N)(IM)
		ν_{13}	147	ν(NH...N)(IM)

continued overleaf

Table 6.1 (*continued*)

Complex	Symmetry species	Mode	Wavenumber (cm^{-1})	Approximate description
	a''	ν_{14}	3731	$\nu_a(NH_3)$
		ν_{15}	3620	$\nu_a(NH_2)$
		ν_{16}	1739	$\delta_a(NH_3)$
		ν_{17}	1334	$\tau(NH_2)$
		ν_{18}	814	$\gamma(OH...N)(IM)$
		ν_{19}	252	$\rho(NH_2)$
		ν_{20}	194	$l(NH_3)(IM)$
		ν_{21}	140	$\gamma(NH...N)(IM)$
NH$_2$OH.OH$_2$^{c}	a	ν_1	3972	$\nu(OH)(W)$ (free)
		ν_2	3763	$\nu(OH)(H)$
		ν_3	3705	$\nu(OH)(W)$ (bonded)
		ν_4	3634	$\nu_a(NH_2)(H)$
		ν_5	3532	$\nu_s(NH_2)(H)$
		ν_6	1710	$\delta(HOH)(W)$
		ν_7	1701	$\delta(NH_2)(H)$
		ν_8	1549	$\delta(NOH)(H)$
		ν_9	1330	$\tau(NH_2)(H)$
		ν_{10}	1214	$\omega(NH_2)(H)$
		ν_{11}	968	$\nu(NO)(H)$
		ν_{12}	733	$\delta(OH...N)(IM)$
		ν_{13}	690	$\gamma(OH...O)(IM)$
		ν_{14}	384	$\rho(HOH)(W)(IM)$
		ν_{15}	259	$\tau(OH)(W)(IM)$
		ν_{16}	234	$\rho(NH_2)(H)$
		ν_{17}	217	$\tau(HOH)(W)(IM)$
		ν_{18}	213	$\nu(H...O)(IM)$

[a] Yeo and Ford (1991a).
[b] Yeo and Ford (1991b).
[c] Yeo and Ford (1991c).
IM—intermolecular mode; W—water; H—hydroxylamine.

various combinations of level of theory and basis set have been well documented (Besler *et al.*, 1988; Thomas *et al.*, 1993). At the MP2 level of theory, and with the 6–31G** basis set, 10 per cent is not an unrealistic expectation of the degree of agreement between theory and experiment. However, approximately the same level of overestimation may be expected for calculations on complexes and on their component monomers. Therefore by cancellation of errors, the computed shifts in the monomer wavenumbers resulting from complexation may be found to be quite close to those measured experimentally, for example by matrix isolation infrared spectroscopy. The computed wavenumber shifts of the modes of HOH.NH$_3$, NH$_2$OH.NH$_3$ and NH$_2$OH.OH$_2$ from the corresponding monomer mode positions are presented in Table 6.2 (Yeo and Ford 1991a–c).

The HOH.NH$_3$ complex has been studied in nitrogen and argon matrices by Nelander and Nord (1982) and Yeo and Ford (1991d), and in neon and krypton by Engdahl and Nelander (1989). The experimental complex-monomer wavenumber shifts of HOH.NH$_3$ (Engdahl and Nelander, 1989; Yeo and Ford, 1991d) are shown in Table 6.3. In Tables 6.2 and 6.3 the shift of the bonded water OH stretching mode of the complex is measured from the position of the $\nu_s(OH_2)$ mode of the monomer and that of the free water OH

Table 6.2 Computed complex–monomer wavenumber shifts (MP2/6–31G**) of the 1:1 water–ammonia, hydroxylamine–ammonia and hydroxylamine–water complexes

Monomer unit	Symmetry species	Mode	Wavenumber shift (cm^{-1})		
			HOH.NH$_3$[a]	H$_2$NOH.NH$_3$[b]	NH$_2$OH.OH$_2$[c]
H$_2$O	a_1	ν_s(OH$_2$)	−167		−190
		δ(OH$_2$)	60		27
	b_1	ν_a(OH$_2$)	−45		−62
NH$_3$	a_1	ν_s(NH$_3$)	3	−11	
		δ_s(NH$_3$)	36	−9	
	e	ν_a(NH$_3$)(a')	0	−13	
		(a'')	2	10	
		δ_a(NH$_3$)(a')	−13	19	
		(a'')	−4	11	
NH$_2$OH	a'	ν(OH)		−278	−129
		ν_s(NH$_2$)		−13	1
		δ(NH$_2$)		−2	−3
		δ(NOH)		144	119
		ω(NH$_2$)		28	22
		ν(NO)		9	14
	a''	ν_a(NH$_2$)		−16	−2
		τ(NH$_2$)		−13	−17
		τ(OH)		406	282

[a] Yeo and Ford (1991a).
[b] Yeo and Ford (1991b).
[c] Yeo and Ford (1991c).

Table 6.3 Experimental complex–monomer wavenumber shifts of the 1:1 water–ammonia complex observed in cryogenic matrices

Monomer unit	Mode	Wavenumber shift (cm^{-1})			
		N$_2$[a]	Ne[b]	Ar[a]	Kr[b]
H$_2$O	ν_s(H$_2$O)	−223	−209.1	−205	−195.5
	δ(H$_2$O)	−[c]	29.4	−[c]	−[c]
	ν_a(H$_2$O)	−31	−34.1	−25	−28.2
NH$_3$	ν_s(NH$_3$)	−[c]	−1.4	−[c]	−[c]
	δ_s(NH$_3$)	76	67.2	61	69.0
	ν_a(NH$_3$)	−[c]	−[c]	−[c]	−[c]
		−[c]	−[c]	−[c]	−[c]
	δ_a(NH$_3$)	−[c]	−[c]	−[c]	−[c]
		−[c]	−[c]	−[c]	−[c]

[a] Yeo and Ford (1991d).
[b] Engdahl and Nelander (1989).
[c] Not observed.

from the monomer ν_a(OH$_2$). Owing to the decoupling of the water monomer modes on complexation, the symmetric/antisymmetric distinction in the complex is lost. For convenience in measuring and discussing the shifts the comparisons are made between the corresponding complex and monomer modes whose wavenumbers are closer in value.

The prediction of Table 6.2 is that the bonded OH mode of complexed water shifts by almost 200 cm^{-1} to the red, the free OH by a much smaller amount in the same direction, and the bending vibration by over 50 cm^{-1} to the blue. Simultaneously, the shifts of the ammonia fragment are all predicted to be small, except for the symmetric NH$_3$ bending, which is expected to be displaced almost 50 cm^{-1} to higher values. Comparing the experimental results of Table 6.3 with the predictions of Table 6.2, the bonded OH shift is approximately −200 cm^{-1}, depending on the matrix (even larger than the predicted shift), the free OH moves 25 to 35 cm^{-1} to lower wavenumber, slightly less than the computed value, and the HOH bending, which was only observed in neon (Engdahl and Nelander, 1989), by about 30 cm^{-1} to the blue, half of the predicted shift. Only the symmetric bending mode of the ammonia sub-unit was observed in these matrices; its shift is about twice that predicted. While the calculated shifts are not quantitatively very close to those observed experimentally, their relative orders of magnitude, and particularly their directions, are in remarkably good agreement.

Nelander and Nord (1982), Engdahl and Nelander (1989) and Yeo and Ford (1991d) have observed a number of the intermolecular modes of this complex in cryogenic matrices. In fact, Engdahl and Nelander (1989) were able to make a complete assignment in their spectrum in neon of all six intermolecular modes. Several naming conventions have been proposed in the literature for describing the intermolecular modes of hydrogen-bonded complexes (Bertie and Falk, 1973; Knözinger and Schrems, 1987; Yeo and Ford, 1992). The positions of the low-wavenumber bands observed by Engdahl and Nelander (1989) and Yeo and Ford (1991d) are presented in Table 6.4, along with the assignments of Engdahl and Nelander (1989) and those given in Table 6.1 (Yeo and Ford 1991a). Differences in these assignments are attributed simply to questions of semantics. Table 6.4 shows that agreement with the computed wavenumbers is quite good for the a' modes and for v_{13}. The wavenumber of the ammonia librational mode is severely underestimated, however, and that of the water torsional vibration is overestimated by about 50 per cent. Although the intermolecular modes are usually predicted with far lower accuracy than their intramolecular counterparts, since low-frequency vibrations are typically highly anharmonic, the agreement for HOH.NH$_3$ is fairly satisfactory.

Table 6.4 Experimental intermolecular wavenumbers of the 1:1 water–ammonia complex observed in cryogenic matrices

Assignment[a]	Assignment[b]	Wavenumber (cm^{-1})			
		N$_2$[c]	Ne[b]	Ar[c]	Kr[b]
$v_{13}(a'')$	Out-of-plane shear	600	661.9	638	631.1
$v_8\ (a')$	In-plane shear	440	429.5	419	416.5
$v_{14}(a'')$	Out-of-plane bend	420	411.2	402	389.9
$v_9\ (a')$	Stretch	−[d]	201.8	−[d]	197.1
$v_{10}(a')$	In-plane bend	−[d]	179.6	−[d]	193.2
$v_{15}(a'')$	Torsion	−[d]	19.5	−[d]	−[d]

[a] From Table 6.1.
[b] Engdahl and Nelander (1989).
[c] Yeo and Ford (1991d).
[d] Not observed.

COMPUTED VIBRATIONAL SPECTRA

The GAUSSIAN family of programs is capable of taking a computed force constant matrix from a parent isotopomer, and calculating the wavenumbers of any isotopic variant obtained by substituting one or more atoms by a different isotope. This is carried out using the READISO keyword. Nelander and coworkers (Nelander and Nord, 1982; Engdahl and Nelander, 1989) obtained the spectra of a number of isotopomers of HOH.NH$_3$, and assigned many of the bands of each species. Table 6.5 shows the computed isotopic wavenumbers of HOD.NH$_3$, DOH.NH$_3$, DOD.NH$_3$, HOH.ND$_3$,

Table 6.5 Computed wavenumbers (MP2/6−31G**) of the deuterium-substituted isotopic variants of the 1:1 water−ammonia complex[a]

Symmetry species	Mode[b]	Wavenumber (cm^{-1})			
		HOH.NH$_3$	HOD.NH$_3$	DOH.NH$_3$	DOD.NH$_3$
a'	ν_1	3989	3975	2884	2912
	ν_2	3728	2720	3748	2698
	ν_3	3721	3721	3721	3721
	ν_4	3567	3567	3567	3567
	ν_5	1743	1477	1572	1269
	ν_6	1715	1721	1722	1721
	ν_7	1158	1158	1158	1158
	ν_8	460	411	396	373
	ν_9	202	198	196	194
	ν_{10}	172	163	164	155
a''	ν_{11}	3723	3723	3723	3723
	ν_{12}	1724	1724	1724	1724
	ν_{13}	712	543	712	543
	ν_{14}	161	152	158	149
	ν_{15}	30	30	23	23

Symmetry species	Mode[b]	Wavenumber (cm^{-1})			
		HOH.ND$_3$	HOD.ND$_3$	DOH.ND$_3$	DOD.ND$_3$
a'	ν_1	3989	3975	2884	2911
	ν_2	3728	2720	3747	2699
	ν_3	2738	2738	2738	2738
	ν_4	2549	2549	2549	2549
	ν_5	1737	1477	1572	1269
	ν_6	1248	1248	1248	1247
	ν_7	882	882	882	882
	ν_8	435	379	365	335
	ν_9	190	187	187	185
	ν_{10}	136	132	132	127
a''	ν_{11}	2740	2740	2740	2740
	ν_{12}	1251	1251	1251	1251
	ν_{13}	692	515	692	514
	ν_{14}	124	120	121	117
	ν_{15}	29	29	22	22

[a] Yeo (1991).
[b] The mode numbering of each isotopomer corresponds exactly to that of HOH.NH$_3$, hence the wavenumbers of some isotopomer modes are found to be out of sequence.

HOD.ND$_3$, DOH.ND$_3$ and DOD.ND$_3$, as well as those of the parent HOH.NH$_3$ (Yeo, 1991). The calculated isotopic patterns are fully consistent with those observed in the spectra in cryogenic matrices (Nelander and Nord, 1982; Engdahl and Nelander, 1989).

The infrared spectrum of the NH$_2$OH.NH$_3$ complex has been studied in nitrogen and argon matrices by Yeo and Ford (1991e). The shifts of the monomer sub-unit bands in each matrix are reported in Table 6.6. Again referring to the computed shifts given in Table 6.2, the OH stretching mode of the hydroxylamine fragment is predicted to undergo a massive shift of almost 300 cm^{-1} to the red, and the corresponding NOH bending of about 150 cm^{-1} to the blue. The torsional mode of the OH group is also predicted to shift to the blue by a very large amount. This is because this mode in the monomer is relatively free, while in the complex the motion is constrained due to the effect of the OH...N hydrogen bond, increasing the value of the torsional force constant considerably. The shifts of the ammonia sub-unit are much more modest, being between 10 and 20 cm^{-1} in either direction, in general. The experimental spectrum (Yeo and Ford, 1991e) faithfully reproduces the magnitudes and directions of both the OH stretching and the NOH bending shifts of the hydroxylamine component (-345 and 108 cm^{-1} in nitrogen, for example), as shown in Table 6.6. The experimental ammonia shifts (about -65 cm^{-1} for the two stretching modes and about 60 cm^{-1} for the symmetric bending) are rather larger than those calculated, and because the computed NH$_3$ shifts are quite small, some of their signs are opposite to those determined experimentally. Surprisingly, the OH torsional mode was not observed in either matrix (Yeo and Ford, 1991e), even though it was

Table 6.6 Experimental complex-monomer wavenumber shifts of the 1:1 hydroxylamine–ammonia complex observed in cryogenic matrices

Monomer	Mode	Wavenumber shift (cm^{-1})	
		N$_2$[a]	Ar[a]
NH$_2$OH	ν(OH)	-345	-297
	ν_s(NH$_2$)	–[b]	–[b]
	δ(NH$_2$)	–[b]	–[b]
	δ(NOH)	108	121
	ω(NH$_2$)	13	12
	ν(NO)	10	8
	ν_a(NH$_2$)	–[b]	–[b]
	τ(NH$_2$)	–[b]	–[b]
	τ(OH)	–[b]	–[b]
NH$_3$	ν_s(NH$_3$)	-62	-77
	δ_s(NH$_3$)	69	55
	ν_a(NH$_3$)	-54	–[b]
		–[b]	–[b]
	δ_a(NH$_3$)	–[b]	–[b]
		–[b]	–[b]

[a] Yeo and Ford (1991e).
[b] Not observed.

Table 6.7 Experimental complex-monomer wavenumber shifts of the 1:1 hydroxylamine–water complex observed in cryogenic matrices

Monomer	Mode	Wavenumber shift (cm^{-1})	
		N$_2$[a]	Ar[a]
NH$_2$OH	ν(OH)	−139	−93
	ν_s(NH$_2$)	6	−[b]
	δ(NH$_2$)	−[b]	−[b]
	δ(NOH)	54	10
	ω(NH$_2$)	18	19
	ν(NO)	19	12
	ν_a(NH$_2$)	−[b]	−[b]
	τ(NH$_2$)	−[b]	−[b]
	τ(OH)	−63	−[b]
H$_2$O	ν_s(H$_2$O)	−211	−201
	δ(H$_2$O)	33	−[b]
	ν_a(H$_2$O)	−38	−[b]

[a] Yeo and Ford (1991f).
[b] Not observed.

predicted to be in a relatively accessible region of the spectrum, and its intensity was computed to be quite high (Yeo and Ford, 1991b).

The third complex of this set, NH$_2$OH.OH$_2$, has also been studied in nitrogen and argon matrices by Yeo and Ford (1991f). The experimental complex-monomer wavenumber shifts are collected in Table 6.7. Since both water and hydroxylamine function as proton donors in this complex, both bonded OH stretching modes are predicted to suffer large red displacements, of 190 cm^{-1} for water and 129 cm^{-1} for hydroxylamine (see Table 6.2). The hydroxylamine NOH bending vibration is expected to shift to the blue by over 100 cm^{-1}, as in NH$_2$OH.NH$_3$. Again, the wavenumber of the OH torsional mode is expected to increase significantly as a result of the presence of the OH...O hydrogen bond. The remaining water sub-unit shifts are relatively small. In both nitrogen and argon matrices the bonded water OH shift is about −200 cm^{-1}, remarkably close to the predicted value, while the hydroxylamine OH stretching displacements in each matrix lie on either side of the computed value. The free water OH stretching mode, which was observed only in nitrogen, shifts by only about two-thirds of the predicted value, while the HOH bending shift in nitrogen is very well reproduced. The hydroxylamine NOH bending mode shifts only about half of the predicted difference in nitrogen, while the shift in argon is even smaller. The signs and relative magnitudes of the wavenumber differences of the remaining modes are in fairly good agreement.

In summary, the calculated spectra, measured by their intramolecular wavenumber shifts, of all three complexes are generally in excellent semi-quantitative agreement with the observed spectra in nitrogen and argon, and other matrices, allowing for the standard overestimation of the wavenumbers. The computational technique is clearly potentially a very useful source of information which may be used to aid the assignments and interpretation of infrared spectra measured in cryogenic matrices.

6.3.2 The Complexes of Water with Hydrogen Cyanide, Acetonitrile and Cyanogen

This series of 1:1 complexes provides another example of a set of related adducts displaying quite different structural features. Although the CH group is not a common hydrogen bond proton donor, since hydrogen cyanide is a stronger acid than water, the complex between HCN and H_2O should show evidence of a CH...O hydrogen bond, with the acidic proton of HCN interacting with an oxygen lone pair of the water molecule. As there are two equivalent lone pairs on the oxygen atom of the water molecule, the potential surface of such a complex should show two equivalent minima, corresponding to structures in which the HCN molecule approaches the O atom in an approximately tetrahedral direction, on either side of the H_2O plane. A totally planar structure, with the HCN molecule interacting equally strongly with both lone pairs, represents a transition state on the potential energy surface, connecting the two equivalent non-planar minima. The gas phase structure of the $NCH.OH_2$ complex, determined by microwave spectroscopy, has been interpreted as being 'effectively planar' (Fillery-Travis et al., 1984) and 'pseudoplanar' (Gutowsky et al., 1992), in which the water molecule executes a large-amplitude librational motion out of the heavy atom plane, representing an exchange of the two equivalent water molecule positions referred to above. The complex has been studied by ab initio molecular orbital theory. While the predictions at the self-consistent field (SCF) level with small basis sets favoured a planar equilibrium C_{2v} structure (Vishveshwara, 1978; Hinchliffe, 1986), with larger basis sets, including polarization and diffuse functions (Turi and Dannenberg, 1993), evidence for both a planar C_{2v} and a non-planar C_s structure was found. Tshehla and Ford (1994) found that, at the MP2 level, with the 6–31G** basis set, only the C_s structure is a genuine minimum on the potential energy surface. This structure is illustrated in Figure 6.2(a).

There appears to be no experimental evidence for the structure of the water–acetonitrile complex. An OH...N hydrogen bonded model is proposed, with water as the proton donor, on the basis of the lack of a sufficiently acidic hydrogen atom in the acetonitrile molecule to cause it to act as the proton donor towards the water oxygen. Tshehla (1996) indeed found such a structure to be a genuine minimum; moreover, the structure in which the free water OH group eclipses one of the methyl hydrogens, in a C_s adduct, is slightly favoured over the alternative staggered arrangement, but rotation of the non-bonded OH about the O...NCC axis is virtually free. This structure is shown in Figure 6.2(b).

The structure of the water–cyanogen complex is not the expected classically hydrogen bonded one, with a water OH group donating a proton to one of the cyanogen N atoms, as in $HOH.NCCH_3$, discussed above. Lee et al. (1992), by the use of MBER spectroscopy, showed this complex to have a planar T-shaped structure of C_{2v} symmetry, with the oxygen atom oriented towards the C_2N_2 molecule and the water C_2 axis bisecting the CC bond, as shown in Figure 6.2(c). This complex thus represents the third different type of structure found in this set of adducts. Tshehla and Ford (1994) confirmed the planar C_{2v} structure, at the MP2/6–31G** level, and showed that the alternative perpendicular C_{2v} arrangement represents a saddle point, with a rotational barrier of 2.65 kJ mol^{-1}.

The computed vibrational spectrum of $NCH.OH_2$ (Tshehla and Ford, 1995) is presented in Table 6.8, with the predicted complex-monomer wavenumber shifts listed in Table 6.9. No experimental data appear to be available for this complex, although the spectrum

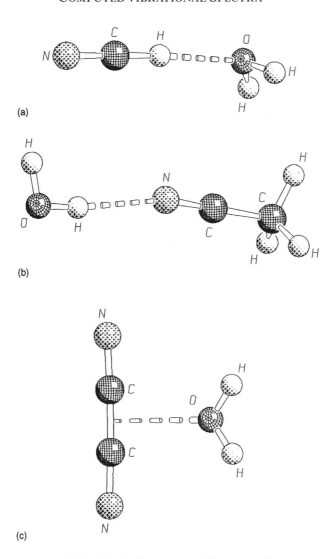

Figure 6.2 Structures of the (a) hydrogen cyanide–water, (b) water–acetonitrile and (c) cyanogen–water complexes

of the isostructural complex between water and acetylene, which is isoelectronic with hydrogen cyanide, has been studied in argon matrices (Andrews *et al.*, 1982; Engdahl and Nelander, 1983). The predicted shifts of the water sub-unit bands of NCH.OH$_2$ (Table 6.9) are minimal, none exceeding 10 cm^{-1}, and all being to lower wavenumber. This confirms the fact that water does not function as the proton donor in this complex. The large predicted shifts of the HCN fragment, however, particularly the red shift of 143 cm^{-1} of the bonded CH stretching mode, testify to the identity of this sub-unit as the proton donor. The degenerate bending mode of the HCN monomer splits into a pair of bands in the complex, one of a' and one of a'' symmetry, each with a blue shift of approximately 200 cm^{-1}. Engdahl and Nelander (1983) found red shifts of 62.3 and

Table 6.8 Computed wavenumbers (MP2/6–31G**) of the 1:1 hydrogen cyanide–water, water–acetonitrile and cyanogen–water complexes

Complex	Symmetry species	Mode	Wavenumber (cm^{-1})	Approximate description
NCH.OH$_2$[a]	a'	ν_1	3887	$\nu_s(OH_2)$
		ν_2	3390	$\nu(CH)$
		ν_3	2039	$\nu(CN)$
		ν_4	1676	$\delta(OH_2)$
		ν_5	872	$\delta(HCN)$
		ν_6	209	$\omega(OH_2)(IM)$
		ν_7	165	$\nu(H...O)(IM)$
		ν_8	118	$\delta(CH...O)(IM)$
	a''	ν_9	4020	$\nu_a(OH_2)$
		ν_{10}	929	$\gamma(HCN)$
		ν_{11}	292	$\tau(OH_2)(IM)$
		ν_{12}	122	$\gamma(CH...O)(IM)$
HOH.NCCH$_3$[b]	a'	ν_1	4005	$\nu(OH)$ (free)
		ν_2	3868	$\nu(OH)$ (bonded)
		ν_3	3254	$\nu_a(CH_3)$
		ν_4	3152	$\nu_s(CH_3)$
		ν_5	2244	$\nu(CN)$
		ν_6	1709	$\delta(HOH)$
		ν_7	1532	$\delta_a(CH_3)$
		ν_8	1462	$\delta_s(CH_3)$
		ν_9	1089	$\rho(CH_3)$
		ν_{10}	948	$\nu(CC)$
		ν_{11}	354	$\delta(CCN)$
		ν_{12}	254	$\delta(OH...N)(IM)$
		ν_{13}	131	$\nu(H...N)(IM)$
		ν_{14}	29	$\delta(H...NC)(IM)$
	a''	ν_{15}	3255	$\nu_a(CH_3)$
		ν_{16}	1532	$\delta_a(CH_3)$
		ν_{17}	1088	$\rho(CH_3)$
		ν_{18}	417	$\gamma(H...NC)(IM)$
		ν_{19}	350	$\gamma(CCN)$
		ν_{20}	34	$\tau(H...N)(IM)$
		ν_{21}	2	$l(HOH)(IM)$
C$_2$N$_2$.OH$_2$[a]	a_1	ν_1	3886	$\nu_s(OH_2)$
		ν_2	2267	$\nu_s(CN)$
		ν_3	1672	$\delta(HOH)$
		ν_4	868	$\nu(CC)$
		ν_5	245	$\delta(CCN)$
		ν_6	122	$\nu(O...X)(IM)$
	a_2	ν_7	533	$\gamma(CCN)$
		ν_8	117	$\tau(OH_2)(IM)$
	b_1	ν_9	246	$\gamma(CCN)$
		ν_{10}	39	$\omega(OH_2)(IM)$
	b_2	ν_{11}	4024	$\nu_a(OH_2)$
		ν_{12}	2069	$\nu_a(CN)$
		ν_{13}	519	$\delta(CCN)$
		ν_{14}	174	$\rho(OH_2)(IM)$
		ν_{15}	24	$l(NCCN)(IM)$

[a] Tshehla and Ford (1995).
[b] Tshehla (1996).
IM—intermolecular mode; X—mid-point of the CC bond.

Table 6.9 Computed complex-monomer wavenumber shifts (MP2/6−31G**) of the 1:1 hydrogen cyanide−water, water−acetonitrile and cyanogen−water complexes

Monomer unit	Symmetry species	Mode	Wavenumber shift (cm^{-1})		
			NCH.OH$_2$[a]	HOH.NCCH$_3$[b]	C$_2$N$_2$.OH$_2$[a]
H$_2$O	a_1	ν_s(OH$_2$)	−4	−23	−5
		δ(OH$_2$)	−6	27	−10
	b_1	ν_a(OH$_2$)	−10	−25	−6
HCN	Σ^+	ν(CH)	−143		
		ν(CN)	0		
	Π	δ(HCN)(a')	147		
		(a'')	204		
CH$_3$CN	a_1	ν_s(CH$_3$)		2	
		ν(CN)		12	
		δ_s(CH$_3$)		0	
		ν(CC)		4	
	e	ν_a(CH$_3$)(a')		3	
		(a'')		4	
		δ_a(CH$_3$)(a')		−3	
		(a'')		−3	
		ρ(CH$_3$)(a')		0	
		(a'')		−1	
		δ(CCN)(a')		−1	
		(a'')		−5	
C$_2$N$_2$	Σ_g^+	ν_s(CN)			10
		ν(CC)			0
	Σ_u^-	ν_a(CN)			8
	Π_g	δ(CCN)(a_2)			7
		(b_2)			−7
	Π_u	δ(CCN)(a_1)			8
		(b_1)			9

[a] Tshehla and Ford (1995).
[b] Tshehla (1996).

48.7 cm^{-1} for the bonded CH stretching mode of HCCH.OH$_2$, relative to the matrix split ν_a(CH) band of acetylene monomer, and a pair of bands blue-shifted by 49 and 56 cm^{-1} from the infrared active degenerate CCH bending mode of the monomer. While these shifts cannot be compared directly with those predicted for NCH.OH$_2$, their signs and relative magnitudes are the same. Tshehla and Ford (1994) found NCH.OH$_2$ to be more than twice as strongly bound as HCCH.OH$_2$ (−22.29 versus −10.26 kJ mol^{-1}), therefore the fact that the observed shifts of HCCH.OH$_2$ (Engdahl and Nelander, 1983) are only approximately half of the predicted ones of NCH.OH$_2$ (Table 6.9) finds a ready explanation.

The infrared spectra of the HOH.NCCH$_3$ and C$_2$N$_2$.OH$_2$ complexes have not been studied experimentally, as far as we are aware. In the first case, both water monomer OH stretching modes are red-shifted, by almost the same amount (indicating virtually no decoupling of these modes relative to the monomer), while the bending vibration is blue-shifted, again by the same amount. It is rather surprising that the lower of the two OH stretching modes does not undergo a larger shift, as it does in the conventionally hydrogen-bonded species discussed above. The small shift is explained by the relative weakness of the hydrogen bonded interaction in HOH.NCCH$_3$ (about -15 kJ mol^{-1}) (Tshehla 1996). The weakness of the binding is confirmed by the small wavenumber shifts of the corresponding acetonitrile modes; apart from the blue shift of the CN stretching mode (12 cm^{-1}), none of the other shifts exceeds 10 cm^{-1} in either direction. Since C$_2$N$_2$.OH$_2$ is not hydrogen bonded, the water vibrations are not expected to suffer significant wavenumber perturbations from the monomer values. Table 6.9 shows this indeed to be the case. The largest shift is that of the bending mode, and it is shifted to lower wavenumber, in the opposite direction to that normally observed for hydrogen-bonded water molecules, reinforcing the conclusion that this species is stabilized by types of forces other than hydrogen bonding. The displacements of the cyanogen modes are also minimal, with the symmetric CN stretching vibration having the largest shift, of 10 cm^{-1} to the blue. These vibrational shift data suggest a weak interaction, and Tshehla and Ford (1994) in fact computed the complex to be stabilized by only a little over -12 kJ mol^{-1}.

6.3.3 The Methanol–Dimethylamine–Dimethyl ether–Trimethylamine System

The four 1:1 complexes formed between methanol and dimethylamine as proton donors, and dimethyl ether and trimethylamine as proton acceptors, provide examples of the OH...O, OH...N, NH...O and NH...N types of hydrogen bond. As such, they illustrate the relative propensities of the OH and NH groups to donate protons and of the oxygen and nitrogen atoms to accept them. This particular series was chosen for study because the remaining valences of the O and N atoms in each molecule are occupied by methyl groups. This ensures that no additional potential sites exist for hydrogen bonding interaction, other than the major one, e.g. the OH...O hydrogen bond in the complex between methanol and dimethyl ether, and moreover, any differences calculated in the properties of the various complexes which might be due to variations in the nature of the attached alkyl groups are eliminated. Because the complexes in this set are considerably larger than those discussed earlier, use of a level of theory which takes account of the effects of electron correlation was precluded, and the computations were carried out at the RHF level only. Even so, the number of basis functions processed in the calculations ranged from 125 for CH$_3$OH.O(CH$_3$)$_2$ to 185 for (CH$_3$)$_2$NH.N(CH$_3$)$_3$. The restriction on the level of theory that could be accommodated in the computations for these complexes had a bonus; the GAUSSIAN programs perform calculations of the Raman band activities and depolarization ratios as part of the normal mode analysis at the RHF level, which are not available at the MP2 level. Thus the vibrational features of the complexes characteristic of the Raman spectra were accessible, and will be discussed below.

The structural and electronic properties of CH$_3$OH.O(CH$_3$)$_2$, CH$_3$OH.N(CH$_3$)$_3$, (CH$_3$)$_2$NH.O(CH$_3$)$_2$ and (CH$_3$)$_2$NH.N(CH$_3$)$_3$ were reported by Bricknell et al. (1995). The optimized structures are illustrated in Figure 6.3. The computed wavenumbers of some of the intramolecular modes of the four complexes are reported in Table 6.10 (Bricknell, 1996). Owing to the large number of normal modes, many of which are associated with the methyl group vibrations, and are therefore not directly perturbed by the hydrogen bonding interaction, the modes listed in the table, and the following discussion, will be

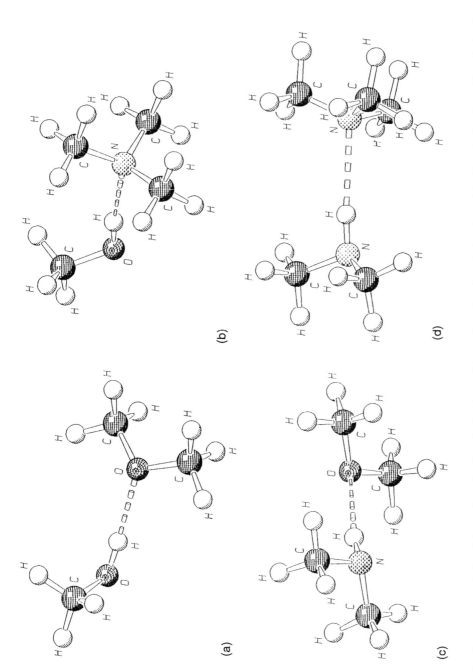

Figure 6.3 Structures of the (a) methanol–dimethyl ether, (b) methanol–trimethylamine, (c) dimethylamine–dimethyl ether and (d) dimethylamine–trimethylamine complexes

Table 6.10 Computed wavenumbers (RHF/6–31G**) of some selected intramolecular modes of the 1:1 methanol–dimethyl ether, methanol–trimethylamine, dimethylamine–dimethyl ether and dimethylamine–trimethylamine complexes

Complex	Monomer fragment	Complex mode	Approximate description	Wavenumber (cm^{-1})
CH$_3$OH.O(CH$_3$)$_2$[a]	CH$_3$OH	ν_1	ν(OH)	4118
		ν_{12}	δ(COH)	1540
		ν_{16}	ν(CO)	1197
	(CH$_3$)$_2$O	ν_{17}	ν_s(COC)	1031
		ν_{18}	δ(COC)	445
		ν_{34}	ν_a(COC)	1224
CH$_3$OH.N(CH$_3$)$_3$[a]	CH$_3$OH	ν_1	ν(OH)	4020
		ν_{16}	δ(COH)	1568
		ν_{21}	ν(CO)	1204
	(CH$_3$)$_3$N	ν_{22}	ν_a(NC$_3$)	1144
		ν_{23}	ν_s(NC$_3$)	897
		ν_{24}	δ_a(NC$_3$)	455
		ν_{25}	δ_s(NC$_3$)	416
		ν_{44}	ν_a(NC$_3$)	1142
		ν_{46}	δ_a(NC$_3$)	454
(CH$_3$)$_2$NH.O(CH$_3$)$_2$[a]	(CH$_3$)$_2$NH	ν_1	ν(NH)	3784
		ν_{19}	ν_s(CNC)	1018
		ν_{20}	δ_s(CNH)	902
		ν_{22}	δ(CNC)	417
		ν_{45}	ν_a(CNC)	1196
		ν_{46}	δ_a(CNH)	1131
	(CH$_3$)$_2$O	ν_{18}	ν_s(COC)	1035
		ν_{21}	δ(COC)	443
		ν_{44}	ν_a(COC)	1225
(CH$_3$)$_2$NH.N(CH$_3$)$_3$[a]	(CH$_3$)$_2$NH	ν_1	ν(NH)	3760
		ν_{24}	ν_s(CNC)	1022
		ν_{25}	δ_s(CNH)	924
		ν_{28}	δ(CNC)	418
		ν_{53}	ν_a(CNC)	1196
		ν_{56}	δ_a(CNH)	1132
	(CH$_3$)$_3$N	ν_{23}	ν_a(NC$_3$)	1146
		ν_{26}	ν_s(NC$_3$)	898
		ν_{27}	δ_a(NC$_3$)	456
		ν_{29}	δ_s(NC$_3$)	3998
		ν_{55}	ν_a(NC$_3$)	1147
		ν_{57}	δ_a(NC$_3$)	456

[a] Bricknell (1996).

confined to those group vibrations which involve the hydrogen bonds directly. These are the OH and NH stretching, the COH and CNH bending, and the CO and CNC stretching modes of the proton donors, and the COC and CNC stretching and bending modes of the acceptors. Table 6.11 lists the complex-monomer wavenumber shifts for these modes. This table shows that by far the largest perturbations occur in the spectra of the proton donors. Among the proton acceptors, only for the symmetric COC stretching mode of the (CH$_3$)$_2$O fragment and the symmetric NC$_3$ bending vibration of the (CH$_3$)$_3$N sub-unit do the absolute shifts exceed 10 cm^{-1}. For the proton donors, however, the displacements are

Table 6.11 Computed selected complex-monomer wavenumber shifts (RHF/6−31G**) of the 1:1 methanol–dimethyl ether, methanol–trimethylamine, dimethylamine–dimethyl ether and di-methylamine–trimethylamine complexes

Monomer unit	Complex mode	Wavenumber shift (cm^{-1})[a]			
		$CH_3OH.$ $O(CH_3)_2$	$CH_3OH.$ $N(CH_3)_3$	$(CH_3)_2NH.$ $O(CH_3)_2$	$(CH_3)_2NH.$ $N(CH_3)_3$
CH_3OH	$\nu(OH)$	−75	−173	−	−
	$\delta(COH)$	49	77	−	−
	$\nu(CO)$	40	47	−	−
$(CH_3)_2NH$	$\nu(NH)$	−	−	1	−23
	$\nu_s(CNC)$	−	−	7	11
	$\delta_s(CNH)$	−	−	71	93
	$\delta(CNC)$	−	−	1	2
	$\nu_a(CNC)$	−	−	2	2
	$\delta_a(CNH)$	−	−	8	9
$(CH_3)_2O$	$\nu_s(COC)$	−11	−	−7	−
	$\delta(COC)$	2	−	0	−
	$\nu_a(COC)$	−4	−	−3	−
$(CH_3)_3N$	$\nu_a(NC_3)(a')$	−	−8	−	−6
	$\nu_s(NC_3)$	−	−3	−	−2
	$\delta_a(NC_3)(a')$	−	−4	−	−3
	$\delta_s(NC_3)$	−	35	−	17
	$\nu_a(NC_3)(a'')$	−	−10	−	−5
	$\delta_a(NC_3)(a'')$	−	−5	−	−3

[a] Bricknell (1996).

quite substantial, particularly those of the OH stretching mode. The red-shifts of 75 and 173 cm^{-1} for $CH_3OH.O(CH_3)_2$ and $CH_3OH.N(CH_3)_3$, respectively, indicate that N is a better acceptor of OH protons than is O. Likewise the blue shifts of the COH bending mode (49 cm^{-1} for $CH_3OH.O(CH_3)_2$ and 77 cm^{-1} for $CH_3OH.N(CH_3)_3$) are in the same order, confirming the relative abilities of O and N to engage in hydrogen bonds. This pattern is even extended to the CO stretching mode shifts. The same result is observed in the case of $(CH_3)_2NH.O(CH_3)_2$ and $(CH_3)_2NH.N(CH_3)_3$; with reference to the $\nu(NH)$ shifts, the perturbation in $(CH_3)_2NH.N(CH_3)_3$ is quite small, but in $(CH_3)_2NH.O(CH_3)_2$ this mode undergoes virtually no change at all. The $\delta_s(CNH)$ shifts confirm that N is a better acceptor of NH protons than is O, while the $\delta_a(CNH)$ mode is hardly affected in either case. The stretching and bending vibrations of the CNC group of $(CH_3)_2NH$ are quite insensitive to the nature of the proton acceptor. A further deduction to be drawn from the results of Table 6.11 is that, both for O and N as acceptor atoms, the OH stretching mode undergoes larger shifts than the NH, while $\delta_s(CNH)$ is displaced more than $\delta(COH)$.

A relationship has been observed between the wavenumber of the highest frequency intermolecular mode of a 1:1 hydrogen-bonded complex and the strength of the hydrogen bond (Kurnig et al., 1987). Table 6.12 lists the computed wavenumbers of the hydrogen bond stretching, and the in- and out-of-plane hydrogen bond bending vibrations of the four complexes of this group. In the cases of $CH_3OH.O(CH_3)_2$ and $CH_3OH.N(CH_3)_3$ the highest wavenumber intermolecular mode is the out-of-plane hydrogen bond bending,

Table 6.12 Computed wavenumbers (RHF/6–31G**) of some selected intermolecular modes of the 1:1 methanol–dimethyl ether, methanol–trimethylamine, dimethylamine–dimethyl ether and dimethylamine–trimethylamine complexes

Complex	Wavenumber (cm^{-1})		
	ν(H...A)	δ(AH...B)	γ(AH...B)
CH$_3$OH.O(CH$_3$)$_2$[a]	82	149	653
CH$_3$OH.N(CH$_3$)$_3$[a]	106	157	723
(CH$_3$)$_2$NH.O(CH$_3$)$_2$[a]	65	109	64
(CH$_3$)$_2$NH.N(CH$_3$)$_3$[a]	64	120	60

[a] Bricknell (1996).

γ(AH...B). The values of these wavenumbers are 653 and 723 cm^{-1}, respectively, in agreement with the evidence provided by the intramolecular wavenumber shifts. For (CH$_3$)$_2$NH.O(CH$_3$)$_2$ and (CH$_3$)$_2$NH.N(CH$_3$)$_3$ the highest wavenumber intermolecular mode is the in-plane bending, δ(AH...B), and their wavenumbers (109 and 120 cm^{-1} respectively) bear out the earlier deductions. A similar corroboration of Kurnig et al.'s (1987) relationship has been observed by Yeo and Ford (1994) in the H$_2$O–NH$_3$–NH$_2$OH system.

Since the four complexes in this group all have qualitatively the same structure, it would be instructive to examine the ways in which the intensities of various modes involved in the hydrogen bonds change on complexation. The relative intensity perturbations should mirror the changes in the wavenumbers of the same modes discussed above. The computed intensities of the intramolecular modes (Bricknell, 1996) are given in Table 6.13.

Since the infrared intensities in the spectrum of a given molecule typically span several orders of magnitude, a more appropriate way of indicating changes in those intensities on complexation is by determining the ratios of the intensities rather than their differences, as was done for the wavenumbers. The complex/monomer intensity ratios are also given in Table 6.13. This table indicates that, apart from the OH and NH stretching modes, with very few other exceptions the ratios lie in a range from 0.5 to 2.0. These exceptions include the modes derived from the antisymmetric NC$_3$ bending vibration of (CH$_3$)$_3$N, which is virtually zero in the monomer, and the symmetric and antisymmetric CNC stretching and CNC bending modes of (CH$_3$)$_2$NH. The modes of the complexes which correlate with these monomer vibrations are all computed to be very weak.

In the context of the OH and NH stretching intensities, significant increases are observed on hydrogen bond formation, entirely consistent with the generalizations described by Pimentel and McClellan (1960). The proportional increase for ν(OH) is greater in CH$_3$OH.N(CH$_3$)$_3$ than in CH$_3$OH.O(CH$_3$)$_2$ and that for ν(NH) is larger in (CH$_3$)$_2$NH.N(CH$_3$)$_3$ than in (CH$_3$)$_2$NH.O(CH$_3$)$_2$. This confirms the evidence from the wavenumber shifts that N is a better proton acceptor than O. For the same proton acceptor, (CH$_3$)$_2$O or (CH$_3$)$_3$N, the enhancement of the NH stretching intensity is found to be greater than that for the OH. This is probably due more to the fact that the ν(NH) intensity in the (CH$_3$)$_2$NH monomer is very low (0.7 km mol^{-1}) (Bricknell, 1996) than to any apparent superiority of NH as a proton donor compared with OH, since the evidence provided by the wavenumber shifts indicates the reverse to be the case.

Table 6.13 Computed infrared band intensities (RHF/6−31G**) of some selected intermolecular modes of the 1:1 methanol–dimethyl ether, methanol–trimethylamine, dimethylamine–dimethyl ether and dimethylamine–trimethylamine complexes, and their ratios with respect to the corresponding monomer intensities

Complex	Monomer fragment	Complex mode	Intensity (km mol^{-1})	Intensity ratio[a]
$CH_3OH \cdot O(CH_3)_2$[b]	CH_3OH	ν_1	373.8	8.80
		ν_{12}	44.1	1.01
		ν_{16}	91.1	1.31
	$(CH_3)_2O$	ν_{17}	67.2	1.64
		ν_{18}	5.97	1.81
		ν_{34}	45.9	1.34
$CH_3OH \cdot N(CH_3)_3$[b]	CH_3OH	ν_1	570.3	13.42
		ν_{16}	33.6	0.77
		ν_{21}	102.6	1.47
	$(CH_3)_3N$	ν_{22}	20.2	0.85
		ν_{23}	35.6	1.77
		ν_{24}	0.16	5.16
		ν_{25}	22.8	1.93
		ν_{44}	25.8	1.09
		ν_{46}	0.16	5.16
$(CH_3)_2NH \cdot O(CH_3)_2$[b]	$(CH_3)_2NH$	ν_1	50.8	73.62
		ν_{19}	1.56	0.28
		ν_{20}	134.0	1.02
		ν_{22}	4.37	0.56
		ν_{45}	0.037	0.61
		ν_{46}	6.64	0.65
	$(CH_3)_2O$	ν_{18}	50.6	1.23
		ν_{21}	3.69	1.11
		ν_{44}	38.3	1.12
$(CH_3)_2NH \cdot N(CH_3)_3$[b]	$(CH_3)_2NH$	ν_1	90.3	130.87
		ν_{24}	0.024	0.00
		ν_{25}	110.4	0.84
		ν_{28}	3.23	0.41
		ν_{53}	0.005	0.08
		ν_{56}	4.94	0.48
	$(CH_3)_3N$	ν_{23}	23.8	1.00
		ν_{26}	40.2	2.00
		ν_{27}	0.041	1.32
		ν_{29}	19.8	1.68
		ν_{55}	23.2	0.98
		ν_{57}	0.015	0.48

[a] Ratio = complex/monomer intensity.
[b] Bricknell (1996).

The same set of four complexes will be used to illustrate the effects of hydrogen bonding on the computed Raman spectra of the component monomer species. The Raman data, specifically the Raman scattering intensity and the depolarization ratio of each mode, are gathered together in Table 6.14 (Ford and Bricknell, 1994). In order to quantify the effects of hydrogen bonding on the Raman data, the complex/monomer intensity ratios and the differences between the depolarization ratios for the corresponding totally symmetric

Table 6.14 Computed Raman intensities and depolarization ratios (RHF/6–31G**) of some selected intramolecular modes of the 1:1 methanol–dimethyl ether, methanol–trimethylamine, dimethylamine–dimethyl ether and dimethylamine–trimethylamine complexes

Complex	Monomer fragment	Complex mode	Raman intensity (10^{-12} m^4 kg^{-1})	Depolarization ratio
$CH_3OH.O(CH_3)_2$[a]	CH_3OH	ν_1	6.94	0.3209
		ν_{12}	0.12	0.7132
		ν_{16}	0.16	0.6444
	$(CH_3)_2O$	ν_{17}	0.72	0.3204
		ν_{18}	0.02	0.0081
		ν_{34}	0.36	0.75
$CH_3OH.N(CH_3)_3$[a]	CH_3OH	ν_1	10.16	0.3374
		ν_{16}	0.17	0.7058
		ν_{21}	0.44	0.3760
	$(CH_3)_3N$	ν_{22}	0.58	0.7498
		ν_{23}	0.77	0.1265
		ν_{24}	0.05	0.7499
		ν_{25}	0.02	0.0005
		ν_{44}	0.58	0.75
		ν_{46}	0.005	0.75
$(CH_3)_2NH.O(CH_3)_2$[a]	$(CH_3)_2NH$	ν_1	7.49	0.2925
		ν_{19}	0.54	0.3176
		ν_{20}	0.21	0.4562
		ν_{22}	0.01	0.0437
		ν_{45}	0.42	0.75
		ν_{46}	0.10	0.75
	$(CH_3)_2O$	ν_{18}	0.76	0.3368
		ν_{21}	0.02	0.0134
		ν_{44}	0.33	0.75
$(CH_3)_2NH.N(CH_3)_3$[a]	$(CH_3)_2NH$	ν_1	10.33	0.3026
		ν_{24}	0.54	0.3342
		ν_{25}	0.29	0.2276
		ν_{28}	0.01	0.0890
		ν_{53}	0.41	0.75
		ν_{56}	0.14	0.75
	$(CH_3)_3N$	ν_{23}	0.56	0.7500
		ν_{26}	0.69	0.1332
		ν_{27}	0.01	0.7450
		ν_{29}	0.01	0.0107
		ν_{55}	0.55	0.75
		ν_{57}	0.005	0.75

[a] Ford and Bricknell (1994).

modes of the complexes and the monomers have been determined (Bricknell, 1996), as was done for the infrared intensities and the wavenumber shifts (see Tables 6.11 and 6.13). These changes on complexation are presented in Table 6.15. Again the major perturbations are in the properties of the proton donors. The OH and NH intensities consistently increase, by factors of between 1.4 and 2.5. The $\delta(COH)$ modes decrease in intensity as a result of hydrogen bond formation, while the $\delta_s(CHN)$ modes change erratically. Conversely, the $\nu(CO)$ mode intensities are enhanced quite substantially, while those of the $\nu_s(CNC)$ modes are reduced to three-quarters of their value in the $(CH_3)_2NH$ monomer.

Table 6.15 Computed Raman intensity ratios and depolarization ratio differences (RHF/6–31G**) of some selected intramolecular modes of the 1:1 methanol–dimethyl ether, methanol–trimethylamine, dimethylamine–dimethyl ether and dimethylamine–trimethylamine complexes relative to those of the monomers

Complex	Monomer fragment	Complex mode	Raman intensity ratio[a]	Depolarization ratio difference[b]
$CH_3OH \cdot O(CH_3)_2$[c]	CH_3OH	ν_1	1.68	0.0261
		ν_{12}	0.60	−0.0055
		ν_{16}	1.78	−0.0682
	$(CH_3)_2O$	ν_{17}	0.96	−0.0755
		ν_{18}	0.67	−0.0077
		ν_{34}	1.13	0.00
$CH_3OH \cdot N(CH_3)_3$[c]	CH_3OH	ν_1	2.45	0.0426
		ν_{16}	0.85	−0.0129
		ν_{21}	4.89	−0.3366
	$(CH_3)_3N$	ν_{22}	0.95	−0.0002
		ν_{23}	1.00	−0.0338
		ν_{24}	0.60	−0.0001
		ν_{25}	0.66	−0.0453
		ν_{44}	0.95	0.00
		ν_{46}	1.67	0.00
$(CH_3)_2NH \cdot O(CH_3)_2$[c]	$(CH_3)_2NH$	ν_1	1.49	0.0129
		ν_{19}	0.75	−0.0074
		ν_{20}	0.88	−0.1041
		ν_{22}	0.50	−0.0830
		ν_{45}	2.10	0.00
		ν_{46}	2.50	0.00
	$(CH_3)_2O$	ν_{18}	1.01	−0.0591
		ν_{21}	0.67	−0.0024
		ν_{44}	1.03	0.00
$(CH_3)_2NH \cdot N(CH_3)_3$[c]	$(CH_3)_2NH$	ν_1	2.05	0.0230
		ν_{24}	0.75	0.0092
		ν_{25}	1.21	−0.3327
		ν_{28}	0.05	−0.0377
		ν_{53}	2.05	0.00
		ν_{56}	3.50	0.00
	$(CH_3)_3N$	ν_{23}	0.92	0.0000
		ν_{26}	0.90	−0.0271
		ν_{27}	1.67	−0.0050
		ν_{29}	0.33	−0.0351
		ν_{55}	0.90	0.00
		ν_{57}	1.67	0.00

[a] Ratio = complex/monomer intensity.
[b] Difference = complex − monomer depolarization ratio.
[c] Bricknell (1996).

While these observations are quite consistent among the four complexes, the significance of the directions and magnitudes of the changes is not clear, and further theoretical work is clearly indicated to enable a sensible interpretation of these results to be made.

With the exception of the in-plane hydrogen bond bending wavenumber, the other spectroscopic properties discussed for this set of four hydrogen-bonded complexes vary with the strength of the hydrogen bond, determined by the computed enthalpy of

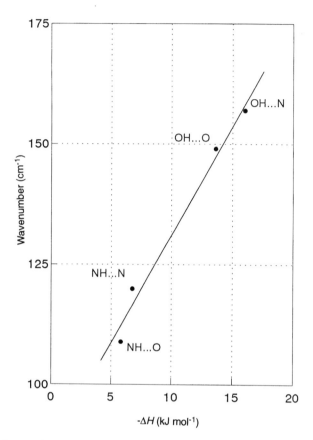

Figure 6.4 Plot of the in-plane hydrogen bond bending wavenumbers of $CH_3OH.O(CH_3)_2$, $CH_3OH.N(CH_3)_3$, $(CH_3)_2NH.O(CH_3)_2$ and $(CH_3)_2NH.N(CH_3)_3$ against the computed hydrogen bond enthalpies

hydrogen bond formation (Bricknell *et al.*, 1995), in such a manner that plots of these properties versus the hydrogen bond enthalpy separate into distinct correlation lines, one for the OH...B and one for the NH...B complexes. A plot of the wavenumber of the $\delta(AH...B)$ mode for each of the four complexes against the computed hydrogen bond enthalpies, however, indicates a sensible linear relationship which includes all four complexes, as shown in Figure 6.4. This observation suggests that this particular intermolecular mode may be used as a good diagnostic parameter in measuring the strength of a hydrogen bond experimentally, and it represents a refinement of the proposal of Kurnig *et al.* (1987) referred to above.

6.3.4 The Hydrogen Fluoride–Hydrogen Chloride–Methane–Silane System

This system of four 1:1 molecular complexes again illustrates a number of different bonding types, between two pairs of otherwise very similar component molecules. The greater polarity of the HF bond, compared with HCl, and the reversal of the bond polarity from $C^{\delta-}$—$H^{\delta+}$ in methane to $Si^{\delta+}$—$H^{\delta-}$ in silane, result in the appearance of four quite different equilibrium structures for these complexes (Govender and Ford, 1995a). The complexes containing HCl are genuinely hydrogen bonded, with HCl as a conventional proton donor. In ClH.CH$_4$, however, the proton acceptor atom is the carbon atom, and in ClH.SiH$_4$ the site of proton acceptance is one of the hydrogen atoms of silane. Both complexes belong to the C$_{3v}$ point group (see Figures 6.5(b) and 6.5(d)). In the CH$_4$.HF complex, one of the methane hydrogen atoms donates to the fluorine atom while a secondary interaction, between the HF hydrogen atom and the methane carbon atom, induces a distortion of the complex structure from C$_{3v}$ to C$_s$. A secondary interaction is also present in SiH$_4$.HF, where the primary attraction is between the F and Si atoms. This is a fairly common phenomenon, associated with the known ability of silicon-containing compounds to form five-coordinate species (Boal and Ozin, 1973). In this case the secondary interaction occurs between the $H^{\delta+}$ atom of HF and one of the $H^{\delta-}$ atoms of SiH$_4$, also resulting in a complex of C$_s$ symmetry. The equilibrium structures of CH$_4$.HF and SiH$_4$.HF are shown in Figures 6.5(a) and 6.5(c) (Govender and Ford, 1995a). These structures are consistent with those of CH$_4$.HF (Legon *et al.*, 1990) and ClH.CH$_4$ (Legon *et al.*, 1990; Ohshima and Endo, 1990) observed in the gas phase.

The computed wavenumbers of the four complexes are listed in Table 6.16, along with the assignments of their normal modes (Govender and Ford, 1995b). Table 6.17 shows the wavenumber shifts calculated from these values and those of the corresponding modes of the component monomers (Govender and Ford, 1995b). Once again, the major wavenumber changes occurring on hydrogen bond formation are those of the proton donor molecules, i.e. the HF and HCl stretching vibrations. In the case of the axial complexes, in which only a single interaction is present, the greater v(HCl) shift is found in ClH.SiH$_4$, indicating that the ClH...H hydrogen bond is stronger than the ClH...C, other factors being equal. This observation is consistent with the fact that, while the interaction energies of both complexes are very close, -2.19 kJ mol^{-1} for ClH.SiH$_4$ and -1.93 kJ mol^{-1} for ClH.CH$_4$ (Govender and Ford, 1995a), that of ClH.SiH$_4$ is fractionally higher. The displacements of the CH$_4$ and SiH$_4$ modes are usually smaller; only the v_s(SiH$_4$) mode of ClH.SiH$_4$ has a shift comparable with that of v(HCl). The v_s(CH$_4$) mode shift of ClH.CH$_4$ is also significant, and in this complex the antisymmetric CH$_4$ stretching vibration also undergoes a moderate red shift.

In the CH$_4$.HF and SiH$_4$.HF complexes the largest shifts are those of the HF stretching vibrations; in these cases, too, the shift in SiH$_4$.HF exceeds that in CH$_4$.HF. These shifts, however, are not consistent with the computed interaction energies; these are -3.08 and -0.14 kJ mol^{-1} for CH$_4$.HF and SiH$_4$.HF respectively (Govender and Ford, 1995a). This is because in these two complexes a second interaction is present, and the CH...F and Si...F interactions account for only a part of the total interaction energy. None of the shifts of the CH$_4$ and SiH$_4$ sub-units exceeds 20 cm^{-1}; their signs, mode for mode, agree with one another quite well. Clearly, in this set of complexes, neither the spectra of the CH$_4$ nor those of the SiH$_4$ fragments are particularly sensitive to the molecular interaction.

The spectra of ClH.CH$_4$ (Barnes, 1983) and of CH$_4$.HF (Davis and Andrews, 1987a,b) and SiH$_4$.HF (Davis and Andrews 1987a, 1989) have been studied in argon matrices. In ClH.CH$_4$, Barnes (1983) observed a red shift of the HCl stretching mode of 16 cm^{-1}, fortuitously exactly equal to that calculated by Govender and Ford (1995a,b). In CH$_4$.HF Davis and Andrews (1987a,b) found a red shift of 23 cm^{-1} relative to the HF monomer;

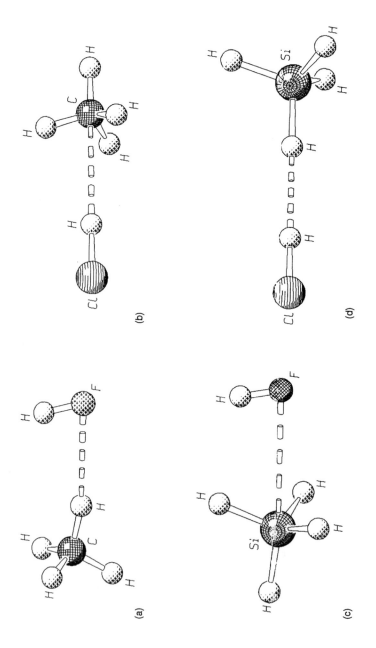

Figure 6.5 Structures of the (a) methane–hydrogen fluoride, (b) hydrogen chloride–methane, (c) silane–hydrogen fluoride and (d) hydrogen chloride–silane complexes

Table 6.16 Computed wavenumbers (MP2/6−31G**) of the 1:1 methane–hydrogen fluoride, hydrogen chloride–methane, silane–hydrogen fluoride and hydrogen chloride–silane complexes

Complex	Symmetry species	Mode	Wavenumber (cm^{-1})	Approximate description
CH$_4$.HF[a]	a'	ν_1	4175	ν(HF)
		ν_2	3299	ν_s(CH$_2$)
		ν_3	3275	ν(CH) (free)
		ν_4	3134	ν(CH) (bonded)
		ν_5	1633	δ(HCH)
		ν_6	1417	δ(CH$_2$)
		ν_7	1404	ω(CH$_2$)
		ν_8	243	δ(HF...H)(IM)
		ν_9	144	l(HF)(IM)
		ν_{10}	110	ν(H...F)(IM)
	a''	ν_{11}	3273	ν_a(CH$_2$)
		ν_{12}	1632	ρ(CH$_2$)
		ν_{13}	1415	τ(CH) (free)
		ν_{14}	156	l(CH$_4$)(IM)
		ν_{15}	14	l(HF)(IM)
ClH.CH$_4$[a]	a_1	ν_1	3277	ν(CH)
		ν_2	3127	ν_s(CH$_3$)
		ν_3	3112	ν(HCl)
		ν_4	1400	δ_s(CH$_3$)
		ν_5	74	ν(C...H)(IM)
	e	ν_6	3270	ν_a(CH$_3$)
		ν_7	1624	δ_a(CH$_3$)
		ν_8	1407	ρ(CH$_3$)
		ν_9	124	δ(ClH...C)(IM)
		ν_{10}	78	l(CH$_4$)(IM)
SiH$_4$.HF[a]	a'	ν_1	4167	ν(HF)
		ν_2	2361	ν_s(SiH$_2$)
		ν_3	2341	ν(SiH) (free)
		ν_4	2326	ν(SiH) (bonded)
		ν_5	1029	δ(HSiH)
		ν_6	983	δ(SiH$_2$)
		ν_7	977	ω(SiH$_2$)
		ν_8	271	δ(HF...Si)(IM)
		ν_9	161	l(HF)(IM)
		ν_{10}	90	ν(Si...F)(IM)
	a''	ν_{11}	2369	ν_a(SiH$_2$)
		ν_{12}	1021	τ(SiH) (free)
		ν_{13}	981	ρ(SiH$_2$)
		ν_{14}	175	l(SiH$_4$)(IM)
		ν_{15}	99	l(HF)(IM)
ClH.SiH$_4$[a]	a_1	ν_1	3097	ν(HCl)
		ν_2	2349	ν_s(SiH$_3$)
		ν_3	2325	ν(SiH)
		ν_4	969	δ_s(SiH$_3$)
		ν_5	54	ν(H...H)(IM)
	e	ν_6	2360	ν_a(SiH$_3$)
		ν_7	1014	δ_a(SiH$_3$)
		ν_8	974	ρ(SiH$_3$)
		ν_9	179	δ(ClH...H)(IM)
		ν_{10}	13	δ(SiH...H)(IM)

[a]Govender and Ford (1995b).
IM—intermolecular mode.

Table 6.17 Computed complex-monomer wavenumber shifts (MP2/6–31G**) of the 1:1 methane–hydrogen fluoride, hydrogen chloride–methane, silane–hydrogen fluoride and hydrogen chloride–silane complexes

Monomer unit	Symmetry species	Mode	Wavenumber shift[a] (cm^{-1})				
			CH$_4$.HF	SiH$_4$.HF		ClH.CH$_4$	ClH.SiH$_4$
HF		ν(HF)	−22	−30		−	−
HCl		ν(HCl)	−	−		−16	−31
CH$_4$	a_1	ν_s(CH$_4$)	−9	−		−16	−
	e	δ_s(CH$_4$)(a′)	10	−	(e)	1	−
		(a″)	9	−		−	−
	t_2	ν_a(CH$_4$)(a′)	10	−	(a_1)	−12	−
		(a′)	−14	−	(e)	−19	−
		(a″)	−16	−		−	−
		δ_a(CH$_4$)(a′)	2	−	(a_1)	−2	−
		(a′)	15	−	(e)	5	−
		(a″)	13	−		−	−
SiH$_4$	a_1	ν_s(SiH$_4$)	−	−19		−	−20
	e	δ_s(SiH$_4$)(a′)	−	12		−	−3
		(a″)	−	4		−	−
	t_2	ν_a(SiH$_4$)(a′)	−	6	(a_1)	−	−6
		(a′)	−	−14	(e)	−	5
		(a″)	−	14		−	−
		δ_a(SiH$_4$)(a′)	−	11	(a_1)	−	−3
		(a′)	−	5	(e)	−	2
		(a″)	−	9		−	−

[a] Govender and Ford (1995b).

this is to be compared with the predicted shift of 22 cm^{-1} in the same direction (see Table 6.17). These workers also measured a red shift of 65 cm^{-1} for ν(HF) of SiH$_4$.HF and a blue shift of 51 cm^{-1} from the ν_a(SiH$_4$) wavenumber of the SiH$_4$ monomer (Davis and Andrews, 1987a, 1989), although to which of the three distinct complex modes this corresponds was not specified. Govender and Ford's (1995b) computed SiH$_4$.HF displacements were −30 cm^{-1} for ν(HF) and 6 and 14 cm^{-1} for the two SiH stretching modes blue-shifted from ν_a(SiH$_4$). While for CH$_4$.HF and ClH.CH$_4$ the predicted shifts are in remarkably good agreement with the experimental ones, for SiH$_4$.HF the computed values are only about a half or a third of those determined experimentally. Davis and Andrews (1987a,b, 1989) have also (1987a, 1989) reported bands in the intermolecular mode region in argon, at 130 cm^{-1} in CH$_4$.HF and 165 cm^{-1} in SiH$_4$.HF, which they have assigned to the HF librational mode. The predicted positions of the in-plane librational modes are 144 and 161 cm^{-1} respectively (Govender and Ford, 1995b). For such low-lying vibrations, where agreement is not generally found to be very close, as pointed out above, the degree of harmony in the cases of these two bands of CH$_4$.HF and SiH$_4$.HF is unusually good.

6.3.5 The Complexes of Water with Carbon Dioxide and Nitrous Oxide

The final system we wish to discuss is that of the pair of complexes formed between water and the isoelectronic analogues carbon dioxide and nitrous oxide. This system was chosen in order to sound a note of caution, by illustrating the fact that not every 1:1 molecular complex lends itself to smooth geometry optimization followed by normal mode analysis leading to the identification of a genuine minimum on a particular potential energy surface. Peterson and Klemperer (1984), by means of MBER spectroscopy, determined the gas-phase structure of the carbon dioxide-water complex to be a planar C_{2v} species, containing a C...O interaction perpendicular to the CO_2 axis, rather similar to that formed between water and cyanogen, discussed above (Lee *et al.*, 1992). This structure is illustrated in Figure 6.6(a); it has been confirmed on the basis of the matrix isolation infrared spectroscopic work of Fredin *et al.* (1975) and Tso and Lee (1985a,b), and by a variety of *ab initio* theoretical studies (Jönsson *et al.*, 1975, 1976, 1977, 1978; George *et al.*, 1982; Nguyen and Ha, 1984; Liang and Lipscomb, 1986; Sokalski, 1986; Reed *et al.*, 1986; Hurst *et al.*, 1986; Nguyen *et al.*, 1987; Damewood *et al.*, 1989; Block *et al.*, 1992; Cox *et al.*, 1993). Both Damewood *et al.* (1989) and Cox *et al.* (1993) examined a number of plausible complex structures, using different basis sets. Cox *et al.* observed that, while the experimental planar C_{2v} structure was found to have the lowest energy of all the species considered, using the MP2/6−31G** combination, it was found to have two negative eigenvalues. Some alternative structures, e.g. the bifurcated C_{2v}, cyclic C_{2v} and 'straddled' C_s isomers, had higher absolute energies but had only one negative eigenvalue. Particularly noteworthy was the fact that they could not achieve a successful geometry optimization starting with a 'linear' hydrogen-bonded structure; this structure invariably collapsed to the planar C_{2v} form. This finding contrasts with that of Block *et al.* (1992), who were able, using a larger basis set, to force convergence for a linear C_s hydrogen-bonded structure, which they found to be a genuine minimum. The usual explanation for behaviour of this type is that the potential energy surface of the complex is very flat in the region of the minimum, and either the level of theory or the flexibility of the basis set, or both, are inadequate for ensuring a successful optimization.

In the case of the nitrous oxide-water complex, Cox *et al.* (1993) found two genuine minima, a lower-energy planar N...O bonded form, similar to the C_{2v} $CO_2.OH_2$ complex but with one of the water OH bonds tilted significantly towards the nitrous oxide O atom, apparently stabilized by a secondary H...O interaction (see Figure 6.6(b)), and a higher-energy 'linear' OH...N hydrogen bonded form. A similar *ab initio* study by Sadlej and Sicinski (1990) also identified the same two stable structures, while Slanina (1991)

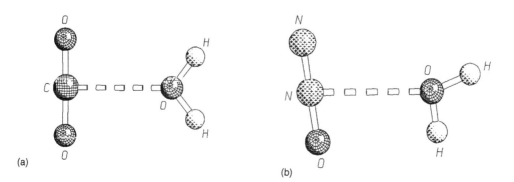

Figure 6.6 Structures of the (a) carbon dioxide-water and (b) nitrous oxide-water complexes

distinguished the N...O bonded structure as the more stable of the two at low temperature, but observed that the OH...N hydrogen-bonded isomer became predominant at higher temperatures. The N...O bonded structure has been confirmed experimentally in the gas phase by MBER spectroscopy by Zolandz *et al.* (1992), who proposed a planar T-shaped arrangement similar to that found in $CO_2.OH_2$ (Peterson and Klemperer, 1984), with a pronounced tilt of one of the OH bonds of about 20° towards the NO bond, and of the NNO axis of about 9° from the perpendicular to the N...O bond.

The computed wavenumbers, including the two negative ones, for the b_1 water wagging mode and the b_2 carbon dioxide libration of $CO_2.OH_2$, are given in Table 6.18 (Cox *et al.*, 1993). Table 6.19 lists the complex-monomer wavenumber shifts, which are all marginal, for both sub-units. Both water-stretching modes increase minimally in wavenumber on complexation, and the bending mode suffers a small wavenumber decrease, as was the case in $C_2N_2.OH_2$ (see Table 6.9). Furthermore, the CO_2 stretching modes undergo small blue shifts, and the degenerate bending vibration splits into two modes, shifted by 8 cm^{-1} to either side of the monomer value. These small shifts are consistent with a weak interaction; the computed interaction energy (Cox *et al.*, 1994) is only -9.06 kJ mol^{-1}. Table 6.20 shows the wavenumber shifts determined experimentally for $CO_2.OH_2$ in nitrogen (Fredin *et al.*, 1975) and oxygen matrices (Tso and Lee, 1985a,b). The predicted

Table 6.18 Computed wavenumbers (MP2/6–31G**) of the 1:1 carbon dioxide–water and nitrous oxide–water complexes

Complex	Symmetry species	Mode	Wavenumber (cm^{-1})	Approximate description
$CO_2.OH_2$[a]	a_1	ν_1	3896	$\nu_s(OH_2)$
		ν_2	1674	$\delta(OH_2)$
		ν_3	1340	$\nu_s(CO_2)$
		ν_4	634	$\delta(CO_2)$
		ν_5	125	$\nu(C...O)(IM)$
	a_2	ν_6	149	$\tau(OH_2)(IM)$
	b_1	ν_7	650	$\gamma(CO_2)$
		ν_8	(negative)	$\omega(OH_2)(IM)$
	b_2	ν_9	4035	$\nu_a(OH_2)$
		ν_{10}	2459	$\nu_a(CO_2)$
		ν_{11}	186	$\rho(OH_2)(IM)$
		ν_{12}	(negative)	$l(CO_2)(IM)$
$N_2O.OH_2$[a]	a'	ν_1	4033	$\nu_a(OH_2)$
		ν_2	3902	$\nu_s(OH_2)$
		ν_3	2281	$\nu(NN)$
		ν_4	1692	$\delta(OH_2)$
		ν_5	1303	$\nu(NO)$
		ν_6	591	$\delta(NNO)$
		ν_7	150	$\rho(OH_2)(IM)$
		ν_8	90	$\nu(N...O)(IM)$
		ν_9	26	$l(NNO)(IM)$
	a''	ν_{10}	591	$\gamma(NNO)$
		ν_{11}	150	$\tau(OH_2)(IM)$
		ν_{12}	31	$\omega(OH_2)(IM)$

[a]Cox *et al.* (1993).
IM—intermolecular mode.

Table 6.19 Computed complex-monomer wavenumber shifts (MP2/6−31G**) of the 1:1 carbon dioxide–water and nitrous oxide–water complexes

Monomer unit	Symmetry species	Mode	Wavenumber shift (cm^{-1})	
			$CO_2.OH_2$[a]	$N_2O.OH_2$[a]
H_2O	a_1	$\nu_s(OH_2)$	1	7
		$\delta(OH_2)$	−9	9
	b_1	$\nu_a(OH_2)$	1	−1
CO_2	Σ_g^+	$\nu_s(CO_2)$	3	−
	Π_u	$\delta(CO_2)(a_1)$	−8	−
		(b_1)	8	−
	Σ_u^-	$\nu_a(CO_2)$	4	−
N_2O	Σ^+	$\nu(NN)$	−	25
		$\nu(NO)$	−	10
	Π	$\delta(NNO)(a')$	−	7
		(a'')	−	7

[a] Cox et al (1993).

Table 6.20 Experimental complex-monomer wavenumber shifts of the 1:1 carbon dioxide–water complex observed in nitrogen and oxygen matrices

Monomer unit	Symmetry species	Mode	Wavenumber shift (cm^{-1})		
			N_2[a]	O_2[b]	O_2[c]
H_2O	a_1	$\nu_s(OH_2)$	−2.3	−[d]	−3.5
		$\delta(OH_2)$	0.9	−12.1	−8.9
	b_1	$\nu_a(OH_2)$	−1.9	−1.0	2.9
CO_2	Σ_g^+	$\nu_s(CO_2)$	−[e]	−[e]	−[e]
	Π_u	$\delta(CO_2)(a_1)$	−6.0		−7.9
		(b_1)	3.0		1.2
	Σ_u^-	$\nu_a(CO_2)$	2		0.8[f]

[a] Fredin et al. (1975).
[b] Tso and Lee (1985a).
[c] Tso and Lee (1985b).
[d] Not observed.
[e] Monomer mode is infrared inactive.
[f] Average of matrix-split bands.

small shifts of the water sub-unit are confirmed; all shifts are less than 10 cm^{-1} except that of $\delta(OH_2)$ in oxygen (Tso and Lee, 1985a). The signs of the shifts are mostly in agreement, although there is some discrepancy between matrices, and in the case of the $\nu_a(OH_2)$ mode in oxygen, between methods of preparation of the sample (Tso and Lee, 1985a,b). The predicted splitting of 16 cm^{-1} between the components of the $\delta(CO_2)$ band is larger than those observed experimentally, 9.0 cm^{-1} (Fredin et al., 1975) and 9.1 cm^{-1} (Tso and Lee, 1985b). Neither of the two groups which studied the matrix spectra of $CO_2.OH_2$ was able to assign the upper and lower components of the split $\delta(CO_2)$ band of the CO_2 fragment to the a_1 or the b_1 mode in the complex (Fredin et al., 1975, Tso and Lee, 1985b). Tso and Lee, in fact, noted that it would be interesting to identify these modes through an *ab initio* calculation. Both Block et al. (1992) and Cox et al. (1993)

agree that the out-of-plane (b_1) CO_2 bending vibration of $CO_2.OH_2$ occurs at the higher wavenumber.

The computed spectrum of the more stable N...O bonded isomer of $N_2O.OH_2$ is also reported in Table 6.18, and the relevant complex–monomer wavenumber shifts in Table 6.19 (Cox *et al.*, 1993). As was the case for the $CO_2.OH_2$ complex, the water wavenumber shifts are very small. In this case the lower wavenumber stretching mode is blue–shifted by 7 cm^{-1} and the higher wavenumber counterpart shifts by 1 cm^{-1} to lower value. The bending mode undergoes a small blue shift of 9 cm^{-1}. These minimal shifts are indicative, once again, of a very weak interaction, calculated to be -6.40 kJ mol^{-1} (Cox *et al.*, 1994). The nitrous oxide wavenumber perturbations are slightly larger than those of CO_2 in $CO_2.OH_2$, and are all to the blue. Unexpectedly, the NN stretching mode suffers a larger blue shift than the NO stretching, even though the secondary interaction occurs through the oxygen, rather than through the terminal nitrogen atom. There do not appear to be any experimental matrix isolation infrared data on this complex in the literature.

6.4 SUMMARY

In this chapter we have attempted to illustrate the utility of *ab initio* molecular orbital calculations in predicting the vibrational spectra of the binary molecular complexes formed between a variety of small molecules. For consistency, we have selected several series of complexes for which the computations were carried out at the same level of theory and using the same basis set. These groups of complexes were associated through the presence of common proton donors or acceptors, or through common electron donors or acceptors, as appropriate. Within each set the calculations show that by varying the components of the complexes, the nature of the interaction could change drastically, with consequent variations in the predicted wavenumber shifts and intensity changes of the infrared bands of the adducts. Where possible, the computed spectra have been compared alongside the spectra measured experimentally in cryogenic matrices. Allowing for the standard overestimation of the computed wavenumbers, the essential features of the experimental spectra, almost without exception, are faithfully reproduced. One set of related complexes was studied at a lower level of theory, on account of the comparative complexity of their structures. This enabled the Raman spectra of these complexes to be examined, the computation of which is not available at higher levels of theory in the computer package we have described. While the interpretation of the changes in the Raman spectra on complexation is not as well established as that of the infrared spectral perturbations at this stage, a number of consistent variations in the spectra are reported, which parallel those of the infrared changes in several ways. The potential for the extension of the routine calculation of infrared spectra to include Raman spectra is obvious, and promises to be a valuable tool for experimental Raman spectroscopists in the interpretation of their data.

ACKNOWLEDGMENTS

The author wishes to thank the Foundation for Research Development and the University of Natal Research Fund for ongoing financial support, and Dr Geoffrey Yeo, Messrs

Tankiso Tshehla, Bradley Bricknell and Maganthran Govender, Professor Leslie Glasser and Miss Amanda Cox for assistance in generating the data discussed in this chapter.

REFERENCES

Andrews, L. (1988) *Faraday Discuss. Chem. Soc.*, No. 86, 37.
Andrews, L. (1990) in *Vibrational Spectra and Structure*, Vol. 18, edited by Durig, J. R., Elsevier, Amsterdam, pp. 183–216.
Andrews, L., Johnson, G. L. and Kelsall, B. J. (1982) *J. Phys. Chem.*, **86**, 3375.
Barnes, A. J. (1983) *J. Mol. Structure*, **100**, 259.
Bertie, J. E. and Falk, M. V. (1973) *Can. J. Chem.*, **51**, 1713.
Besler, B. H., Scuseria, G. E., Scheiner, A. C. and Schaefer III, H. F. (1988) *J. Chem. Phys.*, **89**, 360.
Binkley, J. S., Whiteside, R. A., Krishnan, R., Seeger, R., DeFrees, D. J., Schlegel, H. B., Topiol, S., Kahn, L. R. and Pople, J. A. (1980) *GAUSSIAN-80*, Carnegie–Mellon Quantum Chemistry Publishing Unit, Carnegie-Mellon University, Pittsburgh.
Binkley, J. S., Frisch, M. J., DeFrees, D. J., Raghavachari, K., Whiteside, R. A., Schlegel, H. B., Fluder, E. M. and Pople, J. A. (1982) *GAUSSIAN-82*, Carnegie-Mellon Quantum Chemistry Publishing Unit, Carnegie-Mellon University, Pittsburgh.
Block, P. A., Marshall, M. D., Pedersen, L. G. and Miller, R. E. (1992) *J. Chem. Phys.*, **96**, 7321.
Boal, D. and Ozin, G. A. (1973) *Can. J. Chem.*, **51**, 609.
Bricknell, B. C. (1996) PhD thesis, University of Natal, Durban.
Bricknell, B. C., Letcher, T. M. and Ford T. A. (1995) *S. Afr. J. Chem.*, in press.
Chin, S. and Person, W. B. (1984) *J. Phys. Chem.*, **88**, 553.
Cox, A. J., Ford, T. A. and Glasser, L. (1993) in *Structures and Conformations of Non-Rigid Molecules*, edited by Laane, J., Dakkouri, M., van der Veken, B. and Oberhammer, H., Kluwer, Dordrecht, pp. 391–408.
Cox, A. J., Ford, T. A. and Glasser, L. (1994) *J. Mol. Structure (Theochem)*, **312**, 101.
Damewood, Jr, J. A., Kumpf, R. A. and Muhlbauer, W. C. F. (1989) *J. Phys. Chem.*, **93**, 7640.
Davis, S. R. and Andrews, L. (1987a) *J. Chem. Phys.*, **86**, 3765.
Davis, S. R. and Andrews, L. (1987b) *J. Am. Chem. Soc.*, **109**, 4768.
Davis, S. R. and Andrews, L. (1989) *J. Phys. Chem.*, **93**, 1273.
Del Bene, J. E. (1988) *J. Phys. Chem.*, **92**, 2874.
Diercksen, G. H. F., Kramer, W. P. and von Niessen, W. (1972) *Theoret. Chim. Acta*, **28**, 67.
Engdahl, A. and Nelander, B. (1983) *Chem. Phys. Letters*, **100**, 129.
Engdahl, A. and Nelander, B. (1989) *J. Chem. Phys.*, **91**, 6604.
Fillery-Travis, A. J., Legon, A. C. and Willoughby, L. C. (1984) *Proc. Roy. Soc. (London), Ser. A*, **396**, 405.
Ford, T. A. and Bricknell, B. C. (1994) in *Proceedings of the XIVth International Conference on Raman Spectroscopy, Hong Kong, 1994*, edited by Yu, N.-T. and Li, X.-Y., Wiley, Chichester, pp. 232–233.
Francl, M. M., Pietro, W. J., Hehre, W. J., Binkley, J. S., Gordon, M. S., DeFrees, D. J. and Pople, J. A. (1982) *J. Chem. Phys.*, **77**, 3654.
Fredin, L., Nelander, B. and Ribbegard, G. (1975) *Chem. Scripta*, **7**, 11.
Frisch, M. J., Binkley, J. S., Schlegel, H. B., Raghavachari, K., Melius, C. F., Martin, R. L., Stewart, J. J. P., Bobrowicz, F. W., Rohlfing, C. M., Kahn, L. R., DeFrees, D. J., Seeger, R., Whiteside, R. A., Fox, D. J., Fluder, E. M. and Pople, J. A. (1984) *GAUSSIAN-86*, Carnegie–Mellon Quantum Chemistry Publishing Unit, Carnegie-Mellon University, Pittsburgh.
Frisch, M. J., Head-Gordon, M., Schlegel, H. B., Raghavachari, K., Binkley, J. S., Gonzalez, C., DeFrees, D. J., Fox, D. J., Whiteside, R. A., Seeger, R., Melius, C. F., Baker, J., Martin, R., Kahn, L. R., Stewart, J. J. P., Fluder, E. M., Topiol, S. and Pople, J. A. (1988) *GAUSSIAN-88*, Gaussian, Inc., Pittsburgh.
Frisch, M. J., Head-Gordon, M., Trucks, G. W., Foresman, J. B., Schlegel, H. B., Raghavachari, K., Robb, M. A., Binkley, J. S., Gonzalez, C., DeFrees, D. J., Fox, D. J., Whiteside, R. A., Seeger, R.,

Melius, C. F., Baker, J., Martin, R., Kahn, L. R., Stewart, J. J. P., Topiol, S. and Pople, J. A. (1990) *GAUSSIAN-90*, Gaussian, Inc., Pittsburgh.
Frisch, M. J., Trucks, G. W., Head-Gordon, M., Gill, P. M. W., Wong, M. W., Foresman, J. B., Johnson, B. G., Schlegel, H. B., Robb, M. A., Replogle, E. S., Gomperts, R., Andres, R. L., Raghavachari, K., Binkley, J. S., Gonzalez, C., Martin, R. L., Fox, D. J., DeFrees, D. J., Baker, J., Stewart, J. J. P. and Pople, J. A. (1992) *GAUSSIAN-92*, Gaussian, Inc., Pittsburgh.
Frisch, M. J., Trucks, G. W., Schlegel, H. B., Gill, P. M. W., Johnson, B. G., Robb, M. A., Cheeseman, J. R., Keith, T., Petersson, G. A., Montgomery, J. A., Raghavachari, K., Al-Laham, M. A., Zakrzewski, V. G., Ortiz, J. V., Foresman, J. B., Cioslowski, J., Stefanov, B. B., Nanayakkara, A., Challacombe, M., Peng, C. Y., Ayala, P. Y., Chen, W., Wong, M. W., Andres, J. L., Replogle, E. S. Gomperts, R., Martin, R. L., Fox, D. J., Binkley, J. S., DeFrees, D. J., Baker, J., Stewart, J. J. P., Head-Gordon, M., Gonzalez, C. and Pople, J. A. (1995) *GAUSSIAN-94*, Gaussian, Inc., Pittsburgh.
Frisch, M. J., Yamaguchi, Y., Schaefer III, H. F. and Binkley, J. S. (1986) *J. Chem. Phys.*, **84**, 531.
George, P., Bock, C. W. and Trachtman, M. (1982) *J. Comput. Chem.*, **3**, 283.
Govender, M. G. and Ford, T. A. (1995a) *S. Afr. J. Chem.*, **48**, 98.
Govender, M. G. and Ford, T. A. (1995b) *J. Mol. Structure (Theochem)*, **338**, 141.
Gutowsky, H. S., Germann, T. C., Augspurger, J. D. and Dykstra, C. E. (1992) *J. Chem. Phys.*, **96**, 5808.
Hehre, W. J., Radom, L., Schleyer, P. von R. and Pople, J. A. (1986) *Ab initio Molecular Orbital Theory*, Wiley, New York, pp. 21, 22.
Herbine, P. and Dyke, T. R. (1985) *J. Chem. Phys.*, **83**, 3768.
Hinchliffe, A. (1986) *J. Mol. Structure (Theochem)*, **136**, 193.
Howard, B. J. and Brown, J. M. (1992) in *Applied Laser Spectroscopy*, edited by Andrews, D. L., VCH, New York, pp. 185-225.
Hurst, G. J. B., Fowler, P. W., Stone, A. J. and Buckingham, A. D. (1986) *Intern. J. Quantum Chem.*, **29**, 1223.
Jönsson, B., Karlström, G. and Wennerström, H. (1975) *Chem. Phys. Letters*, **30**, 58.
Jönsson, B., Karlström, G. and Wennerström, H. (1976) *Chem. Phys. Letters*, **41**, 317.
Jönsson, B., Karlström, G. and Wennerström, H. (1978) *J. Am. Chem. Soc.*, **100**, 1658.
Jönsson, B., Karlström, G., Wennerström, H., Forsen, S., Roos, B. and Almlöf, J. (1977) *J. Am. Chem. Soc.*, **99**, 4628.
Kerns, R. C. and Allen, L. C. (1978) *J. Am. Chem. Soc.*, **100**, 6587.
Knözinger, E. and Schrems, O. (1987) in *Vibrational Spectra and Structure*, Vol. 16, edited by Durig, J. R., Elsevier, Amsterdam, pp. 141-225.
Komornicki, A. and Jaffe, R. L. (1979) *J. Chem. Phys.*, **71**, 2150.
Krishnan, R., Schlegel, H. B. and Pople, J. A. (1980) *J. Chem. Phys.*, **72**, 4654.
Kurnig, I. J., Szczesniak, M. M. and Scheiner, S. (1987) *J. Chem. Phys.*, **87**, 2214.
Latajka, Z. and Scheiner, S. (1990) *J. Phys. Chem.*, **94**, 217.
Lee, S., Suni, I. I. and Klemperer, W. (1992) *J. Chem. Phys.*, **96**, 5577.
Legon, A. C. (1995) *Chem. Phys. Letters*, **247**, 24.
Legon, A. C., Roberts, B. P. and Wallwork, A. L. (1990) *Chem. Phys. Letters*, **173**, 107.
Liang, J.-Y. and Lipscomb, W. N. (1986) *J. Am. Chem. Soc.*, **108**, 5051.
Millen, D. J. and Mines, G. W. (1977) *J. Chem. Soc., Faraday Trans. II*, **73**, 369.
Miller, R. E. (1988) *Science*, **240**, 447.
Miller, R. E. (1990) *Accounts Chem. Research*, **23**, 10.
Møller, C. and Plesset, M. S. (1934) *Phys. Rev.*, **46**, 618.
Mulliken, R. S. (1955) *J. Chem. Phys.*, **23**, 1997.
Nelander, B. and Nord, L. (1982) *J. Phys. Chem.*, **86**, 4375.
Nguyen, M. T. and Ha, T.-K. (1984) *J. Am. Chem. Soc.*, **106**, 599.
Nguyen, M. T., Hegarty, A. J. and Ha, T.-K. (1987) *J. Mol. Structure*, **150**, 319.
Ohshima, Y. and Endo, Y. (1990) *J. Chem. Phys.*, **93**, 6256.
Person, W. B., Brown, K. G., Steele, D. and Peters, D. (1981) *J. Phys. Chem.*, **85**, 1998.
Peterson, K. I. and Klemperer, W. (1984) *J. Chem. Phys.*, **80**, 2439.
Pimentel, G. C. and McClellan, A. L. (1960) *The Hydrogen Bond*, Freeman, San Francisco, pp. 70, 71.

Pople, J. A., Krishnan, R., Schlegel, H. B. and Binkley, J. S. (1979) *Intern. J. Quantum Chem., Symp.*, **13**, 325.
Reed, A. E., Weinhold, F., Curtiss, L. A. and Pochatko, D. J. (1986) *J. Chem. Phys.*, **84**, 5687.
Sadlej, J. and Sicinski, M. (1990) *J. Mol. Structure (Theochem)*, **204**, 1.
Saykally, R. J. (1989) *Accounts Chem. Research*, **22**, 295.
Schlegel, H. B., Binkley, J. S. and Pople, J. A. (1984) *J. Chem. Phys.*, **80**, 1979.
Slanina, Z. (1991) *J. Mol. Structure (Theochem)*, **231**, 215.
Sokalski, W. A. (1986) *J. Mol. Structure*, **138**, 77.
Steele, D., Person, W. B. and Brown, K. G. (1981) *J. Phys. Chem.*, **85**, 2007.
Thomas, J. R., DeLeeuw, B. J., Vacek, G., Crawford, T. D., Yamaguchi, Y. and Schaefer III, H. F. (1993) *J. Chem. Phys.*, **99**, 403.
Tokhadze, K. G. and Tkhorzhevskaya, N. A. (1992) *J. Mol. Structure*, **270**, 351.
Tshehla, T. M. (1996) PhD thesis, University of Natal, Durban.
Tshehla, T. M. and Ford, T. A. (1994) *Bull. Polish Acad. Sci., Chem.*, **42**, 397.
Tshehla, T. M. and Ford, T. A. (1995) *S. Afr. J. Chem.*, **48**, 127.
Tso, T.-L. and Lee, E. K. C. (1985a) *J. Phys. Chem.*, **89**, 1612.
Tso, T.-L. and Lee, E. K. C. (1985b) *J. Phys. Chem.*, **89**, 1618.
Turi, L. and Dannenberg, J. J. (1993) *J. Phys. Chem.*, **97**, 7899.
Vishveshwara, S. (1978) *Chem. Phys. Letters*, **59**, 26.
Wilson, E. B., Decius, J. C. and Cross, P. C. (1955) *Molecular Vibrations*, McGraw-Hill, New York.
Yamaguchi, Y., Frisch, M., Gaw, J., Schaefer III, H. F. and Binkley, J. S. (1986) *J. Chem. Phys.*, **84**, 2262.
Yeo, G. A. (1991) PhD thesis, University of the Witwatersrand, Johannesburg.
Yeo, G. A. and Ford, T. A. (1991a) *Can. J. Chem.*, **69**, 632.
Yeo, G. A. and Ford, T. A. (1991b) *Chem. Phys. Letters*, **178**, 266.
Yeo, G. A. and Ford, T. A. (1991c) *J. Mol. Structure (Theochem)*, **235**, 123.
Yeo, G. A. and Ford, T. A. (1991d) *Spectrochim. Acta*, **47A**, 485.
Yeo, G. A. and Ford, T. A. (1991e) *Spectrochim. Acta*, **47A**, 919.
Yeo, G. A. and Ford, T. A. (1991f) *Vibr. Spectrosc.*, **2**, 173.
Yeo, G. A. and Ford, T. A. (1992) *J. Mol. Structure*, **266**, 183.
Yeo, G. A. and Ford, T. A. (1994) *Spectrochim. Acta*, **50A**, 5.
Zolandz, D., Yaron, D., Peterson, K. I. and Klemperer, W. (1992) *J. Chem. Phys.*, **97**, 2861.

7 Conventional and Unconventional Hydrogen-bonded Ionic Clusters

C. A. DEAKYNE

Eastern Illinois University, USA

7.1 INTRODUCTION

Recently there has been a rapid increase of interest in the structures and properties of gas-phase clusters as a result of their relevance to nucleation phenomena, catalysis, combustion, biochemistry, atmospheric chemistry, plasma chemistry and many other areas. For example, proton-bound clusters are of fundamental importance to condensed-phase chemistry and are also found in the earth's lower ionosphere (Arijs *et al.*, 1980; Eisele, 1986, 1988). Solvated protons found in the troposphere and stratosphere include $(H_2O)_nH^+$, $(CH_3CN)_n(H_2O)_mH^+$, and $(NH_3)_n(H_2O)_mH^+$ (Arijs *et al.*, 1980; Eisele, 1986, 1988). The negative ion content of the atmosphere is also dominated by cluster ions of the type $(H_2SO_4)_n(HNO_3)_mHSO_4^-$ (Arnold and Fabian, 1980). Information on the properties of these ionic clusters is needed to determine the abundance of trace neutral compounds in the atmosphere (Eisele, 1988) and to discover how the clusters enter into the atmospheric chemistry.

Work on gaseous solvated ions is also of interest since it 'bridges the gap' between studies of gas-phase and condensed-phase chemistry (Castleman and Keesee, 1988). By adding solvent molecules to gas-phase ions, one may approach the conditions within the condensed phase while working with a system of greatly reduced complexity. Also, of course, the interactions and forces, e.g. hydrogen bonding, that determine the structures and properties of finite clusters are present in the bulk phases of matter. Analysis of the binding forces in cluster ions provides information on ion–molecule interactions in electrolytic solutions, enzyme active sites, DNA double helices, and K^+ transport channels (Burda *et al.*, 1996; Dougherty, 1996; Meot-Ner, 1988).

The novel and intriguing properties exhibited by cluster ions that have no condensed-phase analogs provide another motivation for their study. Two such properties are the 'magic number' mass-spectral distributions and the 'size-selective' chemistry observed for cluster ions. A specific cluster size is termed a 'magic number' if it is associated with a discontinuity in the smooth cluster ion intensity distribution. 'Magic numbers' are believed to have a structural origin related to the extra stability of completed solvation shells (El-Shall *et al.*, 1991; Garvey *et al.*, 1994) or cyclic structures (Wei *et al.*, 1991a,b). Certain reactions that clusters experience can be 'switched on' and 'switched off' depending on the size of the cluster. Some of these reactions appear to require participation by four or more chemical species and are, therefore, unlikely to occur outside the confines of a cluster. The special environment of a cluster ion allows the requisite number and types of species to be brought together with the correct orientation for the reaction to proceed (Winkel and Stace, 1994).

For all the above reasons, neutral and ionic clusters have been the subject of many computational and experimental investigations. The consequent wealth of information on clusters makes it necessary to narrow the focus of this chapter, and we have chosen to discuss conventional *ab initio* and experimental studies of, primarily, gas-phase hydrogen-bonded ionic clusters appearing in the literature since 1994. Even so, it is not possible to be comprehensive, and we have covered representative topics of current interest in this area.

While the focus of this chapter is on conventional *ab initio* calculations, there is also interest in the study of kinetics, thermodynamics, phase coexistence, and related cluster phenomena via Monte Carlo or molecular dynamics calculations (see, for example, Smith, 1994). The application of density functional theory to the investigation of clusters is also receiving considerable attention lately (Hobza *et al.*, 1995; Kim and Jordan, 1994; Pudzianowski, 1996). Two features of density functional theory that make it attractive are that it includes some correlation effects and it is much less computationally demanding than traditional molecular orbital methods. Because of its speed, there is much promise in interfacing density functional theory with Monte Carlo and/or molecular dynamics calculations. Indeed, the work of Car, Parrinello and coworkers has made significant strides in this area (Laasonen *et al.*, 1993). Nevertheless, questions about the applicability and performance of this methodology are still in the process of being answered.

One of the most widely used experimental methods in the study of gas-phase clusters is mass spectrometry (Castleman and Wei, 1994). This method provides high selectivity and sensitivity after ionization of the clusters. Developments in multiple or tandem mass spectrometric techniques coupled with developments in ion sources have greatly enhanced the power of mass spectrometry for determining structures and analyzing selected classes of compounds. Supersonic expansion techniques are commonly used to produce beams of neutral clusters. Coexpansion and pick-up source techniques have been devised for the preparation of clusters of mixed composition. The clusters are then ionized by, for example, electron impact ionization, chemical ionization or multiphoton ionization. Alternatively, cluster ions can be generated directly through the coexpansion of ions and neutral ligands. Other approaches to the production of cluster ions are fast atom (or ion) bombardment (FAB), matrix-assisted laser desorption ionization (MALDI) and electrospray ionization (ESI). The latter approaches (and others) have been successfully implemented to yield singly- and multiply-charged clusters and large biologically interesting systems such as proteins and oligonucleotides (Biemann and Papayannopoulos, 1994). Chemical and physical properties of selected ions can then be probed by photoexcitation or dissociation, metastability, or collision-induced dissociation investigations or by ion–molecule reactions. In this way information can be gained on structures, conformations, and dynamics and on reactions undergone within the cluster (intracluster reactions) or with other species. Technological advances in the sources for flow-tube reactors have also allowed ion–molecule reactions involving relatively large clusters to be studied under well-defined thermal reaction conditions. Traditionally, information on bond dissociation energies has been determined from van't Hoff plots (Meot-Ner and Speller, 1986) but lately this information has been determined also from analysis of photoelectron spectra (Snodgrass *et al.*, 1995), kinetic energy release distributions (KERDs) (Lifshitz and Feng, 1995), and thresholds for collision-induced dissociation (More *et al.*, 1996). For more details, see the excellent reviews on these techniques and their applications that have appeared in the recent literature (Biemann and Papayannopoulos, 1994; Bleske, 1995; Castleman and Wei, 1994; de Hoffmann, 1996; Freeman and March, 1996; Garvey *et al.*, 1994; Gross, 1994; McLafferty, 1994; Syage, 1995; Topp, 1995; Vincenti, 1995).

Both theorists and experimentalists are interested in the following questions concerning the buildup of a cluster ion.

1. How much does clustering affect the geometry and charge distribution of the subunits?
2. What is the relationship between structure, charge transfer and energetics for cluster ions?
3. In what proportion do electrostatic and delocalization interactions contribute to the bonding in a particular cluster?
4. Do clusters build up through the filling of distinct solvent shells?
5. Can clusters with a given composition assume several isomeric structures?
6. What is the relationship between the pronounced mass-spectral abundances of certain ionic clusters ('magic numbers') and the structures of the clusters?
7. What types of reactions do ionic clusters undergo?
8. To what degree does clustering simulate bulk solvation?

Additional questions specific to protonated cluster ions are the following:

9. What combination of solvent molecules about the proton best stabilizes the proton?
10. Starting with a protonated solute molecule, can a group of solvent molecules 'pull away' the proton from the solute molecule?
11. In complexes of mixed solvents is the protonated solute molecule the molecule with the highest proton affinity?

Each of the articles discussed in this chapter attempts to answer one or more of these questions.

7.2. COMPUTATIONAL DETAILS

7.2.1 General aspects

In the supermolecule approach, the interaction energy is calculated by subtracting the sum of the total energies of the subunits from the total energy of the complex:

$$\Delta E = E_{X\cdots Y} - (E_X + E_Y) \tag{7.1}$$

As a guide to understanding the nature of molecular interactions, the interaction energy can be partitioned into physically intuitive components, e.g. electrostatic, exchange repulsion, polarization, charge transfer, and dispersion, and various procedures have been developed for this purpose. (See, for example, Cybulski *et al.*, 1990; Glendening, 1996; Glendening and Streitwieser, 1994; Hayes and Stone, 1984; Kitaura and Morokuma, 1976; Morokuma, 1971; Rybak *et al.*, 1991; Stone, 1993). Clearly, in order to obtain accurate interaction energies, the level of theory chosen must yield reliable electric properties (multipole moments and polarizabilities). In addition, any system for which dispersion makes a significant contribution must be calculated at a correlated level of theory. Often second-order Møller-Plesset theory (MP2) (Møller and Plesset, 1934; Pople *et al.*, 1976) is sufficient, as the higher-order terms frequently are much smaller in magnitude and alternate in sign (Deakyne, 1986; Del Bene, 1989; Frisch *et al.*, 1985a; Magnusson, 1994; More *et al.*, 1996). The electrostatic and dispersion contributions are the most slowly convergent with respect to the size of the basis set (Cybulski *et al.*, 1990; Szczesniak *et al.*, 1993). Polarization functions of higher symmetry than *d* are required to obtain accurate dispersion energies. Placing a set of *sp* functions at the center of the intermolecular bond also improves both the electrostatic and dispersion energies (Szczesniak *et al.*, 1993; van Duijneveldt-van de Rijdt and van Duijneveldt, 1992; van Mourik and van Duijneveldt, 1995).

Regardless of the model utilized to calculate total energies, if interaction energies are evaluated with the supermolecule approach the problem of basis set superposition error (BSSE) must be addressed. BSSE is the nonphysical charge delocalization and stabilization of the complex (supermolecule) caused by the different sizes of the subunit and supermolecule basis sets. (In the complex, each fragment has a portion of the basis set asssociated with the other fragment accessible to it). An efficacious way of reducing BSSE is to select a basis set that contains diffuse functions (Del Bene, 1996; Del Bene and Shavitt, 1994; Feller *et al.*, 1994). The customary way of correcting for BSSE is by the Boys–Bernardi counterpoise method, i.e. with the use of ghost functions (Boys and Bernardi, 1970). For each subunit, the counterpoise correction to the interaction energy is equal to the difference between the total energy of the subunit calculated with the supermolecule set of basis functions and the total energy of the subunit calculated with the subunit set of basis functions. In each case, the geometry of the fragment in the optimized complex is used. Unless otherwise noted, counterpoise corrections have been included in the *ab initio* calculations described below. However, Cook *et al.* (1993) have pointed out that the use of ghost functions with correlation treatments leads to special problems that do not occur with self-consistent field (SCF) treatments. Counterpoise calculations create a set of spurious orbitals that are not important at the SCF level but cause an overcorrection at the correlated level. Pudzianowski's (1995) analysis of the effect of ghost functions on the compositions and eigenvalues of the occupied and virtual molecular orbitals of the proton donor NH_4^+ and the acceptor NH_3 lends support to Cook *et al.*'s contention. As noted below, Pudzianowski, Del Bene (1992) and Wehbeh and Allen (personal communication) all recommend not correcting for BSSE when interaction energies are obtained with the protocols they have advanced.

In all the work discussed in this chapter, the geometries of the subunits and the complex have been fully optimized. The authors have also invariably computed harmonic vibrational frequencies. The latter calculation ensures that the optimized structure is in fact a minimum on the potential energy surface (no imaginary frequencies) and allows the calculation of zero-point energies, thermal vibrational energies, and thermodynamic quantities. Since it has been found that correlated energies appear to be remarkably insensitive to the geometry used to compute them (Frisch et al., 1985a, 1986; Meot-Ner et al., 1996a; More et al., 1996) rather than reoptimizing structures at every calculational level of interest, single-point energies are computed. The conventional notation of, for example, MP2/6−31+G(d)//HF/6−31G, indicates that the total energy has been obtained at the MP2/6−31+G(d) level of theory with the geometry obtained at the HF/6−31+G level (Binkley and Pople 1977; Francl et al., 1982; Hehre et al., 1972b). For some complexes, e.g. $(H_2O)H_3O^+$, the proton-transfer potential changes from a double- to a single-well type when electron correlation is taken into account (Del Bene, 1988; Meot-Ner et al., 1996a; Pudzianowski, 1995). Thus, optimizing geometries at the SCF level and calculating interaction energies at a correlated level should be done with caution.

7.2.2 Calculational Protocols

Wehbeh and Allen (personal communication) have examined 84 calculational levels with the objective of identifying the simplest level that yields consistently accurate results. Forty-two basis sets have been considered at both the SCF and MP2 levels of theory. Each basis set was derived from the 6−31G or the 6−311G (Krishnan et al., 1980b) basis by adding various combinations of p- and d-type polarization functions (one or two sets) (Frisch et al., 1985b; Hariharan and Pople, 1972; Krishnan et al., 1980a) and s and sp-type diffuse functions (Clark et al., 1983; Spitznagel et al., 1982). The basis sets were augmented with functions on both hydrogen and nonhydrogen atoms or with functions on only nonhydrogen atoms. Optimized geometries were obtained at all 84 levels of approximation for HF, H_2O, HCN, (HF)HF, (HCN)HF, and (H_2O)HF. In each case HF is the proton donor. Calculated data on $(H_2O)H_2O$ obtained by Frisch et al. (1986) were also utilized in the analysis. These dimers were chosen because accurate experimental data are available for their binding energies, vibrational frequencies and structures. Since neither the SCF nor the MP2 energies converged to the experimental binding energies, Wehbeh and Allen used a different strategy for choosing among the calculational levels. They looked for the level with a low BSSE that most nearly matches the available experimental values for the binding energies as well as the following monomer and dimer electronic properties: bond lengths, bond angles, dipole moments, vibrational frequencies, and ir intensities. They found that a d-type polarization function on the heavy atoms is essential and that diffuse functions on the heavy atoms lower the BSSE and improve the binding energies. Their overall conclusion is that hydrogen-bond energies should be calculated at the HF/6−31+G(d) (Ditchfield et al., 1971; Hariharan and Pople, 1973) level without correcting for BSSE. Other researchers have also noted that, due to the opposing effects on the binding energy of basis set expansion and inclusion of electron correlation, Hartree−Fock theory and moderately-sized basis sets often predict the structures and energies of hydrogen-bonded complexes well (Del Bene, 1989; Frisch et al., 1985a; Meot-Ner et al., 1996a; More et al., 1996). One advantage of the recommended level of theory is that the 6−31+G(d) basis set is small enough to allow relatively large systems to be studied. One disadvantage is that any complex for which electron correlation effects are important (with respect to the binding energy and/or the geometry) will be improperly represented.

The 6−31+G(d) basis set appears to be the minimum basis set with enough flexibility to describe the long- and short-range forces that contribute to complexation. (It should be mentioned, however, that the 6−31G(d) basis set (Francl et al., 1982; Hariharan and Pople, 1972) is often adequate when one is primarily interested in the trend in interaction energies for a series of related complexes (Boyd and Choi, 1986; Carroll and Bader, 1988; Deakyne et al., 1994; Koch and Popelier, 1995)). Hashimoto and Morokuma (1994, 1995) and Feller and coworkers (Feller et al., 1994; Glendening and Feller, 1995; More et al., 1996) also have had success with the 6−31+G(d) basis set in their investigations of metal cations complexing with the neutral ligands water, ammonia, or dimethyl ether. The 6−31+G(d) results compared favorably with results from more extensive basis sets and with experimental results. These researchers did find, however, that accounting for electron correlation is particularly important for obtaining accurate binding energies for the larger systems ($n \geq 4$), for which ligand–ligand interactions can be significant. The majority of the effect is recovered at the MP2 level.

Surveying the literature on charged hydrogen-bonded dimers for a set of *ab initio* benchmark calculations against which density functional methods could be evaluated, Pudzianowski (1995) found that no such set existed at a consistently high level of theory for the representative ions he had chosen. Therefore, he obtained MP2/6−311++G(d,p) (Krishnan et al., 1980a) equilibrium geometries and binding energies for $(H_2O)H_3O^+$, $(H_2O)NH_4^+$, $(H_2O)CH_3NH_3^+$, $(NH_3)NH_4^+$, $(NH_3)CH_3NH_3^+$, $(H_2O)OH^-$, $(H_2O)CH_3O^-$, $(H_2O)HCOO^-$, $(H_2O)CN^-$, and $(H_2O)HCC^-$. A comparison of the optimal structures found in this work with those in the recent extant literature led to the conclusion that geometry optimizations carried out at the correlated level with a basis set containing polarization functions on all atoms and diffuse functions on the nonhydrogen atoms show convergence with more sophisticated models. Changing from a double-split to a triple-split valence basis set has a considerably smaller effect on the geometries (and energies). This conclusion is in agreement with the optimization level recommended by Del Bene, i.e. MP2/6−31+G(d,p) (see below) (Del Bene, 1992; Del Bene and Shavitt, 1994).

All but two of the MP2/6−311++G(d,p)//MP2/6−311++G(d,p) complexation enthalpies are within 2 kcal mol^{-1} of the experimental $\Delta H°$ values (Pudzianowski, 1995). The exceptions are the two NH_3 dimers for which the errors are $\approx$2.2 kcal mol^{-1}. Assessment of the effect of expanding the basis set to 6−311++G(2d,2p) (Krishnan et al., 1980a) versus extending the correlation treatment to MP4 indicated that the former improvement was more efficacious for these systems. In contrast, for the $(H_2O)OH^-$ ion, for which the error is 1.6 kcal mol^{-1}, the latter improvement is more effective. Del Bene and Shavitt (1994) have reported the need for similar procedural enhancements for these systems. For those ions studied by both groups, the two sets of calculated complexation enthalpies are comparable (Del Bene and Shavitt, 1994; Pudzianowski, 1995).

Another observation from Pudzianowski's work is that each of the 'linear' complexes has a hydrogen-bond stretching vibrational mode between about 200 and 350 cm^{-1} (Pudzianowski, 1995). For complexes with the same electron donor, the magnitude of the frequency correlates directly with the acid strength of the proton donor. No such relationship between base strength and frequency is observed for complexes with the same proton donor. More information on these trends would be useful since this vibrational mode should contribute to proton transfer (Pudzianowski, 1995).

For some time now, Del Bene and coworkers have actively pursued the methodological requirements for calculating quantitatively accurate interaction energies (cf. Chapter 5, this volume). The bulk of their work has concentrated on neutral and ionic hydrogen-bonded complexes, and in a recent article (Del Bene and Shavitt, 1994) systematic changes in the Dunning correlation-consistent polarized split-valence basis sets (cc-pVXZ, X = D (double-split), T (triple-split), and Q (quadruple-split)) were evaluated with respect to convergence of computed interaction energies (Dunning, 1989; Kendall et al., 1992). The effect of including electron correlation via full many-body fourth-order Møller–Plesset perturbation theory MP4 versus the infinite-order coupled cluster method CCSD + T(CCSD) (Purvis and Bartlett, 1982; Urban et al., 1985) was also compared. The cc-pVXZ and aug-cc-pVXZ basis sets were modified to form the cc-pVXZ+, cc-pVQZ'+ and aug'-cc-pVXZ bases. The cc-pVXZ+ sets contain diffuse sp functions on the nonhydrogen atoms (Del Bene, 1992). The cc-pVQZ'+ basis is cc-pVQZ+ minus the g orbitals on the heavy atoms and the f orbitals on the hydrogens. The aug'-cc-pVXZ sets contain s, p and d diffuse functions on the nonhydrogen atoms only. The aug'-cc-pVTZ basis set was modified further by removing the diffuse f functions (aug'-cc-pVTZ($-f$)) or the diffuse d and f functions (aug'-cc-pVTZ($-df$) from the heavy atoms. MP2/6–31+G(d,p) (Dill and Pople, 1975; Hariharan and Pople, 1973) optimal geometries were used for the $(NH_3)NH_2^-$, $(NH_3)NH_3$, $(NH_3)NH_4^+$, $(H_2O)OH^-$, $(H_2O)H_2O$, $(H_2O)H_3O^+$, $(HF)F^-$, $(HF)HF$, and $(HF)H_2F^+$ complexes and their constituent monomers. The results of this work can be summarized as follows.

First, the recommended calculational level for neutral and cationic hydrogen-bonded complexes is MP4/aug'-cc-pVTZ//MP2/6–31+G(d,p) and for anionic hydrogen-bonded complexes is CCSD + T(CCSD)/aug'-cc-pVTZ//MP2/6–31+G(d,p). Second, diffuse f functions can be omitted from the aug'-cc-pVTZ basis without affecting the quantitative accuracy. In fact, for the neutral and positively charged systems the diffuse d functions may not be required either. Third, for the neutral dimers the aug'-cc-pVDZ and aug'-cc-pVTZ energies are close, so the former basis may be useful for larger systems but additional evaluation is needed.

An analogous study on the effect of diffuse functions, electron correlation, and BSSE on Li^+ affinities has also been reported by Del Bene (1996). Complexes between Li^+ ;and NH_3, H_2O, HF, HCN, CO, PH_3, H_2S, and HCl were examined. Optimum geometries for the complexes were obtained at the MP2/6–31+G(d,p) level. MP2, MP4 and CCSD(T) single-point energies were computed with the cc-pVXZ, aug'-cc-pVXZ, aug'-cc-pVTZ($-f$), and aug'-cc-pVTZ($-df$) basis sets, where X = D, T and Q. The CCSD(T) method is an infinite-order coupled-cluster method with noniterative incorporation of triple excitations (Purvis and Bartlett, 1982; Urban et al., 1985). Many of the conclusions from this work were identical to those given above. The exceptions are: (1) diffuse f functions are required in the basis set for some of the anionic bases; (2) CCSD(T) calculations are recommended for some of the anionic bases; and (3) although the trends in computed and experimental binding energies are similar, the computed values are consistently about 1–2 kcal mol^{-1} lower than the experimental values. BSSE corrections are small (< 1 kcal mol^{-1}) when diffuse functions are included in the basis set, and the uncorrected MP4/aug'-cc-pVTZ Li^+ interaction energies are in better agreement with those from higher levels of theory than are the corrected energies.

Therefore, Del Bene (1996) did not correct the calculated MP4/aug′-cc-pVTZ Li$^+$ affinities for BSSE.

7.3 CONVENTIONALLY HYDROGEN-BONDED IONIC SYSTEMS

7.3.1 Proton-Bound Clusters

A conventional hydrogen bond is one for which both the proton- and electron-donating atoms are *n*-donors. Typical strengths of conventional neutral hydrogen bonds are in the 2–16 kcal mol^{-1} range but for conventional ionic hydrogen bonds the range rises to 9–45 kcal mol^{-1}. Clustering enthalpies of $(HCN)_n(NH_3)_mH^+$ have been measured by pulsed high-pressure mass spectrometry and the interactions analyzed by *ab initio* calculations. Deakyne *et al.* (1994) considered five isomeric forms for complexes with $n + m \leq 5$. One form has all the ligands in the inner shell and the other four have one ligand in the outer shell, with a NCH...NCH, NCH...NH$_3$, NH$_3$...NCH or NH$_3$...NH$_3$ interaction. The isomers are illustrated in Scheme 7.1 for the $(HCN)_2(NH_3)_3H^+$ system, with the appropriate designation to show the connectivity. Here Am and Ac represent ammonia and hydrogen cyanide, respectively. Clusters with two ligands bonded to other ligands rather than to the core ion were examined also, but were found to be much less stable.

Optimum geometries were obtained at the SCF level with the 3–21G (Binkley *et al.*, 1980) and 6–31G(*d*) basis sets. In order to confirm that all the equilibrium structures are minima (no imaginary frequencies), normal mode vibrational frequencies were calculated from the 3–21G geometries. Zero-point vibrational energies (ZPEs) and the temperature dependence of the vibrational correction were determined from the computed frequencies. The vibrational frequencies were corrected by the usual factor of 0.89 (Hehre *et al.*, 1986). The zero-point energies were corrected by the factor of 0.91 suggested by Grev *et al.* (1991). Torsional modes with corrected energies ≤ 500 cm^{-1} were treated as pure rotations.

Total energies and complexation energies of the clusters were computed at the SCF and MP2 levels with the 6–31+G(*d*) and 6–31+G(2*d*,2*p*) basis sets and the 6–31G(*d*) geometries. The stabilization energies were obtained from the following equation:

$$\Delta E_{n-1,n} = E_T((HCN)_n(NH_3)_mH^+) - E_T((HCN)_{n-1}(NH_3)_mH^+) - E_T(HCN) \quad (7.2)$$

The complexation energies were converted into enthalpies of reaction via equation (7.3).

$$\Delta H_{n-1,n} = \Delta E_{n-1,n} + \Delta ZPE + \Delta E^t_{298} + \Delta E^r_{298} + \Delta E^v_{298} + \Delta PV \quad (7.3)$$

The translational (ΔE^t_{298}) and rotational (ΔE^r_{298}) temperature terms are classically equal to $\pm 1/2 RT$ for each degree of freedom gained or lost in the reaction. The vibrational temperature term is given by the following equation (Pitzer, 1961):

$$\Delta E^v_{298} = RT \left[\sum_{p,i} \frac{u_i}{e^{u_i} - 1} - \sum_{r,j} \frac{u_j}{e^{u_j} - 1} \right] \quad (7.4)$$

where *p* represents products, *r* represents reactants, and $u = hc\nu/kT = 4.826 \times 10^{-3}\nu$ at 298 K. Here ν is the vibrational frequency in cm^{-1}. The pressure-volume work term

CONVENTIONALLY HYDROGEN-BONDED IONIC SYSTEMS 225

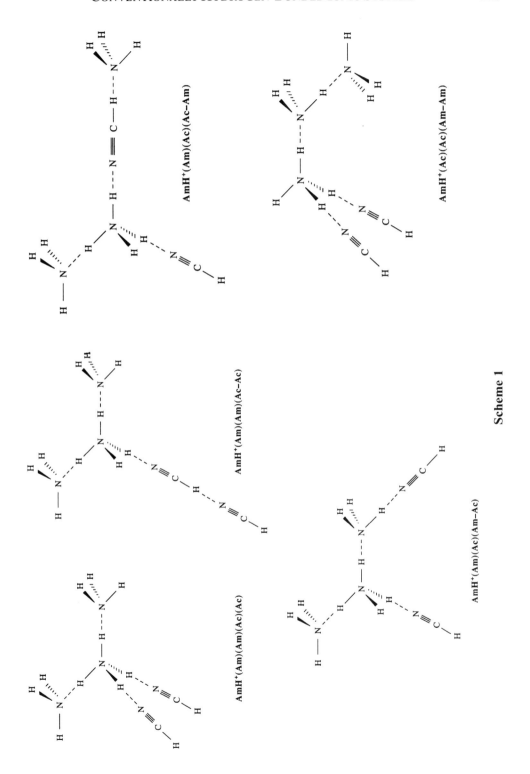

Scheme 1

(ΔPV) is equal to $\pm RT$ for each mole of gas produced or consumed, assuming ideal behavior.

For each cluster ion with all of the ligands bonded to the central ion, the basis set superposition error was estimated via the full counterpoise approach of Boys and Bernardi (1970). The superposition error was determined for both the proton donor and acceptor molecules at the MP2/6−31+G(d)//HF/6−31G(d) and MP2/6−31+G($2d,2p$)//HF/6−31G(d) levels of theory. The same BSSE correction was applied to each isomeric form of a particular cluster ion. The BSSE of ≈ 1 kcal mol^{-1} is independent of calculational level and cluster size and type.

The HF/6−31G(d) optimized values of selected intermolecular parameters for the (HCN)(NH$_3$)$_3$H$^+$ and (HCN)$_2$(NH$_3$)$_3$H$^+$ complexes are given in Table 7.1. These cluster ions have been chosen for illustrative purposes, since the results for these ions are representative of the results for the remaining ions in this work and for ions in other work (Deakyne, 1986; Deakyne *et al.*, 1986; Hirao *et al.*, 1982; Newton, 1977; Newton and Ehrenson, 1971). r_1 is the distance between the proton-donating atom and the hydrogen-bonded proton; r_2 is the distance between the hydrogen-bonded proton and the electron-donating atom. The intermolecular C—H...N and H...N—H hydrogen bond angles, which

Table 7.1 Selected intermolecular geometrical parameters of (HCN)(NH$_3$)$_3$H$^+$ and (HCN)$_2$(NH$_3$)$_3$H^{+a}

Cluster	Hydrogen bond	r_1[b]	r_2[c]	<N—H...N[d]	<H...N—C[d]
(HCN)(NH$_3$)$_3$H$^+$					
AmH$^+$(Am)(Am)(Ac)[e]	AmH$^+$...Ac	1.016	2.043	178.2	178.5
C$_s$	AmH$^+$...Am	1.034	1.899	179.0	
AmH$^+$(Am)(Ac—Am)	AmH$^+$...Ac	1.025	1.901	178.2	177.9
C$_s$	AmH$^+$...Am	1.043	1.837	180.0	
	Ac...Am	1.086	1.995		
(HCN)$_2$(NH$_3$)$_3$H$^+$					
AmH$^+$(Am)(Am)(Ac)(Ac)	AmH$^+$...Ac	1.014	2.095	179.7	178.4
C$_{2v}$	AmH$^+$...Am	1.029	1.937	178.8	
AmH$^+$(Am)(Am)(Ac—Ac)	AmH$^+$...Ac	1.018	1.990	178.3	178.1
C$_s$	AmH$^+$...Am	1.032	1.908	178.0	
	Ac...Ac	1.070	2.150		179.2
AmH$^+$(Am)(Ac)(Ac—Am)	AmH$^+$...Ac	1.020	1.954	176.9	179.0
C$_1$	AmH$^+$...Ac	1.016	2.039	176.5	179.6
	AmH$^+$...Am	1.035	1.884	179.8	
	Ac...Am	1.084	2.014		
AmH$^+$(Am)(Ac)(Am—Ac)	AmH$^+$...Ac	1.015	2.065	177.3	179.3
C$_1$	AmH$^+$...Am	1.040	1.842	177.9	
	AmH$^+$...Am	1.031	1.916	179.8	
	Am...Ac	1.007	2.307	174.7	165.6
AmH$^+$(Ac)(Ac)(Am—Am)	AmH$^+$...Ac	1.016	2.050	174.4	178.9
C$_s$	AmH$^+$...Am	1.048	1.798	178.5	
	Am...Am	1.013	2.181	174.7	

[a] Data from Deakyne *et al.* (1994). Am = NH$_3$ and Ac = HCN. When two hydrogen bonds of the same type exist for a cluster, the data for the ligand that forms an inner and outer shell hydrogen bond are given first.
[b] N—H or C—H bond length in Å.
[c] H...N bond length in Å.
[d] Bond angle in degrees.
[e] AmH$^+$(Am)(Am—Ac) is not a minimum on the HF/6−31G(d) potential energy surface for (HCN)(NH$_3$)$_3$H$^+$.

have not been included in the table, range from 179.5° to 179.9° and from 106.5° to 115.5°, respectively. Several patterns in the geometrical data can be discerned. First, all the structures have NH_4^+ as the core ion (most of the positive charge is retained by the NH_4 moiety). Second, as a result of the weaker basicity of HCN (Meot-Ner and Sieck, 1991) r_1 is shorter and r_2 is longer for the NH_4^+...NCH hydrogen bond compared to the NH_4^+...NH_3 hydrogen bond. The same relative bond lengths are seen for analogous neutral hydrogen bonds, e.g. NCH...NCH versus NCH...NH_3. Third, when solvent molecules are added to the inner shell of the core ion, the other r_2 distances increase and the other r_1 distances decrease, thereby weakening the hydrogen bonds. Fourth, when an inner shell solvent molecule acts as both a proton donor and a proton acceptor, the hydrogen bond between NH_4^+ and the molecule becomes stronger (r_1 lengthens and r_2 shortens), at the expense of the other hydrogen bonds involving NH_4^+. Points three and four are consistent with cooperative interactions between the hydrogen bonds (Deakyne, 1986; Hirao et al., 1982; Karpfen and Yanovitskii, 1994; Newton, 1977; Newton and Ehrenson, 1971). The more stable the newly formed hydrogen bond, the larger are the observed changes in the bond distances (cf. Table 7.2).

Table 7.2 presents the calculated and experimental enthalpies ($\Delta H°$), entropies ($\Delta S°$), and free energies ($\Delta G°$) of reaction for the clustering reactions of the $(HCN)(NH_3)_4H^+$, $(HCN)_2(NH_3)_3H^+$, and $(HCN)_4(NH_3)H^+$ systems. $\Delta E_{n-1,n}$ is also tabulated. Again the thermodynamic data for the selected complex ions are representative of the data for all the ions examined in this work. The isomer with all the ligands bound to NH_4^+ is the most stable isomer for all the complexes. The binding energies for the ions with one ligand in the outer shell decrease in the order NCH...NCH > NCH...NH_3 > NH_3...NCH > NH_3...NH_3. The relative energies of the clusters do not depend on the size of the basis nor on whether electron correlation is included in the calculations. The experimental stabilization energy of $(NH_3)_2H^+$ is 3.5 kcal mol^{-1} greater than that of

Table 7.2 Thermodynamic quantities for the clustering reactions $(HCN)_{n-1}(NH_3)_m H^+ + HCN \rightarrow (HCN)_n(NH_3)_m H^{+a}$

Cluster	$-\Delta E^b$	$-\Delta H°_{calc}{}^c$	$-\Delta H°_{exp}{}^c$	$-\Delta S°_{calc}{}^d$	$-\Delta S°_{exp}{}^d$	$-\Delta G°_{calc}{}^c$	$-\Delta G°_{exp}{}^c$
$(HCN)_4(NH_3)H^+$							
$AmH^+(Ac)(Ac)(Ac)(Ac)$	12.1	10.4	11.0	27.4	20.9	2.2	4.8
$AmH^+(Ac)(Ac)(Ac-Ac)$	9.4	7.5		24.9		0.1	
$(HCN)_2(NH_3)_3H^+$							
$AmH^+(Am)(Am)(Ac)(Ac)$	11.6	10.4	11.2	24.3	(20.0)	3.2	5.2
$AmH^+(Am)(Am)(Ac-Ac)$	9.5	7.9		25.7		0.2	
$AmH^+(Am)(Ac)(Ac-Am)$	8.9	7.4		23.3		0.5	
$AmH^+(Am)(Ac)(Am-Ac)$	7.8	6.6		25.1		−0.8	
$AmH^+(Ac)(Ac)(Am-Am)$	5.8	5.0		25.5		−2.6	
$(HCN)(NH_3)_4H^+$							
$AmH^+(Am)(Am)(Am)(Ac)$	11.3	10.0		22.4		3.3	
$AmH^+(Am)(Am)(Ac-Am)$	9.3	7.9		19.5		2.1	
$AmH^+(Am)(Am)(Am-Ac)$	7.6	6.5		21.1		0.2	
$AmH^+(Am)(Ac)(Am-Am)$	6.3	5.3		23.8		−1.8	

[a] Data from Deakyne et al. (1994).
[b] MP2/6-31+G(2d, 2p)//HF/6-31G(d) complexation energies corrected for BSSE. Am = NH_3 and Ac = HCN.
[c] $\Delta H°$ and $\Delta G°$ in kcal mol^{-1}.
[d] $\Delta S°$ in cal mol^{-1} K^{-1}. Value in parentheses is estimated.

$(HCN)(NH_3)H^+$, consistent with the much larger proton affinity of NH_3 (Meot-Ner and Sieck, 1991). However, replacing NH_3 molecules with HCN molecules has very little effect on the stabilities of the three- to six-membered clusters provided the cluster has at least two NH_3 ligands. Indeed, related structures with higher HCN content tend to have higher stabilities. Evidently, the larger dipole moment of HCN counterbalances the better hydrogen-bonding capacity of NH_3. With two exceptions, $(HCN)_{1,2}(NH_3)_2H^+$, the calculated and experimental $\Delta H°$ values agree to within ± 1.5 kcal mol^{-1}. Agreement between the calculated and experimental $\Delta S°$ and $\Delta G°$ values is less good at ± 6.5 cal K^{-1} mol^{-1} and ± 3 kcal mol^{-1}, respectively.

For many of the $(X)_n(NH_3)_mH^+$ cluster ions studied to date, e.g. $(NH_3)_mH^+$ (Payzant et al., 1973; Tang and Castleman, 1975), $(CH_3CN)_n(NH_3)_mH^+$, $(CH_3CHO)_n(NH_3)_mH^+$, and $((CH_3)_3N)_n(NH_3)_mH^+$ (Tzeng et al., 1991; Wei et al., 1991b,d)), it has been inferred from the mass spectrometric data that the ions have distinct solvent shells. In contrast, plots of $\Delta H°_{n-1,n}$ versus n for $(H_2O)_n(NH_3)_mH^+$ (Meot-Ner and Speller, 1986) and $(HCN)_n(NH_3)_mH^+$ (Deakyne et al., 1994) decrease smoothly with n showing no evidence of a transition between shells upon addition of a fifth ligand molecule. This suggests that a second shell might start to build up before the first shell is completed. The *ab initio* results for the $(H_2O)_n(NH_3)H^+$ (Deakyne, 1986), $(HCN)_n(NH_3)H^+$ and $(HCN)(NH_3)_mH^+$ ions support this conclusion (Deakyne et al., 1994). According to the calculations on $(HCN)_n(NH_3)_mH^+$, the entropy term stabilizes the isomer with the NCH...NH_3 interaction compared to the other isomers, and the relative $\Delta G°$ values ($\partial \Delta G° \leq 2$ kcal mol^{-1}) for the $(HCN)(NH_3)_mH^+$ complexes, $m \geq 4$, suggest that mixtures of this isomer and the most stable isomer will be present at equilibrium (Table 7.2). Likewise, when $n \geq 4$, the relative $\Delta G°$ values for the $(HCN)_n(NH_3)H^+$ clusters suggest that an equilibrium mixture may contain the most stable isomer and the isomer with a HCN group bonded to HCN. Forms with ligands in the outer shell become competitive with the form with all the ligands in the inner shell for the following reason. As $n + m$ increases the hydrogen bond strength decreases; however, the drop is much larger for the isomers for which the ligand is added to the inner shell than for those for which the ligand is added to the outer shell. When $n + m$ increases from 3 to 5, the drop is 5–6 kcal mol^{-1} for the former clusters but is 2 kcal mol^{-1} or less for the latter complexes.

If there is in fact a mixture of cluster populations, the experimental van't Hoff plot represents a composite formed by two equilibria, and the experimental $\Delta H°$ and $\Delta S°$ data derived from the plot are different from the actual values for the formation of either cluster. It should be noted, however, that the presence of isomeric clusters cannot cause very large errors. If one form predominates in the mixture, the measured quantity would be correct for this form. If several isomers have essentially the same stability, the measured value will be close to that shared value.

Other recent experimental investigations of proton-bound clusters include collision-induced dissociation (CID, or collision-assisted dissociation (CAD)) and metastable fragmentation experiments on the clusters in order to obtain information on their structures, energetics, and reactivities. One particularly intriguing result from these experiments is that the dominant dissociation channel sometimes changes with cluster size ('size-selective' chemistry). The switch in dominant pathway can correlate with clusters of pronounced mass-spectral abundance and is interpreted as being indicative of a structural change within the cluster. Garvey and coworkers (Garvey et al., 1994) have recently reviewed their (and others') work on $(M)_n(H_2O)H^+$ cluster ions, where

M = ROH, ROCH$_2$CH$_2$OH, and CH$_3$OCH$_2$CH$_2$OCH$_3$. The methodology utilized to produce and detect the clusters, the structures and reactivities of the clusters, and the origin and importance of the 'magic number' clusters are all described in the article. For each M there is a value of n for which the ligand dissociation channel changes from predominantly water to predominantly M (Garvey *et al.*, 1994; Stace and Moore, 1983; Stace and Shukla, 1982). This change in dissociation behavior is ascribed to a switch in core ion from MH$^+$ to H$_3$O$^+$. Briefly, the structural implications of the observed 'magic numbers' and changing branching ratios are interpreted as follows. First, although it is generally assumed that the proton is associated with the solvent of higher proton affinity, Garvey *et al.* (1994) suggest that as long as the central ion is completely solvated, maximizing the number of hydrogen bonds or forming a hydrogen-bonded network is more important than the relative proton affinities of the solvent molecules in determining the structure. This suggestion is supported by the proton switching found in protonated complexes of water with acetonitrile (Deakyne *et al.*, 1986), acetone (Tzeng *et al.*, 1990), diethyl ether and glyme (Meot-Ner *et al.*, 1994). Second, fused five-membered ring geometries will be favored since they minimize the number of nonhydrogen-bonded acidic hydrogens, presumably without substantially reducing the strength of the individual hydrogen bonds. The authors point out, however, that bent hydrogen bonds and unfavorable entropic effects do argue against the latter conclusion. Theoretical calculations on the (M)$_n$(H$_2$O)$_m$H$^+$ systems would aid in resolving questions about the above interpretation of the experimental results.

Garvey and coworkers' (Garvey *et al.*, 1994) proposal about the preference for fused five-membered rings is based, in part, on earlier work on (H$_2$O)$_n$H$^+$. A fused five-membered ring structure has been postulated to account for the enhanced abundance of (H$_2$O)$_{21}$H$^+$ (Kassner and Hagen, 1976; Wei *et al.*, 1991a). In a titration experiment with trimethylamine, Castleman and coworkers (Wei *et al.*, 1991a) showed that no more than ten trimethylamine molecules can bond to (H$_2$O)$_{21}$H$^+$. The maximum number of titratable protons in (H$_2$O)$_{20}$H$^+$ is eleven. These results suggest that only ten or eleven 'dangling' hydrogens are present in these ions, and the cyclic structure proposed to account for the number of 'dangling' hydrogens is the clathrate-like pentagonal dodecahedron (Wei *et al.*, 1991a). (H$_2$O)$_{20}$H$^+$ is postulated to be pentagonal dodecahedron with a mobile proton on the surface; (H$_2$O)$_{21}$H$^+$ is postulated to be pentagonal dodecahedron with a hydronium ion in the center of the cage. It should be noted, however, that Smith and Dang (1994) have carried out classical molecular dynamics simulations utilizing a polarizable interaction model to examine the structures and energies of (H$_2$O)$_n$Cs$^+$, $n = 18-22$, as a function of temperature. The simulations have revealed several other cyclic structures (containing four-, five-, six- and/or seven-membered rings) with the requisite number of nonhydrogen-bonded hydrogens and stabilities similar to that of the pentagonal dodecahedron. The solid- and liquid-like regime incremental binding energies suggest that 'magic numbers' may be a phenomenon limited to the solid-like cluster regime, and density of states arguments suggest that this phenomenon is not entropic in origin, as proposed by Castleman and coworkers (Shi *et al.*, 1993) but enthalpic in origin (Smith and Dang, 1994).

If the ten nonhydrogen-bonded hydrogens in the cyclic structure can be replaced by methyl groups without significant loss of stability, clathrate (or other cyclic) structures can be formed in mixed protonated water/methanol clusters, also. Zhang and Castleman (1994) explored this possibility by investigating reactions between protonated water clusters and methanol under equilibrium conditions at low temperatures in a flow tube reactor. In a series of switching reactions, mixed $(H_2O)_n(CH_3OH)_mH^+$ complexes were formed, with magic numbers at $m + n = 21$, $m = 0 - 5$, under all methanol flow rates. The product distributions for ions with $m > 5$ were not analyzed. In an additional set of experiments both water and methanol were added in the source and the same pronounced abundances were observed. The agreement between these experiments and preceding molecular beam experiments (Shi *et al.*, 1992) indicates that the clathrate (or other cyclic) structures result from an inherent stability rather than from kinetic effects (Zhang and Castleman, 1994).

Karpas *et al.* (1994) have performed collision-induced dissociation investigations of protonated neat alcohol clusters and alcohol/water heteroclusters in a tandem mass spectrometer. The protonated clusters were produced in an atmospheric pressure ionization (API) corona discharge source, thereby avoiding the 'boiling off' of ligand molecules that occurs when large cold neutral clusters are ionized with molecular beam techniques. The alcohols examined were ethanol, *n*-propanol, isopropanol, *n*-butanol, isobutanol, *sec*-butanol, and *tert*-butanol. The CID spectra of $(ROH)_n(H_2O)_mH^+$, when $n + m < 9$, demonstrate that loss of water is favored for these clusters, indicating that the alcohol is the preferred protonation site. This result is in agreement with the results of Stace and coworkers (Stace and Moore, 1983; Stace and Shukla, 1982), since the switch in preferred ligand dissociation channel does not occur for mixed protonated water/alcohol ions with $m = 1$ until $n \geq 8$. Due to the way the experiments are carried out, Karpas *et al.* (1994) neither anticipated nor observed 'magic numbers', which leads them to question the generality of the concept of 'magic numbers'.

In contrast to the possible change in core ion for the mixed protonated alcohol/water complexes, protonated alcohol/ammonia complexes appear to have only an NH_4^+ core ion. Xia and Garvey (1995) have carried out a multiphoton ionization study of $(ROH)_m(NH_3)_nH^+$, R = CH_3, C_2H_5 and C_3H_7, in which they utilized a reflectron time-of-flight mass spectrometer to examine the product ions of competitive metastable decomposition pathways. When $n = 1$, the dominant decomposition channel is the loss of a single alcohol molecule. Loss of ammonia does not occur regardless of the value of m, suggesting that the core ion is NH_4^+. As the number of ammonia and alcohol molecules and the polarizability of the alcohol molecules increase, loss of an ammonia molecule becomes increasingly prevalent. The switch in the dominant decomposition pathway from alcohol loss to ammonia loss ranges from $n = 2$ to $n = 4$, depending on the alcohol and the value of m. Since the dipole moments and polarizabilities of the alcohols are greater than those of the ammonia, the NH_4^+...alcohol electrostatic interactions are expected to be greater than the NH_4^+...ammonia electrostatic interaction (Xia and Garvey, 1995). This inequality explains the loss of ammonia for the 1-propanol complexes when $n \geq 2$ but not the loss of alcohol for the methanol and ethanol complexes when $n \leq 4$. Xia and Garvey rationalize the latter trend by suggesting that ammonia molecules are close to the core ion in the smaller clusters (due to the larger proton affinity of ammonia) but get pushed into outer shells as the number of alcohol molecules increases. Assessment of the viability of these explanations awaits theoretical calculations.

Lifshitz and coworkers have investigated proton-bound hydrazine clusters (Feng *et al.*, 1995) and proton-bound acetic acid clusters (Lifshitz and Feng, 1995) by a pressure- and temperature-variable ion source in combination with tandem mass spectrometry. The two major cluster series obtained from the hydrazine/water sample were

$(N_2H_4)_nH^+$, $n = 1 - 8$, and $(N_2H_4)_n(H_2O)_mH^+$, $n = 1 - 6$, $m = 1 - 4$. Minor series include $(N_2H_4)_n^+$, $n = 1 - 8$. Pronounced mass-spectral abundances were observed for $(N_2H_4)_4H^+$ and for $(N_2H_4)_n(H_2O)_mH^+$, $n + m = 5$. In addition, for the neat protonated hydrazine complexes, there is a break in the plot of metastable decay fractions versus n and a maximum in the average kinetic energy releases at $n = 4$. The primary CAD channels show N_2H_4 loss for the $(N_2H_4)_nH^+$ series, H_2O loss for the $(N_2H_4)_n(H_2O)_mH^+$ series when $n + m \leqslant 5$, and N_2H_4 loss for the $(N_2H_4)_n(H_2O)_mH^+$ series when $n + m > 5$. The latter two results are consistent with the $n + m = 5$ clusters being 'magic numbers' and are indicative of a structural change at $n + m = 5$. One possible interpretation of these results is that the $n + m = 5$ systems are cyclic (Feng et al., 1995). The dominant decomposition pathway also changes for the $(N_2H_4)_4(H_2O)H^+$ ion when the source temperature is changed. The loss of N_2H_4 at high source temperatures and H_2O at low source temperatures suggests that either the core ion changes from $N_2H_5^+$ to H_3O^+ as the temperature is raised or that $(N_2H_4)_4(H_2O)H^+$ has a cyclic structure at low temperatures from which the ligand with lower proton affinity is preferentially lost.

Additional information on the structures and energetics of the $(N_2H_4)_nH^+$ clusters was obtained via *ab initio* (HF/6−31G) and semiempirical (PM3 (Stewart, 1989) and AM1 (Dewar et al., 1985)) calculations (Feng et al., 1995). Optimizations were carried out for $n = 1 - 4$ at both computational levels and for $n = 5, 6$ at the semiempirical level only. The $(N_2H_4)_nH^+$ ions have a $N_2H_5^+$ core with the ligands preferentially associated with the protonated nitrogen atom. Thus, this nitrogen atom is the more effective proton donor, which is consistent with the Mulliken charge of $+0.748$ on the NH_3^+ moiety and accounts for the 'magic number' at $n = 4$. Binding affinities were determined from the computed total energies or enthalpies of formation and from the experimental kinetic energy release distributions (KERDs), following the procedure developed by Klots (1993). As is usual for conventionally hydrogen-bonded complexes, the binding energies decrease (from ≈ 32 to ≈ 8 kcal mol^{-1}) with increasing n. There is no significant drop in energy between $n = 4$ and 5 but the energies do level off for $n \geqslant 5$.

In their study of $(CH_3COOH)_nH^+$, Lifshitz and Feng (1995) found two loss pathways: loss of a single CH_3COOH and of intact CH_3COOH dimers. The former evaporation channel is observed preferentially for the small clusters, the latter for the large clusters. The crossover occurs at $n = 7$. Similar results were observed for $(HCOOH)_nH^+$ at $n = 6$ (Feng and Lifshitz, 1994). For both series of ions, incremental binding affinities were deduced from the KERDs by applying Klots' (1993) approach. The values obtained in this work for $(CH_3COOH)_nH^+$, $n = 3 - 5$, agree well with those obtained by Meot-Ner (1992) via high-pressure mass spectrometry. Using thermodynamic cycles, Lifshitz and Feng (1995) evaluated the dissociation energy $D(RCOOH)_2$ of the neutral product dimers and concluded that the dimers are evaporated in the cyclic rather than open-chain form.

In order to verify this conclusion, Zhang and Lifshitz (1996) performed *ab initio* calculations on $(HCOOH)_nH^+$, $n = 1 - 6$, and $(CH_3COOH)_nH^+$, $n = 1 - 5$. After confirming that the 4−31G# basis set (polarization functions on the oxygens only) reproduces geometries and relative isomer energies obtained with larger sets for the proton-bound formic acid dimer, the remaining structures were optimized at the HF/4−31G# level. Single-point calculations were performed at the HF/6−31+G(d)//HF/4−31G# level. Three types of structures were found, namely chains terminated at both ends by a cyclic dimer unit, chains terminated at one end by a cyclic dimer unit, and chains terminated at both ends by monomer units. In each case the added proton is near the center of the chain. The calculated and experimental binding energies are in good agreement for the acetic acid series of ions (Lifshitz and Feng, 1995; Meot-Ner, 1992). In order to determine the crossover point from monomer loss to dimer loss, it was assumed that dimer loss occurs only from the first two structures and monomer loss occurs only from the last two. As was found experimentally (Feng and Lifshitz, 1994; Lifshitz and Feng, 1995), the calculated evaporation enthalpies for the formic acid series indicate that monomer evaporation is favored for $n < 6$ and dimer evaporation is favored for $n \geqslant 6$.

7.3.2 Intracluster Reactions

Uncatalyzed intracluster reactions are another interesting facet of cluster ion chemistry. The mixed protonated water/alcohol complexes studied by Garvey's and Castleman's groups are generated from neat alcohol expansions via intracluster reactions (Garvey *et al.*, 1994; Morgan *et al.*, 1989; Zhang *et al.*, 1991). The production of the mixed complexes is another example of 'size-selective' chemistry for protonated clusters, i.e. Castleman and coworkers have shown that the 10-mer is the smallest neat protonated methanol cluster for which the reaction occurs (Morgan *et al.*, 1989; Zhang *et al.*, 1991). Two pathways have been proposed for this process. Methanol clusters react only by the first pathway; the second pathway is more facile for larger, branched alcohols. Here X, Y, Y' and Z are functional groups (Garvey *et al.*, 1994; Morgan *et al.*, 1989; Zhang *et al.*, 1991):

$$(ROH)_n H^+ \rightarrow (ROH)_{n-3}(H_2O)H^+ + ROH + R_2O \tag{7.5}$$

$$(ROH)_n(CHYZ - CXY' - OH)H^+ \rightarrow (ROH)_n(H_2O)H^+ + CYZ = CXY' \tag{7.6}$$

Similar intracluster ion–molecule reactions in $(CH_3OH)_n Na^+$ and $(CH_3OH)_n Cs^+$ have been reported by Lisy and coworkers (reaction (7.7)) (Draves and Lisy, 1990; Selegue and Lisy, 1994). Their tandem mass spectrometric investigation of these clusters demonstrates that only the dimethyl ether is lost in the first step of the reaction:

$$(CH_3OH)_n M^+ \rightarrow (CH_3OH)_{n-2}(H_2O)M^+ + (CH_3)_2O$$

$$\rightarrow (CH_3OH)_{n-2-m}(H_2O)M^+ + mCH_3OH \tag{7.7}$$

The heterogeneous alcohol complexes originate from collisions of Na^+ or Cs^+ ions with large, neutral methanol clusters produced in a supersonic expansion. The collision and exothermic solvation processes supply enough internal energy to the clusters for the intracluster reaction to proceed. This intracluster reaction is also 'size-selective'. The first clusters for which the reaction occurs are $(CH_3OH)_8 Na^+$ and $(CH_3OH)_{12}Cs^+$, i.e. in each case two methanol molecules have been added to the second solvation shell before the reaction occurs. As n increases, the reaction becomes more facile and multiple dehydrations within a given cluster are detected. The mechanism of the intracluster reaction was probed by performing isotopic substitution experiments with CH_3OD, CD_3OH, and CD_3OD. The protons or deuterons in the water product are supplied exclusively by the hydroxyl group. These results led Lisy and coworkers (Selegue and Lisy, 1994) to suggest a mechanism by which two methanol molecules in the second shell are hydrogen-bonded to an inner shell methanol, one to the hydroxyl oxygen and one to the hydroxyl hydrogen.

The two outer shell methanols are oriented such that, rather than forming a hydrogen bond, the oxygen of one of the outer shell methanols is directed toward the methyl group of the other. A backside S_N2 attack of the oxygen on the methyl group and subsequent proton transfers along the hydrogen bonds yield the observed products. As the number of methanols in the second shell rises, multiple reaction sites can be formed and the activation barrier is likely to be lowered. Preliminary molecular dynamics calculations on the reaction between Cs^+ and neutral methanol clusters indicate that the ion is solvated on a much faster time scale than the ionic cluster is dehydrated, lending support to the proposed mechanism (Selegue and Lisy, 1994).

Watanabe and Iwata (1996) have studied several stable isomeric forms of $(H_2O)_n B^+$, $n = 1 - 4$, and $(H_2O)_m Al^+$, $m = 1 - 5$, and the spontaneous intracluster reactions that connect some of the observed isomers. $(H_2O)_n M^+$ and $(H_2O)_{n-1} HOMH^+$ were examined in detail via *ab initio* calculations. Both surface forms, in which the cation is situated atop the solvent cluster, and interior forms, in which the cation is surrounded by solvent molecules, were explored. Geometries were optimized with the RHF/6−31G(d) and MP2/6−31G(d) models, and there is little difference in the two sets of equilibrium structures with the exception of the water–water hydrogen bonds. The O...H distances are about 0.1 Å shorter in the MP2 optimum structures. Watanabe and Iwata (1996) base all their conclusions on the MP2 data. For the singly solvated boron complexes, the $HBOH^+$ insertion product is 78 kcal mol^{-1} more stable than $(H_2O)B^+$ and 79 kcal mol^{-1} more stable than H_2BO^+. The latter isomer is an ion complex between BO^+ and H_2. The activation barrier for the production of the $HBOH^+$ from $(H_2O)B^+$ is 34 kcal mol^{-1}. In contrast, $(H_2O)Al^+$ is the most stable singly solvated aluminium system by 12 kcal mol^{-1} with respect to $HAlOH^+$ and 106 kcal mol^{-1} with respect to H_2AlO^+. The barrier height for the production of the $HAlOH^+$ is 68 kcal mol^{-1}. The most remarkable aspect of the aluminium insertion product is that the charge on the aluminium, as computed by natural population analysis (Reed *et al.*, 1985, 1988), is +1.90. Thus, the aluminium is oxidized in $HAlOH^+$. The corresponding charge on the boron is +1.19.

Surface structures are preferred for both the $(H_2O)_n Al^+$ and $(H_2O)_n B^+$ systems, but the difference in energy between the interior and surface structures is larger for the boron complexes. The smaller radius of B^+ leads to greater repulsion between the water ligands in the interior structures, and in fact, no structure with three or four waters attached to the B^+ was found for $(H_2O)_4 B^+$. Networks of hydrogen bonds among the water molecules stabilize the surface structures. All boron complexes of the type $(H_2O)_{n-1}(H_2O)B^+$, which have only one water molecule hydrating the boron, spontaneously transfer a proton to form complexes of the type $(H_2O)_{n-2}(BOH)H_3O^+$. Of particular interest is the $(H_2O)(H_2O)B^+$ complex, for which the presence of the boron leads to the spontaneous dissociation of the $(H_2O)_2$ group to form H_3O^+ and BOH. When $n \geq 3$, some of the $(H_2O)_{n-2}(BOH)H_3O^+$ clusters continue reacting with no barrier to form the insertion products $(H_2O)_{n-1} HOBH^+$ with a linear $HOBH^+$ moiety. No such reactions are observed for the aluminium complexes or for any other atomic ion complexes studied thus far. This unique behavior of the boron complexes is attributed to the tendency of the boron atom to be *sp* hybridized and to the larger electronegativity of the boron atom. Overall, the most stable isomers are of the type $(H_2O)_{n-1} HOMH^+$ with a bent (*cis*) $HOMH^+$ group. Some of these clusters also undergo spontaneous proton transfer (acid-base) reactions as depicted in equations (7.8)–(7.10). The product ions in these reactions have covalent M—OH bonds.

$$(H_2O)_2HOBH^+ \rightarrow (HB(OH)_2)H_3O^+ \tag{7.8}$$

$$(H_2O)_3HOBH^+ \rightarrow (HB(OH)_2)H_5O_2^+ \tag{7.9}$$

$$(H_2O)_3HOAlH^+ \rightarrow (HAl(OH)_2)H_5O_2^+ \tag{7.10}$$

Stace et al. (1994) have investigated the unimolecular decay patterns of $(D_2O)_nNO^+$, $n = 2 - 9$, ions in a tandem mass spectrometer. The ions were formed by electron impact ionization of NO/H_2O clusters prepared by a pick-up technique. The aim of the work was to provide information on reactions that generate proton hydrates in the D region of the ionosphere. NO^+ has been postulated to be involved in a sequence of reactions leading to the formation of the hydrates. The final two reactions in the series are believed to be the following, but the latter reaction has never been directly observed:

$$(H_2O)_2NO^+ + H_2O \rightarrow (H_2O)_3NO^+ \tag{7.11}$$

$$(H_2O)_3NO^+ + H_2O \rightarrow (H_2O)_3H^+ + HNO_2 \tag{7.12}$$

Approaching the problem indirectly, $(H_2O)_4NO^+$ can be viewed as a collision intermediate in reaction (7.12), which can then rearrange to form the transition state for the production of the desired products (reaction (7.13)):

$$(H_2O)_4NO^+ \rightarrow (H_2O)_3H^+ + HNO_2 \tag{7.13}$$

The dominant dissociation channel for the smaller $(D_2O)_nNO^+$ ions was loss of a water molecule to form the $n-1$ ion. When $n=4$, however, both decay channels, i.e. formation of the $n-1$ ion and of the proton hydrate, are observed with comparable fragmentation percentages. When $n=5$, reaction (7.13) (for D_2O) predominates. Thus, the switch in the preferred dissociation channel correlates with the reaction sequence given above.

7.3.3 Polyethers

Another active research area is the complexing of cations by polyethers, including crown ethers. The interest in polyethers arises from the similar nature of polyethers and the polar environments of protein interiors (Warshel, 1981), from the possible application of crown ethers to separating radionuclides from contaminated streams and soils (Chiarizia et al., 1992; Du et al., 1993), and from the use of crown ethers in the treatment of cancerous tissue (see, for example, Kozak et al., 1985). When the cation contains more than one acidic hydrogen, the complexes are stabilized by multiple cation–ether intermolecular hydrogen bonds as well as by interactions between the cation and the bond dipoles of unbound ether groups. Meot-Ner et al. (1994, 1996b) have examined, via pulsed high-pressure mass spectrometry, complexes of H^+, $(H_2O)_nH^+$, $n = 1 - 3$, and NH_4^+ with monodentate ligands and with acyclic and cyclic polydentate ligands. More specifically, in the earlier paper the ligands were n-propyl ether, glyme (G1, $CH_3OCH_2CH_2OCH_3$), diglyme (G2, $CH_3(OCH_2CH_2)_2OCH_3$), and triglyme (G3, $CH_3(OCH_2CH_2)_3OCH_3$). The experimental binding energies were compared with literature values for acetone, 12-crown-4, 15-crown-5, and 18-crown-6. In the subsequent article, the ligands were acetone, $CH_3OCH_2OCH_3$, glyme and diglyme. In this article, gas-phase and condensed-phase NH_4^+ data were compared with data for K^+, H_3O^+, and Na^+. Acetone and n-propyl ether were utilized as monodentate models of the polydentate ethers.

The thermochemical trends observed for the complexation of H^+, H_3O^+ and NH_4^+ to the ethers are exemplified by the results obtained when four oxygen groups are available to bind to the H^+ (Meot-Ner et al., 1994, 1996b). $\Delta H°$ for the association reaction is $-225.0\,\text{kcal\,mol}^{-1}$ for (12-crown-4)H^+ and is more negative by $5.0\,\text{kcal\,mol}^{-1}$ for G3H^+, by $11.0\,\text{kcal\,mol}^{-1}$ for (G1)$_2H^+$, and by $24.7\,\text{kcal\,mol}^{-1}$ for $(CH_3CO)_4H^+$. The increased flexibility of the ligands allows the geometries of the $[OH...O]^+$ (or $[NH...O]^+$) hydrogen bonds to optimize and leads to a more favorable alignment of the bond dipoles of the free ligands with the cation. However, concomitant destabilizing entropy effects make the range of $\Delta G°$ values for these systems much more narrow than the range of $\Delta H°$ values. In fact, for some analogous series of complexes the relative association free energies are rearranged with respect to the relative association enthalpies. The atypically large negative $\Delta S°$ values for the water and ammonia complexation reactions led to the conclusions that multiple hydrogen bonds are formed with the first ligand, the proton is associated with the water or ammonia rather than the ether, and, when not prevented by steric constraints, the second water forms a bridged structure in which each water molecule interacts with an ether oxygen.

In order to gain more insight into the structures of these complexes, Scheiner's group carried out *ab initio* calculations on the model systems dihydroxyethane ($HO(CH_2)_2OH$, DHE), protonated dihydroxyethane (DHEH^+), (DHE)(H_2O)H^+, and (DHE)(H_2O)$_2H^+$, and on G1, G1H^+, (G1)(H_2O)H^+, (G1)(H_2O)$_2H^+$ (Meot-Ner et al., 1994), (G1)(NH_3)H^+, (G1)Na^+, and (G1)K^+ (Meot-Ner et al., 1996b). Equilibrium geometries were obtained at the SCF level with use of the 6–31G basis set for Na, the 3–21G basis set for K, and the 4–31G basis set for all the other atoms (Hehre et al., 1972a). DHE, DHEH^+ and G1H^+ contain asymmetric internal hydrogen bonds, although the hydrogen bond is weak in the neutral system due to its constrained molecular skeleton (Meot-Ner et al., 1994). When a water is added to DHEH^+ or G1H^+, a symmetric ring structure is formed with the proton on the H_2O moiety, despite the smaller proton affinity of H_2O compared to those of the polyethers. In (DHE)(H_2O)$_2H^+$ and (G1)(H_2O)$_2H^+$, a ring structure is retained with the second water hydrogen bonded to the H_3O^+ group and to a glyme oxygen, such that a network of hydrogen bonds is present. An additional open structure with the second water hydrogen bonded only to the first water was considered for (G1)(H_2O)$_2H^+$, but it was found to be $8.0\,\text{kcal\,mol}^{-1}$ less stable than the ring form. Interestingly, the proton remains on G1 in the open form. The calculated complexation energies for G1 are weaker than the corresponding energies for DHE, particularly for the addition of the first H_2O, indicative of a less efficient interaction with the second ether group in the former complexes. Perhaps the most significant implication of the combined experimental and computational data is that these complexes may serve as model building blocks for the development of the proton wires postulated to be involved in proton transport in biomembranes (Nagle and Tristram-Nagle, 1983).

According to the calculated structures, the proton is on the ammonia group in (G1)(NH_3)H^+ and the (G1)(NH_3)H^+, (G1)Na^+ and (G1)K^+ complexes all have the cation interacting with both ether oxygens (Meot-Ner et al., 1996b). The calculations reproduce the relative gas-phase binding energies observed experimentally, i.e. $H_3O^+ > Na^+ > NH_4^+ \approx K^+$. Although the amount of electron density transferred from the ether to the cation is small (< 0.2 e), the relative complexation energies do correlate with this parameter. The relative complexation energies of H_3O^+, Na^+, NH_4^+, and K^+ in the condensed phase are also analyzed in the article. All the additional data provide further evidence for the thermochemical similarity of the binding of NH_4^+ and K^+ cations in gas- and condensed-phase environments (Liebman et al., 1991).

The interactions between 18-crown-6 and Li^+, Na^+, K^+, Rb^+ and Cs^+ have been studied with use of *ab initio* calculations by Glendening et al. (1994). $(H_2O)_nM^+$ complexes were examined also in order to probe the selectivity of the cations in aqueous solution. Another motivation of this work was to furnish information that would aid in the development of reliable classical force fields. Geometry optimizations were carried out for two conformations of each system at the SCF level with the 3–21G and '6–31+G(d)' basis sets. MP2 energies were calculated for the SCF geometries. The standard 3–21G basis set is available for all atoms but Cs, for which Glendening et al. developed a 3–21G-type basis set. For the higher-level '6–31+G(d)' calculations, the standard 6–31+G(d) basis was employed for H, Li, O, and Na and the standard 6–31G(d) for C. For K, Rb and Cs the Hay–Wadt valence basis set with effective core potentials (Wadt and Hay, 1985) was contracted to $(5s, 5p)/[3s, 3p]$ and a set of d functions was added. The core electrons in the $n - 1$ shell were treated explicitly.

The C_i conformation of 18-crown-6, with four of the ether oxygens pointing toward the center of the ring and two of them pointing upward, was found to be more stable than the D_{3d} conformation, with all six oxygens pointing toward the center of the ring, by about 5 kcal mol^{-1}. The former conformation is stabilized by intramolecular C—H...O interactions and is found in crystalline 18-crown-6 (Dunitz et al., 1974); the latter is dominant in polar solvents (Live and Chan, 1976). In the complexes the K^+ and Na^+ ions are located inside a D_{3d} cavity, the cavity puckers around the Li^+ ion forming a complex with a distorted octahedral arrangement of the oxygen atoms, and the Rb^+ and Cs^+ cations sit atop the crown ether forming a complex with C_{3v} symmetry. In general, the calculated geometries are in good accord with reported crystal structures (Dunitz et al., 1974) but the calculated binding affinities differ from the experimental values reported by Katritzky et al. (1992) by as much as 30 kcal mol^{-1}. In contrast, the binding affinity of -71.5 kcal mol^{-1} for (18-crown-6)K^+ obtained by the calculations is in excellent agreement with the value of -71 ± 3 kcal mol^{-1} estimated by Meot-Ner et al. (1996b). Some general trends were observed for the association energies (Glendening et al., 1994):

1. As the size of the cation increases, the gas-phase association energies decrease. However, the sequence changes to $K^+ > Rb^+ \approx Na^+ > Cs^+ > Li^+$ when the cations are solvated by as few as four water molecules. The experimental ordering for the fully solvated cations is identical for the first four ions. (No data are available for Li^+.)
2. The HF/3–21G bonding energies are not representative of the values computed at the higher levels of theory.
3. Inclusion of electron correlation increases the complexation energies.
4. A natural energy decomposition analysis (Glendening and Streitwieser, 1994) of the binding affinity indicates that the dominant stabilizing component is the electrostatic term but the charge transfer term is also important. The cation also weakly polarizes the crown ether, all of which should be taken into account in the potential function used for computer simulations.

7.3.4 Negatively Charged Clusters

The development of electrospray mass spectrometry by Fenn and coworkers (Fenn et al., 1985; Yamashita and Fenn, 1984) has made it possible to produce multiply charged positive and negative ions in the gas phase. Such ions include transition metal– and alkaline earth metal–ligand complexes, peptides and proteins, and multiply deprotonated nucleic acids. For example, researchers have utilized this technique in their studies of the gas-phase sequences and conformations of singly- and multiply-protonated proteins (Clemmer et al., 1995; Jones et al., 1994; Ogorzalek Loo and Smith, 1994), of proton-transfer rates and gas-phase basicities of singly- and multiply-protonated peptides and proteins (Kaltashov et al., 1995; Ogorzalek Loo et al., 1994), of ion–ligand interactions with methylmercury (II) (Canty et al., 1994), and of the molecular masses and sequences of modified oligonucleotides (Barry et al., 1995). Kebarle and coworkers have utilized this method in their measurements of the free energies of hydration of singly- and multiply-charged negative ions, and some of their work is described below.

One way of probing whether an ion has an intramolecular hydrogen bond is to determine its clustering free energies with water. The presence of the hydrogen bond leads to lower hydration energies and the binding of fewer water molecules compared to open systems (Gao et al., 1989; Klassen et al., 1994). With this in mind, Blades et al. (1996) investigated the hydration free energies of the orthophosphate, diphosphate, D-ribose 5-phosphate, adenosine 5′-monophosphate, and adenosine 5′-diphosphate ions. The $-\Delta G^\circ_{0,1}$ values decrease from 8.4 kcal mol^{-1} for $H_2PO_2^-$ to 7.9 kcal mol^{-1} for $(OH)HPO_2^-$ to 7.6 kcal mol^{-1} for $(OH)_2PO_2^-$ to 5.8 kcal mol^{-1} for D-ribose 5-phosphate. Moreover, under the same conditions the D-ribose 5-phosphate ion adds one fewer water molecule than the orthophosphate ion $(OH)_2PO_2^-$. The decrement in dissociation free energies for the first three ions can be rationalized by the electron withdrawing effected by each additional OH group. The larger decrease for the D-ribose 5-phosphate anion is attributed to an internal hydrogen bond. Other possible explanations for this decrease can be ruled out. The negative charge is not delocalized over a greater number of equivalent oxygens, and the electron-withdrawing effect of the ribose group should be comparable to that of a hydroxy group. A qualitative estimate of the strength of this hydrogen bond was made by the following analysis. The $HCO_2(CH_2)_nCO_2^-$ anions, $n = 3-8$, have been shown to be internally hydrogen bonded (Blades et al., 1995). The $\partial \Delta G^\circ_{0,1}$ value for the D-ribose 5-phosphate anion and the orthophosphate anion (1.8 kcal mol^{-1}) is smaller than the $\partial \Delta G^\circ_{0,1}$ value for the propionate anion $C_2H_5CO_2^-$ and the α, ω-dicarboxylate anion $HCO_2(CH_2)_5CO_2^-$ (3.8 kcal mol^{-1}); thus, the intramolecular hydrogen bond is weaker for the ribose phosphate. On the basis of the $-\Delta G^\circ_{0,1}$ results for adenosine 5′-monophosphate, adenosine 3′-monophosphate, and adenosine 5′-diphosphate, these anions are also predicted to be stabilized by internal hydrogen bonding (Blades et al., 1996).

Kebarle and colleagues have now examined ion–water clustering reactions for several series of doubly charged anions, i.e. anions of some oxo acids of C, S and P (Blades et al., 1995, 1996; Blades and Kebarle, 1994). General observations from this work are:

1. Hydration free energies $(-\Delta G^\circ)$ for doubly charged ions are greater than those for singly charged ions.
2. The ion peaks of maximum intensity occur at much higher solvent numbers for doubly charged ions than for singly charged ions.

3. Separating the two negative charges with intermediate groups reduces the hydration free energies.
4. Greater charge delocalization within a given charged group reduces the hydration free energies.
5. Successive waters alternate between charge sites so $\Delta G°$ does not decrease smoothly with increasing solvent number.

$(H_2O)_n X^-$ clusters, where $X =$ F, Cl, Br, I, and $n = 1 - 6$ or $1 - 8$ (I only), have been examined by Combariza and Kestner (1994) and Combariza et al. (1993, 1994a,b). Geometry optimizations were performed at the SCF level for all clusters and at the MP2 level for $n = 1 - 3$ (F) or $n = 1$ (Cl, Br, I). MP2 single-point energies were calculated for the remaining systems. The 6−31+G(d) basis set was used for the water molecules and the McClean and Chandler (1980) double-zeta basis set augmented with a set of diffuse sp functions and three sets of d functions was used for fluorine and chlorine. For bromine and iodine the basis sets and effective core potentials developed by Christiansen and coworkers were utilized (LaJohn et al., 1987). The outermost core d-orbitals were included in the valence region, and sets of 3 and 1d-functions were constructed from the four original primitives. Two sets of diffuse sp functions and a third set of d functions were added to the basis set. Structures with X^- at the surface of the cluster and with X^- in the interior of the cluster were characterized. The interplay between the $X^-...H_2O$ and $H_2O...H_2O$ interactions determines the structures of the clusters. For the $(H_2O)_n F^-$ systems (Combariza and Kestner, 1994) the calculations indicate that as the number of water molecules increases the $F^-...H_2O$ interactions weaken and the $H_2O...H_2O$ interactions become more important. Based on the ΔMP2 + ZPE energies, with the exception of $n = 4$, the more stable isomer is a surface isomer. However, the differences in stability are small and will be affected by entropy corrections (which favor the interior form by about 10–20 cal mol^{-1} K^{-1}), so integral and differential ionization energy shifts were utilized to predict the preferred structures. The integral shift is the difference in the vertical ionization energies of the nth cluster ion and the unsolvated F^-; the differential shift is the difference in the vertical ionization energies of the nth and $(n - 1)$th cluster ions. Plots of the integral shift and the differential shift versus the number of water molecules show a change in slope at $n = 4, 5$, respectively, which, in conjunction with the entropy effects, is interpreted as indicative of a change in preference from surface to interior states at $n = 4, 5$. Combariza et al. (1993, 1994a) obtained different results for the larger halide anions Cl^-, Br^-, and I^- for which the $X^-...$ water interactions are weaker and the number and strength of the $H_2O...H_2O$ interactions are particularly important. Based on plots of ionization energy shifts versus n and on a comparison with experimental vertical ionization energies (see below) (Markovich et al., 1994), no internal states were found for $n < 5$ but there was some evidence for $I^-(H_2O)_n$ that the transition from exterior to interior states occurs at $n = 6$.

Markovich et al. (1994) have measured the photoelectron spectroscopy (PES) of $(H_2O)_n Cl^-$, $n = 1 - 7$, $(H_2O)_n Br^-$, $n = 1 - 16$, and $(H_2O)_n I^-$, $n = 1-60$. The photoelectron spectra of the iodide clusters suggest that a first solvation layer is formed that contains six water molecules. This conclusion is based on a sharp drop in the incremental increase in E_{stab}, the vertical binding energy of the cluster minus the electron affinity of the unsolvated anion, when $n = 7$. The smaller increase in the value of $E_{stab}(n) - E_{stab}(n - 1)$ observed with increasing n for each of these ions is consistent with a weaker interaction between the outer shell water molecules and the anion. Although the patterns for the iodide ion are most dramatic, when E_{stab} is normalized with respect to the appropriate $n = 1$ cluster for each of the halide ions, the trends in E_{stab} versus n and in $E_{stab}(n) - E_{stab}(n - 1)$ versus n are similar for all three anions. The above observations and an analysis of $E_{stab}(n)$ for large $(H_2O)_nI^-$ in terms of a classical continuous dielectric model led the authors (Markovich *et al.*, 1994) to tentatively propose that all three of these cluster ions are surface rather than internally solvated. This suggestion agrees with several molecular dynamics (Dang and Garrett, 1993; Perera and Berkowitz, 1993) and Monte Carlo (Jorgensen and Severance, 1993) simulations but disagrees with Combariza *et al.*'s (1993, 1994a) *ab initio* evidence for the I^- clusters. However, as Markovich *et al.* point out, none of the studies are conclusive.

$(H_2O)_nF^-$ clusters, $n = 1 - 3$, have also been studied by Xantheas and Dunning (1994) but their focus was on determining transition state as well as equilibrium structures, vibrational frequencies, magnitudes of the many-body interaction terms, and the change in these properties with cluster size. Xantheas and Dunning optimized geometries at the HF and MP2 levels of theory with the aug-cc-pVDZ basis set ($n = 1 - 3$) and with the aug-cc-pVTZ basis set ($n = 1$). $(H_2O)F^-$ was also optimized at the MP4 level with both basis sets. For the larger clusters, the MP2/aug-cc-pVDZ geometries were employed to calculate MP4/aug-cc-pVDZ single-point energies. Unlike Combariza and Kestner (1994), Xantheas and Dunning (1994) found substantial differences in the optimal MP2 and HF structures of these clusters, in particular a shorter distance and stronger interaction between the waters at the MP2 level, which could affect MP2 single-point energies obtained with HF geometries. Another deviation in the two sets of results is that Cambariza and Kestner's $\Delta MP2 + ZPE$ energy changes agree better with Kebarle and coworkers' experimental data (Arshadi *et al.*, 1970; Kebarle *et al.*, 1968), whereas Xantheas and Dunning's ΔH_{298} values agree better with Hiraoka and coworkers' experimental data (Hiraoka *et al.*, 1988). Again the most stable structures computed for these small clusters are the surface structures; the symmetric internal structures are transition states. The minima–minima and minima–transition state energy differences are all small. As the authors point out (Xantheas and Dunning, 1994) with such flat potential energy surfaces identifying global minima may not be meaningful, and dynamical effects must be taken into account in any consideration of the properties of these clusters. Moreover, the interaction potential utilized in the simulation must include three-body terms (Xantheas and Dunning, 1993) and account for halide to water electron density transfer (Combariza *et al.*, 1994a) in order to replicate the subtle variations in energy among the configurations.

7.4 UNCONVENTIONALLY HYDROGEN-BONDED IONIC SYSTEMS

There is an increasing interest in the nature of unconventional ionic and neutral hydrogen bonds. Included in this category are CH...Y (Aakeröy and Seddon, 1993; Green, 1974; Taylor and Kennard, 1982, 1984), XH...C (Del Bene, 1993; Legon, 1992; Legon *et al.*, 1992a–c; Legon and Thorn, 1992; Meot-Ner *et al.*, 1996a) and XH...π interactions (Al-Juaid *et al.*, 1991; Atwood *et al.*, 1991; Caldwell and Kollman, 1995; Dougherty, 1996; Dougherty and Stauffer, 1990; Kearney *et al.*, 1993; Kim *et al.*, 1994; Lee *et al.*,

1995; Viswamitra et al., 1993). Metal ion–π interactions have also been investigated in some detail recently (Caldwell and Kollman, 1995; Dougherty, 1996; Kearney et al., 1993; Mecozzi et al., 1996). In these cases the term 'hydrogen bonding' indicates only that the interaction involves partially charged hydrogen atoms and polar and/or polarizable ligands. It does not imply that there is a significant amount of n-donation into the bond. For ionic systems, the strength of these interactions varies from 8 to 25 kcal mol^{-1}.

7.4.1   CH...Y and XH...C Hydrogen Bonds

Strong evidence for the existence of CH...Y hydrogen bonds has been provided by crystallographic and spectroscopic studies. Here Y is most frequently O but can be C, F, N, Cl, Br, or I (Abdul-Sada et al., 1986, 1990; Chuen and Sammes, 1983; Desiraju, 1991; Taylor and Kennard, 1982, 1984). This type of interaction occurs frequently in the solid state (Desiraju, 1991; Steiner and Saenger, 1993; Taylor and Kennard, 1982) and has been shown to have an important influence on packing motifs (Aakeröy and Seddon, 1993; Desiraju et al., 1993; Desiraju, 1991; Reetz et al., 1993). For example, CH...O interactions help form networks of hydrogen bonds that are an important factor in the packing diagrams found for some cyclopenta[a]phenanthrenes (Desiraju et al., 1993) and for complexes of the proton sponge 1,8-bis(dimethylamino)naphthalene with 1,2-dichloromaleic acid (Wozniak et al., 1995). Since C—H...Y interactions are particularly common when the C—H group is adjacent to an aliphatic or aromatic nitrogen (Taylor and Kennard, 1982), C—H...O hydrogen bonds also occur frequently in biomolecules such as carbohydrates (Steiner and Saenger, 1992) and nucleosides (Jeffrey and Saenger, 1991; Saenger 1984). They have an impact on the conformations of small molecules (Adcock and Zhang, 1995; Gil et al., 1995; Tsuzuki et al., 1993) and on the tertiary structure of macromolecules (Jeffrey and Saenger, 1991; Saenger, 1984). In addition, the C—H group may act as the proton donor in the coordination of water molecules (Kim et al., 1994; Meot-Ner and Deakyne, 1985; Steiner and Saenger, 1992, 1993). In crystals the role of the C—H group appears to be to complete the tetrahedral coordination of the water molecule (Steiner and Saenger, 1993).

One of the general characteristics of a conventional X—H...Y hydrogen bond is that the X—H bond lengthens upon hydrogen bond formation, leading to a decrease in the X—H stretching frequency v_{XH} and the X—H force constant k_{XH}. If the C—H...Y interaction is a hydrogen bond analogous, albeit smaller, effects should be observed in spectroscopic investigations of these bonds. The C—H...N bond in the 1, 3, 5-trichlorobenzene complex with pyridine leads to a 35 cm^{-1} lowering of the C—H stretch (Allerhand and Schleyer, 1963). Comparable spectral shifts have been reported for the C—H bonds in acetylenes as a result of C—H...O interactions (DeLaat and Ault, 1987; Desiraju and Murty, 1987; Peterson and Klemperer, 1984). Further support for the existence of C—H...Y hydrogen bonds comes from Koch and Popelier's (1995) analysis of the neutral intermolecular C—H...O interactions in the formaldehyde/chloroform, acetone/chloroform, formaldehyde/benzene, and acetone/1,1-dichloroethane complexes and of the intramolecular C—H...O interactions in 3'-azido-3'-deoxythymidine (AZT). Koch and Popelier have proposed a set of seven criteria that can be employed to confirm the presence of a hydrogen bond. The criteria are based on the theory of 'atoms in molecules' (AIM)

(Bader, 1990) and characterize hydrogen bonding solely from the total charge density. Each of the above C—H...O interactions obey all seven criteria. Since it has been shown that ionic hydrogen bonds have similar, but more pronounced, characteristics to those of related neutral hydrogen bonds (Carroll *et al.*, 1988; Deakyne, 1987; Desmeules and Allen, 1980; Platts and Laidig, 1995), these studies also provide evidence for the existence of C—H...Y ionic hydrogen bonds.

[C—H...O]$^+$, [C—H...N]$^+$, [N—H...C]$^+$, and [C—H...C]$^+$ interactions have been examined by Meot-Ner *et al.* (1996a). Hydrogen-bond dissociation enthalpies (ΔH°_{exp}) in protonated dimer ions containing isocyanides were measured by pulsed high-pressure mass spectrometry, and the interactions were analyzed by *ab initio* calculations. The systems studied experimentally are (1) CH_3NC with $(CH_3)_nNH_{4-n}^+$, $n = 0 - 3$, (2) CH_3NCH^+ with CH_3CN, CH_3NC, H_2O, CH_3OH, and $(CH_3)_2O$, and (3) C_2H_5NC with $C_2H_5NCH^+$ and $CH_3NH_3^+$. Strong bonding (19–25 kcal mol^{-1}) is observed when the carbon lone pair is the electron donor, i.e. in the complexes of the isocyanides with the protonated amines and the protonated isocyanides. The bonding is weaker (14–21 kcal mol^{-1}) in the complexes of the oxygen bases with CH_3NCH^+. Plots of ΔH°_{exp} versus the proton affinity difference of the neutral subunits are linear for the $R_3NH^+...CNR$ and $CH_3NCH^+...O$ type complexes, and the intercepts yield intrinsic bond strengths ($\Delta PA = 0$) of 26.5 kcal mol^{-1} and 23.1 kcal mol^{-1}, respectively.

The dimers examined *ab initio* are $(HCN)NH_4^+$, $(HNC)NH_4^+$, $(CH_3CN)NH_4^+$, $(CH_3NC)NH_4^+$, $(HCN)CH_3NH_3^+$, $(HNC)CH_3NH_3^+$, $(CH_3CN)CH_3NH_3^+$, $(CH_3NC)CH_3NH_3^+$, $(CH_3CN)CH_3CNH^+$, $(CH_3NC)CH_3CNH^+$, $(CH_3CN)CH_3NCH^+$, and $(CH_3NC)CH_3NCH^+$ (Meot-Ner *et al.*, 1996a). Geometry optimizations were carried out at four calculational levels, HF/3−21G, HF/6−31G(d), MP2/6−31+G(d), and MP2/6−31+G(d,p). Two significant differences were obtained in the results from these four levels of theory. First, $(CH_3NC)CH_3CNH^+$ is a minimum on the SCF potential energy surfaces but not on the MP2 potential energy surfaces. Second, $(CH_3CN)CH_3CNH^+$ changes from an asymmetric to a symmetric complex at the MP2/6−31+G(d,p) level. MPn, $n = 2 - 4$, and infinite-order quadratic configuration interaction (QCISD(T)) (Wiese and Weiss, 1968) single-point energies were obtained with the HF/6−31G(d) and MP2/6−31+G(d,p) geometries. Increasingly flexible basis sets, i.e. 6−31+G(d), cc-pVDZ+, 6−31+G(d,p), aug'-cc-pVDZ, aug-cc-pVDZ, 6−31+G(2d,2p), 6−311++G(2d,2p), and cc-pVTZ+, were employed in the single-point calculations. The computed binding energies were not corrected for BSSE.

Normal-mode vibrational frequencies were obtained at all four levels of theory for which optimal structures were computed. The HF/3−21G and HF/6−31G(d) vibrational modes and zero-point energies were corrected by the usual factors of 0.89 and 0.91, respectively. The MP2/6−31+G(d) and MP2/6−31+G(d,p) vibrational frequencies and ZPEs were not scaled. Each cyano and isocyano dimer has a hydrogen-bond stretching mode in the 160–365 cm^{-1} range, which is similar to the range of roughly 200–350 cm^{-1} found by Pudzianowski (1995). Clustering enthalpies were determined using equations (7.3) and (7.4). For this set of reactions, ΔE^t_{298} is $-3/2$ RT, ΔE^r_{298} is $-$RT for the linear bases and $-3/2$ RT for the nonlinear bases, and ΔPV is $-$RT. Torsional vibrational modes characterized by scaled frequencies $\leqslant 500$ cm^{-1} were treated as pure rotations. The results indicate that for those complexes for which the shape of the proton-transfer potential energy curve is unaffected by the level of calculation, the HF/6−31G(d) and MP2/6−31+G(d) models can be utilized to convert clustering energies into clustering enthalpies if (1) utilizing the MP2/6−31+G(d,p) model is not computationally feasible and (2) only trends in clustering enthalpies are desired.

Calculated interaction energies as a function of level of theory are listed in Table 7.3 for two representative sets of the $XH^+ + Y \rightarrow (Y)XH^+$ reactions (Meot-Ner et al., 1996a). The entries are arranged in order of increasing basis set size. Trends in association energies within each series of complexes and between the two series of complexes hold for every basis set considered. SCF hydrogen-bond energies are larger when complexation occurs through the nitrogen but by no more than 2.5 kcal mol^{-1}.

Table 7.3 Calculated clustering energies for the reactions $XH^+ + Y \rightarrow (Y)XH^{+a}$

Complex	Basis	HF[b]	MP2[c]	MP4[c]	QCISD(T)[c]
$(CH_3CN)NH_4^+$	6−31G(d)	−25.5	−27.1[b]		
	6−31+G(d)	−24.8	−26.3[b]		
	cc-pVDZ+		−26.7	−26.6	−26.5
	6−31+G(d,p)		−26.4		
	aug'-cc-pVDZ		−27.2	−27.0	−26.9
	aug-cc-pVDZ		−27.9	−27.7	−27.6
	6−31+G(2d,2p)	−25.2	−27.6		
	6−311++G(2d,2p)	−25.1	−27.6		
	cc-pVTZ+		−27.7		
$(CH_3NC)NH_4^+$	6−31G(d)	−24.2	−28.8[b]		
	6−31+G(d)	−22.1	−26.1[b]		
	cc-pVDZ+		−27.6	−26.5	−26.1
	6−31+G(d,p)		−27.0		
	aug'-cc-pVDZ		−28.0	−26.9	−26.6
	aug-cc-pVDZ		−28.6	−27.6	−27.2
	6−311+G(2d,2p)	−22.5	−27.9		
	6−31++G(2d,2p)	−22.8	−28.4		
	cc-pVTZ+		−28.5		
$(CH_3CN)CH_3NCH^+$	6−31G(d)	−22.8	−24.0[b]		
	6−31+G(d)	−21.9	−23.0[b]		
	cc-pVDZ+		−23.6	−23.9	−23.7
	6−31+G(d,p)		−23.0		
	aug'-cc-pVDZ		−24.1	−24.3	−24.1
	aug-cc-pVDZ		−24.8		
	6−31+G(2d,2p)	−22.5	−24.3		
	6−311++G(2d,2p)	−22.3	−24.2		
	cc-pVTZ+		−24.2		
$(CH_3NC)CH_3NCH^+$	6−31G(d)	−21.5	−25.2[b]		
	6−31+G(d)	−19.4	−22.5[b]		
	cc-pVDZ+		−24.3	−23.8	−23.1
	6−31+G(d,p)		−23.5		
	aug'-cc-pVDZ		−24.9	−24.5	−23.8
	aug-cc-pVDZ		−25.4		
	6−31+G(2d,2p)	−19.9	−24.5		
	6−311++G(2d,2p)	−20.1	−24.9		
	cc-pVTZ+		−24.8		

[a]Data from Meot-Ner et al. (1996a). Energies in kcal mol^{-1}.
[b]HF/6−31G(d) geometries.
[c]MP2/6−31+G(d,p) geometries.

Complexation through the carbon is preferentially stabilized when electron correlation is taken into account, yielding hydrogen-bond strengths that are essentially equal for analogous cyanide and isocyanide systems. The MP2 interaction energies have converged to within 0.1 kcal mol^{-1} for all dimers examined. Compared to the QCISD(T) correlation correction, the MP2 correction is substantially overestimated (by 1–2 kcal mol^{-1}) for the isocyanides. These clusters provide additional examples of binary systems for which the MP4/aug'-cc-pVDZ and, for the cyano complexes, MP2/6–311++G(2d,2p) models reproduce the results of larger models (cf. Section 7.2.2).

Since the QCISD(T) correction (Table 7.3) is essentially independent of the basis set employed to calculate it, effective QCISD(T)/cc-pVTZ+//MP2/6–31+G(d,p) hydrogen-bond energies were determined via an additivity approximation (equation (7.14)) based on MP2/cc-pVTZ+//MP2/6–31+G(d,p) energies:

$$\Delta E_{\text{eff}} = \Delta E(\text{MP2/cc-pVTZ+}) + (\Delta E(\text{QCISD(T)}) - \Delta E(\text{MP2})) \quad (7.14)$$

The second term on the right-hand side of equation (7.14) was estimated separately for each $XH^+ + Y \rightarrow (Y)XH^+$ reaction by finding the average of the difference in the QCISD(T) and MP2 energies for that reaction. Using ΔE_{eff} to determine clustering enthalpies, calculated and experimental ΔH° values agree within the standard uncertainty of ±1.5 kcal mol^{-1} for only four of the nine complexes for which experimental data were measured. (Examples are listed in Table 7.4.) Two other complexes have $\Delta H^\circ_{\text{calc}}$ and $\Delta H^\circ_{\text{exp}}$ values that agree to ±2 kcal mol^{-1}. It can be argued, based on the consistency of the theoretical data and on patterns in the effect of methyl substitution of the proton donor on hydrogen-bond enthalpies (Keesee and Castleman, 1986), that some of the experimental clustering enthalpies are too low (Meot-Ner et al., 1996a).

Two other results from the calculations on these complexes are the following. First, the separation between proton donor and acceptor differs by about 0.1 Å for all corresponding cyano and isocyano complexes, consistent with the larger van der Waals radius of isocyano C compared to cyano N (Legon, 1992). Second, the relative importance of the electrostatic and delocalization components of the dissociation energy is different for the two sets of complexes, with delocalization effects being more important for the isocyanides (Meot-Ner et al., 1996a).

Complexes with [C—H...O]$^+$ hydrogen bonds have been invoked as intermediates

Table 7.4 Clustering enthalpies for the reactions $XH^+ + Y \rightarrow (Y)XH^{+a}$

Complex	$-\Delta E^b$	$-\Delta E_{\text{eff}}^c$	$-\Delta H^\circ_{\text{calc}}$	$-\Delta H^\circ_{\text{exp}}$
(CH$_3$CN)NH$_4^+$	27.7	27.4	26.8	27.6^d
(CH$_3$NC)NH$_4^+$	28.5	27.2	26.8	24.0
(CH$_3$CN)CH$_3$NH$_3^+$	25.4	25.0	24.4	24.5^d
(CH$_3$NC)CH$_3$NH$_3^+$	26.0	24.7	24.2	23.8
(CH$_3$CN)CH$_3$CNH$^+$	33.3	31.9	32.8	30.2^d
(CH$_3$CN)CH$_3$NCH$^+$	24.2	24.2	23.3	19.1
(CH$_3$NC)CH$_3$NCH$^+$	24.8	23.7	23.4	25.2

^{a}Data from Meot-Ner et al. (1996a). ΔE and ΔH in kcal mol^{-1}.
^{b}MP2/cc-pVTZ+//MP2/6–31+G(d,p) values.
$^c\Delta E$ adjusted for the estimated QCISD(T) correction, i.e. the average difference in the MP2 and QCISD(T) energies for the given complex.
^{d}Data from Keesee and Castleman (1986).

in the dissociation of radical cations of general formula $HOCH(R_1)C(=O)R_2^{·+}$. In particular, Terlouw and colleagues have probed the mechanism of the breakdown of ionized acetol $HOCH_2C(=O)CH_3^{·+}$, acetoin $HOCH(CH_3)C(=O)CH_3^{·+}$, methyl glycolate $HOCH_2C(=O)OCH_3^{·+}$, and methyl lactate $HOCH(CH_3)C(=O)OCH_3^{·+}$ to $R_1CO^·$ (George et al., 1994; Suh et al., 1995a,b). Analysis of the metastable ion (MI) mass spectra and collision-induced dissociation (CID) mass spectra of deuterium labeled reactants and products has led to the proposal that the mechanism proceeds by a double hydrogen transfer via a $[C-H...O]^+$ rather than an $[O-H...O]^+$ hydrogen-bonded intermediate. *Ab initio* calculations have been carried out on the reactants and on some of the proposed intermediates and transition states.

Consider the methyl glycolate radical cation (Suh et al., 1995b). The products of its dissociation are protonated methyl formate $HC(OH)OCH_3^+$ and $HCO^·$. Labeling experiments, which can be utilized because the reactant and product ions experience little or no isotopic scrambling, show that the following transformation occurs:

$$H^{18}O^{13}CH_2C(=O)OCH_3^{·+} \rightarrow HC(OH)OCH_3^+ + H^{13}C^{18}O^· \quad (7.15)$$

These results led to two proposed mechanisms for the dissociation which can be distinguished by deuterium labeling of the hydroxyl hydrogen. Mechanism (7.16) involves an $[O-D...O]^+$ hydrogen-bonded intermediate, similar to the postulated intermediate in the loss of $HOCH_2^·$ from ionized methyl acetate $CH_3C(=O)OCH_3^{·+}$ (Heinrich et al., 1987). Mechanism (7.17) involves a $[C-D...O]^+$ hydrogen-bonded intermediate:

$$DOCH_2C(=O)OCH_3^{·+} \rightarrow [CH_3OC(=O)...O(D)CH_2]^{·+}$$
$$\rightarrow [CH_3OC=O...D...OCH_2]^{·+} \rightarrow HC(OD)OCH_3^+ + HCO^· \quad (7.16)$$

$$DOCH_2C(=O)OCH_3^{·+} \rightarrow [CH_3OC(=O)...O(D)CH_2]^{·+}$$
$$\rightarrow [CH_3OC(=O)-D...OCH_2]^{·+} \rightarrow DC(OH)OCH_3^+ + HCO^· \quad (7.17)$$

The CID mass spectrum of the product cation is almost identical to that of the reference ion formed via protonation of $DC(=O)OCH_3$. ^{18}O, ^{13}C labeling studies of the methyl glycolate radical cation also provide support for reaction (7.17). Further clarification of the mechanism is obtained from the *ab initio* calculations (George et al., 1994; Suh et al., 1995b).

UHF/6–31G(d) equilibrium structures were identified, and MP*n* (up to third order) single-point energies were then computed with these geometries (UMP3/6–31G(d)//UHF/6–31G(d)). Based on the calculations, a direct transfer of a hydrogen atom from the $R_1C(H)O$ molecule (formaldehyde in equation (7.17)) to the $R_2C(=O)-H^{·+}$ moiety ($CH_3OC(=O)-D^{·+}$ in equation (7.17)) is energetically too costly to be feasible. Rather a rearrangement and charge transfer within the $[CH_3OC(=O)-H...OCH_2]^{·+}$ system is suggested, (equation (7.18)). Once the aldehyde becomes charged, a proton is donated to the $R_2(H)C=O$ group and the complex breaks

$$[CH_3OC(=O)-H...OCH_2]^{·+} \rightarrow [CH_3OC(H)=O...H-C(O)H]^{·+}$$
$$\rightarrow HC(OH)OCH_3^+ + HCO^· \quad (7.18)$$

down to the desired products.

7.4.2 XH...π Hydrogen Bonds

Much of the evidence for the existence and importance of XH...π interactions comes from crystallographic studies (Al-Juaid *et al.*, 1991; Atwood *et al.*, 1991; Datta *et al.*, 1980; Harel *et al.*, 1996; Viswamitra *et al.*, 1993) but there is gas-phase (Deakyne and Meot-Ner, 1985; Dougherty, 1996; Gross *et al.*, 1995; Meot-Ner and Deakyne, 1985; Sunner *et al.*, 1981) and solution-phase evidence as well (Kearney *et al.*, 1993; Schneider, 1991). (If the proton donor is positively charged, X—H...π hydrogen bonds belong to the general class of interactions termed cation–π interactions.) XH...π interactions gain stability from the centers of high electron density on the proton acceptors. The bonds can be quite strong, as exemplified by the binding enthalpies of -38, -19, -19, and -9 kcal mol^{-1} measured in the gas phase for Li$^+$ (Taft *et al.*, 1990), K$^+$ (Sunner *et al.*, 1981), NH$_4^+$ (Deakyne and Meot-Ner, 1985) and (CH$_3$)$_4$N$^+$ (Meot-Ner and Deakyne, 1985), respectively, with benzene. In fact, benzene competes favorably with water in binding these ions. The gas-phase binding enthalpies with water are -34, -18, -20, and -9 kcal mol^{-1} for Li$^+$ (Woodin and Beauchamp, 1978), K$^+$ (Sunner *et al.*, 1981), NH$_4^+$ (Meot-Ner, 1984) and (CH$_3$)$_4$N$^+$ (Meot-Ner and Deakyne, 1985), respectively. Interactions of this type are believed to determine crystal packing motifs, aid in substrate/ligand binding, and establish ion selectivity in K$^+$ channels (Dougherty, 1996; Viswamitra *et al.*, 1993).

Recent studies (Harel *et al.*, 1996; Sussman *et al.*, 1991) have shown that the neurotransmitter acetylcholine (ACh, [CH$_3$C(=O)OCH$_2$CH$_2$N(CH$_3$)$_3$]$^+$) traverses a narrow, 20 Å deep gorge lined with aromatic residues after leaving the aqueous environment in the synaptic cleft and before reaching the active site in acetylcholinesterase (AChE). Some of these aromatic residues, along with three water molecules, are also involved in binding the choline end of the ACh molecule in the active site. Meot-Ner (1994) has argued qualitatively that the aromatic residues in the gorge may assist in desolvating ACh, while providing a weakly bonding environment that does not prevent diffusion of ACh to the active site. In order to investigate this suggestion more fully, the interactions between ACh and polar and aromatic solvents have been probed via high-pressure mass spectrometry and *ab initio* calculations (Deakyne *et al.*, 1996).

Equilibrium geometries of ACh, (H$_2$O)ACh, (H$_2$O)$_2$ACh, (C$_6$H$_6$)ACh, and (C$_6$H$_6$)$_2$ACh have been obtained at the HF/6–31G(d) level of theory. Five local minima have been identified for ACh, including the conformations found in solution and in crystals of acetylcholine salts (Table 7.5) (Culvenor and Ham, 1966; Datta *et al.*, 1980; Svinning and Sørum, 1975). The primary features that distinguish one conformation from another are the N$_1$C$_3$C$_4$O$_5$ and C$_3$C$_4$O$_5$C$_6$ torsional angles. The C$_2$N$_1$C$_3$C$_4$ and C$_4$O$_5$C$_6$C$_8$ dihedral angles are approximately 180° for each conformer. (See Table 7.5 for atomic numbering.) Positive *gauche* torsional angles are represented by the letter *g*, negative *gauche* torsional angles by *g'*, and *trans* torsional angles by *t*. For the (H$_2$O)ACh complexes, three interaction sites were examined for each of the five ACh conformations. The first site is the carbonyl oxygen, the second is the ester oxygen, and the third is the quaternary ammonium group. No stable complexes of the second type were found. When the second water was added to form the (H$_2$O)$_2$ACh clusters, it was bound to either the first water or the quaternary ammonium moiety. This led to four complexes for each ACh conformation—one with a C=O...H$_2$O...H$_2$O interaction, one with a quaternary ammonium ...H$_2$O...H$_2$O interaction, one with a C=O...H$_2$O and a quaternary ammonium...H$_2$O

Table 7.5 Selected dihedral angles of acetylcholine[a]

```
                    O7
                    ||
      N1      C4    C6
     / \    / \    / \
   C2   C3    O5    C8
```

System[b]	< N$_1$C$_3$C$_4$O$_5$	< C$_3$C$_4$O$_5$C$_6$
ACh, gg	67.3	81.9
ACh, gg′	78.7	−113.5
ACh, gt	62.6	171.1
ACh, tg	−160.5	80.6
ACh, tt[c]	180.0	180.0
AChBr[d]	78.4	78.9
AChTPB[e]	66.7	154.6
Solution[f]	gauche	trans

[a] Bond angle in degrees.
[b] ACh: acetylcholine, g: gauche, g′: gauche′, t: trans.
[c] C$_s$ symmetry.
[d] Acetylcholine bromide (Svinning and Sørum, 1975).
[e] Acetylcholine tetraphenylborate (Datta et al., 1980).
[f] (Culvenor and Ham, 1966).

interaction, and one with two quaternary ammonium...H$_2$O interactions. In the ACh/benzene cluster ions the quaternary ammonium function was the only binding site considered. The HF/6−31G(d) potential energy surfaces of all of these ions are extremely flat, and we have made no attempt other than starting with geometries far different from the equilibrium geometries to locate either the global minimum or other local minima. We do not believe that the conclusions obtained from this work would be altered by attempting to do so. Moreover, since the order of some of the local minima changes when energies are converted to enthalpies and, especially, to free energies, the idea of an 'optimal structure' may have little meaning for these complexes (Table 7.6).

MP2/6−31G(d) energies were computed for each local minimum, since Kim and coworkers have found that electron correlation has a key role in the binding of (CH$_3$)$_4$N$^+$ to benzene (see below) (Kim et al., 1994; Lee et al., 1995). Accounting for electron correlation reorders the relative stabilities of some of the ACh systems and increases complexation energies by as much as 3 kcal mol^{-1}. The MP2/6−31G(d) data and equation (7.3) were utilized to calculate reaction enthalpies. The translational and rotational temperature terms were computed classically, the vibrational temperature term was computed from standard statistical formulas (equation (7.4)) (Pitzer, 1961) and ΔPV was computed assuming ideal gas behavior. Since the primary interest in this work was the relative energies of ACh conformers and their complexes, no correction factor was applied to the vibrational frequencies or to the zero-point energies and the binding energies were not corrected for BSSE.

All conformers except the tt conformer of ACh are stabilized by internal hydrogen bonding. C—H...O contacts are classified as hydrogen bonds when the H...O distance is ≤ 2.7 Å and the C—H...O bond angle is > 90° (Steiner and Saenger, 1992). There are C—H...O hydrogen bonds between the ester oxygen and a quaternary methyl hydrogen,

Table 7.6 Relative thermodynamic quantities[a]

System[b]	ΔE	ΔH	ΔS	ΔG
ACh, gg	0.6	0.5	2.9	0.0
ACh, gg'	0.0	0.0	0.0	0.4
ACh, gt	1.5	1.0	3.0	0.8
ACh, tg	1.8	1.9	3.0	1.3
ACh, tt	4.5	4.4	4.6	3.3
(H_2O)ACh, gg, qa[c]	2.2	2.0	10.3	0.0
(H_2O)ACh, gg', qa[c]	1.7	1.7	4.7	1.3
(H_2O)ACh, gt, br[d]	0.0	0.0	0.0	1.0
(H_2O)ACh, tg, br[d]	2.6	2.3	1.6	3.2
(C_6H_6)ACh, gg, qa[c]	0.0	0.0	1.5	0.0
(C_6H_6)ACh, gg', qa[c]	0.8	0.9	0.0	1.4
(C_6H_6)ACh, tt, qa[c]	4.8	4.7	6.2	3.3
$(H_2O)_2$ACh, gg', qa, w[c,e]	0.4	0.6	0.2	0.5
$(H_2O)_2$ACh, gt, br, qa[c,d]	0.0	0.0	0.0	0.0
$(H_2O)_2$ACh, tg, qa, qa[c]	3.5	3.5	8.9	0.8
$(H_2O)_2$ACh, tt, br, w[d,e]	2.5	2.6	0.8	2.5

[a] MP2/6-31G(d)//HF/6-31G(d) data. ΔE, ΔH, and ΔG in kcal mol^{-1}, ΔS in cal mol^{-1} K^{-1}.
[b] ACh: acetylcholine, g: *gauche*, g': *gauche'*, t: *trans*.
[c] Water or benzene ligand hydrogen bonded to quaternary ammonium.
[d] Water bridged between carbonyl oxygen and quaternary ammonium.
[e] Second water bonded to first water.

between the carbonyl oxygen and a quarternary methyl hydrogen, and between the carbonyl oxygen and a CH_2 hydrogen. The internal solvation is preserved when water or benzene complexes with ACh. One complex for each of the five forms of ACh is displayed in Figure 7.1. When the ligand binding site is the quaternary ammonium group, both water and benzene interact with hydrogens on two or three different methyl groups. Similar structures have been reported previously for $(CH_3)_4N^+$/water and $(CH_3)_4N^+$/benzene clusters (Kim et al., 1994; Meot-Ner and Deakyne, 1985). For four ACh conformations, when water interacts with the carbonyl oxygen, it is bridged between the carbonyl oxygen and the choline moiety. The water oxygen forms a O—H...O hydrogen bond with the carbonyl group and C—H...O hydrogen bonds with two to four hydrogens on different CH_3 or CH_2 groups. No local minimum was found for the analogous complex with ACh in the gg' conformation; the ACh converts to the gt conformation. When a second water molecule solvates a bridged water molecule, the bridge is widened by the first water molecule moving away from the carbonyl oxygen and toward the quaternary ammonium function. When a second water molecule solvates a water molecule bound to the quaternary ammonium group, cooperative effects are observed in that the C—H...O hydrogen bonds shorten. In all cases, the H...O bond distances in the O—H...O hydrogen bonds range from 1.9 to 2.5 Å; those in the C—H...O hydrogen bonds range from 2.3 to 2.7 Å. The hydrogen bond angles vary from 96° to 177°.

Despite the expected flexibility of ACh, solution NMR studies indicate that the gt form predominates in aqueous solution (Culvenor and Ham, 1966) and X-ray diffraction studies indicate that the gt and gg forms predominate in crystals of acetylcholine salts (Table 7.5) (Datta et al., 1980; Svinning and Sørum, 1975). Clearly the conformation of ACh is influenced by its environment, since these and other forms of ACh are of similar stability in the gas phase (Table 7.6). In fact, the calculations suggest that there will be

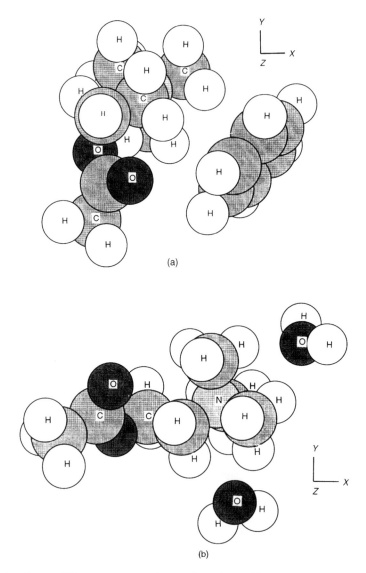

Figure 7.1 Space-filling representations of (C_6H_6)ACh and $(H_2O)_2$ACh clusters. (a) (C_6H_6)ACh, gg, qa: acetylcholine is in the *gauche, gauche* conformation and the benzene is bound to the quaternary ammonium group. (See text.) (b) $(H_2O)_2$ACh, gg′, qa, qa: acetylcholine is in the *gauche, gauche′* conformation, both waters are bound to the quaternary ammonium group. (c) $(H_2O)_2$ACh, gt, br, qa: acetylcholine is in the *gauche, trans* conformation, the first water is bridged between the carbonyl and quaternary ammonium groups and the second water is bound to the quaternary ammonium group. (d) $(H_2O)_2$ACh, tg, qa, w: acetylcholine is in the *trans, gauche* conformation, the first water is bound to the quaternary ammonium group and the second water is bound to the first water. (e) $(H_2O)_2$ACh, tt, br, w: acetylcholine is in the *trans, trans* conformation, the first water is bridged between the carbonyl and quaternary ammonium groups and the second water is bound to the first water.

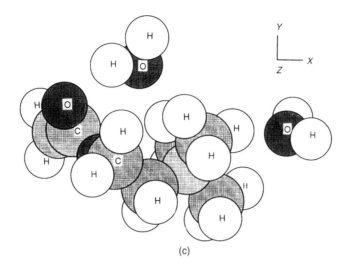

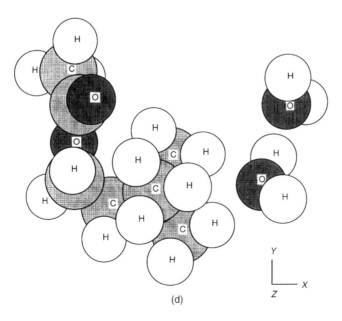

Figure 7.1 (*continued*)

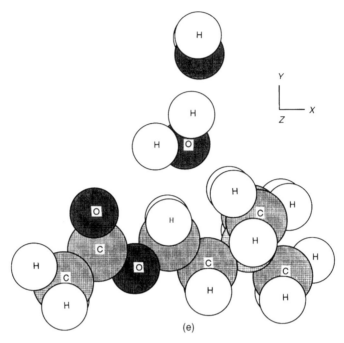

Figure 7.1 (*continued*)

mixed populations at equilibrium for ACh and its complexes with water and benzene ($\partial \Delta G \leq 2$ kcal mol^{-1}). For some of these systems the relative ΔH and ΔG values would predict different equilibrium compositions, underscoring the importance of the entropy contribution to the free energy.

Consider those clusters included in Table 7.6 for which only the choline moiety of ACh is solvated. The strength of the interaction is nearly independent (within 1 kcal mol^{-1}) of the conformational form of the ACh. The evidence for this conformational independence is that the difference in the enthalpies and free energies of the solvated ACh systems in Table 7.6 parallel those of the unsolvated ACh, and that the distances between the solvent and quaternary ammonium are similar for all ACh conformations. The only exception is the (C$_6$H$_6$)ACh complex with the *gg* form of ACh, which is stabilized with respect to the other analogous complexes. These observations hold for all the one- and two-ligand systems studied thus far. So, if, in fact, ACh is in the *tt* conformation in the active site of AChE (Harel *et al.*, 1996; Sussman *et al.*, 1991) conversion of ACh to this rare conformation would appear to have little effect on the free energy of stabilization of the acylation transition state. Now consider those clusters for which a water binds to the carbonyl oxygen. In this case, the calculations suggest that the only competitive clusters are those for which the ACh is in the *gt* conformation. Although these gas-phase data are limited to only two water molecules, if the results can be extrapolated to the bulk solution they do provide a possible explanation for the predominance of the *gt* form of ACh in aqueous solution (Culvenor and Ham, 1966). Finally, if the ACh changes conformation when it moves from the aqueous environment into the gorge, the conformational change would help the desolvation process.

The experimental measurements indicate that ACh binds to the ligands examined fairly

Table 7.7 Thermodynamic quantities for the clustering reactions $(Y)_{n-1}ACh + Y \rightarrow (Y)_n ACh^{a,b}$

Complex	ΔE	$\Delta H°_{calc}$	$\Delta H°_{exp}$	$\Delta S°_{calc}$	$\Delta S°_{exp}$	$\Delta G°_{calc}$	$\Delta G°_{exp}$
$(H_2O)ACh^c$	−12.7	−11.1	−8.0	−21.0	(−22.)	−4.9	−2.9[d]
$(H_2O)_2ACh$	−11.8	−10.2		−22.6		−3.7	
$((CH_3)_2CO)ACh$			−13.2		−21.7		−6.7
$(C_6H_6)ACh$	−10.7	−9.2		−22.1		−2.7	
$(C_6H_6)_2ACh$	−9.8	−8.4		−18.2		−3.0	
$(C_6H_5CH_3)ACh$			−8.1		−15.5		−3.5
$(C_6H_5OCH_3)ACh^c$			−12.8		(−24.)		−5.8[e]

[a] MP2/6−31G(d)//HF/6−31G(d) data. ΔE, ΔH, and ΔG in kcal mol^{-1}, ΔS in cal mol^{-1} K^{-1}. $T = 298$ K unless otherwise noted.
[b] In order to determine the calculated thermodynamic values for a given association reaction, the ACh conformation of the product cluster with the most stable free energy was used for both reactant and product.
[c] $\Delta G°_{exp}$ could be measured at only one temperature; $\Delta S°_{exp}$ is estimated.
[d] $T = 229$ K.
[e] $T = 267$ K.

weakly; the binding enthalpies are in the 8–13 kcal mol^{-1} range (Table 7.7). The measured values are consistent with those observed for other [C—H...O]$^+$ and [C—H...π]$^+$ interactions of blocked ions but are significantly weaker than those for conventional ionic hydrogen bonds (Keesee and Castleman, 1986; Meot-Ner and Deakyne, 1985) and those for [Na...π]$^+$ (Guo et al., 1990) and [K...π]$^+$ interactions (Sunner et al., 1981). In particular, the ACh/ligand complexation energy is within approximately 1 kcal mol^{-1} of the $(CH_3)_4N^+$/ligand complexation energy (Meot-Ner and Deakyne, 1985). In order to estimate thermochemical quantities from the calculated total energies, we used the total energy of the product complex with the most stable free energy for the given reaction and of the reactant complex (or ACh molecule) with the ACh conformation corresponding to that of the product complex. There is reasonable agreement between the estimated computed thermochemical data and the experimental data (Table 7.7). The ACh/water and ACh/aromatic binding enthalpies are comparable but smaller than those for ACh and more polar oxygen or amide groups, supporting the contention that the aromatic residues in the cleft provide an effective but moderately stabilizing environment for the transit of ACh along the cleft. The incremental bond strengths of the first and second ligand molecules are similar, which contrasts with conventional ionic hydrogen bonds for which incremental bond strengths usually decrease by a factor of approximately 0.3 (Meot-Ner, 1984). Similar effects were observed experimentally and calculationally for $(CH_3)_4N^+$ (Meot-Ner and Deakyne, 1985), making the latter ion a good model for ACh.

Sussman and coworkers (Harel et al., (1996) have shown that the choline end of the transition state analog inhibitor m-(N,N,N-trimethylammonio)-2,2,2-trifluoroacetophenone interacts with three water molecules and two aromatic residues (tryptophan and phenylalanine) in the choline binding pocket of AChE. These researchers have estimated that these interactions contribute about 6 kcal mol^{-1} to the total free energy of stabilization of the acylation transition state. This value was obtained by assuming that the individual contributions are additive. The gas-phase results described herein suggest that both the assumption of additivity and the net magnitude of the contributions are reasonable.

The interactions between benzene and the cations NH_4^+ and $(CH_3)_4N^+$ have been analyzed by Kim et al. (1994). The binding forces in these clusters were then compared with those in the corresponding water clusters. Part of their motivation for studying the interactions in the benzene/tetramethylammonium ion complex was their relevance to the interactions between the choline moiety of ACh and AChE (Dougherty and Stauffer, 1990; Harel et al., 1996). HF/6−311+G(d,p) and MP2/6−311+G(d,p) optimum geometries and association energies were computed. For the $(CH_3)_4N^+$ systems the benzene and water interact with three methyl hydrogens, each from a different methyl group. For $(C_6H_6)NH_4^+$, two N−H hydrogens are directed toward the benzene ring. The calculated association enthalpies are in good agreement with experimental data (Deakyne and Meot-Ner, 1985; Meot-Ner and Deakyne, 1985). The cation−dipole, cation−quadrupole, cation-polarizability and polarizability−dipole components of the complexation energy were determined from a multipole expansion. The electron correlation energy was computed as the difference in the correlated and SCF complexation energies for the MP2 equilibrium structure (ΔE(MP2//MP2) − ΔE(HF//MP2)). The main binding force in the water clusters is the cation−dipole interaction, whereas the cation−quadrupole, cation−polarizability, and dispersion interactions all make significant contributions to the binding in the benzene clusters. The large electron correlation effect in these systems is attributed to a novel $\pi-\sigma^*$ through-space interaction brought about by a transfer of electron density from the π-LUMO of the benzene to the σ^*_{C-H} orbitals of the cations. This phenomenon is not observed at the HF level, and it is less important for $(C_6H_6)NH_4^+$ than for $(C_6H_6)(CH_3)_4N^+$ since the overlap is reduced by a tilting of the relevant orbitals in the former complex. In agreement with these results, Caldwell and Kollman (1995) have found that a nonadditive molecular mechanical model, in which electrostatic, polarization, and dispersion terms are included, yields much more accurate cation−π interaction energies than two-body additive models.

The above work by Kim et al. was followed up with a study of benzene/$(CH_3)_nNH_{4-n}^+$ complexes, $n = 0 - 4$, in which the authors focused on the $\pi-\sigma^*$ through-space interaction described in the earlier article (Lee et al., 1995). HF/6−31G(d) and MP2/6−31G(d) equilibrium structures indicate that when $n = 1$ and 2, two N—H hydrogens point toward the benzene. When $n = 3$, the N—H hydrogen is directed toward the center of the benzene ring. The cation orientations of the remaining complexes are identical to those reported previously (Kim et al., 1994). Since the protonated methylamine group is tilted in the benzene/$(CH_3)NH_3^+$ complex, this complex is the only system without a rotational symmetry axis. For the N—H...π hydrogen bonds, the distance between the hydrogen atom and the benzene plane ranges from 2.0 to 2.3 Å; the corresponding distance for the C—H...π hydrogen bond is 2.6 Å. The N—H and C—H bonds involving the hydrogen-bonded protons lengthen when the hydrogen bond is formed, and the elongations are larger for the MP2 geometries than the HF geometries. This observation is interpreted as a manifestation of the $\pi-\sigma^*$ through-space interaction, which plays an important role in stabilizing these complexes. The calculated $\Delta H°$s range from −18 to −16 kcal mol^{-1} and are underestimated with respect to the experimental values, but the calculated and experimental $\Delta G°$s are in reasonably good accord (Deakyne and Meot-Ner, 1985; Meot-Ner and Deakyne, 1985). With the exception of the benzene/$(CH_3)_3NH^+$ complex, the calculated dissociation enthalpies decrease by 0.5−1.0 kcal mol^{-1} with the addition of each methyl group. The experimental trend in dissociation enthalpies is similar.

Dougherty and coworkers have studied cation−π interactions extensively and were early advocates of the importance of this type of interaction in binding ACh and related quaternary ions to biological binding sites (Dougherty, 1996; Dougherty and Stauffer, 1990; Kearney et al., 1993; Kumpf and Dougherty, 1993; Mecozzi et al., 1996). They have designed a series of water-soluble, cyclophane host molecules as models of biological receptors. Since aromatic rings make up the 'walls' of the host molecule, the host provides a hydrophobic environment. The group has used NMR and circular dichroism to measure the binding affinity of a large number of cationic guests to these synthetic receptors in aqueous media, and have demonstrated that even when guests are quite soluble in water, the cation–π interaction can lead to strong binding energies and novel binding selectivities. Effective guests include alkyliminiums, tetraalkylammoniums, trialkylsulfoniums, alkylated guanidiniums and neutral azulenes. Brominating the host improves binding affinities, but exchanging two benzenes in the host with furans or thiophenes does not. Methoxy substitution of the host apparently causes a conformational change that inhibits binding. Adding an organic solvent to an aqueous medium reduces binding constants, presumably by diminishing the difference in the hydrophobicity of the host and aqueous environments. Obviously a number of factors determine the strength of cation–π interactions, and the authors have carried out *ab initio* calculations in order to gain more insight into these factors (Kearney *et al.*, 1993; Mecozzi *et al.*, 1996). Recognizing that no one intermolecular force, e.g. electrostatic, induction, charge transfer, or dispersion, can account for the complex nature of the cation–π interaction, Dougherty and coworkers contend that, to first order, the major contributor to the interaction energy is the electrostatic term and, in particular, the cation-permanent quadrupole term (Dougherty, 1996; Kumpf and Dougherty, 1993). (The second most significant contributor is most likely the cation-induced dipole term.) In a recent article they have also correlated variations in binding energies with variations in the electrostatic component (Mecozzi *et al.*, 1996).

Mecozzi *et al.* have examined complexes of Na^+ bound to the π-face of eleven aromatic compounds, including benzene, eight substituted benzenes, pyridine, and naphthalene, at the SCF level with the 6–31G(d,p) basis set. The effect of including electron correlation in the calculations was checked for three of the complexes but both levels of theory yielded the same conclusions. The simpler Na^+ ion was chosen rather than the experimentally more relevant NH_4^+ ion, since trends for the two ions have been shown to be identical (Kearney *et al.*, 1993). The electrostatic contribution to the complexation energy was evaluated by exchanging a dummy probe atom for the Na^+ in the optimized complex and determining the electrostatic potential at that position. When the binding energy is plotted versus electrostatic potential for the eleven complexes, the plot obtained is linear with a slope of one and an intercept of about -12 kcal mol^{-1}. (The correlation coefficient is 0.991.) These results imply that a complex with no electrostatic component will have a binding energy of about -12 kcal mol^{-1}, and indeed, the calculated binding energy of 1,3,5-trifluorobenzene is -12.4 kcal mol^{-1}. Most importantly, for this series of complexes (with the possible exception of naphthalene) the electrostatic contribution accounts for essentially all the variation in the binding energies; the sum of the other contributions is constant at about 12 kcal mol^{-1}. Thus, analyses of trends in the electrostatic term can aid in designing new systems.

7.5 SUMMARY

Key results from the articles discussed in this chapter are summarized in Table 7.8.

Table 7.8 Summary of key results

Reference	System	Methodology	Key results
Wehbeh and Allen	(HF)HF, (HCN)HF, (H_2O)HF	84 *ab initio* calculational levels	Recommended calculational protocol: HF/6−31+G(d), no correction for BSSE
Pudzianowski (1995)	5 protonated dimers, 5 anionic dimers	MP2/6−311++G(d,p), MP2/6−31++G($2d,2p$)	Recommended calculational protocol: MP2/6−311++G(d,p)//MP2/6−31+G(d,p) for all systems without nitrogen, MP2/6−311++G($2d,2p$)//MP2/6−31+G(d,p) for systems with nitrogen, no correction for BSSE
Del Bene (1996), Del Bene and Shavitt (1994)	(L)Li$^+$, L: neutral or anionic base, 3 protonated dimers, 3 anionic dimers, 3 neutral dimers	Series of correlation consistent basis sets, Møller–Plesset and coupled cluster calculations	Recommended calculational protocol: MP4/aug′-cc-pVTZ//MP2/6−31+G(d,p) for cationic and neutral complexes, CCSD(T)/aug′-cc-pVTZ//MP2/6−31+G(d,p) for negative complexes
Deakyne *et al.* (1994)	(HCN)$_n$(NH$_3$)$_m$H$^+$	High-pressure mass spectometry; MP2/6−31+G($2d,2p$)//HF/6−31G(d) calculations	NH$_4^+$ core ion for all complexes, may have mixtures of isomeric clusters for $n+m \geqslant 5$
Garvey *et al.* (1994)	(L)$_n$(H$_2$O)H$^+$, L: oxygen-containing species	Tandem mass spectrometry, molecular beam source	Review, mixed protonated clusters formed via intracluster reactions, 'size-selective' ligand dissociations, core ion switches from MH$^+$ to H$_3$O$^+$, fused five-membered ring geometries favored, maximizing number of hydrogen bonds can be more important than relative proton affinities in determining structure
Wei *et al.* (1991a)	(H$_2$O)$_{21}$H$^+$	Tandam mass spectrometry, molecular beam source	Number of nonhydrogen-bonded hydrogens consistent with clathrate-like pentagonal dodecahedron structure

Table 7.8 (*continued*)

Reference	System	Methodology	Key results
Smith and Dang (1994)	$(H_2O)_nCs^+$	Classical molecular dynamics simulations	Other cyclic structures competitive in stability with pentagonal dodecahedron, 'magic numbers' limited to solid-like cluster regime
Zhang and Castleman (1994)	$(H_2O)_n(MeOH)_mH^+$	Flow tube reactor	'Magic numbers' at $n + m = 21$, clathrate-like (or other cyclic) structure results from inherent stability not kinetic effects
Karpas *et al.* (1994)	$(ROH)_nH^+$, $(ROH)_n(H_2O)_mH^+$	Tandem mass spectrometry, atmospheric pressure ionization corona discharge source	H_2O loss favored for $n + m < 9$, alcohol preferred protonation site, neither expected nor saw 'magic numbers', question universality of concept of 'magic numbers'
Xia and Garvey (1995)	$(ROH)_n(NH_3)_mH^+$	Tandem mass spectrometry, molecular beam source	NH_4^+ core ion for all complexes, switch in dominant decomposition pathway, NH_3s close to core ion in small clusters but pushed to periphery in large clusters
Feng *et al.* (1995)	$(N_2H_4)_nH^+$, $(N_2H_4)_n(H_2O)_mH^+$	Tandem mass spectrometry, pressure- and temperature-variable source; HF/6−31G, AM1 and PM3 calculations for $(N_2H_4)_nH^+$	'Magic numbers' for $(N_2H_4)_4H^+$ and $(N_2H_4)_n(H_2O)_mH^+$, $n + m = 5$, $N_2H_5^+$ core ion with protonated nitrogen preferentially solvated
Lifshitz and Feng (1995)	$(MeCOOH)_nH^+$	Tandem mass spectrometry, pressure- and temperature-variable source	Evaporation channel shifts from loss of single MeCOOH to intact MeCOOH dimers at $n = 7$, dimers evaporate in cyclic form
Zhang and Lifshitz (1996)	$(MeCOOH)_nH^+$, $(HCOOH)_nH^+$	HF/6−31+G(d)// HF/4−31G# calculations	Stable isomeric chains terminated by monomers and/or dimers, evaporation channel switches from monomeric to dimeric HCOOH at $n = 6$

continued overleaf

Table 7.8 (continued)

Reference	System	Methodology	Key results
Selegue and Lisy (1994)	$(MeOH)_n Na^+$, $(MeOH)_n Cs^+$	Tandem mass spectrometry, molecular beam source	'Size-selective' intracluster reaction forms Me_2O, postulated reaction mechanism: $S_N 2$ attack followed by proton transfer along hydrogen bonds between two second shell ligands and one first shell ligand
Watanabe and Iwata (1996)	$(H_2O)_n B^+$, $(H_2O)_n Al^+$	MP2/6–31G(d) calculations	Overall surface forms of $(H_2O)_{n-1} HOMH^+$ most stable, spontaneous intracluster insertion (for B only) and proton transfer reactions observed
Stace et al. (1994)	$(D_2O)_n NO^+$	Tandem mass spectrometry, pick-up technique	'Size-selective' intracluster reaction forming $(D_2O)_n D^+$ switched on at $n = 4$, support for postulated reaction sequence for production of proton hydrates in atmosphere
Meot-Ner et al. (1994)	$(L)_n H^+$, $(L)_n (H_2O)_m H^+$, L: acetone, monodentate ether, polydentate ether	High-pressure mass spectrometry, HF/4–31G calculations	Entropy plays important role in relative binding free energies of flexible versus inflexible ligands, hydrogen-bonded rings formed with proton on H_2O, complexes model proton wires
Meot-Ner et al. (1996b)	$(L)_n M^+$, L: acetone, polydentate ether, M^+: NH_4^+, Na^+, K^+	High-pressure mass spectrometry, HF/4–31G calculations	All cations interact with both ether oxygens, additional gas-phase and condensed-phase evidence for similarity of binding of NH_4^+ and K^+
Glendening et al. (1994)	$(18\text{-crown-}6)M^+$, $(H_2O)_n M^+$, M: Li, Na, K, Cs, Rb	MP2/'6–31+G(d)'//HF/'6–31+G(d)' calculations	C_i symmetry for 18-crown-6, gas-phase binding energies in order $Li^+ > Na^+ > K^+ > Cs^+ > Rb^+$, changes to order in bulk solution with 4 water ligands

Summary

Table 7.8 (*continued*)

Reference	System	Methodology	Key results
Blades et al. (1996)	$(H_2O)_nX^-$, $(H_2O)_nX^{2-}$, X^-, X^{2-}: phosphate anion	Tandem mass spectrometry, electrospray ionization	Relative dissociation free energy $\Delta G^\circ_{0,1}$ values indicate presence of intramolecular hydrogen bond in D-ribose 5'-phosphate, adenosine 5'-monophosphate, adenosine 3'-monophosphate, and adenosine 5'-diphosphate
Blades and Kebarle (1994), Blades et al. (1995)	$(H_2O)_nX^{2-}$, X^{2-}: anions of C, S, and P oxo acids	Tandem mass spectrometry, electrospray ionization	Physically separating negatively charged groups reduces hydration free energies, successive waters solvate alternate charged groups
Combariza and Kestner (1994)	$(H_2O)_nF^-$	H_2O: MP2/6–31+G(d)//HF/6–31+G(d), F: MP2/DZP+//HF/DZP+ calculations	Switch from surface to interior structures at $n = 4, 5$, entropy important factor in relative free energies
Combariza et al. (1994)	$(H_2O)_nX^-$, X: Cl, Br, I	H_2O: MP2/6–31+G(d)//HF/6–31+G(d), Cl: MP2/DZP+//HF/DZP+, Br, I: MP2/'TZP+'//HF/'TZP+' calculations	Only surface states stable for $n < 5$, switch to internal states for $(H_2O)_6I^-$
Markovich et al. (1994)	$(H_2O)_nX^-$, X: Cl, Br, I	Photoelectron spectroscopy	Six ligands in first solvation shell for $(H_2O)_nI^-$, surface structures predicted for all X at all n
Xantheas and Dunning (1994)	$(H_2O)_nF^-$	MP4/aug-cc-pVDZ//MP2/aug-cc-pVDZ calculations	Important to optimize geometries at correlated level, surface structures more stable, need three-body terms in interaction potential for simulations
Meot-Ner et al. (1996a)	$(L)Me_nNH_{4-n}{}^+$, $(L)(RCNH^+)$, $(L)RNCH^+$, L: RCN, RNC, Me_mOH_{2-m}	High-pressure mass spectrometry, effective QCISD(T)/cc-pVTZ+//MP2/6–31+G(d,p) calculations	Binding energies 14–25 kcal mol^{-1}, electron correlation preferentially stabilizes isocyano complexes, binding energies of analogous cyano and isocyano complexes similar

continued overleaf

Table 7.8 (continued)

Reference	System	Methodology	Key results
Suh et al. (1995)	HOCH$_2$C(=O)OMe$^{·+}$	Tandem mass spectrometry, multiphoton ionization, MP2/6–31G(d)//HF/6–31G(d) calculations	Breakdown to HC(OH)OMe$^+$ and HCO· proceeds via [C—H...O]$^+$ intermediate, calculations help clarify reaction mechanism
Deakyne et al. (1996)	(H$_2$O)$_n$ACh, (C$_6$H$_6$)$_n$ACh	High-pressure mass spectrometry, MP2/6–31G(d)//HF/6–31G(d) calculations	Binding energies of water and benzene differ by ≈ 1 kcal mol^{-1}, first and second ligands similar binding energies, aromatic groups may help desolvate ACh in AChE cleft
Kim et al. (1994)	(L)NH$_4^+$, (L)Me$_4$N$^+$, L: H$_2$O, C$_6$H$_6$	MP2/6–311+G(d,p) calculations	Large electron correlation effect in benzene complexes due to novel π–σ^* through-space interaction, cation–quadrupole and cation–polarizability terms also important
Lee et al. (1995)	(C$_6$H$_6$)Me$_n$NH$_{4-n}^+$	MP2/6–31G(d)	π–σ^* through-space interaction stabilizes all complexes
Mecozzi et al. (1996)	(L)Na$^+$	HF/6–31G(d,p)	Electrostatic component accounts for all variation in binding energies, sum of all other components constant at 12 kcal mol^{-1}

REFERENCES

Aakeröy, C. B. and Seddon, K. R. (1993) *Chem. Soc. Rev.*, **22**, 397.
Abdul-Sada, A. K., Al-Juaid, S., Greenway, A. M., Hitchcock, P. B., Howells, M. J., Seddon, K. R. and Welton, T. (1990) *Struct. Chem.*, **1**, 391.
Abdul-Sada, A. K., Greenway, A. M., Hitchcock, P. B., Mohammed, T. J., Seddon, K. R. and Zora, J. A. (1986) *J. Chem. Soc. Chem. Commun.*, 1753.
Adcock, J. L. and Zhang, H. (1995) *J. Org. Chem.*, **60**, 1999.
Al-Juaid, S. S., Al-Nasr, A. K. A., Eaborn, C. and Hitchcock, P. B. (1991) *J. Chem. Soc. Chem. Commun.*, 1482.
Allerhand, A. and Schleyer, P. v. R. (1963) *J. Am. Chem. Soc.*, **85**, 1715.
Arijs, E., Nevejans, D. and Ingels, J. (1980) *Nature*, **288**, 684.
Arnold, F. and Fabian, R. (1980) *Nature*, **283**, 55.
Arshadi, M., Yamdagni, R. and Kebarle, P. (1970) *J. Phys. Chem.*, **74**, 1475.
Atwood, J. L., Hamada, F., Robinson, K. D., Orr, G. W. and Vincent, R. L. (1991) *Nature*, **349**, 683.
Bader, R. F. W. (1990) *Atoms in Molecules. A Quantum Theory*, Clarendon, Oxford.

Barry, J. P., Vouros, P., Van Schepdael, A. and Law, S.-J. (1995) *J. Mass Spectrom.*, **30**, 993.
Biemann, K. and Papayannopoulos, I. A. (1994) *Acc. Chem. Res.*, **27**, 370.
Binkley, J. S. and Pople, J. A. (1977) *J. Chem. Phys.*, **66**, 879.
Binkley, J. S., Pople, J. A. and Hehre, W. J. (1980) *J. Am. Chem. Soc.*, **102**, 939.
Blades, A. T., Ho, Y. and Kebarle, P. (1996) *J. Phys. Chem.*, **100**, 2443.
Blades, A. T. and Kebarle, P. (1994) *J. Am. Chem. Soc.*, **116**, 10761.
Blades, A. T., Klassen, J. S. and Kebarle, P. (1995) *J. Am. Chem. Soc.*, **117**, 10563.
Bleske, E. J. (1995) *J. Chem. Soc. Faraday Trans.*, **91**, 1.
Boyd, R. J. and Choi, S. C. (1986) *Chem. Phys. Lett.*, **129**, 62.
Boys, S. F. and Bernardi, F. (1970) *Mol. Phys.*, **19**, 553.
Burda, J. V., Sponer, J. and Hobza, P. (1996) *J. Phys. Chem.*, **100**, 7250.
Caldwell, J. W. and Kollman, P. A. (1995) *J. Am. Chem. Soc.*, **117**, 4177.
Canty, A. J., Colton, R., D'Agostino, A. and Traeger, J. C. (1994) *Inorg. Chim, Acta*, **223**, 103.
Carroll, M. T. and Bader, R. F. W. (1988) *Mol. Phys.*, **65**, 695.
Carroll, M. T., Chang, C. and Bader, R. F. W. (1988) *Mol. Phys.*, **63**, 387.
Castleman, A. W., Jr and Keesee, R. G. (1988) *Science*, **241**, 36.
Castleman, A. W., Jr and Wei, S. (1994) *Annu. Rev. Phys. Chem.*, **45**, 685.
Chiarizia, R., Horwitz, E. P. and Dietz, M. L. (1992) *Solvent Extraction Ion Exchange*, **10**, 337.
Chuen, L. and Sammes, M. (1983) *J. Chem. Soc. Perkin Trans.*, 1303.
Clark, T., Chandrasekhar, J., Spitznagel, G. W. and Schleyer, P. v. R. (1983) *J. Comput. Chem.*, **4**, 294.
Clemmer, D. E., Hudgins, R. R. and Jarrold, M. F. (1995) *J. Am. Chem. Soc.*, **117**, 10141.
Combariza, J. E. and Kestner, N. R. (1994) *J. Phys. Chem.*, **98**, 3513.
Combariza, J. E., Kestner, N. R. and Jortner, J. (1993) *Chem. Phys. Lett.*, **203**, 423.
Combariza, J. E., Kestner, N. R. and Jortner, J. (1994a) *J. Chem. Phys.*, **100**, 2851.
Combariza, J. E., Kestner, N. R. and Jortner, J. (1994b) *Chem. Phys. Lett.*, **221**, 156.
Cook, D. B., Sordo, J. A. and Sordo, T. L. (1993) *Int. J. Quantum Chem.*, **48**, 375.
Culvenor, C. C. J. and Ham, N. S. (1966) *Chem. Commun.*, 537.
Cybulski, S. M., Chalasinski, G. and Moszynski, R. (1990) *J. Chem. Phys.*, **92**, 4357.
Dang, L. X. and Garrett, B. C. (1993) *J. Chem. Phys.*, **99**, 2972.
Datta, N., Mondal, P. and Pauling, P. (1980) *Acta Crystallogr.*, **B36**, 906.
de Hoffmann, E. (1996) *J. Mass Spectrom.*, **31**, 129.
Deakyne, C. A. (1986) *J. Phys. Chem.*, **90**, 6625.
Deakyne, C. A. (1987) *Molecular Structure and Energetics*, **4**, 105.
Deakyne, C. A., Knuth, D. M., Speller, C. V., Meot-Ner (Mautner), M. and Sieck, L. W. (1994) *J. Mol. Struct. (Theochem)*, **307**, 217.
Deakyne, C. A. and Meot-Mer (Mautner), M. (1985) *J. Am. Chem. Soc.*, **107**, 474.
Deakyne, C. A., Meot-Ner (Mautner), M., Campbell, C. L., Hughes, M. G. and Murphy, S. P. (1986) *J. Chem. Phys.*, **84**, 4958.
Deakyne, C. A., Meot-Ner (Mautner), M. and Sieck, L. W. (1996), in preparation.
Del Bene, J. E. (1988) *J. Phys. Chem.*, **92**, 2874.
Del Bene, J. E. (1989) *J. Comput. Chem.*, **10**, 603.
Del Bene, J. E. (1992) *Int. J. Quantum Chem. Quantum Chem. Symp.*, **26**, 527.
Del Bene, J. E. (1993) *J. Am. Chem. Soc.*, **115**, 1610.
Del Bene, J. E. (1996) *J. Phys. Chem.*, **100**, 6284.
Del Bene, J. E. and Shavitt, I. (1994) *J. Mol. Struct. (Theochem)*, **307**, 27.
DeLaat, A. M. and Ault, B. S. (1987) *J. Am. Chem. Soc.*, **109**, 4232.
Desiraju, G. R., Kashino, S., Coombs, M. M. and Glusker, J. P. (1993) *Acta Crystallogr.*, **B49**, 880.
Desiraju, G. R. (1991) *Acc. Chem. Res.*, **24**, 290.
Desiraju, G. R. and Murty, B. N. (1987) *Chem. Phys. Lett.*, **139**, 360.
Desmeules, P. J. and Allen, L. C. (1980) *J. Chem. Phys.*, **72**, 4731.
Dewar, M. J. S., Zoebisch, E. G., Healy, E. F. and Stewart, J. J. P. (1985) *J. Am. Chem. Soc.*, **107**, 3902.
Dill, J. D. and Pople, J. A. (1975) *J. Chem. Phys.*, **62**, 2921.
Ditchfield, R., Hehre, W. J. and Pople, J. A. (1971) *J. Chem. Phys.*, **54**, 724.
Dougherty, D. A. (1996) *Science*, **271**, 163.

Dougherty, D. A. and Stauffer, D. A. (1990) *Science*, **250**, 1558.
Draves, J. A. and Lisy, J. M. (1990) *J. Am. Chem. Soc.*, **112**, 9006.
Du, H. S., Wood, D. J., Elshani, S. and Wai, C. M. (1993) *Talanta*, **40**, 173.
Dunitz, J. D., Dobler, M., Seiler, P. and Phizackerley, R. P. (1974) *Acta Crystallogr.*, **B30**, 2733.
Dunning, T. H., Jr (1989) *J. Chem. Phys.*, **90**, 1007.
Eisele, F. (1986) *J. Geophys. Res.*, **91**, 7897.
Eisele, F. (1988) *J. Geophys. Res.*, **93**, 716.
El-Shall, M. S., Olafsdottir, S. R., Meot-Ner (Mautner), M. and Sieck, L. W. (1991) *Chem. Phys. Lett.*, **185**, 193.
Feller, D., Glendening, E. D., Kendall, R. A. and Peterson, K. A. (1994) *J. Chem. Phys.*, **100**, 4981.
Feng, W. Y., Aviyente, V., Varnali, T. and Lifshitz, C. (1995) *J. Phys. Chem.*, **99**, 1776.
Feng, W. Y. and Lifshitz, C. (1994) *J. Phys. Chem.*, **98**, 6075.
Fenn, J. B., Mann, M., Meng, C. K., Wong, S. F. and Whitehouse, C. M. (1985) *Science*, **246**, 64.
Francl, M. M., Pietro, W. J., Hehre, W. J., Binkley, J. S., Gordon, M. S., DeFrees, D. J and Pople, J. A. (1982) *J. Chem. Phys.*, **77**, 3654.
Freeman, G. R. and March, N. H. (1996) *J. Phys. Chem.*, **100**, 4331.
Frisch, M. J., Del Bene, J. E., Binkley, J. S. and Schaefer, H. F., III (1986) *J. Chem. Phys.*, **84**, 2279.
Frisch, M. J., Pople, J. A. and Del Bene, J. E. (1985a) *J. Phys. Chem.*, **89**, 3664.
Frisch, M. J., Pople, J. A. and Binkley, J. S. (1985b) *J. Chem. Phys.*, **80**, 3265.
Gao, J., Keczera, K., Tidor, B. and Karplus, M. (1989) *Science*, **244**, 1069.
Garvey, J. F., Herron, W. J. and Vaidyanathan, G. (1994) *Chem. Rev.*, **94**, 1999.
George, M., Kingsmill, C. A., Suh, D., Terlouw, J. K. and Holmes, J. L. (1994) *J. Am. Chem. Soc.*, **116**, 7807.
Gil, F. P. S. C., Amorim da Costa, A. M. and Teixeira-Dias, J. J. C. (1995) *J. Phys. Chem.*, **99**, 8066.
Glendening, E. D. (1996) *J. Am. Chem. Soc.*, **118**, 2473.
Glendening, E. D. and Feller, D. (1995) *J. Phys. Chem.*, **99**, 3060.
Glendening, E. D., Feller, D. and Thompson, M. A. (1994) *J. Am. Chem. Soc.*, **116**, 10657.
Glendening, E. D. and Streitwieser, A. (1994) *J. Chem. Phys.*, **100**, 2900.
Green, R. D. (1974) *Hydrogen Bonding by CH Groups*, Wiley Interscience, New York.
Grev, R. S., Janssen, C. L. and Schaefer, H. F., III, (1991) *J. Chem. Phys.*, **95**, 5128.
Gross, J., Harder, G., Vogtle, F., Stephan, H. and Gloe, K. (1995) *Angew. Chem. Int. Ed. Eng.*, **34**, 481.
Gross, M. L. (1994) *Acc. Chem. Res.*, **27**, 361.
Guo, B. C., Purnell, J. W. and Castleman, A. W., Jr (1990) *Chem. Phys. Lett.*, **168**, 155.
Harel, M., Quinn, D. M., Nair, H. K., Silman, I. and Sussman, J. L. (1996) *J. Am. Chem. Soc.*, **118**, 2340.
Hariharan, P. A. and Pople, J. A. (1973) *Theor. Chim. Acta.*, **28**, 213.
Hariharan, P. C. and Pople, J. A. (1972) *Chem. Phys. Lett.*, **66**, 217.
Hashimoto, K. and Morokuma, K. (1994) *J. Am. Chem. Soc.*, **116**, 11436.
Hashimoto, K. and Morokuma, K. (1995) *J. Am. Chem. Soc.*, **117**, 4151.
Hayes, I. C. and Stone, A. J. (1984) *Mol. Phys.*, **53**, 83.
Hehre, W. J., Ditchfield, R. and Pople, J. A. (1972a) *J. Chem. Phys.*, **56**, 2267.
Hehre, W. J., Ditchfield, R. and Pople, J. A. (1972b) *J. Chem. Phys.*, **56**, 2257.
Hehre, W. J., Radom, L., Schleyer, P. v. R. and Pople, J. A. (1986) *Ab Initio Molecular Orbital Theory*, John Wiley, New York.
Heinrich, N., Schmidt, J., Schwarz, H. and Apeloig, Y. (1987) *J. Am. Chem. Soc.*, **109**, 1317.
Hirao, K., Sano, M. and Yamabe, S. (1982) *Chem. Phys. Lett.*, **87**, 181.
Hiraoka, K., Mizuse, S. and Yamabe, S. (1988) *J. Phys. Chem.*, **92**, 3943.
Hobza, P., Sponer, J. and Reschel, T. (1995) *J. Comput. Chem.*, **16**, 1315.
Jeffrey, G. A. and Saenger, W. (1991). *Hydrogen Bonding in Biological Structures*, Springer-Verlag, Berlin.
Jones, J. L., Dongré, A. R., Somogyi, A. and Wysocki, V. H. (1994) *J. Am. Chem. Soc.*, **116**, 8368.
Jorgensen, W. L. and Severance, D. L. (1993) *J. Chem. Phys.*, **99**, 4233.
Kaltashov, I. A., Fabris, D. and Fenselau, C. C. (1995) *J. Phys. Chem.*, **99**, 10046.
Karpas, Z., Eiceman, G. A., Ewing, R. G. and Harden, C. S. (1994) *Int. J. Mass. Spectrom. Ion Processes*, **133**, 47.

Karpfen, A. and Yanovitskii, O. (1994) *J. Mol. Struct. (Theochem)*, **314**, 211.
Kassner, J. L., Jr and Hagen, D. E. (1976) *J. Chem. Phys.*, **64**, 1860.
Katritzky, A. R., Malhotra, N., Ramanathan, R., Kemerait, R. C., Jr, Zimmerman, J. A. and Eyler, J. R. (1992) *Rapid Commun. Mass Spectrom.*, **6**, 25.
Kearney, P. C., Mizoue, L. S., Kumpf, R. A., Forman, J. E., McCurdy, A. and Dougherty, D. A. (1993) *J. Am. Chem. Soc.*, **115**, 9907.
Kebarle, P., Arshadi, M. and Scarborough, J. (1968) *J. Chem. Phys.*, **49**, 817.
Keesee, R. G. and Castleman, A. W., Jr (1986) *J. Phys. Chem. Ref. Data*, **15**, 1011.
Kendall, R. A., Dunning, T. H., Jr and Harrison, R. J. (1992) *J. Chem. Phys.*, **96**, 6796.
Kim, K. and Jordan, K. D. (1994) *J. Phys. Chem.*, **98**, 10089.
Kim, K. S., Lee, J. Y., Lee, S. J., Ha, T.-K. and Kim, D. H. (1994) *J. Am. Chem. Soc.*, **116**, 7399.
Kitaura, K. and Morokuma, K. (1976) *Int. J. Quantum Chem.*, **10**, 325.
Klassen, J. S., Blades, A. T. and Kebarle, P. (1994) *J. Am. Chem. Soc.*, **116**, 12075.
Klots, C. E. (1993) *J. Chem. Phys.*, **98**, 1110.
Koch, U. and Popelier, P. L. A. (1995) *J. Phys. Chem.*, **99**, 9747.
Kozak, R. W., Waldmann, T. A., Atcher, R. W. and Gansow, O. A. (1985) *Trends Biotechnol.*, **4**, 259.
Krishnan, R., Binkley, J. S., Seeger, R. and Pople, J. A. (1980a) *J. Chem. Phys.*, **72**, 650.
Krishnan, R., Frisch, M. J. and Pople, J. A. (1980b) *J. Chem. Phys.*, **72**, 4244.
Kumpf, R. A. and Dougherty, D. A. (1993) *Science*, **261**, 1708.
Laasonen, K., Sprik, M., Parrinello, M. and Car, R. (1993) *J. Chem Phys.*, **99**, 9080.
LaJohn, L. A., Christiansen, P. A., Ross, R. B., Atashroo, T. and Ermler, W. C. (1987) *J. Chem. Phys.*, **87**, 2812.
Lee, J. Y., Lee, S. J., Choi, H. S., Cho, S. J., Kim, K. S. and Ha, T.-K. (1995) *Chem. Phys. Lett.*, **232**, 67.
Legon, A. C. (1992) *J. Chem. Soc. Perkin Trans.*, **2**, 329.
Legon, A. C., Lister, D. G. and Rego, C. A. (1992a) *Chem. Phys. Lett.*, **189**, 221.
Legon, A. C., Lister, D. G. and Warner, H. E. (1992b) *Angew. Chem. Int. Ed. Engl.*, **31**, 202.
Legon, A. C., Lister, D. G. and Warner, H. E. (1992c) *J. Am. Chem. Soc.*, **114**, 8177.
Legon, A. C. and Thorn, J. C. (1992) *J. Mol. Struct.*, **270**, 449.
Liebman, J. F., Romm, M. J., Meot-Ner (Mautner), M., Cybulski, C. M. and Scheiner, S. (1991) *J. Phys. Chem.*, **95**, 1112.
Lifshitz, C. and Feng, W. Y. (1995) *Int. J. Mass Spectrom. Ion Processes*, **146/147**, 223.
Live, D. and Chan, S. I. (1976) *J. Am. Chem. Soc.*, **98**, 3769.
Magnusson, E. (1994) *J. Phys. Chem.*, **98**, 12558.
Markovich, G., Pollack, S., Giniger, R. and Cheshnovsky, O. (1994) *J. Chem. Phys.*, **101**, 9344.
McClean, A. D. and Chandler, G. S. (1980) *J. Chem. Phys.*, **72**, 5639.
McLafferty. F. (1994) *Acc. Chem. Res.*, **27**, 379.
Mecozzi, S., West, A. P., Jr and Dougherty, D. A. (1996) *J. Am. Chem. Soc.*, **118**, 2307.
Meot-Ner (Mautner), M. (1984) *J. Am. Chem. Soc.*, **106**, 1265.
Meot-Ner (Mautner), M. (1988) *J. Am. Chem. Soc.*, **110**, 3071.
Meot-Ner (Mautner), M. (1992) *J. Am. Chem. Soc.*, **114**, 3312.
Meot-Ner (Mautner), M. (1994) *Computational Approaches in Supramolecular Chemistry*, Kluwer Academic, Dordrecht.
Meot-Ner (Mautner), M. and Deakyne, C. A. (1985) *J. Am. Chem. Soc.*, **107**, 469.
Meot-Ner (Mautner), M. and Sieck, L. W. (1991) *J. Am. Chem. Soc.*, **113**, 4448.
Meot-Ner (Mautner), M., Sieck, L. W., Koretke, K. K. and Deakyne, C. A. (1996a), submitted to *J. Am. Chem. Soc.*
Meot-Ner (Mautner), M., Sieck, L. W., Liebman, J. F. and Scheiner, S. (1996b) *J. Phys. Chem.*, **100**, 6445.
Meot-Ner (Mautner), M., Sieck, L. W., Scheiner, S. and Duan, X. (1994) *J. Am. Chem. Soc.*, **116**, 7848.
Meot-Ner (Mautner), M. and Speller, C. V. (1986) *J. Phys. Chem.*, **90**, 6616.
Møller, C. and Plesset, M. S. (1934) *Phys. Rev.*, **46**, 618.
More, M. B., Glendening, E. D., Ray, D., Feller, D. and Armentrout, P. B. (1996) *J. Phys. Chem.*, **100**, 1605.
Morgan, S., Keesee, R. G. and Castleman, A. W., Jr (1989) *J. Am. Chem. Soc.*, **111**, 3841.

Morokuma, K. (1971) *J. Chem. Phys.*, **55**, 1236.
Nagle, J. F. and Tristram-Nagle, S. J. (1983) *Membr. Biol.*, **74**, 1.
Newton, M. D. (1977) *J. Chem. Phys.*, **67**, 5535.
Newton, M. D. and Ehrenson, S. (1971) *J. Am. Chem. Soc.*, **93**, 4971.
Ogorzalek Loo, R. R. and Smith, R. D. (1994) *J. Am. Soc. Mass Spectrom.*, **5**, 207.
Ogorzalek Loo, R. R., Winger, B. E. and Smith, R. D. (1994) *J. Am. Soc. Mass Spectrom.*, **5**, 1064.
Payzant, J. D., Cunningham, A. J. and Kebarle, P. (1973) *Can. J. Chem.*, **51**, 3242.
Perera, L. and Berkowitz, M. L. (1993) *J. Chem. Phys.*, **99**, 4222.
Peterson, K. J. and Klemperer, W. (1984) *J. Chem. Phys.*, **81**, 3842.
Pitzer, K. S. (1961) *Quantum Chemistry*, Prentice Hall, Englewood Cliffs.
Platts, J. A. and Laidig, K. E. (1995) *J. Phys. Chem.*, **99**, 6487.
Pople, J. A., Binkley, J. S. and Seeger, R. (1976) *Int. J. Quant. Chem. Symp.*, **10**, 1.
Pudzianowski, A. T. (1995) *J. Chem. Phys.*, **102**, 8029.
Pudzianowski, A. T. (1996) *J. Phys. Chem.*, **100**, 4781.
Purvis, G. D., III and Bartlett, R. J. (1982) *J. Chem. Phys.*, **76**, 1910.
Reed, A. E., Curtiss, L. A. and Weinhold, F. (1988) *Chem. Rev.*, **88**, 899.
Reed, A. E., Weinstock, R. B. and Weinhold, F. (1985) *J. Chem. Phys.*, **83**, 735.
Reetz, M. T., Hutte, S. and Goddard, R. (1993) *J. Am. Chem. Soc.*, **115**, 9339.
Rybak, S., Jeziorski, B. and Szalewicz, K. (1991) *J. Chem. Phys.*, **95**, 6576.
Saenger, W. (1984). *Principles of Nucleic Acid Structure*, Springer-Verlag, Berlin.
Schneider, H.-J. (1991) *Angew. Chem. Int. Ed. Eng.*, **30**, 1417.
Selegue, T. J. and Lisy, J. M. (1994) *J. Am. Chem. Soc.*, **116**, 4874.
Shi, Z., Ford, J. V., Wei, S. and Castleman, A. W., Jr (1993) *J. Chem. Phys.*, **99**, 8009.
Shi, Z., Wei, S., Ford, J. W. and Castleman, A. W., Jr (1992) *Chem. Phys. Lett.*, **200**, 142.
Smith, D. A. (ed.) (1994) *Modeling the Hydrogen Bond*, ACS Symposium Series 569, American Chemical Society, Washington, DC.
Smith, D. E. and Dang, L. X. (1994) *J. Chem. Phys.*, **101**, 7873.
Snodgrass, J. T., Coe, J. V., Freidhoff, C. B., McHugh, K. M., Arnold, S. T. and Bowen, K. H. (1995) *J. Phys. Chem.*, **99**, 9675.
Spitznagel, G. W., Clark, T., Chandrasekhar, J. and Schleyer, P. v. R. (1982) *J. Comput. Chem.*, **3**, 363.
Stace, A. J. and Moore, C. (1983) *J. Am. Chem. Soc.*, **105**, 1814.
Stace, A. J. and Shukla, A. K. (1982) *J. Am. Chem. Soc.*, **104**, 5314.
Stace, A. J., Winkel, J. F., Lopez Martens, R. B. and Upham, J. E. (1994) *J. Phys. Chem.*, **98**, 2012.
Steiner, T. and Saenger, W. (1992) *J. Am. Chem. Soc.*, **114**, 10146.
Steiner, T. and Saenger, W. (1993) *J. Am. Chem. Soc.*, **115**, 4540.
Stewart, J. J. P. (1989) *J. Comput. Chem.*, **10**, 209.
Stone, A. J. (1993) *Chem. Phys. Lett.*, **211**, 101.
Suh, D., Burgers, P. C. and Terlouw, J. K. (1995a) *Int. J. Mass Spectrom. Ion Processes*, **144**, L1.
Suh, D., Kingsmill, C. A., Ruttink, P. J. A., Burgers, P. C. and Terlouw, J. K. (1995b) *Int. J. Mass Spectrom. Ion Processes*, **146/147**, 305.
Sunner, J., Nishizawa, K. and Kebarle, P. (1981) *J. Phys. Chem.*, **85**, 1814.
Sussman, J. L., Harel, M., Frolow, F., Oefner, C., Goldman, A., Toker, L. and Silman, I. (1991) *Science*, **253**, 872.
Svinning, T. and Sørum, H. (1975) *Acta Crystallogr.*, **B31**, 1581.
Syage, J. A. (1995) *J. Phys. Chem.*, **99**, 5772.
Szczesniak, M. M., Chalasinski, G., Cybulski, S. M. and Cieplak, P. (1993) *J. Chem. Phys.*, **98**, 3078.
Taft, R. W., Anvia, F., Gal, J. F., Walsh, S., Capon, M., Holmes, M. C., Hosn, K., Olumi, G., Vasanwala, R. and Yazdani, S. (1990) *Pure Appl. Chem.*, **62**, 17.
Tang, I. N. and Castleman, A. W., Jr (1975) *J. Chem. Phys.*, **62**, 4576.
Taylor, R. and Kennard, O. (1982) *J. Am. Chem. Soc.*, **104**, 5063.
Taylor, R. and Kennard, O. (1984) *Acc. Chem. Res.*, **17**, 320.
Topp, M. R. (1995) *J. Chim. Phys.*, **92**, 310.
Tsuzuki, S., Uchimaru, T., Tanabe, K. and Hirano, T. (1993) *J. Phys. Chem.*, **97**, 1346.
Tzeng, W. B., Wei, S. and Castleman, A. W., Jr (1991) *J. Phys. Chem.*, **95**, 5757.

References

Tzeng, W. B., Wei, S., Neyer, D. W., Keesee, R. G. and Castleman, A. W., Jr (1990) *J. Am. Chem. Soc.*, **112**, 4097.
Urban, M., Noga, J., Cole, S. J. and Bartlett, R. J. (1985) *J. Chem. Phys.*, **83**, 4041.
van Duijneveldt-van de Rijdt, J. G. C. M. and van Duijneveldt, F. B. (1992) *J. Chem. Phys.*, **97**, 5019.
van Mourik, T. and van Duijneveldt, F. B. (1995) *J. Mol. Struct. (Theochem)*, **341**, 63.
Vincenti, M. (1995) *J. Mass Spectrom.*, **30**, 925.
Viswamitra, M. A., Radhakrishnan, R., Bandekar, J. and Desiraju, G. R. (1993) *J. Am. Chem. Soc.*, **115**, 4868.
Wadt, W. R. and Hay, P. J. (1985) *J. Chem. Phys.*, **82**, 284.
Warshel, A. (1981) *Acc. Chem. Res.*, **14**, 284.
Watanabe, H. and Iwata, S. (1996) *J. Phys. Chem.*, **100**, 3377.
Wehbeh, J. A. and Allen, L. C. personal communication.
Wei, S., Shi, Z. and Castleman, A. W., Jr (1991a) *J. Chem. Phys.*, **94**, 3268.
Wei, S., Tzeng, W. B. and Castleman, A. W., Jr (1991b) *J. Phys. Chem.*, **95**, 585.
Wei, S., Tzeng, W. B. and Castleman, A. W., Jr (1991c) *Chem. Phys. Lett.*, **178**, 411.
Wei, S., Tzeng, W. B., Keesee, R. G. and Castleman, A. W. Jr (1991d) *J. Am. Chem. Soc.*, **113**, 1960.
Wiese, W. L. and Weiss, A. W. (1968) *Phys. Rev.*, **175**, 50.
Winkel, J. F. and Stace, A. J. (1994) *Chem. Phys. Lett.*, **221**, 431.
Woodin, R. L. and Beauchamp, J. L. (1978) *J. Am. Chem. Soc.*, **100**, 501.
Wozniak, K., He, H., Klinowski, J., Jones, W. and Barr, T. L. (1995) *J. Phys. Chem.*, **99**, 14667.
Xantheas, S. S. and Dunning, T. H., Jr (1993) *J. Chem. Phys.*, **99**, 8774.
Xantheas, S. S. and Dunning, T. H., Jr (1994) *J. Phys. Chem.*, **98**, 13489.
Xia, P. and Garvey, J. F. (1995) *J. Phys. Chem.*, **99**, 3448.
Yamashita, M. and Fenn, J. B. (1984) *J. Phys. Chem.*, **88**, 4451.
Zhang, R. and Lifshitz, C. (1996) *J. Phys. Chem.*, **100**, 960.
Zhang, X. and Castleman, A. W., Jr (1994) *J. Chem. Phys.*, **101**, 1157.
Zhang, X., Yang, X. and Castleman, A. W., Jr (1991) *Chem. Phys. Lett.*, **185**, 298.

8 Case Studies in Cooperativity in Hydrogen-bonded Clusters and Polymers*

A. KARPFEN

Institut für Theoretische Chemie und Strahlenchemie der Universität Wien, Austria

8.1 INTRODUCTION

Hydrogen bonding represents one of the most important types of intermolecular interaction and is almost ubiquitous in many fields of organic and inorganic crystal chemistry, in biochemistry and in the physical chemistry of gases and liquids. Numerous books and collections of review articles have been devoted to thorough and comprehensive treatments of this topic. Representative examples are Hadzi and Thompson (1959), Pimentel and McClellan (1960), Hamilton and Ibers (1968), Vinogradov and Linell (1971), Schuster *et al.* (1976), Hobza and Zahradnik (1988), Michl and Zahradnik (1988), and Castleman and Hobza (1994). The characteristic properties of a hydrogen bond A—H···B formed between a proton donor A–H and a proton acceptor B are often (Pimentel and McClellan, 1960; Schuster, 1978) described in an operational manner by enumerating the structural and the various spectroscopically observable features relative to that of the unpertubed A–H monomer. Apart from the fact that new intermolecular modes appear in the vibrational spectra of the complex, and restricting the discussion to properties in the electronic ground state, the formation of a hydrogen bond A—H...B is generally accompanied by the following trends:

1. The intermolecular distance $R(A\cdots B)$ is much shorter than the sum of the van der Waals radii of A and B.

---

*This chapter is dedicated to Professor A. Neckel.

Molecular Interactions. Edited by S. Scheiner
© 1997 John Wiley & Sons Ltd

2. $r(A—H)$ is longer in the complex than in the monomer.
3. The fundamental frequency $\nu(A—H)$ is smaller in the complex than in the monomer.
4. The infrared intensity of $\nu(A—H)$ is larger in the complex than in the monomer.
5. Proton NMR chemical shifts are significantly smaller in the complex than in the monomer.

The extent to which these trends are actually observable depends on the strength of the hydrogen bond. In general, for single hydrogen bonds in vapour-phase or matrix-isolated, neutral dimers, the above-mentioned effects are significantly less pronounced than in the liquid or solid state, in specific neutral oligomers or in charged species.

The aim of this chapter is to focus on these differences and to describe some aspects of the non-additive or cooperative behaviour in hydrogen-bonded systems. To this end, two model systems have been selected which appear particularly suited for a detailed demonstration of the gradual transition from isolated dimers with one hydrogen bond to sufficiently large clusters that resemble the properties of the hydrogen-bonded solids: (1) neutral, chain-like and cyclic clusters of hydrogen cyanide molecules, $(HCN)_n$, and (2) neutral, chain-like and cyclic clusters of hydrogen fluoride molecules, $(HF)_n$. In the latter case, protonated $([(HF)_nH]^+)$ and deprotonated $([(HF)_{n-1}F]^-)$ chain-like clusters are treated as well. One ingredient that is common to both compounds is the formation of strongly anisotropic crystals, i.e. regular, extended, hydrogen-bonded chains, surrounded by other chains at about a distance compatible with the usual van der Waals distances, are the dominating structural feature in the solid state. Hydrogen cyanide forms extended, perfectly linear chains in the solid state (Dulmage and Lipscomb, 1951). In the hydrogen fluoride crystal, zig-zag chains are formed as revealed by X-ray (Atoji and Lipscomb, 1954) and neutron (Johnson *et al.*, 1975) diffraction. Figure 8.1 shows schematic diagrams of the hydrogen cyanide and hydrogen fluoride chains. Quasi-one-dimensional hydrogen-bonded chains are actually quite often encountered in nature. As further representative examples for anisotropic hydrogen-bonded crystals in which hydrogen-bonded chains constitute the dominating structural pattern, we mention the crystal structures of methanol (Tauer and Lipscomb, 1952), formic (Holtzberg *et al.*, 1953; Nahringbauer, 1978; Albinati *et al.*, Thomas, 1978b) and acetic acid (Jones and Templeton, 1958; Nahringbauer, 1970; Jönsson, 1971; Albinati *et al.*, 1978a), hydrogen chloride (Sandor

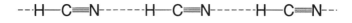

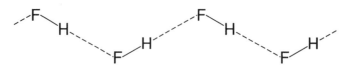

Figure 8.1 Schematic representation of hydrogen cyanide and hydrogen fluoride chains

and Farrow, 1967), imidazole (Martinez-Carrera, 1966) and cyanoacetylene (Shallcross and Carpenter, 1958).

An idea of how strong the cooperative effects are in the above-mentioned hydrogen-bonded crystals and in the corresponding oligomers can be gained most easily by a comparison of the experimental A$\cdots$B intermolecular distances in the isolated dimer and in the crystal, respectively, and by confronting the experimental A—H stretching frequency in the monomer, with that observed in the vapour phase dimer, and in the solid, respectively. Hydrogen cyanide is an example for a system in which only very modest structural and vibrational spectroscopic changes occur upon going from the dimer to larger linear clusters and finally to the molecular crystal. Overall, the experimentally observed reduction of the intermolecular distance in the hydrogen cyanide crystal (Dulmage and Lipscomb, 1951) relative to that in the hydrogen cyanide dimer (Buxton *et al.*, 1981) amounts to about 0.1 Å only. The shifts of the C—H stretching frequency in the hydrogen cyanide dimer (Jucks and Miller, 1988b) and in the hydrogen cyanide crystal (Friedrichs and Krause, 1973) relative to the monomer fundamental (Bendtsen and Edwards, 1974) are -70 cm^{-1} and -182 cm^{-1}, respectively. Thus, the entire difference between dimer and crystal amounts to -112 cm^{-1} only.

In hydrogen fluoride, on the other hand, the corresponding effects are much larger. Among the molecular crystals, hydrogen fluoride is the extreme case. There, the reduction of the intermolecular distance from the dimer (Dyke *et al.*, 1972) to the crystal (Atoji and Lipscomb, 1954; Johnson *et al.*, 1975) is about 0.3 Å, considerably larger than in the hydrogen cyanide case. Whereas the -93 cm^{-1} shift of the F—H stretching frequency in the hydrogen fluoride dimer (Pine and Lafferty, 1983; Pine *et al.*, 1984) relative to the monomer is still comparable to the hydrogen cyanide dimer case, the shift of the F—H stretching frequency from the monomer to the molecular crystal (Kittelberger and Hornig, 1967) is 896 cm^{-1}, thus a dimer to crystal shift of about 800 cm^{-1}. Despite these large shifts, hydrogen fluoride can evidently still be considered as a molecular crystal.

Considerably stronger structural and vibrational spectroscopic changes do occur in ionic hydrogen bonds which, however, cannot reasonably be treated as intermolecular interactions anymore. Quite typical examples are deprotonated species like (FHF$^-$), or protonated oligomers like ((H$_2$O)$_n$H$^+$). For example, the F$\cdots$F distance in the (FHF$^-$) anion is shorter by about 0.5 Å (Kawaguchi and Hirota, 1986) than the vibrationally averaged inermolecular F$\cdots$F distance in the (HF)$_2$ dimer. The frequency of the symmetric stretching motion in (FHF$^-$) is about 600 cm^{-1}, and is hence shifted by about -3350 cm^{-1} relative to the HF monomer stretching frequency. Protonated and deprotonated chains of hydrogen fluoride molecules with varying degree of oligomerization may be considered as model systems for the investigation of defect structures on extended chains which play a role in proton conductivity and lend themselves to a systematic investigation of the spatial extension of such protonated or deprotonated defects and of how the ionic hydrogen bonds at such defect centres influence the surrounding hydrogen bonds.

A further reason for having chosen hydrogen cyanide and hydrogen fluoride as model systems is that the structural and spectroscopic properties of small hydrogen cyanide and hydrogen fluoride clusters beyond the dimers is currently a very actively and intensively investigated topic, both from the experimental side and from theory.

8.2 WEAK COOPERATIVITY: THE CASE OF HYDROGEN CYANIDE

The gas-phase structures and vibrational spectra of small clusters of hydrogen cyanide have been quite extensively investigated (Maroncelli *et al.*, 1985., Hopkins *et al.*, 1985; Anex *et al.*, 1988; Jucks and Miller, 1988a, b; Ruoff *et al.*, 1988a, b, Meyer *et al.*, 1989; Kerstel *et al.*, 1993). Similarly, considerable efforts have been directed towards spectroscopic investigations of hydrogen cyanide in inert matrices at low temperatures (King and Nixon, 1968; Pacansky, 1977; Walsh *et al.*, 1978; Knözinger and Wittenbeck, 1982; Knözinger *et al.*, 1986a, b; Schrems *et al.*, 1987; Langel *et al.*, 1989; Beichert *et al.*, 1995). These and earlier microwave experiments on the linear dimer of hydrogen cyanide molecules (Legon *et al.*, 1977; Brown *et al.*, 1981; Buxton *et al.*, 1982; Campbell and Kukolich, 1983; Fillery-Travis *et al.*, 1983; Georgiou *et al.*, 1985; Wofford *et al.*, 1986) have led to a detailed knowledge of the properties of the dimer and of the two different trimers, the linear and the cyclic species, in particular of the vibrational spectra in the C—H stretching frequency region. Although additional spectral lines were observed in the C—H stretching region, vapour-phase assignments for clusters larger than the trimers have not yet been made with certainty.

Hydrogen cyanide clusters larger than the dimer were also investigated theoretically. *Ab initio* investigations on the structure, the vibrational spectra and on other properties were performed at various methodical levels (Kofranek *et al.*, 1987; Kofranek *et al.*, 1987b; Dykstra, 1987, 1996; Anex *et al.*, 1988; De Almeida *et al.*, 1989; Kurnig *et al.*, 1990; Dykstra, 1990; De Almeida, 1991; King and Weinhold, 1995; King *et al.*, 1995, Beichert *et al.*, 1995; Karpfen, 1996). In addition to the theoretical investigations on clusters, several studies dealt with the properties of infinitely extended, perfectly linear hydrogen cyanide chains (Kertész *et al.*, 1976; Karpfen, 1983; Springborg, 1991, 1995; Suhai, 1994).

As far as the data on hydrogen cyanide clusters are concerned, the theoretical results of a very recent systematic study on linear and cyclic hydrogen cyanide and on the related cyanoacetylene clusters (Karpfen, 1996) are taken advantage of in order to illustrate the modification of and the convergence of various spectroscopic observables from the unperturbed monomer to a molecule embedded in an infinitely extended hydrogen-bonded chain. From the different series of calculations performed in that work, only calculated results using the 6–31G(d,p) basis set (Hehre *et al.*, 1972; Frisch *et al.*, 1984) within the framework of the Møller–Plesset second-order (MP2) (Møller and Plesset, 1934) are discussed. In that work, and for all the calculations on the hydrogen fluoride clusters discussed in the next section, the Gaussian 94 suite of programs (Frisch *et al.*, 1995) was used.

The structures of the linear and cyclic hydrogen cyanide oligomers up to the tetramers are shown schematically in Figure 8.2. In the following, we demonstrate how all the features considered as typical for the hydrogen bond are gradually enhanced upon the formation of increasingly larger, linear and cyclic hydrogen cyanide oligomers and how convergence to the same asymptotic limit is approached. However, before discussing the characteristic hydrogen bond properties mentioned in the previous section, we first turn to the energetics.

The increase in the stabilization energies per hydrogen bond taking place upon enlarging the size of the chain-like and cyclic clusters, and the convergence to the infinite

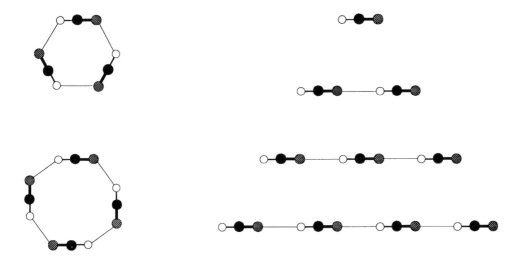

Figure 8.2 Linear and cyclic hydrogen cyanide clusters up to the tetramers

chain limit are most conveniently discussed by defining averaged stabilization energies per hydrogen bond, $\Delta E_a(n)$ and $\Delta E_b(n)$, (Karpfen et al., 1974) as

$$\Delta E_a(n) = E(n) - E(n-1) - E(1) \quad (8.1)$$

and

$$\Delta E_b(n) = (E(n) - n*E(1))/m \quad (8.2)$$

where $E(n)$ is the total energy of the n-mer, $m = n$ for cyclic clusters, and $m = n - 1$ for linear clusters. Analogously, also the corresponding zero-point-energy (ZPE) corrected quantities can be defined. $\Delta E_a(n)$ can be best described as the hydrogen bond energy for the insertion of one molecule in the centre of an already performed chain and, thus, gives higher weight to bulk-like hydrogen bonds, whereas $\Delta E_b(n)$ corresponds to the traditional average hydrogen bond energy, which gives equal weights to bulk and edge hydrogen bonds. As an illustration, the quantities $\Delta E_a(n)$ and $\Delta E_b(n)$, and their ZPE-corrected counterparts, $\Delta E_a^{ZPE}(n)$ and $\Delta E_b^{ZPE}(n)$, are plotted versus $1/n$ in Figure 8.3 for the linear and cyclic hydrogen cyanide clusters up to $n = 6$. In view of the smooth behaviour of all four quantities and the fact that they necessarily must converge to the same limit, about -30 kJ/mol for the stabilization energy and about -25 kJ/mol for the corresponding ZPE-corrected value, further costly calculations on still larger cluster appear superfluous, at least for our purposes. It should be kept in mind that because of the calculational level chosen and because corrections to the interaction energy resulting from the basis set superposition error (BSSE, Boys and Bernardi, 1970) are neglected here, the stabilization energies just discussed are too negative by about 4 kJ/mole.

The five selected properties typical for hydrogen bonding and their behaviour upon increasing the cluster size are collected in Table 8.1 and depicted in Figures 8.4–8.6. The trends in the elongation of the C—H bond and in the reduction of the intermolecular N$\cdots$H bond with increasing chain length or ring size are illustrated in Figures 8.4(a) and 8.4(b) for the case of the shortest hydrogen bond in each cluster by plotting them versus

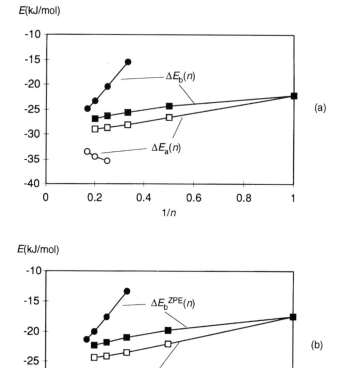

Figure 8.3 Convergence of stabilization energies in linear and cyclic hydrogen cyanide oligomers, $(HCN)_n$ as calculated at the MP2/6–31G(d,p) level. Plot of (a) the stabilization energy per hydrogen bond and (b) the ZPE-corrected stabilization energy per hydrogen bond versus $1/n$ using equations (8.1) and (8.2)

$1/n$, were n is the number of hydrogen bonds in the cluster. The almost perfect correlation between these two quantities for all C—H$\cdots$N hydrogen bonds in linear and cyclic clusters up to the hexamers is shown in Figure 8.4(c). Rough extrapolations to $n = \infty$ lead to values of r(C—H) and r(N$\cdots$H) of about 1.079 ± 0.001 and 2.06 ± 0.002 Å for the infinitely extended polymer. That corresponds to an overall reduction of the C$\cdots$N distance from about 3.29 Å in the dimer to about 3.14 ± 0.02 Å in the polymer, i.e. 0.15 ± 0.02 Å, comparable to the above-mentioned experimental value of about 0.10 Å (Buxton et al., 1981; Dulmage and Lipscomb, 1951).

The trends in the C—H stretching frequencies are illustrated in Figure 8.5(a) by plotting the frequency of the strongest infrared active C—H stretching vibration of each cluster versus $1/n$, where n is the number of C—H bonds in the cluster. In the case of the linear clusters, this always corresponds to the lowest-lying C—H stretching vibration,

Table 8.1 Theoretical characterization of the C—H···N hydrogen bond in linear and cyclic hydrogen cyanide clusters, (HCN)$_n$, as calculated at the MP2/6-31G(d,p) level

n	r(C—H) (Å)	r(N···H) (Å)	ν(C—H) (cm^{-1})	A(C—H) (km/mol)	σ(^{1}H) (ppm)
			Linear		
1	1.0653	–	3529	66	28.9
2	1.0706	2.2203	3455	345	28.2
3	1.0724	2.1762	3426	858	27.7
4	1.0743	2.1347	3400	1113	27.5
5	1.0749	2.1215	3386	1625	27.3
6	1.0755	2.1095	3376	2088	27.2
			Cyclic		
3	1.0681	2.4514	3500	305	27.9
4	1.0712	2.2564	3456	822	27.1
5	1.0731	2.1861	3425	1509	26.9
6	1.0741	2.1558	3406	2248	26.9

which turns out to be the in-phase motion, quite independent of the cluster size. In the case of the cyclic clusters, the only infrared active C—H stretching vibration corresponds to the second-lowest (doubly degenerate) vibration. Approximate extrapolation leads to a value of about 3330 cm^{-1} for both series, which corresponds to an overall reduction of about −200 cm^{-1} from the monomer to the infinitely extended chain. The corresponding experimental red-shift amounts to −182 cm^{-1} (Friedrichs and Krause, 1973).

As a further indication for the non-additive nature of the hydrogen bond, the infrared intensities of the very same modes are shown in Figure 8.5(b). Thereby, the values given in Table 8.1 have been divided by n for the case of the linear clusters and by $2n$ in case of the cyclic clusters; the additional factor of 2 arising from the twofold degeneracy of the infrared active ring modes. With the exception of the *flyer* in case of the linear trimer, well understood because of two almost identical intermolecular distances, and consequently an extraordinary amplification of the intensity of the in-phase mode and an almost complete annihilation of the intensity of the out-of-phase mode, the behaviour with increasing cluster size appears smooth again, although extrapolation is very uncertain in that case and has, therefore, been omitted.

Finally, the NMR σ(^{1}H) shielding constants as evaluated within the framework of the SCF-GIAO approach (Ditchfield, 1974; Wolinsky et al., 1990) at the respective MP2 optimized geometries are plotted versus r(C—H) and r(N···H), respectively. Relative to the HCN monomer, the calculated overall reduction of the chemical shift in the extended polymer amounts to about 2 ppm only. As with the other properties discussed, the chemical shifts of the cyclic and linear oligomers asymptotically approach the same limit. Furthermore, as with the vibrational frequencies, the σ(^{1}H) shielding constants are very sensitive probes of small structural details in the C—H···N hydrogen bonds. Experimental σ(^{1}H) shielding constants in HCN clusters are not yet available.

Summarizing the data on hydrogen cyanide clusters, we obtain comparatively small and modest modifications of the C—H···N hydrogen bond in homomolecular (HCN)$_n$ clusters, in quite reasonable agreement with the available experimental data.

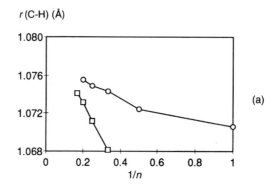

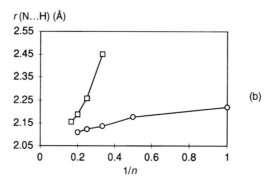

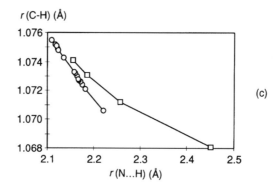

Figure 8.4 Structural convergence in linear and cyclic hydrogen cyanide oligomers, (HCN)$_n$ as calculated at the MP2/6–31G(d,p) level. (a) Plot of r(C—H) versus $1/n$. (b) Plot of r(N···H) versus $1/n$. (c) Correlation between r(C—H) and r(N···H). Open squares from cyclic oligomers, open circles from linear oligomers

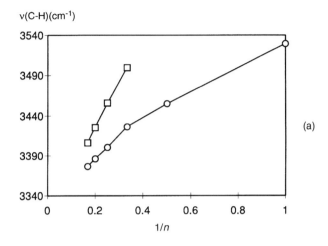

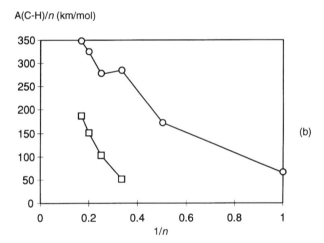

Figure 8.5 Convergence of vibrational spectroscopic data pertinent to the C—H group in linear and cyclic hydrogen cyanide oligomers, $(HCN)_n$. Plot of (a) the frequency of the most intense infrared active C—H stretching vibration and (b) the corresponding infrared intensity. Open squares from cyclic oligomers, open circles from linear oligomers

8.3 MEDIUM COOPERATIVITY: THE CASE OF HYDROGEN FLUORIDE

The number of experimental and theoretical investigations performed on the vapour phase dimer $(HF)_2$ is much too large to discuss or quote individual papers here. Hence, only an exhaustive review (Truhlar, 1990) and a few of the more recent *ab initio* studies (Kofranek *et al.*, 1988; Racine and Davidson, 1993; Peterson and Dunning, 1995; Burcl *et al.*, 1995, Collins *et al.*, 1995; Quack and Suhm, 1996) are mentioned which may be

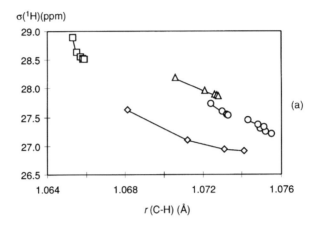

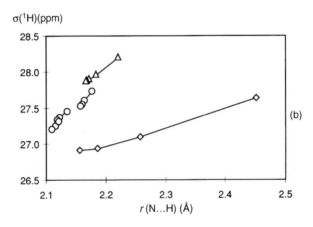

Figure 8.6 Proton NMR chemical shieldings, $\sigma(^1H)$, in linear and cyclic hydrogen cyanide oligomers, (HCN)$_n$, (a) Plot of $\sigma(^1H)$ versus r(C—H); (b) Plot of $\sigma(^1H)$ versus r(N···H). Squares: from hydrogens at the non-hydrogen-bonded edges of the linear clusters; triangles: from hydrogens at the hydrogen-bonded edges of the linear clusters; circles: from hydrogens in the inner hydrogen bonds of the linear clusters; diamonds: from cyclic clusters

also used as sources for further reference. Because of these considerable efforts, (HF)$_2$ is probably one of the best understood hydrogen-bonded dimers. Unlike the C$_s$-symmetric open chain structure of (HF)$_2$, the available experimental (Lisy *et al.*, 1981; Michael and Lisy, 1986; Kolenbrander *et al.*, 1988; Sun *et al.*, 1992; Suhm *et al.*, 1993; Suhm and Nesbitt, 1995) as well as all the theoretical (Karpfen *et al.*, 1981, 1983; Gaw *et al.*, 1984; Heidrich *et al.*, 1985; Liu *et al.*, 1986; Scuseria and Schaefer, 1986; Latajka and Scheiner, 1988; Chalasinski *et al.*, 1989; Karpfen, 1990; Del Bene and Shavitt, 1991; Komornicki *et al.*, 1992; Heidrich *et al.*, 1993; Quack *et al.*, 1993b, Karpfen and Yanovitskii, 1994b; Luckhaus *et al.*, 1995) evidence points to a cyclic C$_{3h}$ configuration as the most stable structure of the vapour phase trimer (HF)$_3$. From the experimental side, the stable structures of the vapour phase oligomers with $n = 4$, 5, and 6 have generally been considered as ring-like clusters (Smith, 1958; Janzen and Bartell, 1969; Dyke *et al.*, 1972; Hinchen and Hobbs, 1979; Lisy *et al.*, 1981; Vernon *et al.*, 1982; Redington, 1981 1982a, b). However, that of the tetramer, (HF)$_4$, and of the pentamer, (HF)$_5$, are currently again very much under debate (Sun *et al.*, 1992; Luckhaus *et al.*, 1995; Huisken *et al.*, 1995a, b). From the theoretical side, the cyclic structures have been treated almost exclusively (Del Bene and Pople, 1971; Liu *et al.*, 1986; Dykstra, 1987, 1990; Zhang *et al.*, 1989; Karpfen, 1990; Quack *et al.*, 1993, Liedl *et al.*, 1995), although alternative structures have been suggested as well (Huisken *et al.*, 1995b,c). The vibrational spectroscopic information as obtained from low-temperature matrix investigations was interpreted in terms of cyclic and chain-like clusters (Andrews and Johnson, 1983, 1984; Andrews *et al.*, 1984; Andrews, 1984; Andrews *et al.*, 1992).

Vibrational spectroscopic investigations on solid hydrogen fluoride, including external pressure effects, have been presented by Lee *et al.* (1986), Jansen *et al.* (1987) and by Pinnick *et al.* (1989). Theoretical investigations on infinitely extended zig-zag chains of hydrogen fluoride molecules have also already been performed (Kertész *et al.*, 1975; Beyer and Karpfen, 1982; Springborg, 1987, 1988). In two recent studies (Karpfen and Yanovitskii, 1994a,b), the mode of convergence of the various ground state properties of chain-like and cyclic hydrogen fluoride oligomers to that of the infinitely extended zig-zag chains has been investigated in some detail within the framework of the SCF approximation. Here, we extend these investigations to the MP2 and to three recently suggested density functional (DF) levels. On the basis of a number of very recent systematic investigations on hydrogen-bonded systems in which the advantages and disadvantages of a large number of DF variants have been thoroughly compared (Sim *et al.*, 1992; Latajka and Bouteiller, 1994; Stanton and Merz, 1994; Latajka *et al.*, 1995; Zhang *et al.*, 1995; Del Bene *et al.*, 1995; Xantheas, 1995; Novoa and Sosa, 1995), the following DF methods were considered as the most promising: (i) the BLYP method, i.e. Becke's (1988) gradient-corrected exchange-functional in combination with the Lee, Yang, Parr non-local correlation functional (Lee *et al.*, 1988; Mielich *et al.*, 1989), (2) the hybrid-B3LYP (Becke, 1993), and (3) the hybrid-B3P86 with Perdew's (1986) gradient-correct exchange functional. Full details of these calculations, in which several extended bases had been applied, will be given elsewhere (Karpfen, in preparation). Here, only a few selected results as obtained applying the 6−311++G(2*d*,*p*) (Frisch *et al.*, 1984; Krishnan *et al.*, 1980; Clark *et al.*, 1983) basis set are presented.

As with the hydrogen cyanide case treated in the previous section, we discuss only the five mentioned properties characteristic for the F—H···F hydrogen bond. The structures of the chain-like and cyclic hydrogen fluoride oligomers up to the tetramers are shown in Figure 8.7. The qualitatitive characterization of the various optimized structures is identical at all four levels of approximation. Among the chain-like clusters, only (HF)$_2$ is a true minimum, whereas the larger linear clusters are either minima, first- or second-order saddle points, in the latter cases with exceedingly small imaginary frequencies corresponding to out-of plane vibrations. Among the cyclic C_{nh}-symmetric clusters, the C_{2h} arrangement is the well-known first-order saddle point between two equivalent C_s minima. The C_{nh}-symmetric (HF)$_n$ clusters with $n = 3 - 5$ are minima. Planar, cyclic, C_{nh}-symmetric (HF)$_6$ is a saddle point of third order, with all imaginary frequencies ($< 20i$) again originating from out-of-plane vibrations.

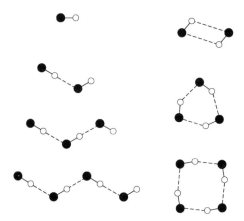

Figure 8.7 The structures of chain-like and cyclic (HF)$_n$ clusters up to the tetramers

Before turning to a detailed discussion of the convergence behaviour in the hydrogen fluoride clusters, one general comment appears to be in order. Unlike the hydrogen cyanide case, in which the dimer minimum corresponds to the fully linear structure, the optimal (HF)$_2$ structure is bent with an FFH angle around 110–120°. The experimental FFF angle of 116° in the zig-zag HF-chains in the solid is also in that range. Hence, the optimal cyclic structures are encountered for $n = 5$ or 6, larger rings will tend to become less stable. Therefore, for eventual extrapolations to the solid, the analysis of the chain-like structures appears to be more appropriate.

The computed stabilization energies and the ZPE-corrected stabilization energies are compiled in Table 8.2. In agreement with practically all the previously published theoretical studies and with the available experimental evidence, the cyclic C_{nh}-symmetric structures are more stable than the chain-like configurations for $n = 3–6$. The three DF variants deliver interaction energies that are in quite good agreement with the MP2 data. The qualitatively identical description of the relative stabilities of cyclic and chain-like clusters as obtained with the four computational levels is illustrated in Figure 8.8, where the stabilization energies per hydrogen bond, $\Delta E_a(n)$ and $\Delta E_b(n)$, are plotted versus $1/n$. The excellent agreement indicates that, as far as the interaction energies are concerned, the much cheaper DF variants produce reliable values.

The structural convergence, i.e. the r(F—H) and the r(F$\cdots$F) distances, are collected in Tables 8.3 and 8.4. The characteristic structural parameters of the shortest hydrogen bond in each of the clusters are depicted in Figure 8.9. Whereas the qualitative picture of the structural distortions with respect to monomer and dimer is still acceptably described with all three DF variants, and is also in qualitative agreement with previous SCF investigations applying the same basis set (Karpfen and Yanovitskii, 1994b), we now observe stronger quantitative differences between MP2 and DF results, in particular for the r(F—H) elongations as computed with all three DF variants and to a lesser degree for the r(F$\cdots$F) contractions. The r(F—H) distances are quite difficult to determine experimentally, and hence only rather approximate crystal values of 0.95 – 0.97 Å as derived from neutron diffraction (Johnson et al., 1975) and 0.95 ± 0.03 Å from NMR data (Habuda and Gagarinsky, 1971) are available. The experimentally derived r(F$\cdots$F)

Table 8.2 Total stabilization energies without (ΔE) and with ZPE-correction (ΔE^{ZPE}) of chain-like and cyclic, C_{nh}-symmetric hydrogen fluoride clusters (HF)$_n$ from $n = 2$ to 6 as obtained at various theoretical levels with the $6-311++G(2d,p)$ basis. (all values in kJ/mol)

n	B3LYP	B3P86	BLYP	MP2
		ΔE chains		
2	−20.8	−20.3	−19.6	−20.6
3	−48.0	−47.5	−45.5	−47.3
4	−79.1	−79.0	−75.4	−77.4
5	−112.2	−112.7	−107.4	−109.3
6	−146.6	−148.0	−140.9	−142.2
		ΔE^{ZPE} chains		
2	−13.3	−12.6	−12.2	−13.3
3	−33.2	−32.4	−31.0	−32.6
4	−56.2	−55.5	−53.3	−55.0
5	−82.3	−82.1	−78.6	−78.6
6	−108.1	−108.5	−103.5	−103.2
		ΔE rings		
2	−16.0	−15.4	−14.9	−15.7
3	−68.1	−69.9	−66.1	−64.8
4	−126.3	−131.2	−124.2	−118.9
5	−174.3	−181.2	−171.5	−164.1
6	−213.9	−222.0	−209.7	−202.3
		ΔE^{ZPE} rings		
2	−9.9	−9.2	−8.9	−9.8
3	−47.2	−48.5	−45.6	−44.4
4	−95.8	−100.2	−95.0	−87.4
5	−135.6	−142.7	−134.9	−124.5
6	−169.1	−177.4	−167.4	−155.3

distances of about 2.53 Å for (HF)$_6$ from electron diffraction (Janzen and Bartell, 1969) and of about 2.50 Å from neutron diffraction of solid HF (Johnson et al., 1975) appear more reliable. The theoretical r(F—H) elongations from the monomer to the cyclic (HF)$_6$ hexamer amount to 0.043, 0.053, 0.050, and 0.034 Å, at the B3LYP, B3P86, BLYP and MP2 levels, respectively. All four methods predict reductions of r(F$\cdots$F) of about 0.28–0.30 Å from the dimer to cyclic (HF)$_6$, somewhat larger than the experimental 0.26 Å. However, uncertainties as to the relative importance of vibrational averaging effects on the structure in vapour phase (HF)$_2$ and in solid HF are actually quite difficult to assess. In agreement with previous experience on hydrogen-bonded clusters, one notices again that all eventual errors that appear already in the dimer description at a given computational level are carried over to the larger clusters.

The consequences of the less good agreement between the computed MP2 and DF structures are also clearly seen in the computed harmonic ν(F—H) vibrational frequencies. These are reported in Table 8.5. A plot of the frequency of the most intense ν(F—H)

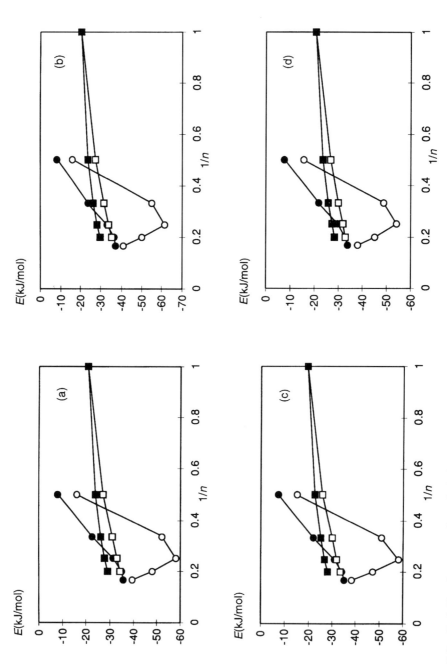

Figure 8.8 Stabilization energies of chain-like and cyclic hydrogen fluoride clusters plotted versus $1/n$. (a) B3LYP, (b) B3P86, (c) BLYP, and (d) MP2. Circles originating from cyclic clusters, squares from chain-like clusters. Filled symbols: $\Delta E_b(n)$; open symbols: $\Delta E_a(n)$

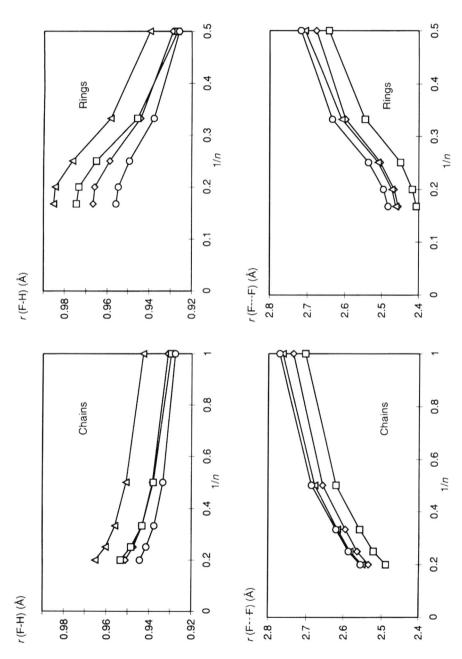

Figure 8.9 Trends in optimized $r(F—H)$ and $r(F\cdots F)$ distances in chain-like and cyclic hydrogen fluoride clusters plotted versus $1/n$. Triangles: B3LYP; squares: B3P86; diamonds: BLYP; circles: MP2

Table 8.3 r(H—F) distances in chain-like and cyclic, C_{nh}-symmetric hydrogen fluoride clusters (HF)$_n$ from $n = 1$ to 6 as obtained at various theoretical levels with the 6–311++G(2d,p) basis. (All values in Å)

n	B3LYP	B3P86	BLYP	MP2
		Chains		
1	0.9241	0.9215	0.9349	0.9221
2	0.9314	0.9299	0.9430	0.9277
	0.9274	0.9249	0.9382	0.9253
3	0.9341	0.9330	0.9460	0.9299
	0.9383	0.9378	0.9506	0.9336
	0.9283	0.9259	0.9391	0.9263
4	0.9358	0.9350	0.9480	0.9313
	0.9431	0.9433	0.9559	0.9377
	0.9414	0.9414	0.9540	0.9360
	0.9290	0.9266	0.9398	0.9267
5	0.9366	0.9359	0.9488	0.9321
	0.9460	0.9469	0.9591	0.9402
	0.9473	0.9484	0.9605	0.9413
	0.9429	0.9432	0.9554	0.9375
	0.9292	0.9269	0.9399	0.9270
6	0.9372	0.9368	0.9496	0.9325
	0.9476	0.9492	0.9612	0.9415
	0.9510	0.9535	0.9652	0.9444
	0.9498	0.9519	0.9638	0.9434
	0.9440	0.9446	0.9569	0.9383
	0.9295	0.9272	0.9402	0.9272
		Rings		
2	0.9288	0.9267	0.9399	0.9261
3	0.9441	0.9457	0.9585	0.9377
4	0.9588	0.9652	0.9764	0.9493
5	0.9657	0.9732	0.9844	0.9548
6	0.9667	0.9744	0.9851	0.9556

infrared mode of each cluster versus $1/n$ is shown in Figure 8.10. As with the previously discussed case of hydrogen cyanide clusters, the strongest infrared active mode in the chains is always the lowest within the group of F—H stretchings and corresponds again to the in-phase mode for all chain lengths and will finally converge to the optically allowed in-phase $k = 0$ phonon. In the rings, the corresponding in-phase mode is optically forbidden, the strongly infrared active mode is always the second-lowest and is always doubly degenerate. Because of the too large elongations of r(F—H) as computed with the DF methods, the corresponding shifts of the ν(F—H) vibrational frequencies in the dimer, and much more so in the longer chains and larger rings are considerably too large. This trend is already visible in case of the cyclic trimer. The experimental shift of the

Table 8.4 $r(F\cdots F)$ distances in chain-like and cyclic, C_{nh}-symmetric hydrogen fluoride clusters $(HF)_n$ from $n = 2$ to 6 as obtained at various theoretical levels with the 6–311++G(2d,p) basis. (All values in Å)

n	B3LYP	B3P86	BLYP	MP2
		Chains		
2	2.7350	2.7030	2.7612	2.7704
3	2.6871	2.6515	2.7126	2.7157
	2.6567	2.6203	2.6785	2.6858
4	2.6629	2.6261	2.6844	2.6920
	2.5964	2.5588	2.6170	2.6209
	2.6265	2.5883	2.6472	2.6579
5	2.6525	2.6169	2.6736	2.6795
	2.5690	2.5274	2.5888	2.5922
	2.5635	2.5214	2.5833	2.5866
	2.6127	2.5735	2.6335	2.6408
6	2.6454	2.6064	2.6654	2.6736
	2.5555	2.5109	2.5733	2.5782
	2.5340	2.4892	2.5506	2.5552
	2.5430	2.5003	2.5624	2.5681
	2.6039	2.5627	2.6241	2.6333
		Rings		
2	2.6757	2.6419	2.7062	2.7156
3	2.5967	2.5447	2.6071	2.6336
4	2.5050	2.4497	2.5120	2.5374
5	2.4664	2.4164	2.4729	2.4957
6	2.4558	2.4062	2.4629	2.4844

vibrationally allowed, doubly degenerate F—H stretching mode with respect to the monomer is -249 cm^{-1} (Michael and Lisy, 1986), in excellent agreement with the MP2 shift of -253 cm^{-1}. The shift of the remaining totally symmetric F—H stretching mode could be derived indirectly by the same authors with the aid of partially deuterated clusters and was found to be -354 cm^{-1}, again extremely close to the -358 cm^{-1} MP2 shift. The computed MP2 shifts for the strongly infrared active F—H stretching mode of cyclic tetramer, pentamer and hexamer of -504, -655, and -719 cm^{-1}, respectively, are also in excellent agreement with the experimental data by Quack et al. (1993a), who reported corresponding shifts of -514, -655, and -716 cm^{-1}. The corresponding DF predictions for the frequency shifts from the monomer to the cyclic trimer, tetramer, pentamer and hexamer are -338, -646, -818, and -881 cm^{-1} (B3LYP), -416, -804, -999, and -1074 cm^{-1} (B3P86), and -389, -731, -912, and -970 cm^{-1} (BLYP), respectively, are all significantly larger, the error being smallest for the B3LYP variant.

Experimentally, the in-phase $k = 0$ F—H stretching frequency in the hydrogen fluoride crystal (Kittelberger and Honig, 1967) is shifted by -896 cm^{-1} relative to the monomer

Table 8.5 Harmonic ν(H—F) vibrational frequencies in chain-like and cyclic, C_{nh}-symmetric hydrogen fluoride clusters (HF)$_n$ from $n = 1$ to 6 as obtained at various theoretical levels with the 6−311++G(2d,p) basis. (All values in cm^{-1})

n		B3LYP	B3P86	BLYP	MP2
			Chains		
1		4093	4136	3937	4134
2		4050	4090	3895	4093
		3940	3956	3769	4022
3		4035	4075	3880	4078
		3896	3904	3721	3987
		3799	3792	3622	3902
4		4026	4065	3872	4072
		3860	3859	3681	3956
		3779	3766	3602	3887
		3672	3646	3491	3795
5		4021	4060	3867	4067
		3838	3835	3661	3938
		3744	3725	3566	3857
		3677	3649	3495	3798
		3563	3517	3375	3700
6		4019	4056	3866	4065
		3827	3823	3647	3929
		3721	3698	3541	3840
		3653	3616	3466	3777
		3592	3545	3403	3725
		3469	3400	3271	3618
			Rings		
2		4028	4061	3866	4081
		4006	4036	3843	4061
3	d^a	3755	3720	3548	3881
		3627	3561	3401	3776
4		3534	3440	3304	3703
	d	3447	3332	3206	3630
		3218	3034	2939	3447
5	d	3418	3313	3185	3600
	d	3275	3137	3025	3479
		3019	2808	2726	3276
6		3423	3320	3200	3601
	d	3376	3265	3150	3561
	d	3212	3062	2967	3425
		2979	2761	2696	3241

[a] Doubly degenerate.

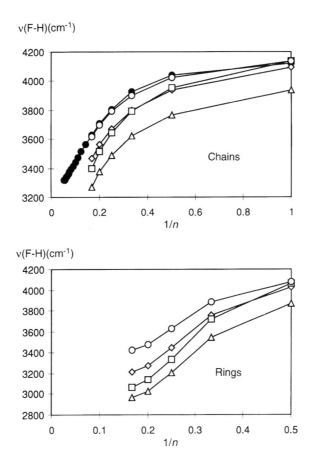

Figure 8.10 Calculated $\nu(F-H)$ vibrational frequencies in chain-like and cyclic hydrogen fluoride clusters plotted versus $1/n$. Triangles: B3LYP; squares: B3P86; diamonds: BLYP; circles: MP2; full circles: SCF/4−31G results (Karpfen and Yanovitskii, 1994a).

frequency, and hence, about -180 cm^{-1} relative to the (HF)$_6$ case. As already mentioned, extrapolation to the infinite chain limit is more difficult in the case of hydrogen fluoride, partly because the optimal structure is not the fully linear arrangement, partly also because of the greater importance of non-additive contributions to the interaction energy, which cause also a slower decay of the two edge effects at both sides of the chains. Hence, chain-like clusters approach the bulk limit considerably slower. Nevertheless, because of the close structural similarity of the F—H···F hydrogen bond in the crystal to that in the cyclic (HF)$_6$ cluster, the infrared inactive in-phase mode of the cyclic (HF)$_6$ cluster should also be close to the in-phase $k = 0$ phonon and could, therefore, give an indication of the theoretical crystal limit. For the latter, we obtain shifts relative to the monomer of -1114, -1375, -1241, and -893 cm^{-1} at the B3LYP, B3P86, BLYP, and MP2 levels, respectively. Again, as with the cyclic (HF)$_n$ clusters discussed before, given the limitations in the basis set applied, the limitations in the accuracy of MP2, and the fact that $r(F\cdots F)$ is still larger in $(HF)_6$ by about 0.03 Å than in solid HF, the MP2 value is somewhat fortuitously close to the experimental monomer to crystal shift of -896 cm^{-1}.

As shown in Table 8.5 and in Figure 8.10, the lowest $\nu(F-H)$ frequencies of chain-like clusters converge slowly, and actually larger clusters should be treated. Since that becomes, however, quickly prohibitive at the computational levels chosen, simpler SCF calculations with smaller basis sets can be a helpful guide in such extrapolation attempts. For curiosity, the corresponding SCF/4−31G frequencies are included in Figure 8.10 up to cluster sizes of $n = 19$ (Karpfen and Yanovitskii, 1994a). The almost complete error cancellation between SCF/4−31G and MP2/6−311++G(2d,p) frequencies which practically lie on top of each other is quite fortuitous.

The trends in the computed infrared intensities of the $\nu(F-H)$ modes are shown in Figure 8.11 and in Table 8.6. The methodical trends upon chain length elongation or upon

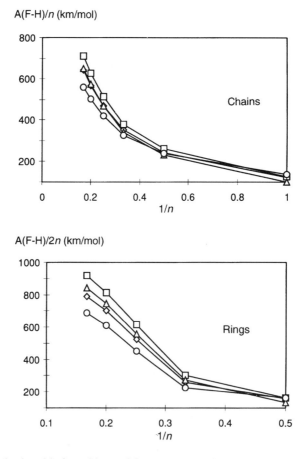

Figure 8.11 Calculated infrared intensities A(F—H) of the $\nu(F-H)$ vibrational frequencies in chain-like and cyclic, C_{nh}-symmetric hydrogen fluoride clusters plotted versus $1/n$. Triangles: B3LYP; squares: B3P86; diamonds: BLYP; circles: MP2

Table 8.6 Infrared intensities A(H—F) of ν(H—F) vibrational frequencies in chain-like and cyclic, C_{nh}-symmetric hydrogen fluoride clusters (HF)$_n$ from $n = 1$ to 6 as obtained at various theoretical levels with the 6−311++G(2d,p) basis. (All values in km/mol)

n	B3LYP	B3P86	BLYP	MP2
		Chains		
1	122	127	100	137
2	147	155	126	141
	488	521	464	475
3	161	169	139	166
	337	382	338	286
	1052	1138	1028	977
4	169	178	146	172
	502	557	503	444
	228	249	218	195
	1874	2057	1878	1679
5	173	182	150	176
	516	573	519	460
	584	654	594	524
	101	112	102	93
	2840	3138	2877	2506
6	176	186	153	178
	537	602	546	479
	561	640	587	502
	684	777	715	627
	18	24	26	20
	3848	4264	3898	3358
		Rings		
2	312	326	271	324
3	1566	1816	1642	1341
4	4178	4910	4452	3612
5	6998	8128	7432	6118
6	9474	11002	10074	8274

increasing the ring size parallel those already observed for the frequencies, with the MP2 and B3P86 intensities as the limiting cases. As with the structural parameters and the vibrational frequencies, the computed infrared intensities of the chain-like clusters allow, in combination with an analysis of the normal modes, a nice separation of the convergence to edge and bulk properties, respectively. An indication for this behaviour can be found in Table 8.6 by inspecting the intensities of the two highest-lying modes up to the chain-like hexamer, corresponding to localized F—H stretches at non-hydrogen-bonded and hydrogen-bonded edges, respectively.

Finally, the proton NMR chemical shifts $\sigma^1(H)$, as computed within the framework of the SCF-GIAO approach at the respective MP2-optimized geometries, are compiled in Table 8.7. As with the structural and vibrational spectroscopic properties, the differences in the chemical shifts from the monomer to cyclic $(HF)_6$ are significantly larger than in the hydrogen cyanide clusters.

The MP2 structural and vibrational spectroscopic data and the SCF proton chemcial shifts for the neutral hydrogen fluoride clusters are summarized in Table 8.8, which can be viewed as an analogue to Table 8.1, where the corresponding trends as evaluated in the series of hydrogen cyanide clusters were shown. For all the five properties discussed, the cooperative effects observed in the hydrogen fluoride system exceed by far those in the hydrogen cyanide system.

Table 8.7 Proton NMR chemical shifts, $\sigma(^1H)$, in chain-like and cyclic, C_{nh}-symmetric hydrogen fluoride clusters $(HF)_n$ from $n = 1$ to 6 as obtained at the GIAO-SCF level at MP2/6–311++G(2d,p) optimized geometries. (All values in ppm)

n	Chains	Rings
1	28.9	
2	26.9 27.9	28.1
3	26.3 25.3 27.6	25.3
4	26.0 24.4 24.7 27.4 27.4	22.8
5	25.8 23.9 23.6 24.4 27.4	21.6
6	25.7 23.7 23.1 23.2 24.2 27.3	21.4

Table 8.8 Theoretical characterization of the F—H···F hydrogen bond in chain-like and cyclic, hydrogen fluoride clusters, (HF)$_n$, as calculated at the MP2/6-311++G(2d,p) level

n	r(F—H) (Å)	r(F···F) (Å)	ν(F—H) (cm^{-1})	A(F—H) (km/mol)	σ(^{1}H) (ppm)
			Linear		
1	0.9221	–	4134	137	28.9
2	0.9253	2.7704	4022	475	26.9
3	0.9336	2.6858	3902	977	25.3
4	0.9377	2.6209	3795	1679	24.4
5	0.9413	2.5866	3700	2506	23.6
6	0.9444	2.5552	3618	3358	23.1
			Cyclic		
2	0.9261	2.7156	4081	324	28.1
3	0.9377	2.6336	3881	1341	25.3
4	0.9493	2.5374	3630	3612	22.8
5	0.9548	2.4957	3479	6118	21.6
6	0.9556	2.4844	3425	8274	21.4

8.4 THE TRANSITION TO STRONG IONIC HYDROGEN BONDS: PROTONATED AND DEPROTONATED CHAIN-LIKE CLUSTERS OF HYDROGEN FLUORIDE

As model systems to describe the gradual transition from medium strong hydrogen bonds to the case of very strong, ionic hydrogen bonds, the series of protonated ([(HF)$_n$H]$^+$) and deprotonated ([(HF)$_{n-1}$F]$^-$) defects on chain-like hydrogen fluoride oligomers have been chosen. The shortest hydrogen-bonded species in these series, the protonated dimer, (H$_3$F$_2$)$^+$, and the deprotonated dimer, (FHF)$^-$, are characterized by very short and extremely strong hydrogen bonds. Upon dressing these species by a successively increasing number of hydrogen fluoride molecules at the edges, the properties of charged defects on otherwise unperturbed hydrogen-bonded chains can be studied and the questions, how far such defects would actually extend, and how the local structural properties of the hydrogen bond at the defect site would decay to that of the hydrogen bond embedded in the neutral chain can be investigated in detail. The model character of these investigations is stressed; the global minima for these charged systems are most probably not extended chains. As with the neutral, chain-like oligomers, we ignore the fact that most of these charged clusters are actually first- or second-order saddle points. For recent detailed *ab initio* SCF investigation on these systems see Karpfen and Yanovitskii (1994a,b). Similar model systems, e.g. protonated and deprotonated defects on H$_2$O chains, have already been used for investigations of various quantities relevant for proton transport in extended systems (Scheiner and Nagle, 1983; Nagle, 1992).

The structures of the series of protonated and deprotonated chain-like hydrogen fluoride oligomers are illustrated in Figure 8.12 by sketching the hexamer geometries. For simplicity, only the results for the even-membered species, i.e. protonated and deprotonated dimers, tetramers and hexamers, are discussed here. The even-numbered members

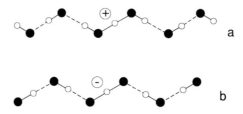

Figure 8.12 The structures of protonated and deprotonated hydrogen fluoride hexamers

of these series have in common that they all have C_{2h} symmetry and a symmetric hydrogen bond in the centre. The odd-membered species are all C_{2v}-symmetric, with strong but asymmetric hydrogen bonds at the centre, and can be viewed best as F^- and HFH^+ ions dressed symmetrically by a successively increasing number of hydrogen fluoride molecules.

In Tables 8.9 and 8.10 the computed $r(F—H)$ and $r(F \cdots F)$ distances are reported, again as obtained with the $6-311++G(2d,p)$ basis set within the framework of the three DF variants and of the MP2 approximation. The qualitative picture that one has to expect

Table 8.9 Calculated $r(H—F)$ distances in protonated, $([(HF)_n H]^+)$, and deprotonated, $([(HF)_{n-1} F]^-)$, chain-like hydrogen fluoride clusters for $n = 2$, 4, and 6 as obtained at various theoretical levels with the $6-311++G(2d,p)$ basis. (All values in Å)

n	B3LYP	B3P86	BLYP	MP2
		Protonated		
2	1.1526	1.1467	1.1683	1.1454
	0.9498	0.9460	0.9597	0.9470
4	1.1443	1.1387	1.1594	1.1379
	0.9995	1.0013	1.0156	0.9914
	0.9371	0.9347	0.9479	0.9352
6	1.1418	1.1360	1.1567	1.1357
	1.0246	1.0289	1.0430	1.0146
	0.9686	0.9710	0.9839	0.9612
	0.9336	0.9314	0.9445	0.9314
		Deprotonated		
2	1.1480	1.1427	1.1645	1.1433
4	1.1402	1.1349	1.1557	1.1355
	0.9781	0.9793	0.9937	0.9725
6	1.1386	1.1332	1.1539	1.1335
	1.0120	1.0162	1.0302	1.0036
	0.9541	0.9554	0.9686	0.9485

Table 8.10 Calculated $r(F \cdots F)$ distances in protonated, $([(HF)_n H]^+)$, and deprotonated, $([(HF)_{n-1} F]^-)$, chain-like hydrogen fluoride clusters for $n = 2$, 4, and 6 as obtained at various theoretical levels with the 6–311++G$(2d,p)$ basis. (All values in Å)

n	B3LYP	B3P86	BLYP	MP2
		Protonated		
2	2.3052	2.2934	2.3366	2.2908
4	2.2886	2.2774	2.3188	2.2758
	2.3931	2.3704	2.4177	2.3902
6	2.2836	2.2720	2.3134	2.2714
	2.3397	2.3186	2.3636	2.3353
	2.4715	2.4383	2.4903	2.4833
		Deprotonated		
2	2.2960	2.2854	2.3290	2.2866
4	2.2804	2.2698	2.3114	2.2710
	2.4223	2.3995	2.4475	2.4206
6	2.2772	2.2664	2.3078	2.2670
	2.3462	2.3242	2.3695	2.3424
	2.5126	2.4792	2.5319	2.5206

upon increasing the chain length of the charged oligomers, is a structural convergence to the defect properties at the centre of the clusters and, provided the clusters are sufficiently large, convergence to an intermediate region with the properties of the neutral bulk, followed by convergence to the two kind of effects, the non-hydrogen-bonded edge in the case of the protonated clusters and to the hydrogen-bonded edge in the case of the deprotonated clusters. The above-mentioned SCF investigations (Karpfen and Yanovitskii, 1994a, b) showed that the extension of both types of charged defects, protonated or deprotonated, for the even- and also for the odd-membered species, is much larger than the hexamers discussed here. Nevertheless, even with this very restricted set of data, the major trends are clearly visible. From the structural details given in Tables 8.9 and 8.10, only that of the deprotonated dimer, the FHF$^-$ anion, can be compared to previous calculations. For a far more extended comparison of DF results with that of other calculations including electron correlation at various levels of sophistication see the recent work of Latajka *et al.* (1995). With the exception of the BLYP method, which produces somewhat larger bond distances, the remaining two DF variants are in close agreement with the MP2 structure. Upon increasing the chain length of the clusters, the length of the central symmetric hydrogen bond remains almost unchanged. However, a small tendency to further shrinking is still visible.

The F—H stretching frequencies and the corresponding infrared intensities reported in Table 8.11 clearly reflect this behaviour. The two lowest-lying F—H stretching modes of each cluster lie in the vicinity of that of the FHF$^-$ anion. However, the intensity of the asymmetric combination increases substantially in the larger clusters. The remaining

Table 8.11 Calculated ν(H—F) vibrational frequencies and infrared intensities in protonated, $([(HF)_nH]^+)$, and deprotonated, $([(HF)_{n-1}F]^-)$, chain-like hydrogen fluoride clusters for $n = 2, 4,$ and 6 as obtained at various theoretical levels with the $6-311++G(2d,p)$ basis. (Frequencies in cm^{-1}, intensities in km/mol (given in parentheses)

n	B3LYP		B3P86		BLYP		MP2	
				Protonated				
2	3729	(0)	3782	(0)	3600	(0)	3781	(0)
	3719	(782)	3772	(788)	3591	(704)	3770	(823)
	1694	(2399)	1761	(2750)	1661	(2475)	1736	(2292)
	576	(0)	586	(0)	546	(0)	594	(0)
4	3910	(0)	3946	(0)	3761	(0)	3946	(0)
	3909	(550)	3945	(564)	3760	(489)	3946	(571)
	2818	(3211)	2801	(3160)	2676	(3050)	2928	(3346)
	2815	(0)	2788	(0)	2667	(0)	2923	(0)
	1402	(3275)	1650	(3533)	1396	(3084)	1416	(3486)
	625	(0)	637	(0)	597	(0)	639	(0)
6	3961	(0)	3995	(0)	3808	(0)	4002	(0)
	3961	(463)	3995	(482)	3808	(411)	4002	(476)
	3278	(0)	3233	(0)	3089	(0)	3412	(0)
	3276	(2526)	3232	(2659)	3088	(2502)	3410	(2438)
	2514	(3375)	2508	(3089)	2394	(3060)	2610	(3640)
	2452	(0)	2418	(0)	2317	(0)	2565	(0)
	1337	(6919)	1405	(7362)	1317	(7468)	1369	(6570)
	647	(0)	665	(0)	620	(0)	659	(0)
				Deprotonated				
2	1323	(3382)	1442	(3378)	1328	(3180)	1338	(3708)
	619	(0)	628	(0)	586	(0)	636	(0)
4	3066	(0)	3046	(0)	2894	(0)	3157	(0)
	3053	(2966)	3037	(3016)	2884	(2916)	3144	(2968)
	1443	(4332)	1544	(4944)	1426	(4614)	1467	(3973)
	657	(0)	667	(0)	622	(0)	672	(0)
6	3488	(0)	3455	(0)	3289	(0)	3600	(0)
	3486	(2049)	3453	(2193)	3287	(2070)	3599	(1937)
	2582	(3614)	2564	(3416)	2442	(3401)	2682	(3759)
	2555	(0)	2515	(0)	2403	(0)	2661	(0)
	1342	(5523)	1461	(4644)	1323	(5549)	1373	(5585)
	669	(0)	682	(0)	635	(0)	686	(0)

F—H stretching vibrations tend to cover the entire region between the chain-like clusters, the molecular crystal and in particular also the ionic crystals, such as KHF$_2$ (Ibers, 1964), the latter containing the FHF$^-$ anion as a structural element.

The proton NMR chemical shifts, $\sigma(^1H)$, shown in Table 8.12 display the expected strong sensitivity to the structural details and to the charged nature of the hydrogen bonds. The overall change of $\sigma(^1H)$ between the HF monomer and the FHF$^-$ anion amounts to

Table 8.12 Proton NMR chemical shifts, $\sigma(^1H)$, in models for protonated, $([(HF)_nH]^+)$, and deprotonated, $([(HF)_{n-1}F]^-)$, chain-like hydrogen fluoride clusters for $n = 2$, 4, and 6 as obtained at the SCF-GIAO level at MP2/6–311++G($2d,p$) optimized geometries. (All values in ppm)

n	Protonated	deprotonated
2	11.2	13.0
	22.9	
4	12.9	14.2
	18.9	21.7
	25.8	
6	11.7	12.5
	15.1	16.5
	20.3	22.9
	26.1	

about 16 ppm at the GIAO-SCF level. The very smooth correlation between $\sigma(^1H)$ and the structural parameters $r(F-H)$ and $r(F\cdots F)$ over the entire series of neutral chains, neutral rings and charged chains is depicted in Figure 8.13(a) and 8.13(b). Finally, in Figure 8.13(c) a purely theoretical correlation between $r(F-H)$ and $r(F\cdots F)$ of a given F—H...F hydrogen bond is displayed, again as obtained from the MP2-optimized geometries of all the neutral and charged fluoride cluster types. The latter curve is actually very similar to that reported by Panich (1995), derived from a comprehensive analysis of available NMR and neutron diffraction data on a large number of solids containing different F—H$\cdots$F hydrogen bonds.

8.5 CONCLUSIONS AND OUTLOOK

In this chapter the influence of cooperative hydrogen bonding on structure, vibrational spectra, and proton NMR chemical shifts, as obtained from *ab initio* investigations including electron correlation at the MP2 level and with three recent DF variants, has been systematically evaluated. Three different scenarios have been developed: the hydrogen cyanide system, a case with comparatively weak cooperative effects, the hydrogen fluoride system with large cooperative effects, and, finally, the charged defects on hydrogen fluoride chains, with the transition to ionic hydrogen bonds. The five properties selected, $r(X-H)$, $r(X\cdots X)$, $\nu(X-H)$, A(X—H), and $\sigma(^1H)$, all vary quite smoothly with the strength of the hydrogen bonds. As far as the still scarce experimental data on vapour phase clusters and on the molecular solids admit a comparison with the theoretical data, the various methods applied perform quite satisfactorily for the interaction energies and for the structures. The main systematic weakness of the DF methods concerns the too

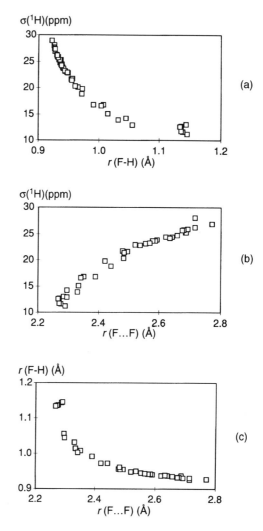

Figure 8.13 Plots of (a) $\sigma(^1H)$ versus $r(F\text{—}H)$; (b) $\sigma(^1H)$ versus $r(F\cdots F)$, and (c) $r(F\text{—}H)$ versus $r(F\cdots F)$ in neutral chain-like and cyclic, C_{nh}-symmetric $(HF)_n$, and in protonated, $([(HF)_n H]^+)$, and deprotonated, $([(HF)_{n-1} F]^-)$, clusters

large shifts of the $\nu(X\text{—}H)$ vibrational frequencies in the region from medium sized vapour phase clusters to the molecular solids. These shifts, as it turned out, can be overestimated by nearly 50 per cent by the BLYP and B3P86 variants in the hydrogen fluoride system. In agreement with previous investigations on single hydrogen bond systems, the B3LYP variant appears the most promising DF variant so far.

In future work, the calculations presented in this chapter should be extended to somewhat larger cluster sizes and, more importantly, they should also be augmented by quantum chemical calculations on idealized periodic systems at identical methodical levels, in order to provide limiting values for the various expectation levels and thus facilitate extrapolation to intermediate cluster sizes.

ACKNOWLEDGMENTS

The calculations were performed on the RISC 6000/550-Cluster and on the Cluster of Digital Alpha Servers (2100 4/275) of the computer centre of the University of Vienna. The author is grateful for ample supply of computer time on these installations.

REFERENCES

Albinati, A., Rouse, K. D. and Thomas, M. W. (1978a) *Acta Cryst.*, **B34**, 2184–2187.
Albinati, A., Rouse, K. D. and Thomas, M. W. (1978b) *Acta Cryst.*, **B34**, 2188–2190.
Andrews, L. and Johnson, G. L. (1983) *Chem. Phys. Lett.*, **96**, 133–138.
Andrews, L. and Johnson, G. L. (1984) *J. Phys. Chem.*, **88**, 425–432.
Andrews, L. (1984) *J. Phys. Chem.*, **88**, 2940–2949.
Andrews, L., Bondybey, V. E. and English, J. H. (1984) *J. Chem. Phys.*, **81**, 3452–3457.
Andrews, L., Davis, S. R. and Hunt, R. D. (1992) *Mol. Phys.*, **77**, 993–1003.
Anex, D. S., Davidson, E. R., Douketis, C. and Ewing, G. E. (1988) *J. Phys. Chem.*, **92**, 2913–2925.
Atoji, M. and Lipscomb, W. N. (1954) *Acta Cryst.*, **7**, 173–175.
Becke, A. D. (1988) *Phys. Rev.*, **A38**, 3098–3100.
Becke, A. D. (1993) *J. Chem. Phys.*, **98**, 5648–5652.
Beichert, P., Pfeiler, D. and Knözinger, E. (1995) *Ber. Bunsenges. Phys. Chem.*, **99**, 1469–1478.
Bendtsen, J. and Edwards, H. G. M. (1974) *J. Raman Spectrosc.*, **2**, 407–421.
Beyer, A. and Karpfen, A. (1982) *Chem. Phys.*, **64**, 343–357.
Boys, S. F. and Bernardi, F. (1970) *Mol. Phys.*, **19**, 553–566.
Brown, R. D., Godfrey, P. D. and Winkler, D. A. (1981) *J. Mol. Spectrosc.*, **89**, 352–355.
Burcl *et al.*, (1995) *J. Chem. Phys.*, **103**, 1498–1507.
Buxton, I. W., Campbell, E. J. and Flygare, W. H. (1981) *Chem. Phys.*, **56**, 399–406.
Campbell, E. J. and Kukolich, S. G. (1983) *Chem. Phys.*, **76**, 225–229.
Castleman, A. W. Jr and Hobza, P. (eds) (1994) *van der Waals Molecules II*, the entire volume of *Chem. Rev.*, **94**, No. 7.
Chalasinski, G., Cybulski, S. M., Szczesniak, M. M. and Scheiner, S. (1989) *J. Chem. Phys.*, **91**, 7048–7056.
Clark, T., Chandrasekhar, J., Spitznagel G. W. and von Ragué Schleyer, P. (1983) *J. Comp. Chem.*, **4**, 294–301.
Collins, C. L., Morihashi, K., Yamaguchi, Y. and Schaefer, H. F. (1995) *J. Chem. Phys.*, **103**, 6051–6056.
De Almeida, W. B., Craw, J. S., and Hinchliffe, A. (1989) *J. Mol. Struct. (Theochem.)*, **184**, 381–389.
De Almeida, W. B. (1991) *Can. J. Chem.*, **69**, 2044–2052.
Del Bene, J. E. and Pople, J. A. (1971) *J. Chem. Phys.*, **55**, 2296–2299.
Del Bene, J. E. and Shavitt, I. (1991) *J. Mol. Struct. (Theochem.)*, **234**, 499–508.
Del Bene, J. E., Person, W. B. and Szczepaniak, K. (1995) *J. Phys. Chem.*, **99**, 10705–10707.
Ditchfield, R. (1974) *Mol. Phys.*, **27**, 789–807.
Dulmage, W. J. and Lipscomb, W. N. (1951) *Acta Cryst.*, **4**, 330–334.
Dyke, T. R., Howard, B. J. and Klemperer, W. (1972) *J. Chem. Phys.*, **56**, 2442–2454.
Dykstra, C. E. (1987) *Chem. Phys. Lett.*, **141**, 159–162.
Dykstra, C. E. (1990) *J. Phys. Chem.*, **94**, 180–185.
Dykstra, C. E. (1996) *J. Mol. Struct. (Theochem.)*, **362**, 1–6.

Fillery-Travis, A. J., Legon, A. C., Willoughby, L. C. and Buckingham, A. D. (1983) *Chem. Phys. Lett.*, **102**, 126–131.
Friedrichs, H. B. and Krause, P. F. (1973) *J. Chem. Phys.*, **59**, 4942–4948.
Frisch, M. J., Pople, J. A. and Binkley, J. S. (1984) *J. Chem. Phys.*, **80**, 3265–3269.
Frisch, M. J., Trucks, G. W., Schlegel, H. B., Gill, P. M. W., Johnson, B. G., Robb, M. A., Cheeseman, J. R., Keith, T. A., Petersson, G. A., Montgomery, J. A., Raghavachari, K., Al-Laham, M. A., Zakrzewski, V. G., Ortiz, J. V., Foresman, J. B., Cioslowski, J., Stefanov, B. B., Nanayakkara, A., Challacombe, M., Peng, C. Y., Ayala, P. Y., Chen, W., Wong, M. W., Andres, J. L., Replogle, E. S., Gomperts, R., Martin, R. L., Fox, D. J., Binkley, J. S., Defrees, D. J., Baker, J., Stewart, J. J. P., Head-Gordon, M., Gonzalez, C. and Pople, J. A. (1995) *GAUSSIAN 94*, Revision A.1, Gaussian, Inc., Pittsburgh.
Gaw, J. F., Yamaguchi, Y., Vincent, M. A., and Schaefer, H. F. III (1984) *J. Amer. Chem. Soc.*, **106**, 3133–3138.
Georgiou, K., Legon, A. C., Millen, D. J. and Mjöberg, P. J. (1985) *Proc. R. Soc. Lond.*, **A399**, 377–390.
Habuda, S. B. and Gagarinsky, Yu. V. (1971) *Acta Cryst.*, **B27**, 1677–1678.
Hadzi, D. and Thompson, H. W. (eds.) (1959) *Hydrogen Bonding*, Pergamon Press, London.
Hamilton, W. C. and Ibers, J. A. (1968) *Hydrogen Bonding in Solids*, Benjamin, New York.
Hehre, W. J., Ditchfield, R. and Pople, J. A. (1972) *J. Chem. Phys.*, **56**, 2257–2261.
Heidrich, D., Köhler, H. -J. and Volkmann, D. (1985) *Intern. J. Quantum Chem.*, **27**, 781–786.
Heidrich, D., Van Eikema Hommes, N. J. R. and Schleyer, P. v. R. (1993) *J. Comp. Chem.*, **14**, 1149–1163.
Hinchen, J. J. and Hobbs, R. H. (1979) *J. Opt. Soc. Am.*, **69**, 1546–1549.
Hobza, P. and Zahradnik, R. (1988) *Intermolecular Complexes, The role of van der Waals Systems in Physical Chemistry and in the Biodisciplines*, Studies in Physical and Theoretical Chemistry 52, Elsevier, Amsterdam.
Holtzberg, F., Post, B. and Fankuchen, I. (1953) *Acta Cryst.*, **6**, 127–130.
Hopkins, G. A., Maroncelli, M., Nibler, J. W. and Dyke, T. R. (1985) *Chem. Phys. Lett.*, **114**, 97–102.
Huisken, F., Kaloudis, M., Kulcke, A., and Voelkel, D. (1995a) *Infrared Phys. Technol.*, **36**, 171–178.
Huisken, F., Kaloudis, M., Kulcke, A., Laush, C., and Lisy, J. M. (1995b) *J. Chem. Phys.*, **103**, 5366–5377.
Huisken, F., Tarakanova, E. G., Vigasin, A. A. and Yukhnevich, G. V. (1995c) *Chem. Phys. Lett.*, **245**, 319–325.
Ibers, J. A. (1964) *J. Chem. Phys.*, **40**, 402–404.
Jansen, R. W., Bertoncini, R., Pinnick, D. A., Katz, A. I., Hanson, R. C., Sankey, O. F. and O'Keeffe, M. (1987) *Phys. Rev.*, **B35**, 9830–9846.
Janzen, J. and Bartell, L. S. (1969) *J. Chem. Phys.*, **50**, 3611–3618.
Johnson, M. W., Sandor, E. and Arzi, E. (1975) *Acta Cryst.*, **B31**, 1998–2003.
Jones, R. E. and Templeton, D. H. (1958) *Acta Cryst.*, **11**, 484–487.
Jönsson, P. -G. (1971) *Acta Cryst.*, **B27**, 893–898.
Jucks, K. W. and Miller, R. E. (1988a) *J. Chem. Phys.*, **88**, 2196–2204.
Jucks, K. W. and Miller, R. E. (1988b) *J. Chem. Phys.*, **88**, 6059–6067.
Karpfen, A., Ladik, J., Russegger, P., Schuster, P. and Suhai, S. (1974) *Theor. Chim. Acta (Berl.)*, **34**, 115–127.
Karpfen, A. Beyer, A., and Schuster, P. (1981) *Intern. J. Quantum Chem.*, **19**, 1113–1119.
Karpfen, A. (1983) *Chem. Phys.*, **79**, 211–218.
Karpfen, A. Beyer, A. and Schuster, P. (1983) *Chem. Phys. Lett.*, **102**, 289–291.
Karpfen, A. (1990) *Intern. J. Quantum Chem. Symp.*, **24**, 129–140.
Karpfen, A. and Yanovitskii, O. (1994a) *J. Mol. Struct. (Theochem.)*, **307**, 81–97.
Karpfen, A. and Yanovitskii, O. (1994b) *J. Mol. Struct. (Theochem.)*, **314**, 211–227.
Karpfen, A. (1996) *J. Phys. Chem.*, **100**, 13474–13486.
Kawaguchi, K. and Hirota, E. (1986) *J. Chem. Phys.*, **84**, 2953–2960.
Kerstel, E. R. Th., Lehmann, K. K., Gambogi, J. E., Yang, X. and Scoles, G. (1993) *J. Chem. Phys.*, **99**, 8559–8570.

Kertész, M., Koller, J. and Azman, A. (1975) *Chem. Phys. Lett.*, **36**, 576–579.
Kertész, M., Koller, J. and Azman, A. (1976) *Chem. Phys. Lett.*, **41**, 146–148.
King, B. F. and Weinhold, F. (1995) *J. Chem. Phys.*, **103**, 333–347.
King, B. F., Farrar, T. C. and Weinhold, F. (1995) *J. Chem. Phys.*, **103**, 348–352.
King, Ch. M. and Nixon, E. R. (1968) *J. Chem. Phys.*, **48**, 1685–1695.
Kittelberger, J. S. and Hornig, D. F. (1967) *J. Chem. Phys.*, **46**, 3099–3108.
Knözinger, E. and Wittenbeck, R. (1982) *Ber. Bunsenges. Phys. Chem.*, **86**, 742–746.
Knözinger, E., Kollhoff, H., Schrems, O. and Langel, W. (1986a) *J. Mol. Struct.*, **141**, 399–404.
Knözinger, E., Kollhoff, H. and Langel, W. (1986b) *J. Chem. Phys.*, **85**, 4881–4889.
Kofranek, M., Karpfen, A. and Lischka, H. (1987a) *Mol. Phys.*, **61**, 1519–1539.
Kofranek, M., Lischka, H. and Karpfen, A. (1987b) *Chem. Phys.*, **113**, 53–64.
Kofranek, M., Lischka, H. and Karpfen, A. (1988) *Chem. Phys.*, **121**, 137–153.
Kolenbrander, K. D., Dykstra, C. E. and Lisy, J. M. (1988) *J. Chem. Phys.*, **88**, 5995–6012.
Komornicki, A., Dixon, D. A. and Taylor, P. R. (1992) *J. Chem. Phys.*, **96**, 2920–2925.
Krishnan, R., Binkley, J. S., Seeger, R. and Pople, J. A. (1980) *J. Chem. Phys.*, **72**, 650–654.
Kurnig, I. J., Lischka, H. and Karpfen, A. (1990) *J. Chem. Phys.*, **92**, 2469–2477.
Langel, W., Kollhoff, H. and Knözinger, E. (1989) *J. Chem. Phys.*, **90**, 3430–3442.
Latajka, Z. and Scheiner, S. (1988) *Chem. Phys.*, **122**, 413–430.
Latajka, Z. and Bouteiller, Y. (1994) *J. Chem. Phys.*, **101**, 9793–9799.
Latajka, Z., Bouteiller, Y. and Scheiner, S. (1995) *Chem. Phys Lett.*, **234**, 159–164.
Lee, C., Yang, W., and Parr, R. G. (1988) *Phys. Rev.*, **B37**, 785–789.
Lee, S. A., Pinnick, D. A., Lindsay, S. M. and Hanson, R. C. (1986) *Phys. Rev.*, **B34**, 2799–2806.
Legon, A. C., Millen, D. J. and Mjöberg, P. J. (1977) *Chem. Phys. Lett.*, **47**, 589–591.
Liedl, K. R. Kroemer, R. T. and Rode, B. M. (1995) *Chem. Phys. Lett.*, **246**, 455–462.
Lisy, J. M., Tramer, A., Vernon, M. F. and Lee, Y. T. (1981) *J. Chem. Phys.*, **75**, 4733–4734.
Liu, S.-Y., Michael, D. W., Dkystra, C. E. and Lisy, J. M. (1986) *J. Chem. Phys.*, **84**, 5032–5036.
Luckhaus, D., Quack, M., Schmitt, U. and Suhm, M. A. (1995) *Ber. Bunsenges. Phys. Chem.*, **99**, 457–468.
Maroncelli, M., Hopkins, G. A., Nibler, J. W. and Dyke, T. R. (1985) *J. Chem. Phys.*, **83**, 2129–2146.
Martinez-Carrera, S. (1966) *Acta Cryst.*, **20**, 783–789.
Meyer, H., Kerstel, E. R. Th., Zhuang, D. and Scoles, G. (1989) *J. Chem. Phys.*, **90**, 4623–4625.
Michael, D. W. and Lisy, J. M. (1986) *J. Chem. Phys.*, **85**, 2528–2537.
Michl, J. and Zahradnik, R. (eds) (1988) *van der Waals Molecules*, the entire volume *Chem. Rev.*, **88**, No. 6.
Mielich, B., Savin, A., Stoll, H. and Preuss, H. (1989) *Chem. Phys. Lett.*, **157**, 200–206.
Møller, C. and Plesset, M. S. (1934) *Phys. Rev.*, **46**, 618–622.
Nagle, J. (1992) in *Proton Transfer in Hydrogen-Bonded Systems*, edited by Bountis, T., NATO-ASI Series **B291**, Plenum, New York, pp. 17–28.
Nahringbauer, I. (1970) *Acta Chem. Scand.*, **24**, 453–462.
Nahringbauer, I. (1978) *Acta Cryst.*, **B34**, 315–318.
Novoa, J. J. and Sosa, C. (1995) *J. Phys. Chem.*, **99**, 15837–15845.
Pacansky, J. (1977) *J. Phys. Chem.*, **81**, 2240–2243.
Panich, A. M. (1995) *Chem. Phys.*, **196**, 511–519.
Perdew, J. P. (1986) *Phys. Rev.*, **B33**, 8822–8824.
Peterson, K. A. and Dunning T. H., Jr (1995) *J. Chem. Phys.*, **102**, 2032–2041.
Pimentel, G. C. and McClellan, A. C. (1960) *The Hydrogen Bond*, Freeman, San Francisco.
Pine, A. S. and Lafferty, W. J. (1983) *J. Chem. Phys.*, **78**, 2154–2162.
Pine, A. S., Lafferty, W. J. and Howard, B. J. (1984) *J. Chem. Phys.*, **81**, 2939–2950.
Pinnick, D. A., Katz, A. I. and Hanson, R. C. (1989) *Phys. Rev.*, **B39**, 8677–8688.
Quack, M., Schmidt, U. and Suhm, M. A. (1993a) *Chem. Phys. Lett.*, **208**, 446–452.
Quack, M., Stohner, J. and Suhm, M. A. (1993b) *J. Mol. Struct.*, **294**, 33–36.
Quack, M. and Suhm, M. A. (1996) *Theoret. Chim. Acta*, **93**, 61–65.
Racine, S. C. and Davidson, E. R. (1993) *J. Phys. Chem.*, **1993**, 6367–6372.
Redington, R. L. (1981) *J. Chem. Phys.*, **75**, 4417–4421.
Redington, R. L. (1982a) *J. Phys. Chem.*, **86**, 552–560.

Redington, R. L. (1982b) *J. Phys. Chem.*, **86**, 561–563.
Ruoff, R. S., Emilsson, T., Chuang, C., Klots, T. D. and Gutowsky, H. S. (1988a) *Chem. Phys. Lett.*, **138**, 553–558.
Ruoff, R. S., Emilsson, T., Klots, T. D., Chuang, C. and Gutowsky, H. S. (1988b) *J. Chem. Phys.*, **89**, 138–148.
Sandor, E. and Farrow, R. F. C. (1967) *Nature*, **213**, 171–172.
Scheiner, S. and Nagle, J. (1983) *J. Phys. Chem.*, **87**, 4267–4272.
Schrems, O., Huth, M., Kollhoff, H., Wittenbeck, R. and Knözinger, E. (1987) *Ber. Bunsenges. Phys. Chem.*, **91**, 261–1266.
Schuster, P., Zundel, G. and Sandorfy, C. (eds.) (1976) *The Hydrogen Bond—Recent Developments in Theory and Experiment*, Vols 1–3, North-Holland, Amsterdam.
Schuster, P. (1978) in *Intermolecular Interactions: From Diatomics to Biopolymers*, edited by Pullman, B., Wiley, Chichester, pp. 363–432.
Scuseria, G. E. and Schaefer, H. F. III (1986) *Chem. Phys.*, **107**, 33–38.
Shallcross, V. F. and Carpenter, G. B. (1958) *Acta Cryst.*, **11**, 490–496.
Sim, F., St-Amant, A., Papai, I. and Salahub, D. R. (1992) *J. Am. Chem. Soc.*, **114**, 4391–4400.
Smith, D. F. (1958) *J. Chem. Phys.*, **28**, 1040–1056.
Springborg, M. (1987) *Phys. Rev. Lett.*, **59**, 2287–2290.
Springborg, M. (1988) *Phys. Rev.*, **B38**, 1483–1503.
Springborg, M. (1991) *Ber. Bunsenges. Phys. Chem.*, **95**, 1238–1255.
Springborg, M. (1995) *Chem. Phys.*, **195**, 143–155.
Stanton, R. V. and Merz, K. M. Jr (1994) *J. Chem. Phys.*, **101**, 6658–6665.
Suhai, S. (1994) *Intern. J. Quantum Chem.*, **52**, 395–412.
Suhm, M. A., Farrell, J. T. Jr, Ashworth, S. H. and Nesbitt, D. J. (1993) *J. Chem. Phys.*, **98**, 5985–5989.
Suhm, A. M. and Nesbitt, D. J. (1995) *Chem. Soc. Rev.*, **24**, 45–53.
Sun, H., Watts, R. O. and Buck, U. (1992) *J. Chem. Phys.*, **96**, 1810–1821.
Tauer, K. J. and Lipscomb, W. N. (1952) *Acta Cryst.*, **5**, 606–612.
Truhlar, D. G. (1990) in *Dynamics of Polyatomic Van der Waals Complexes*, edited by Halberstadt, N. And Janda, K. C., NATO-ASI Series **227**, Plenum, New York, pp. 159–185.
Vernon, M. F., Lisy, J. M., Krajnovich, D. J., Tramer, A., Kwok, H. -S., Shen, Y. R. and Lee, Y. T. (1982) *Faraday Discuss. Chem. Soc.*, **73**, 387–397.
Vinogradov, S. N. and Linell, R. H. (1971) *Hydrogen Bonding*, Van Nostrand-Reinhold, New York.
Walsh, B., Barnes, A. J., Suzuki, S. and Orville-Thomas, W. J. (1978) *J. Mol. Spectrosc.*, **72**, 44–56.
Wofford, B. A., Bevan, J. W., Olson, W. B. and Lafferty, W. J. (1986) *J. Chem. Phys.*, **85**, 105–108.
Wolinsky, K., Hilton, J. F. and Pulay, P. (1990) *J. Am. Chem. Soc.*, **112**, 8251–8260.
Xantheas, S. S. (1995) *J. Chem. Phys.*, **102**, 4505–4517.
Zhang, C., Freeman, D. L. and Doll, J. D. (1989) *J. Chem. Phys.*, **91**, 2489–2497.
Zhang, Q., Bell, R. and Truong, T. N. (1995) *J. Phys. Chem.*, **99**, 592–599.

9 Electrostatic Forces in Molecular Interactions

S. L. PRICE

University College London, UK

9.1 INTRODUCTION

The electrostatic forces between molecules are the easiest intermolecular forces to quantify from *ab initio* wavefunctions of the isolated molecules, as the charge distribution can be well represented by sets of multipoles at each atomic site. This approach combines the flexibility to describe the anisotropic shape of a charge distribution of the central multipole expansion with the spatial extent of an atomic point charge model. It has the advantage over both the central multipole expansion, which is generally used for small polyatomics, and the atomic charge model, which is the usual model for organic molecules, of being able to readily represent the electrostatic effects of non-spherical features such as lone pair and π electron density.

Fortunately, it appears that the electrostatic interactions (when accurately calculated) often dominate the orientation dependence of the total intermolecular potential, particularly for hydrogen bonding and π–π interactions between uncharged molecules. Thus, the combination of such an electrostatic model, plus a fairly crude model for the molecular shape, or other terms in the potential, provides a useful model for finding low-energy structures of assemblies of organic molecules, despite only being theoretically rigorous for one contribution. It has found wide application to van der Waals molecules, and in biological modelling, and has recently been extended for use in modelling molecular crystal structures. This model also finds wide use as a means of scanning complex potential energy surfaces to locate the low-energy regions which can be explored in greater detail by more rigorous methods. Distributed multipole electrostatic models will therefore provide a starting point for the development of more theoretically based force-fields for organic and biochemical interactions.

Molecular Interactions. Edited by S. Scheiner
© 1997 John Wiley & Sons Ltd

9.2 REPRESENTATIONS OF THE ELECTROSTATIC INTERACTION

The electrostatic interaction is strictly defined as the Coulombic interaction between the charge distributions of the molecules, when they have not been distorted by the interaction. Thus, the definition, corresponding to the first-order term in the perturbation theory expansion for intermolecular interactions (with the perturbation operator $H' = \Sigma e_i e_j /(4\pi\varepsilon_0 r_{ij})$) is:

$$U_{\text{estatic}} = \int_{\text{all space}} \frac{\rho^A(\mathbf{r}_1)\rho^B(\mathbf{r}_2)}{4\pi\varepsilon_0 |\mathbf{r}_1 - \mathbf{r}_2|} \, d^3\mathbf{r}_1 \, d^3\mathbf{r}_2 = \langle 0^A 0^B | H' | 0^A 0^B \rangle \quad (9.1)$$

where ρ^A is the charge distribution corresponding to the ground-state wavefunction 0^A of molecule A in isolation. Thus, by definition, the electrostatic energy can be evaluated from the *ab initio* wavefunctions of the isolated molecules, only requiring a good-quality ground state wavefunction. This is in stark contrast to the other contributions to the intermolecular potential: although the polarisation or induction energy (the increased classical Coulombic interaction arising from the mutual distortion of the charge distributions) and the dispersion energy can be related to the polarizabilities of the isolated molecules, these properties require higher-quality wavefunctions for the ground and excited states of the molecule. At short range, when there is significant overlap of the molecular charge distributions, then the intermolecular potential cannot be rigorously analytically described in terms of the properties of the isolated molecules, and *ab initio* supermolecule or perturbation theory calculations at a large number of relative orientations are required to map out this region of the potential energy surface.

Recent increases in computer power, and better computational methods such as the Direct SCF method (Almlöf *et al.*, 1982) have led to a significant increase in the size of molecule for which a wavefunction adequate for a worthwhile estimate of the electrostatic forces can be determined routinely. At the moment, self-consistent field calculations on organic molecules of up to fifty atoms can be considered routine. Larger calculations are possible; we have studied the electrostatic interactions of a derivative of the immunosuppressive drug cyclosporin ($C_{63}H_{113}N_{11}O_{12}$) using an SCF 3 − 21G wavefunction (Price *et al.*, 1989).

However, the electrostatic energy cannot be quickly enough evaluated by direct integration over the charge densities of the two molecules for use in molecular simulations, since an enormous number of relative orientations are sampled. Thus, it is necessary to parameterise a representation of the charge densities to give an analytical expression for the electrostatic energy as a function of the relative orientation of the two molecules. This has traditionally been done by using a central multipole expansion for small, symmetric molecules or an atomic point charge model for organic molecules. These approaches have different strengths and weaknesses, as discussed below. A combination of these two methods in the distributed multipole approach provides the more realistic model whose use is described in this chapter.

9.2.1 The Central Multipole Expansion of the Electrostatic Energy

The traditional method of quantifying the electrostatic forces between small, symmetrical polyatomics, at large separations, is in terms of the total multipole moments of the molecule. The dipole moments of polar molecules such as HCl can be measured experimentally by various methods, such as the Stark effect, and the quadrupole moments of centrosymmetric molecules by induced birefringence experiments, as reviewed in the compilation by Gray and Gubbins (1984). A full expansion of the electrostatic energy in terms of the multipole moments of the individual molecules, and their relative separation, can be derived from equation (9.1). For example, for two linear neutral molecules, without a centre of symmetry, the electrostatic energy can be expressed in terms of the molecular dipole (Q_{10}), quadrupole (Q_{20}), octopole (Q_{30}), etc., as

$$(4\pi\varepsilon_0)U_{\text{estatic}} = Q_{10}^1 Q_{10}^2 R^{-5}[(\mathbf{z}_1.\mathbf{z}_2)R^2 - 3(\mathbf{z}_1.\mathbf{R})(\mathbf{z}_2.\mathbf{R})]$$
$$+ Q_{20}^1 Q_{10}^2 R^{-7}\tfrac{3}{2}[-5(\mathbf{z}_1.\mathbf{R})^2(\mathbf{z}_2.\mathbf{R}) + 2R^2(\mathbf{z}_1.\mathbf{R})(\mathbf{z}_1.\mathbf{z}_2) + R^2(\mathbf{z}_2.\mathbf{R})]$$
$$+ Q_{10}^1 Q_{20}^2 R^{-7}\tfrac{3}{2}[+5(\mathbf{z}_1.\mathbf{R})(\mathbf{z}_2.\mathbf{R})^2 - 2R^2(\mathbf{z}_2.\mathbf{R})(\mathbf{z}_1.\mathbf{z}_2) - R^2(\mathbf{z}_1.\mathbf{R})]$$
$$+ Q_{30}^1 Q_{10}^2 R^{-9}\tfrac{1}{2}[-35(\mathbf{z}_1.\mathbf{R})^3(\mathbf{z}_2.\mathbf{R}) + 15R^2(\mathbf{z}_1.\mathbf{R})^2(\mathbf{z}_1.\mathbf{z}_2)$$
$$+ 15R^2(\mathbf{z}_1.\mathbf{R})(\mathbf{z}_2.\mathbf{R}) - 3R^4(\mathbf{z}_1.\mathbf{z}_2)] \qquad (9.2)$$
$$+ Q_{10}^1 Q_{30}^2 R^{-9}\tfrac{1}{2}[-35(\mathbf{z}_2.\mathbf{R})^3(\mathbf{z}_1.\mathbf{R}) + 15R^2(\mathbf{z}_2.\mathbf{R})^2(\mathbf{z}_1.\mathbf{z}_2)$$
$$+ 15R^2(\mathbf{z}_1.\mathbf{R})(\mathbf{z}_2.\mathbf{R}) - 3R^4(\mathbf{z}_1.\mathbf{z}_2)]$$
$$+ Q_{20}^1 Q_{20}^2 R^{-9}\tfrac{3}{4}[35(\mathbf{z}_1.\mathbf{R})^2(\mathbf{z}_2.\mathbf{R})^2 - 5R^2[(\mathbf{z}_1.\mathbf{R})^2 + (\mathbf{z}_2.\mathbf{R})^2]$$
$$- 20R^2(\mathbf{z}_1.\mathbf{R})(\mathbf{z}_2.\mathbf{R})(\mathbf{z}_1.\mathbf{z}_2) + 2R^4(\mathbf{z}_1.\mathbf{z}_2)^2 + R^4] + \ldots$$

where the molecular orientation is defined in terms of unit vectors $\mathbf{z}_1$ and $\mathbf{z}_2$ along the molecular axes and the intermolecular vector $\mathbf{R}$. This expression is useful at such long range that the dependence on the intermolecular separation R ensures that only the first term is significant, as only the first non-zero multipole moment can generally be determined experimentally. However, the advent of reasonable *ab initio* wavefunctions enables the series of multipole moments to be determined directly from the definition in terms of ground-state expectation values of the operators

$$\hat{Q}_{lk} = \rho r^l C_{lk}(\theta, \phi) \qquad (9.3)$$

where ρ is the charge density operator and $C_{lk}(\theta, \phi)$ is a modified spherical harmonic. (The relationship between this spherical tensor definition and the Cartesian definitions are straightforward (Price et al., 1984).) This soon established that the multipole moments often converged very slowly, even for quite spherical and symmetrical molecules. Diagrammatic representations of the charge distributions corresponding to a point charge, dipole, quadrupole and octopole for a linear molecule are shown in Figure 9.1, which suggest the crude approximations to the molecular shape that are implicit in using only the first, or even first few non-zero moments to represent the molecular charge distribution of molecules. This is quantified in Table 9.1, where the central multipole moments are given for F_2, benzene, uracil and a model peptide. The convergence is very poor for the less spherical molecules. It is clear that many terms in the irregular spherical harmonic expansion of the molecular shape of an organic molecule are needed before even the crudest representation of, for example, the NH_2 group charge distribution within a nucleic acid base is included. Thus, the first term of the central multipole expansion is only useful at intermolecular separations when the differences between a point dipole, and the actual shapes of HF or uracil is unimportant.

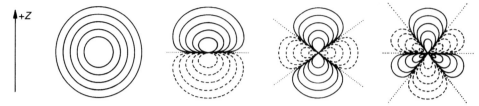

Figure 9.1 The charge distributions corresponding to pure multipolar distributions for a linear molecule. From left to right: a charge, dipole, quadrupole and octopole. The solid contours are positive, dotted zero and dashed negative (build-up of electron) charge density

Table 9.1 The convergence of the total multipole expansion of various molecular charge distributions

Total $Q(l, k)$	Fluorine	Benzene	Total $Q(l, k)$	Uracil	Glycine dipeptide
$Q(2,0)$	0.705	−6.355	$Q(1,1c)$	−1.528	1.276
$Q(4,0)$	13.329	170.042	$Q(1,1s)$	−1.407	0.113
$Q(6,0)$	69.087	−3936.000	$Q(2,0)$	1.725	0.258
$Q(6,6c)$		5819.000	$Q(2,2c)$	−5.079	10.194
$Q(8,0)$	252.695	95 250.000	$Q(2,2s)$	13.565	−18.916
$Q(8,6c)$		−135 100.000	$Q(3,1c)$	32.944	−46.381
$Q(10,0)$	776.042	−2 296 000.000	$Q(3,1s)$	39.521	−24.657
$Q(10,6c)$		3 352 000.000	$Q(3,3c)$	−102.331	−17.357
$Q(12,0)$	2143.190	N/A	$Q(3,3s)$	−10.708	82.370
$Q(14,0)$	5558.447	N/A	$Q(4,0)$	13.862	47.118
$Q(16,0)$	13654.650	N/A	$Q(4,2c)$	120.798	−271.032
			$Q(4,2s)$	−246.918	275.950
			$Q(4,4c)$	−29.694	648.718
			$Q(4,4s)$	−19.502	−88.341

Values of the non-zero central multipole moments Q_{lk} (defined in equation (9.3)) in atomic units (e a_0^l). The real multipole moments Q_{lk_c} and Q_{lk_s} are defined as $\sqrt{2}$ times the real and imaginary parts of the complex multipole Q^*_{l-k} (Price et al., 1984). The results for F_2 are for a large basis set MP2 correlated wavefunction calculated by Amos (1985) for a bondlength of 1.4184 Å. The other calculations are all for a 6−31G** SCF wavefunction for comparative purposes, and so are far from definitive quantum-mechanical results. (The experimental quadrupole moment for benzene is -6.46 ± 0.4 ea_0^2 (Battaglia et al., 1981).) The geometries are taken from Raman spectroscopy for benzene (Stoicheff, 1954), the crystal structure for uracil (Stewart and Jensen, 1969) with adjustment for planarity and hydrogen atom positions for uracil, and constructed from standard bond lengths and angles (Weiner, et al., 1986) for the glycine dipeptide $CH_3.NH(CO.CH_2.NH)CO.CH_3$.

The central multipole expansion is not only poorly convergent at shorter separations, but it is also not valid for many orientations of non-spherical molecules which are found in molecular liquids and crystals. The multipole series is only valid when there is no overlap of the spheres around each molecule which contain the molecular charge distribution ($R > r_{max}^A + r_{max}^B$). Alternatively, if the exponentially decaying effects of the interpenetration of the charge distributions are considered separately, then the convergence sphere only need contain all the nuclei in the molecule (Stone and Price, 1988). These conditions for validity severely limit the use of the central multipole expansion for organic molecules. As can be seen from Figure 9.2, even the less demanding criterion is only just valid for two benzene molecules in van der Waals contact, and will not be met

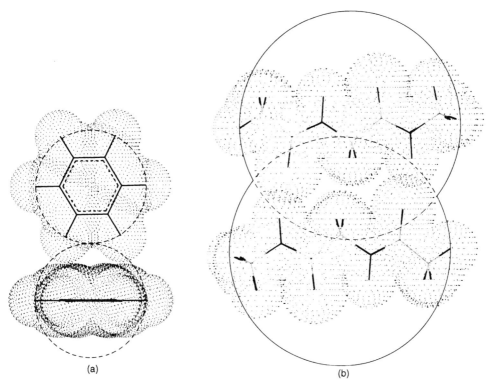

Figure 9.2 The convergence sphere for a central multipole expansion for (a) benzene and (b) the extended model peptide $CH_3.NH(CO.CH_2.NH)CO.CH_3$. The benzene molecules are in van der Waals contact, and the model peptide hydrogen bonded as in an antiparallel β-sheet. The multipole expansion is only just valid for benzene, and completely invalid for the dipeptide because of the overlap of the spheres which contain all the nuclei of each molecule

by less spherical molecules, such as the hydrogen bonded pair of extended peptide fragments.

The importance of the errors associated with the non-validity and poor convergence of the central multipole series in evaluating electrostatic energies are demonstrated in Figure 9.3. This compares the electrostatic energy as a function of separation, as calculated from a central multipole expansion (Table 9.1) and the more exact and detailed distributed multipole representation (see Section 9.2.3) of the same wavefunctions, using all terms up to R^{-5} in either multipole expansion (R is the intermolecular separation for the central expansion, and an atom-atom distance for the distributed multipole model.) For benzene (Figure 9.3(a)), although the expansion is formally valid, the central quadrupole-quadrupole interaction does not distinguish benzene from an X_2 molecule. Hence, it predicts the same energy for a T-shaped dimer, regardless of whether a vertex or edge of one molecule is pointing into the aromatic ring of the other, whereas the difference is significant for intermolecular separations of less than about 7 Å. The localised nature of the central multipole model also produces a significant overestimate of the repulsion between two

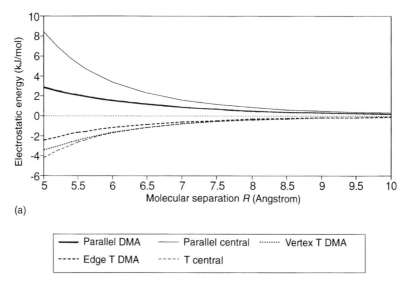

(a)

Figure 9.3 Plot of electrostatic energies as calculated from a central multipole expansion and atomic distributed multipole analysis (bolder lines) of the same SCF 6 − 31G** wavefunctions. All terms up to R^{-5} in each expansion are included. (a) Benzene dimer. The geometries are: parallel—two superimposed molecules displaced by R along the sixfold axis; vertex T—two perpendicular molecules with two vertices of one along the sixfold axis in a bond bisecting mirror plane, with centre of mass separation R; edge T— as above, but with an edge instead of a vertex pointing into the aromatic ring (i.e. two bond mid-points along the sixfold axis). The central multipole model, up to quadrupole–quadrupole terms, gives the same energies for all T-shaped dimers, independent of the rotation around the sixfold axes. In contrast, an atomic point charge model severely overestimates the difference between the vertex and edge T-shaped dimers (Price and Stone, 1987). (b) Uracil dimer. The x and y displacements are parallel and perpendicular to the N···C(=O) axis, starting at the hydrogen-bonded minimum in the DMA electrostatic energy subject to a pseudo hard-sphere model to define accessible orientations, which is illustrated in the insert. The z displacements start from the superposition of inverted molecules, i.e. with a N—H vertically above a C=O and vice versa. (c) The dimer of the model peptide $CH_3.NH(CO.CH_2.NH)CO.CH_3$, in the extended conformation. The origin corresponds to the minimum in the DMA electrostatic energy, subject to a pseudo hard-sphere model, which is illustrated in the insert. The relative displacements of the two molecules shown are either parallel to the molecular long axes or a perpendicular displacement in the plane of the molecules, defined by the motion of the lower molecule along the axes shown

vertically stacked (D_{6h}) molecules, which increases to a sixfold overestimate at van der Waals contact ($R = 3.5$ Å), when the expansion is invalid. Even when multipoles up to $l = 6$ are used (Bauer and Huiszoon, 1982) the central multipole expansion only leads to an appropriate representation of the electrostatic potential of the azabenzenes for molecular separations of greater than 5 Å.

The poor prediction of the electrostatic potential is also obvious for two stacked

REPRESENTATIONS OF THE ELECTROSTATIC INTERACTION 303

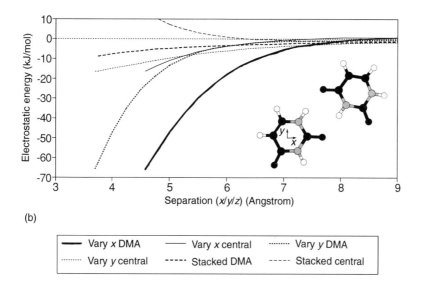

(b)

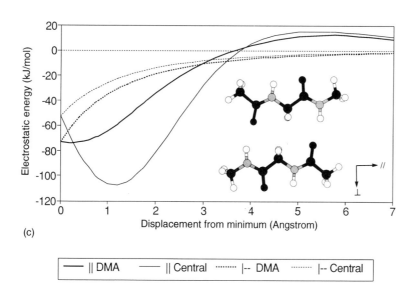

(c)

parallel uracil molecules (Figure 9.3(b)), where the central expansion predicts a drastic repulsion for the inversion related geometry, instead of the attraction that arises from two N—H groups stacked above C=O groups. The other configurations considered for uracil are derived by displacements from the hydrogen-bonded dimer structure shown in Figure 9.3(b). This structure was found as the minimum in the electrostatic energy, calculated from the DMA, within accessible orientations defined by a hard-sphere repulsion potential. It is a very plausible dimer structure, with two hydrogen bonds, and so translations in either direction from this minimum result in a rapid loss in electrostatic energy. The lack of spatial information in the central multipole model means that the electrostatic energies in this hydrogen bonded orientation are grossly underestimated. The minimum sterically accessible electrostatic energy for the central multipole expansion (-52 kJ/mol) is not much less than for the DMA electrostatic model, but is at a totally different implausible geometry with the two uracil molecules tilted and partially overlapping each other.

The model extended dipeptide is also predicted to have a plausible dimer structure involving two hydrogen bonds, shown in Figure 9.3(c), by minimising the DMA electrostatic energy in accessible orientations. The central multipole expansion is just not valid in this orientation (see Figure 9.2) and so it is hardly surprising that the predicted electrostatic energy at the hydrogen-bonded conformation is significantly less than when the two molecules are displaced parallel to their long axes, so that the carbonyl oxygen of one is nearly midway between the CH_2 and N—H groups of the other (Figure 9.3(c)). In all three cases, the two descriptions converge at large separations, giving the same relative energies for the different orientations, but this is only at separations where the electrostatic energy is getting very small.

The total multipole moments provide a clear measure of the polarity of a charge distribution. The first non-zero multipole moment is an experimentally measurable quantity, that is independent of the choice of origin. However, it is a poor model for quantifying the electrostatic forces between molecules unless the molecule is a close approximation to a sphere and a number of terms in the multipole series are used, or you are only evaluating the energy at large separations. Central multipole moments have been successfully used for the electrostatic contribution to van der Waals forces between small polyatomics (Dykstra, 1993). However, validity and convergence problems prevent the central multipole expansion from being used to model organic systems.

9.2.2 The Atomic Point Charge Model

The assignment of a point partial charge to each atomic nucleus clearly gives a representation of the charge distribution that is a far better representation of the spatial extent of the molecule. It also produces a computationally simple model for the electrostatic energy which is readily incorporated into any isotropic atom-atom potential scheme—simply sum $q_i q_j/(4\pi\varepsilon_o R_{ij})$ over each intermolecular pair of charges q_i and q_j separated by R_{ij}. The problem comes in determining the values of the atomic charges.

For simple, symmetric molecules, values of the atomic charges can be deduced from the first total multipole moment of the molecule available from experiment. This can work well. For example, the charges of $+0.153$ e on H and -0.153 e on the bonded C atom and zero on other C atoms deduced by Williams and Starr (1977), by empirical fitting to aromatic hydrocarbon crystal structures, provide a remarkably reasonable estimate of the central out-of-plane quadrupole moment of benzene, naphthalene, anthracene, phenanthrene, pyrene, triphenylene, perylene and coronene (Price, 1985). It can also highlight the inadequacies of the model: a point charge model for nitrogen which corresponds to the known negative quadrupole moment, implies equal negative charges on the nitrogen atoms, and a positive charge in the centre of the N≡N triple bond (Stone and Alderton, 1985). It requires five point charges, including lone pair sites, to get a point charge model for nitrogen that is chemically reasonable and approximates the higher multipole moments well (Murthy *et al.*, 1993). Any point charge model that reproduces the first non-zero central multipole moment includes an implicit estimate of the higher multipole moments. Thus the predicted electrostatic energies will differ from those given by the central multipole expansion except at such large separations that only the first non-zero total multipole moment contributes.

For most organic molecules, the central multipole moments do not provide sufficient information to determine the atomic charges without making further assumptions, which may not be chemically realistic. For example, Bauer and Huiszoon (1982) derived a point charge model for several of the azabenzenes from their central multipole moments by assuming that the electron density around each atom only depended on the nature of the atom (i.e. C or H or N), though the fitted values of q_C, q_H, q_N and the lone pair charge differed between benzene, pyridine, pyridazine, pyrimidine, pyrazine, *s*-triazine and *s*-tetrazine. A later analysis of these charge distributions (Price and Stone, 1983) showed that a better assumption would have been to assume that every nitrogen atom withdrew $p = 0.282 \pm 0.017$ e from each neighbouring C—H group in the aromatic ring to give a charge model that was transferable throughout the series. This just demonstrates that the variation of atomic charge distributions through short-range inductive effects etc makes it difficult to deduce a reasonable set of atomic charges from experimental information on the charge distributions of organic molecules.

Hence, work on atomic charges has concentrated on methods of deriving the charges from the molecular wavefunction. Mulliken charges were initially used, despite the fact that Mulliken (1955) analysis was not devised for estimating intermolecular interactions. It gives charges that are very dependent on the basis set, and so are often chemically unrealistic. Many alternative methods of deriving charges from the wavefunction have been proposed. The most accurate for calculating intermolecular electrostatic energies is to derive the charges by fitting them to reproduce the electrostatic potential, as calculated directly from the wavefunction by integration (see equation 9.1), on a grid of points outside the molecule (Momany, 1978; Cox and Williams, 1981; Singh and Kollman, 1984, and others). The charges will depend on the choice of grid or surface, as discussed by Spackman (1996), and some constraints on the charges for atoms far from the van der Waals surface can help the convergence of the fitting process (Stouch and Williams, 1993). Nevertheless as a comparison of various methods of deriving atomic charges by Wiberg and Rablen (1993) concluded, no atomic point charge model is capable of exactly reproducing the electrostatic potential in the van der Waals region around a wide range of molecules.

The reason for this is the basic assumption of the atomic point charge model, that the molecular charge density can be represented by a superposition of spherical atomic charge densities. This may give a good first approximation to the molecular shape, but it is not the total shape of the charge density that gives rise to the electrostatic forces. There would be no electrostatic potential outside a molecule that was a superposition of neutral spherical atomic charge distributions. Thus the electrostatic forces arise from the rearrangement of the valence electron density (from spherical) on bonding. The atomic point charge model is effectively assuming that on bonding, charge moves from the electropositive atoms to the electronegative atoms so that the charge around each atom remains spherical. This is a travesty of molecular bonding theory. The difference between the molecular charge density and a superposition of neutral atoms, which gives rise to the electrostatic forces, will show markedly non-spherical features, such as bonding density, lone-pair and π electron density, as illustrated for Cl_2 in Figure 9.4. Thus, the atomic point charge model, while giving a good model for the gross molecular shape, is a crude representation of the component of the charge distribution around each atom that gives rise to the electrostatic forces.

9.2.3 Distributed Multipole Models

A more complete alternative to either the central multipole model or the atomic point charge model is to combine the advantages of both approaches and represent the molecular charge distribution by a set of multipoles at each atomic centre. Distributing the mutipole centres around the molecule provides a good model of the molecular shape, and the multipoles at each atom can represent the non-spherical features in the atomic charge distribution, such as lone pairs and π electron density. A comparison of the electron density difference map of chlorine (Figure 9.4) with the distributions for idealised atomic charge models (Figure 9.1) shows that atomic quadrupoles give a good first approximation to the charge density. The model can be refined by atomic dipole and

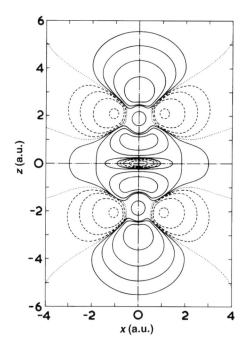

Figure 9.4 The SCF density difference of chlorine, defined by subtracting spherical atomic charge densities from the molecular charge density calculated using a $(7s6p3d)$ basis set. The solid contours are positive, dotted zero and the dashed negative (build-up of electron density). The non-zero contours are at 10^{-3}, 3×10^{-3}, 10^{-2}, 3×10^{-2} and 10^{-1} atomic units. Adapted from Alderton (1983)

octopole moments, to reflect the difference in bonding between the bonding region and the ends of the molecule. This will provide a good representation of the difference density, whereas a central multipole series would have to go to a very high level to include the shape of the lone pair density at all.

Actually quantifying a distributed multipole model requires a method of dividing up the molecular charge density between the sites. This can be done (Cooper and Stutchbury, 1985, Laidig, 1993) rigorously by the Atoms-in-Molecules partitioning of Bader (1990), but this is computationally demanding, and the shapes of the atoms defined by this method can include sharp edges which are not readily represented by a low-order multipole expansion. Hirshfelder partitioning, which is based on differences from a superposition of spherical neutral atoms, has also been used by Ritchie (1985). Other methods have the partitioning defined by the basis set, such as those of Pullman and Perahia (1978) or Vigné-Maeder and Claverie (1988), an optimised method for linear molecules (Wheatley, 1993), or the Culmulative Atomic Multipole Moments of Sokalski and Sawaryn (1987), which extends Mulliken analysis to give the higher moments. The method which is used in most of the work described in this chapter is the Distributed Multipole Analysis (DMA) of Stone (1981), which was explored in detail by Stone and Alderton (1985). This is based on the density matrix of the molecule, expressed in terms of the Gaussian primitives η that comprise the atomic orbital basis set:

$$\rho(r) = \Sigma_{ij} \rho_{ij} \eta_i \eta_j \qquad (9.4)$$

Each contribution to the charge density can be exactly represented (outside itself) by a series of point multipoles at a point determined by the origin and exponents of the two Gaussian primitives involved. If both orbitals are s orbitals, then the contribution is represented by a charge, if two p orbitals, then a charge, dipole and quadrupole at the overlap centre. (The series terminates at the lth multipole, where l is the sum of the angular momenta of the basis functions involved.) Thus, for example, the π electron density in aromatic compounds is represented by out-of-plane quadrupole moments ($l = 2$). If both Gaussian functions are on the same atom, then the analysis generates the corresponding limited multipole series on that atom. If they are on different atoms, then the analysis 'moves' the contribution from the overlap centre to the nearest atom, using the relationship between a point multipole at one site to an infinite series at another point. This series converges rapidly provided the distance moved is not too great. Thus for a molecule which is well described by an sp basis set, the atomic charges, dipoles and quadrupoles are fundamental to the description of the valence density, whereas the higher moments arise from the displacement of the overlap density from the atoms and improve the quantitative accuracy of the description.

All the methods of partitioning should be able to describe the electrostatic potential outside the molecular charge distribution to essentially the accuracy of the wavefunction, provided that sufficient expansion sites and orders of multipole are included. (Some partitioning schemes have multipole sites between as well as on the nuclei, either by definition or as an option.) Thus, Spackman (1986) found that several different distributed multipolar expansions for $(HF)_2$ gave results in excellent agreement provided that terms up to quadrupole were included. The DMA method of moving the contribution to the nearest site optimises this convergence. The use of distributed sites, each representing the charge distribution around an atom (or smaller fragment) also ensures that the expansion is valid for almost all relative orientations where an intermolecular potential model is appropriate (contrast Figure 9.2).

The use of distributed multipoles inevitably involves a large number of parameters. A charge distribution with no symmetry is described by a point charge, three dipole components, five quadrupole moments, seven octopole moments, nine hexadecapole components etc. Although some of these components will be zero if the charge distribution has elements of symmetry, an atomic site usually has less symmetry than the entire molecule.

The electrostatic interaction energies are evaluated using the multipole expansion formulae for each intermolecular pair of sites. Explicit expressions for all terms up to R^{-5} are given by Price et al. (1984), up to R^{-6} by Hättig and Heß (1994), and Stone (1991) has provided a general formulation and discussion of the spherical tensor and Cartesian tensor approaches. The program ORIENT by Stone et al. (1994) incorporates these expressions and is widely used for gas phase modelling of molecular complexes using distributed multipole electrostatic models.

A multipole series should strictly be truncated at a given power ($-(l_A + l_B + 1)$) of the inter-site separation R, rather than order of multipole l_A on molecule A and l_B on molecule B. Hence, the series should be taken up to R^{-5} to include the important quadrupole–quadrupole ($l_A = l_B = 2$) interactions, and thus also the hexadecapole–charge ($l_A = 4, l_B = 0$) and octopole–dipole ($l_A = 3, l_B = 1$) terms.

The contribution of the anisotropic atomic multipoles, which represent the lone pair and π electron density, to the predicted electrostatic potential rapidly become less important as the distance between the molecules increases. This not only results from the inverse power of R but there is also cancellation between the contributions from different multipoles and different atoms. For example, there is generally an atomic dipole component along a bond which opposes the polarity implied by the atomic charges, as shown in the DMAs of the azabenzenes (Price and Stone, 1983). Thus, the accuracy gained by using a distributed multipole model is very dependent on the relative separation and orientation of the molecules, as well as the actual distribution of charge in the molecule.

9.2.4 Limitations to the Accuracy of Distributed Multipole Electrostatic Energies

The accuracy of a distributed multipole representation can always be tested by seeing the effect of increasing the number of interaction sites, for example, by adding sites at the mid-points of bonds, or by increasing the order of multipoles taken into consideration, on the calculated electrostatic energies in the region of interest in the simulation. Similarly, simplifications of the model, such as removing small multipole components, can be evaluated. However, the accuracy of the calculated electrostatic energies is inevitably limited by the quality of the wavefunction and the neglect of penetration effects.

Increasingly high-quality wavefunctions can be used to calculate the distributed multipoles, with near-definitive charge distributions being possible for small polyatomics, and split-valence and better basis set SCF calculations being used for organic molecules. Fortunately, a comparison of the electrostatic potential on a grid of points outside N-acetylalanine N'-methylamide, as calculated from the DMAs of various wavefunctions differing in the basis set and inclusion electron correlation (Price *et al.*, 1992), suggests that the approximations in the wavefunction, for split valence or better basis sets, will primarily produce a scaling of the electrostatic interactions, rather than any significant change in the orientation dependence of the electrostatic interaction. Since a 6-31G** SCF wavefunction generally overestimates molecular total dipole moment by 12–15 per cent (Ryan, 1994), a 0.9 fudge factor is commonly used to allow for the effect of electron correlation on the estimated electrostatic energies. This is based on an average of the ratio of the calculated to experimental total dipole moment of eight small polyatomic molecules, ranging from 0.78 for NH_3 to 0.96 for CH_3CN (Cox and Williams, 1981). However, the practical necessity of making such assumptions is likely to diminish rapidly, as higher-quality wavefunctions become routine for organic molecules.

A second approximation (or matter of definition) is that the electrostatic energies calculated from any of the point multipolar representations of the molecular charge distribution do not include the effect of any interpenetration of the molecular charge distributions. The penetration energy is defined as the difference between the exact electrostatic energy (equation (9.1)) involving the full spatial extent of charge distribution, and a converged distributed point-multipolar representation. The penetration energy decays exponentially with separation, reflecting the decay of the charge distribution with distance. It can be significant at the van der Waals separations involved in hydrogen bonding, but generally does not alter the orientation dependence significantly, as shown by the comparison of the electrostatic energy evaluated by perturbation theory and from DMAs for various van der Waals complexes by Hurst *et al.* (1986). A method of calculating electrostatic energies using Gaussian multipoles, which have an appropriate spatial extent has been developed by Wheatley and Mitchell (1994), from which it is estimated that the penetration energy contributes 25 per cent to the electrostatic energy of the formamide—formaldehyde complex. Nevertheless, the penetration energy is a short-range, exponentially decaying effect which is often empirically absorbed into the repulsion potential.

9.3 THE IMPORTANCE OF THE ELECTROSTATIC ENERGY IN INTERMOLECULAR INTERACTIONS

The distributed multipole model incorporates a very exact description of the molecular charge distribution into the evaluation of the electrostatic energy. Is the increase in accuracy from representing the effects of lone pair and π electrons density worth the extra complexity in the potential model? Even if there is a significant difference, is it worth using such an elaborate model when only crude models, such as the isotropic atom–atom 6-*exp* potential, are available, for the other contributions?

There have been many studies which contrast the accuracy of various atomic charge and distributed multipole models, in addition to the extensive tests performed when the various distributed multipole methods were proposed (see Section 2.3). For example, there are published contour plots of the potential around a water molecule (Dykstra, 1993), the amino acid histidine (Sokalski and Sawaryn, 1992), and variations in the electrostatic energies of nucleic acid bases (Ritchie and Copenhaver, 1995), which confirm the significance of the atomic anisotropy shown in the colour 3D displays of the electrostatic potential around cytosine (Price, 1996b) and electrostatic field around uracil and pyrimidine (Price and Richards, 1991). The analysis of Sokalski *et al.* (1993) of the electrostatic potential around model-blocked peptides shows that it can be considerably distorted around the functional groups capable of forming hydrogen bonds by various models relying on only isotropic atomic charges. It is clear that the difference made in going from an atomic point charge model to a distributed multipole model depends considerably on the molecule and region of the intermolecular potential being sampled, often more than on the precise variation of the different types of model used. The distributed multipole model is most important in the regions close to atoms with marked non-spherical features in their charge distributions, such as lone pair and π electron density.

The numerical differences in the predicted electrostatic properties can be sufficient to qualitatively change the predicted minimum energy structures. Perhaps the most compelling evidence for the need to use accurate electrostatic models comes from the structures of van der Waals complexes. Buckingham and Fowler (1985) showed that the experimental structures of over two dozen van der Waals complexes of small polyatomics could be predicted remarkably well by just minimising the electrostatic energy of the pair of molecules, within the sterically accessible region defined by the van der Waals surfaces of the molecules. The structures of these molecules, which included the bent $(HF)_2$, $HF \cdots H_2O$ and $(Cl_2)_2$, linear $HCN \cdots HCN$, $N \equiv N \cdots HF$ and T-shaped $HCCH \cdots HF$ can be summarised by the empirical rules of Legon and Millen (1987). These stated that the hydrogen bond forms to regions of high electron density in the acceptor, so that the axis of the HX molecule coincides with the supposed lone pair axis on the acceptor atom. When no lone pair density is present, the HX axis is perpendicular to the plane of symmetry of, and bisects, a π-bond. The observation that these structures could be predicted using an electrostatic model for the binding showed that it was not necessary to invoke charge transfer or weak bonding between the molecules in the hydrogen bond as determining the structures. It also showed the dangers of using low order central multipole expansions or atomic point charges to represent the electrostatic forces as these generally omit the anisotropic contributions of the lone pair and π electron density.

This observation that the electrostatic forces were able to predict the structure of the complexes with only the crudest model repulsion potential, suggests that the electrostatic forces dominate the anisotropy of the intermolecular forces, even between neutral molecules. This was confirmed as a generally useful (though not infallible) rule, by Intermolecular Perturbation Theory calculations on all the major contributions to the intermolecular potential by Hurst *et al.* (1986). They showed for several complexes, such as $(HF)_2$, $N_2O \cdots HF$ and $H_2CO \cdots HF$, that the DMA electrostatic energy had a similar anisotropy and minimum position as the total energy of the complex. This is because the anisotropy of the other contributions tends to cancel, with increased overlap of the charge distributions, giving both a higher repulsion energy but also more favourable charge-transfer energy and penetration energies. This cancellation is far from exact, and examination of the few cases where the simple electrostatic model fails to predict the correct structure shows its limitations. One class where the model failed was because both the total and the DMA electrostatic energy changed little, with, for example, the angle around O in $H_2O \cdots H-C \equiv C-H$, so that the structure is a vibrational average over a wide potential well. The second class was exemplified by $HF \cdots ClF$ where the repulsion potential is so anisotropic that the spherical van der Waals radius gave a very poor prediction of the relative separations. The electrostatic model did locate two minima, one corresponding to a bent anti-hydrogen bonded structure H—F···Cl—F, the other to the Legon-Millen rule predicted Cl—F···H—F, but the failure to include the observed anomalously short F···Cl distance meant that the predicted relative energies favoured the hydrogen-bonded structure over the observed structure.

The success of this model seems to extrapolate to larger molecules—for example, it is successful for many van der Waals complexes involving benzene and s-tetrazine (Price and Stone, 1987). The DMA energy also mirrors the total energy (calculated by Intermolecular Perturbation Theory) fairly well for the orientation dependence of the N—H···O=C hydrogen bond in the formamide/formaldehyde complex (Mitchell and Price, 1990). Thus, it seems plausible that for most organic molecules, especially those involving heteroatoms, hydrogen bonding and aromatic rings, the use of an accurate electrostatic model, such as a DMA, will not only give a theoretically justified model for an important to dominant contribution but should also model the orientation dependence of the total interaction fairly well. This assumption has been justified by the success of various organic modelling studies, some of which are described below, which use realistic (DMA) electrostatic models in conjunction with much cruder models for the other contributions. The following illustrations of this point are applications of molecular electrostatic modelling which have engaged the author. There are numerous other applications of molecular electrostatic modelling, many of which have recently been reviewed by Náray-Szabó and Ferenczy (1995).

9.4 MOLECULAR CRYSTAL STRUCTURE MODELLING

The packing adopted by a molecule can determine many of the properties of the solid. An extreme example is the non-linear optical coefficient, which can be high for the isolated molecule, but zero for the crystal if the molecule packs in a centrosymmetric space group. An energetic molecule produces a more effective explosive if the molecules pack densely. The ease with which a pharmaceutical can be tableted depends on mechanical properties which can vary between different crystalline forms (polymorphs) of the same molecule. Hence there are many 'molecular design' problems where the optimisation of a molecular material would be helped if it were possible to predict the crystal structure in advance of synthesis. Such calculations should also predict possible polymorphic forms, as new polymorphs could threaten patents.

We are still a long way from being able to predict molecular crystal structures, as discussed provocatively by Gavezzotti (1994), though recently there has been considerable progress towards this fundamental target. There are three main components to any theoretical technique for predicting molecular crystal structure: a method of simulating the crystal, a method of generating sufficient hypothetical crystal structures so that the most stable structure is likely to be found, and a model for the intermolecular forces which controls the molecular packing.

Lattice energy minimisation has generally been used for studying molecular crystals, mainly for reasons of computational expense, though more experimental data would be required for many organic molecules to justify a more elaborate treatment. The global minimum in the lattice energy should correspond to the 0 K structure. This usually has to be compared with a room-temperature crystal structure, thus neglecting the effect of the molecular motion on both the structure and the thermodynamics. From the relatively few rigid organic molecules where a low-temperature crystal structure is also available, it appears that lattice vectors are usually few per cent larger at room temperature, but this varies between the different lattice vectors, depending on the packing (some even contract on heating) and between different organic molecules (Price, 1996b). Thus, the use of static simulation techniques probably leads to errors of a few per cent in the lattice vectors. The crude treatment of the energetics, implicit in comparing the minimum lattice energy (calculated by summing over the intermolecular pair potentials) with the experimental enthalpy of sublimation, is usually justified because the experimental errors (usually 1–3 kcal/mol) are generally of the same order as the neglected vibrational and entropic contributions. The net result is that discrepancies of up to 3–4 kcal/mol between the lattice energy and the observed heat of sublimation should not cause concern (Pertsin and Kitaigorodsky, 1987).

Various strategies have recently been proposed for trying to ensure that the global minimum in the lattice energy can be found, including a simulated annealing Monte Carlo-based technique (Karfunkel and Gdanitz, 1992). Other methods systematically search through the most commonly observed packing types (in terms of the symmetry relationships between the molecules which define the space group). MOLPAK (Holden *et al.*, 1993) selects promising starting points for energy minimisation on the basis of the density of the hypothetical crystal, whereas PROMET (Gavezzotti, 1991) focuses on the intermolecular energy of a symmetry defined nucleus of molecules within the structure. In the work reported so far, all methods seem successful in finding the minimum in the lattice energy corresponding to the experimental structure. However, all methods often generate other hypothetical structures which are very close in lattice energy. For example, Holden *et al.* (1993) report that a plausible unknown structure for dinitronaphthalene is predicted to be almost 1 kcal/mol more stable than the minimum corresponding to the experimental structure. Is this a possible polymorph, or due to errors in the intermolecular potential?

Thus, a prerequisite to predicting molecular crystal structures is a model for the intermolecular forces which can reproduce the experimental crystal structure, by having a minimum in the lattice energy acceptably close to the experimental structure. It should also be theoretically well based, and capable of reproducing a range of crystal structures of related molecules, so it is reasonable to expect the potential to extrapolate accurately to hypothetical crystal structures.

Molecular crystal modelling has been based on the use of the isotropic atom–atom model potential:

$$U = \sum_{ik} U_{ik} = \sum_{ik} A_{\iota\kappa} \exp(-B_{\iota\kappa} R_{ik}) - C_{\iota\kappa}/R_{ik}^6 \tag{9.5}$$

where atoms i and k are of types ι and κ respectively. The extensive use of this model potential, particularly for hydrocarbons, has been reviewed by Pertsin and Kitaigorodsky (1987). The simple *6-exp* form has been carefully developed by Filippini and Gavezzotti (1993), by considering a wide range of crystal structures, using the equivalent form

$$U = \sum_{ik} U_{ik} = \sum_{ik} \frac{\varepsilon_{\iota\kappa}}{(\lambda - 6)} \left[6 \exp \lambda \exp\left(-\frac{\lambda}{R_{\iota\kappa}^0} R_{ik}\right) - \frac{\lambda (R_{\iota\kappa}^0)^6}{R_{ik}^6} \right] \tag{9.6}$$

where the values of the minimum energy separation $R_{\iota\kappa}^0$ between atoms of type ι and type

κ were largely guided by the shortest contacts in crystal structures, λ was fixed, and the well depths $\varepsilon_{\lambda\kappa}$ were determined by empirical fitting to crystal structures and heats of sublimation. This set of potentials for C, H, N, O, Cl and S atoms was remarkably successful in reproducing the observed heats of sublimation, but the success in reproducing the crystal structures was more varied. The scheme was extended (Gavezzotti and Filippini, 1994) to mono-functional hydrogen bonded structures, with particularly deep potential wells for the heteroatom/polar hydrogen atom potentials, emphasising that the potentials were absorbing the electrostatic contribution.

The alternative approach of adding a point charge electrostatic model to the isotropic *6-exp* repulsion–dispersion model has been used, though only for limited families of molecules containing heteroatoms (Pertsin and Kitaigorodsky, 1987). Williams and Cox (1984) developed a set of *6-exp* parameters for C, H, and N by fitting to the crystal structures of a range of azahydrocarbons, which were complemented by O parameters derived by fitting to oxohydrocarbons (Cox *et al.*, 1981). However, it is notable that Williams and Weller (1983) found it empirically necessary to include off-nuclear lone pair sites in order to both reproduce the electrostatic potential around the azabenzenes and their crystal structures satisfactorily.

Further evidence that molecular crystal structure could be sensitive to the detailed form of the electrostatic model came from a study of the crystal structures adopted by diatomic X_2 molecules (Price, 1987). This compared the minimum lattice energy for various types of packing (space groups) which are adopted by different diatomics at low temperatures, for a range of model potentials. It was found that the symmetry of the predicted structure (global lattice energy minimum) could depend on whether the same total quadrupole moment of the molecule was represented by atomic dipoles or atomic quadrupole moments, for the same repulsion–dispersion potential, thus rationalising the low temperature crystal structures of the diatomics.

Thus, there was some limited evidence that some crystal structures were sufficiently sensitive to the detailed form of the electrostatic potential, that it was worth considering the introduction of *ab initio*-based distributed multipole electrostatic models into crystal structure simulation. The main barrier to using such model potentials for modelling the crystal structures of organic molecules was the need to adapt a simulation program to handle the anisotropic interactions. This required evaluating the forces and torques arising from all the multipole interactions between every intermolecular pair of atoms, summing to give a net force and torque on each rigid molecular unit, and also evaluating the derivatives with respect to changing the unit cell shape. Some second derivatives were also required for the modified Newton–Raphson minimisation process. A second complication was the need to sum the slowly converging charge–dipole and dipole–dipole contributions to the lattice energy, which was done by a modified Ewald summation method. These developments were implemented into a static lattice simulation program DMAREL (Willock *et al.*, 1995).

Accurate models for the electrostatic forces are most likely to be needed for polar molecules, where the crystal structure contains hydrogen-bonding and π–π interactions, and the molecular shape is sufficiently simple that the molecules can be reasonably close packed in a wide range of structures. Thus, to test the value of a distributed multipole electrostatic model in reproducing molecular crystal structures (Coombes *et al.*, 1996), a dataset of forty organic compounds including various combinations of aromatic and heterocyclic rings, amine, nitro and amide groups was selected from molecules where there was a reasonable quality room temperature crystal structure in the Cambridge Structural Database (Allen *et al.*, 1991). The molecular structure was taken from the experimental crystal structure, apart from standardisation of the hydrogen atom positions. The model intermolecular potential comprised an *ab initio*-based distributed multipole model for the electrostatic contribution with an empirically fitted atom-atom *6-exp* repulsion–dispersion potential (equation (9.5)). The DMA electrostatic model was derived from a 6-31G** SCF wavefunction, with all the multipoles scaled by a factor of 0.9 to approximately allow for the omission of electron correlation in the wavefunction. The isotropic atom–atom parameters for H, C, O and N were derived by empirical fitting to azahydrocarbon and oxohydrocarbon crystal structures (Williams and Cox, 1984; Cox *et al.*, 1981) in the presence of a point charge electrostatic model. The parameters for the polar hydrogen atoms were estimated from a fit to an Intermolecular Perturbation Theory surface for formamide/formaldehyde in the region of the N—H···O=C hydrogen bond (Mitchell and Price, 1990). The heteroatomic parameters were fixed using the traditional combining rules:

$$A_{\iota\kappa} = (A_{\iota\iota}A_{\kappa\kappa})^{1/2} \qquad B_{\iota\kappa} = \tfrac{1}{2}(B_{\iota\iota} + B_{\kappa\kappa}) \qquad C_{\iota\kappa} = (C_{\iota\iota}C_{\kappa\kappa})^{1/2} \qquad (9.7)$$

It is worth noting that these parameters, although the most appropriate available in the literature, had not been applied to hydrogen-bonded molecular crystals before, or indeed to any molecule containing both N and O atoms.

The ability of this model potential to reproduce the room temperature crystal structures is illustrated in Figure 9.5. For most of the molecular crystals, the r.m.s. per cent error in the cell lengths is within the error that could be associated with the neglect of thermal effects. Indeed, the errors are comparable to those obtained with model potentials specifically fitted to a small family of molecules, in the few cases where such results have been published (Coombes *et al.*, 1996).

Many of the molecular crystals that are reproduced with larger errors contain an NH_2 group that could be partially sp^3 hybridised, judging by the C—N bondlength. Thus, there is uncertainty in the hydrogen atom position from possible dynamic motion (as well as the usual problems in locating H atoms in X-ray structures). This can be an important factor, since the r.m.s. error for *m*-nitroaniline was reduced from 5.1 per cent (Figure 9.5) to 1.4 per cent when an idealised sp^2 geometry for the amine group was used.

The importance of the electrostatic model in determining these crystal structures was

---

Figure 9.5 Differences between the calculated and the experimental room-temperature crystal structure, using a distributed multipole model for the electrostatic forces plus literature repulsion–dispersion potential (as defined in the text), for a range of organic molecules. The molecules include examples of the nitrobenzenes, nitroanilines, amides, nitrogen-containing heterocycles including nucleic acid bases to purines, and azabenzenes. (The molecules with abbreviated names are: TATNBZ 1,3,5-triamino-2,4,6-trinitrobenzene; JOWWIB 3-amino-5-nitro-1-2-4-triazole; DAMTRZ20 3,5-diamino-1H-1,2,4-triazole; FITXIP 3,6-diamino-1,2,4,5-tetrazine.) * denotes molecules whose NH_2 groups may be partially sp^3 hybridised, and so there is uncertainty associated with the hydrogen atom positions. The error is presented as the r.m.s. per cent error in the independent cell edges

MOLECULAR CRYSTAL STRUCTURE MODELLING

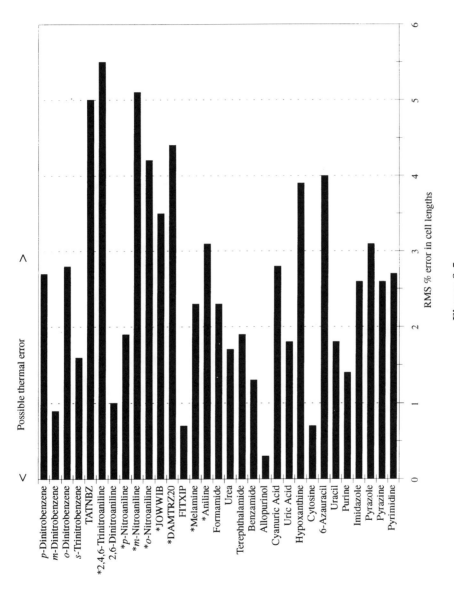

Figure 9.5

tested by removing first the 0.9 scaling factor and then the anisotropic multipoles to leave an atomic point charge model. Although most minimum energy structures are slightly closer to the experimental structure when the electrostatic forces are scaled to allow for the effect of electron correlation, the effect is quite minor. One exception is indazole, which is not included in Figure 9.5, as the scaling of the electrostatic contribution increases the r.m.s. error from 5.6 per cent to 9.3 per cent. Clearly this structure is so sensitive to the balance of the electrostatic forces that the uncertainty in the crude scaling factor is important in this unusual case. The electrostatic contribution to the lattice energy depends on the square of the scaling factor, and since the electrostatic contribution is more than 20 per cent of the total lattice energy (approaching 100 per cent for some heavily hydrogen bonded systems), the predicted lattice energy will be quite sensitive to the quality of wavefunction used. The agreement between the calculated lattice energies and the experimental heats of sublimation is generally acceptable, within the large error associated with such comparisons, and is generally worse when the scaling is removed.

The effect of removing the anisotropic multipoles on the predicted crystal structure was generally very significant, giving errors of greater than 4 per cent in many cases, with nine with errors of greater than 10 per cent, corresponding to qualitatively different structures. Thus, these crystal structures are sensitive to the representation of the electrostatic forces. Although a carefully derived atomic point charge model may be adequate for some crystal structures, just as a simple steric model is adequate for molecules which are so irregularly shaped that there is only one reasonably close packed crystal structure, it would appear that an accurate electrostatic model is necessary for modelling the crystal structures of a wide range of typical polar and hydrogen bonded organic molecules.

Thus, a realistic electrostatic model, in combination with a simple *6-exp* repulsion–dispersion potential, is capable of reproducing crystal structures of a wide range of organic molecules, sufficiently well that the remaining errors are comparable to those involved in using a static minimisation model. Since the dominant electrostatic contribution is derived from theoretical *ab initio* calculations, the potential should extrapolate reliably to hypothetical crystal structures, and thus be useful in developing methods of predicting molecular crystal structures.

9.5 RATIONALISING INTERMOLECULAR BONDING PREFERENCES

In modelling the interactions between organic fragments, whether in crystal structures, protein secondary structure or drug-receptor interactions, there is a need to understand and classify the interactions between various functional groups. Work on identifying the strength and directionality of specific hydrogen bonds, or identifying other interactions as able to significantly stabilise a supramolecular structure, is generally based on empirical observations of known molecular or protein crystal structures, using the Cambridge Structural Database (Allen *et al.*, 1991) or Brookhaven Data Bank (Bernstein *et al.*, 1977), respectively. These observations are required for crystal engineering and *de novo* protein-ligand design work, aimed at producing better molecular materials such as pharmaceuticals. Such insights are more powerful, once they have been understood in

9.5.1 'Lone Pair Directionality'—Electrostatic or 'Specific Interactions'?

The success of an accurate electrostatic plus hard-sphere model for accounting for the structures of hydrogen-bonded complexes of small polyatomics such as HF (see Section 9.2) demonstrated the importance of the electrostatic contribution in determining the preference for hydrogen bonding to the lone pair direction in such van der Waals molecules. Is the electrostatic contribution of the non-spherical features in the charge distribution sufficient to produce a preference for the 'lone pair' direction for hydrogen bonds in larger systems?

A statistical survey of N—H···O═C hydrogen bonds in molecular crystals by Taylor *et al.* (1983) demonstrated a statistical preference for such hydrogen bonds to form with N—H···O linear, the N—H coplanar with the C═O, and in the conventionally viewed lone pair direction. Thus, the empirical evidence is consistent with the intermolecular forces favouring hydrogen bond formation in the lone pair direction. Mitchell and Price (1989) studied the electrostatic contribution to this observed orientational preference by optimising the electrostatic interaction (calculated from a DMA), within accessible orientations, for 29 van der Waals complexes held together by N—H···O═C hydrogen bonds. These complexes involved two molecules chosen from formamide, acetamide, N-methylacetamide, succinimide, formaldehyde, acetaldehyde, acetone and uracil, designed to model the functional groups most commonly involved in such hydrogen bonds. A statistical analysis of the hydrogen bond geometries in the resulting 121 gas-phase minima produced results consistent with the crystal structure analysis. The hydrogen atom lies close to the N···O line, with 87 per cent of the complexes being planar to within 5°, and most hydrogen bonds were formed within 15° of the lone pair direction. However, most of the minima corresponded to one of the five hydrogen-bonding motifs depicted in Figure 9.6. This shows that most of the structures corresponded to two van der Waals contacts between the model molecules, either the double hydrogen bond motif (I) of the amides or one hydrogen bond and a secondary C—H···O contact in motifs II-IV. The single contact minima (motif V) were generally less stable, and also generally had larger C═O···H angles than the lone pair direction. Since the N—H···O═C geometry which is favoured by the electrostatic interaction is remarkably similar to the preferred geometry in crystal structures, the electrostatic interaction must play a major role in determining the angular geometry of this hydrogen bond. However, the electrostatic interaction favours multiple contacts in the van der Waals complex, as would be found in the crystal structures, and neither show the properties of 'an isolated N—H···O═C hydrogen bond'. The observed preference for bonding in the lone pair direction is not a simple consequence of 'lone pair directionality' but arises because this geometry will often allow the formation of a secondary electrostatically favourable contact between neighbouring functional groups.

This result is consistent with Intermolecular Perturbation Theory Calculations on the formamide/formaldehyde complex (Mitchell and Price, 1990). These show that the variation energy for a linear NH···O hydrogen bond in the lone pair plane is slight, varying by only 1 kJ/mol between the two minima in approximately the lone pair

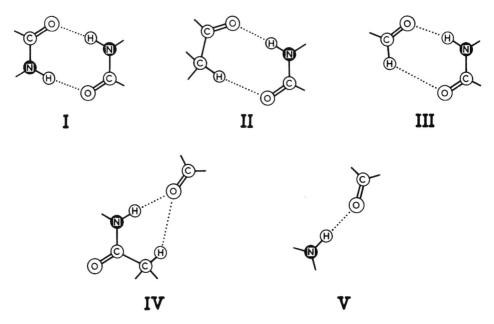

Figure 9.6 Some commonly occurring structural motifs which define many of the minimum electrostatic energy structures for van der Waals complexes involving N—H···O=C hydrogen bonds. The dotted lines indicate electrostatically favourable hydrogen bonding type intermolecular contacts with H···O distances less than 3 Å

positions. This reflects the fact that the electrostatic potential around the oxygen atom varies much less with angle when calculated from the full DMA than from just the charge component, as the effect of lone-pair density is to counteract the contributions from the neighbouring atoms. Thus the energetic preference for the lone pair direction is much less marked than the preference for the lone pair plane, or for a near-linear N—H···O bond.

A more straightforward example of hydrogen bond directionality is shown in organic heterocycles (Nobeli *et al.*, in preparation). Figure 9.7 shows the results of a search of the Cambridge Structural Database for hydrogen bonds to pyridine fragments. There is a preference for the lone pair plane ($\theta = 0°$) and direction ($\phi = 0°$), though there is quite a scatter. The minimum energy for a model system of pyridine with methanol shows a minimum in the intermolecular energy (calculated from a DMA electrostatic model, plus a *6-exp* potential) with a hydrogen bond in the lone pair direction. Intermolecular Perturbation Theory calculations of the main contributions to the intermolecular energy of pyridine with methanol (Figure 9.8) show a minimum in the lone pair direction, consistent with the crystal structure results (Figure 9.7). As usual (cf. Section 9.2), the electrostatic energy has a similar orientation dependence to the total intermolecular energy until there is a marked increase in the overlap of the charge distributions and the repulsion energy around $\phi = 40°$. The total energy is also paralleled by the electrostatic energy calculated from a DMA (i.e. the penetration term has little orientation dependence). However, when the electrostatic energy is calculated from just an atomic charge model or a central multipole model, the orientation dependence is quite different, and does not show the energetic preference for hydrogen bonding in the lone pair direction

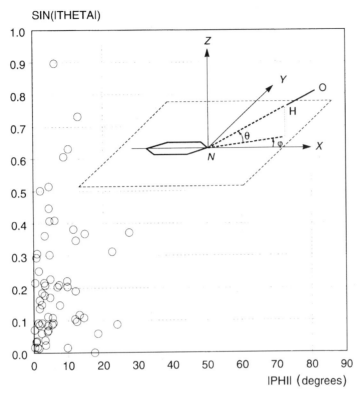

Figure 9.7 The angular geometries of the hydrogen bonds with an C—O—H donor and a pyridine fragment acceptor in molecular crystal structures in the Cambridge Structural database (Version 1996). The distance between the (standardised) hydrogen atom and the N atom is less than 2.64 Å. The plots have been transformed so that equal areas correspond to equal probability for a random distribution of hydrogen atoms about the nitrogen

shown by the total intermolecular energy, the DMA electrostatic energy, and the experimental data.

These examples of hydrogen bonds show that the electrostatic contribution is dominant in explaining the observed orientation dependence, but that this cannot be predicted simply on the basis of lone pair directions in organic molecules (unlike small polyatomics c.f. Section 9.2), as the contributions of neighbouring functional groups is also significant.

The role of the lone pair density in the intermolecular Cl···Cl interaction has also caused some debate. An analysis of Cl···Cl intermolecular interactions in molecular crystal structures by Nyburg and Faerman (1985) showed an effective van der Waals 'radius' of only 1.58 Å in the Cl—C bond direction, compared with 1.78 Å for the perpendicular side-on direction. The common existence of Cl···Cl intermolecular contacts shorter than the sum of the conventional van der Waals contact radii had been taken as evidence for a specific attractive interaction, a weak form of intermolecular bonding involving the lone pair density. Desiraju and Parathasay (1989) claimed that the frequency of such close Cl···Cl interactions in the crystal structures of chlorinated hydrocarbons was evidence that the interaction must be attractive. Unfortunately, the sample contained a disproportionate number of such highly chlorinated structures that Cl···Cl contacts were inevitable, which negates the argument. Intermolecular Perturbation Theory calculations on the methyl chloride dimer confirmed the opposing view, that the 'short intermolecular contacts' were mainly due to anisotropy in the repulsive wall around the Cl atom (Price et al., 1994). Nevertheless, it appears that Cl···Cl contacts can be mildly attractive in some orientations, due to the anisotropy of the electrostatic forces arising from the lone pair density, in conjunction with the attractive dispersion energy.

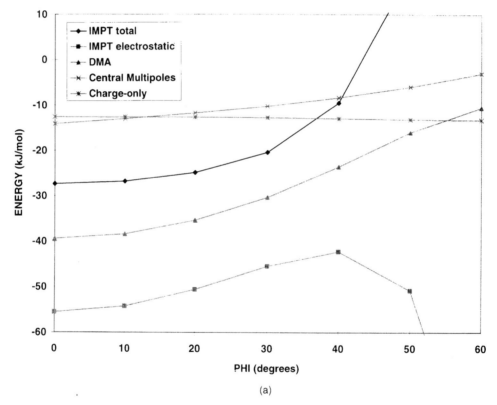

(a)

Figure 9.8 The variation of intermolecular energy terms for a methanol hydrogen bonded to pyridine, for angular variations either (a) in the lone pair plane ($\theta = 0°$) or (b) perpendicular to it ($\phi = 0°$) for a fixed N···H distance of 1.85 Å. The total intermolecular energy ◆, calculated using Intermolecular Perturbation Theory to second order with a 6-31G** basis set, is contrasted with various models for the electrostatic component, namely (a) ■ the IMPT component, which includes the penetration energy, (b) ▲ the DMAs of MP2 6-31G** wavefunctions for the molecules, (c) × the central multipole expansions of the same wavefunctions and (d) ✷ the Mulliken charges of the same wavefunctions, scaled to reproduce the total dipole moments. For (b) and (c) all terms in the expansion up to R^{-5} were included

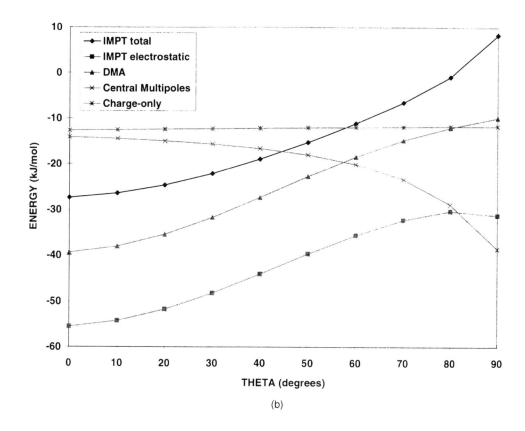

(b)

However, this attraction is not great, certainly not sufficient to significantly affect the position of the repulsive wall. Thus, in this counter-example, the lone pair density affects the intermolecular separation of the chlorine atoms through its effect on the shape of the repulsive wall, which is sufficient to affect the orientation dependence from the electrostatic forces (c.f. the failure of the electrostatic plus hard-sphere model for H—F···Cl—F, Section 9.3).

9.6 BIOLOGICAL MODELLING

Many of the processes in molecular biology depend fundamentally on the intermolecular forces which play a major role determining the secondary structure and ligand binding properties of proteins, and DNA recognition and replication. Many of these interactions, such as hydrogen bonding, are predominantly electrostatic in nature, and hence there is a role for accurate electrostatic modelling in helping to understand the fundamental processes of life. There have been many studies of biological systems involving electrostatic models (Náray-Szabó and Ferenczy, 1995). By using theoretically well-based models such as distributed multipoles, it removes one of the many uncertainties in such studies, and also establishes whether such models are necessary for protein modelling, as suggested by the following examples.

9.6.1 Van der Waals Complexes as Models for Interactions Within Proteins

The study of van der Waals complexes has often been justified as providing information about intermolecular forces which is needed for understanding biological processes. Since accurate modelling of the electrostatic interaction is necessary for understanding the structures of van der Waals complexes, it should also be useful in understanding the interactions within proteins and DNA, and thereby disentangling the roles of various compacting interactions.

Analysis of protein crystal structures shows that phenylalanine side-chains avoid stacking with the phenyl rings directly above each other, like a sandwich. This behaviour is analogous to that observed for the benzene dimer, which does not adopt the sandwich structure predicted by a simple repulsion–dispersion model. Accurate electrostatic models predict, in accord with the developing experimental evidence, that the van der Waals dimer of benzene could be either T-shaped or displaced parallel (Price and Stone, 1987), with a wide range of approximately equi-energetic orientations. The sandwich structure is destabilised by the electrostatic repulsion between the π electrons. This can be simply modelled by additional charges to model the π electron density in order to account for the relative orientation of phenylalanine side-chains (Hunter et al., 1991). This is just one example where a DMA electrostatic model (for the toluene dimer) gives a far better prediction of the experimental side-chain distribution than other models which are widely used in biochemical modelling (Mitchell et al., 1993).

A counter-example, where the van der Waals complex has a different structure from an analogous protein system, is provided by weak hydrogen bonds to aromatic rings. The van der Waals complex of water and benzene involves hydrogen bonding with the water above the aromatic ring (Suzuki et al., 1992), consistent with the Legon–Millen rules and as predicted by all realistic models for the benzene–water complex. However, an analysis of the relative positions of water molecules and phenylalanine residues, based on 48 high-resolution, well-refined protein structures, shows that the dominant interaction is with the edge of the ring and not with the π electrons (Flanagan et al., 1995). A distributed multipole-based model for water interacting with a phenyl alanine-dipeptide showed that indeed a minimum energy site exists above the aromatic ring, such that a solvent molecule can interact with the π electrons, but only when the site is not blocked by the main chain atoms, or disturbed by them. The highly polar protein backbone is sufficiently close to the phenylalanine ring in many probable protein conformations, that it has to be included in the modelling studies.

A second example is the N–H aromatic interaction. The ammonia–benzene van der Waals complex also shows hydrogen bonding to the benzene π electron density (Rodham et al., 1993). The amino-aromatic hydrogen bond is found in some proteins, but only about 10 per cent of interactions between sp^2 hybridised nitrogen atoms and phenylalanine or tyrosine rings have the nitrogen positioned above the ring, and of these, the hydrogen bonds are outnumbered by about 5:2 by stacked orientations with the N–H parallel to the ring (Mitchell et al., 1994). Since a statistical distribution would have far more hydrogen-bonded than stacked geometries, the amino-aromatic hydrogen bonds appear to be less favoured than stacked interactions. The electrostatic energies for the stacked type of geometry were sufficiently close to those of the hydrogen-bonded geometry to warrant more detailed calculations. Intermolecular Perturbation Theory calculations on guanidinium/benzene, formamide/benzene and imidazolium/benzene models for relative orientations representative of those found in protein structures for arginine, asparagine or glutamine, and histidine with N—H groups above phenylalanine, confirmed that the hydrogen-bonded structures were more stable, but that the stacked geometries were surprisingly not much less stable. Indeed, in all three cases, there was at least one experimental hydrogen bonded structure less favourable than at least one stacked structure. However, the stacked sp^2 nitrogen atoms are more accessible to form conventional hydrogen bonds with other parts of the protein or water than those involved in a N—H$\cdots$aromatic hydrogen bond. Thus, the remarkable rarity of amino-aromatic hydrogen bonds in proteins appears to be the consequence of the ability of the stacked conformation to be stabilised simultaneously by the aromatic interaction and conventional hydrogen bonds. The overall conclusions seems to be that weaker hydrogen bonds to π electron density are generally out-competed by stronger interactions within proteins. Thus extrapolating from the structures of van der Waals dimers, or calculations on small model systems, to protein structures requires considerable care.

9.6.2 Electrostatic Modelling of Nucleic Acid Bases—the Variability of Hydrogen Bonds with Environment

Accurate electrostatic models are also useful in considering the effect of molecular environment on hydrogen bond strength. Since the electrostatic contribution is both large, and long range, there can be considerable variation in the interaction of a given donor and acceptor group with the molecular environment. One example of the use and need for such a model as an improvement on the intuitive counting of hydrogen bonds (or its refinement as embodied in some empirical force-fields) is demonstrated by the interactions of pairs of nucleic acid bases.

Nucleic acid bases have lone pair, π electron and other non-spherical features in the valence electron density, which combined with the established importance of using distributed multipoles to represent hydrogen bonding interactions, points to the need for an accurate model for the dominant contribution. Thus, Price *et al.* (1993) found the sterically accessible minima in the electrostatic energy for each of the ten pairings of guanine, adenine, thymine and cytosine, using the DMAs of a 6-31G** SCF wavefunction for each nucleic acid base with a methyl group at the point of attachment to the sugar in DNA. Twenty-nine of the thirty possible multiple-hydrogen-bonded dimers predicted on geometric grounds were found as electrostatic energy minima. The results agreed semi-quantitatively with minimal basis set SCF calculations on these base-pair structures by Hobza and Sandorfy (1987), in that the doubly hydrogen-bonded pairs ranged from being almost as stable as the Watson–Crick triply hydrogen-bonded GC structure to the absence of a minimum in the case of one dimer predicted on geometric grounds. This is a cytosine dimer structure where the formation of the two hydrogen bonds would result in two nitrogen lone pairs being in close proximity. Our calculations also found some minima with one formal hydrogen bond, and a close CH$\cdots$O intermolecular contact that were more stable than some of the weaker doubly hydrogen bonded structures. Thus, accurate electrostatic models are valuable in estimating the sometimes large effect of neighbouring functional groups on hydrogen bond strengths, and for scanning the potential energy surface to find low energy structures which are stabilised by interactions other than conventional hydrogen bonds.

324   ELECTROSTATIC FORCES IN MOLECULAR INTERACTIONS

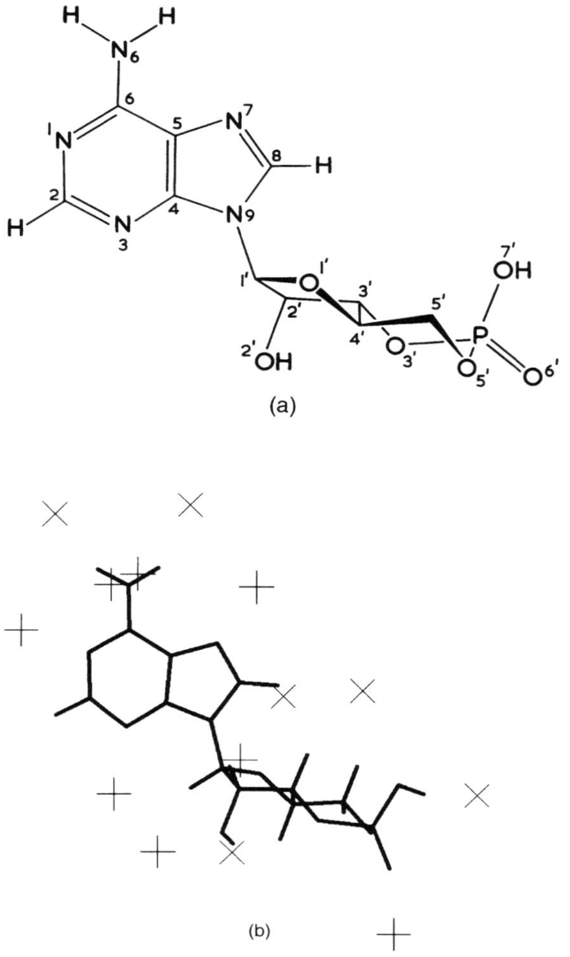

Figure 9.9 (a) The natural substrate, adenosine-3'-5'-cyclic monophosphate (cAMP), and various inhibitors of phosphodiesterase III. Despite the lack of steric and chemical similarity, electrostatically and sterically plausible relative binding orientations were found for all these inhibitors by overlaying three or four of their electrostatic potential extrema with some of those of cAMP illustrated in (b). (b) A 2D projection of the positions of the maxima ($\times$) and minima ($+$) in the electrostatic potential (1.4 Å and 0.5 Å from the van der Waals surface respectively) around cAMP, the natural substrate for phosphodiesterase III

9.6.3 Electrostatic Similarity—a Tool for Protein–ligand Design

Accurate electrostatic models can also be used in drug design to exploit the hypothesis that molecules which bind in the same protein binding site must have similar electrostatic properties. Conceptually, it is obvious that two molecules interacting with the same polar groups of the binding site, will only both bind if any strong electrostatic interactions with the binding site are favourable in both cases. However, actually applying this hypothesis, even qualitatively, to a selection of structurally diverse unsymmetrical ligands is not straightforward, as it involves establishing possible relative orientations of the three-dimensional ligands and the substrate in the unknown binding site. Hence the pharmaceutical industry have considerable interest in an automated method of comparing molecules which will predict the affinity of the ligands for the binding site when the 3D structure of the binding site is unknown. A wide range of molecular matching methods have been proposed (e.g. Perkins *et al.*, 1995), enough to be the subject of a full review, based on various weightings of electrostatic, hydrogen bonding, steric and hydrophobic properties, and considering various regions of space associated with the molecules.

Apaya *et al.* (1995) have developed a novel approach, which focuses only on the electrostatic similarity in the regions where strong interactions with the binding site are plausible, and uses an accurate DMA *ab initio* representation of the electrostatic interactions. It is appropriate for a series of ligands, such as the inhibitors of phosphodiesterase III shown in Figure 9.9(a), where the shapes of the molecules are very different, and the ligands will only occupy part of the enzyme cavity that would be occupied by the natural substrate. For the substrate and ligands, the minima in the electrostatic potential 0.5 Å and the maxima 1.4 Å from the van der Waals surface are found, using the DMA of an *ab initio* wavefunction. These distances are chosen as the approximate separation from the surface of strongly interacting binding site atoms, namely a polar hydrogen at the positions of the minima, which are usually associated with a hydrogen bond acceptors in the ligand, and an oxygen at the maxima which are often near polar hydrogens in the ligand. Figure 9.9(b) shows (in 2D projection) the positions of the many such electrostatic extrema associated with the natural substrate, cyclic adenosine mono-phosphate. Each inhibitor and the substrate are then aligned by minimising the r.m.s. separation of three or four corresponding sign electrostatic extrema in the two molecules. For all the inhibitors of phosphodiesterase III, this process results in a sterically plausible overlay and also matches the sign of significant electrostatic potential in other regions. This method of overlaying electrostatic extrema has also proved a useful tool in selecting the relative conformations of agonists and antagonists of the adenosine A_1 receptor (van der Wenden *et al.*, 1995), rationalising the activity of β-lactam antibiotics (Frau and Price, 1996) and glycolate oxidase inhibitors (Apaya, 1996).

Although there are many assumptions built into the method, not least the rigidity of the unknown binding site and the dominance of the electrostatic forces, the approach does seem to provide useful guidance. This can be partially rationalised in that the method does match up hydrogen-bonding sites, with more subtlety than a geometric rule, and also matches other strong electrostatic interactions, such as those above aromatic rings. However, despite the many uncertainties involved in such drug design methods, it was clear that in both the phophodiesterase and adenosine A_1 systems, the electrostatic similarity is more apparent when the electrostatic properties are calculated accurately, than in the previous, typical medicinal chemistry modelling studies, when they were calculated by atomic charges derived from semi-empirical wavefunctions. Molecular electrostatics plays such an important role in the modelling of biological systems (Náray-Szabó and Ferenczy, 1995), that the choice of model charge distribution is a key assumption in many studies. Distributed multipole models are an intermediate model between point charge models and full integration over molecular wavefunctions, offering better accuracy than the former, at significantly less computational cost than the latter.

9.7 FUTURE OUTLOOK FOR BIOLOGICAL MODELLING—THE ISSUE OF TRANSFERABILITY

The electrostatic term is generally a significant contribution to biological interaction energies, and studies such as those described above show that it is often worth calculating it more accurately than is possible with an atomic charge model. The limitations of the atomic point charge model are becoming accepted by the modelling community, and this has led to the development of extended point charge models, where non-nuclear charge sites represent the electrostatic effects of lone pair and π electron density. Sokalski and Sawaryn (1992) represent their atomic multipoles, up to quadrupole, by a set of nine charges about each atom, for calculating the interactions between protein fragments. Rauhut and Clark (1993) calculate the positions and magnitudes of the eight charges that represent the atomic electron cloud directly from the natural atomic orbitals of the molecular wavefunction. Vinter (1994) has derived extended point charge electron distributions more empirically, using fewer sites, to improve the intermolecular geometry predictions of the COSMIC force-field, building on the π electron charges of Hunter *et al.* (1991). Such extended point charge representations of the molecular charge distribution are also capable of providing an accurate representation of the electrostatic potential outside the molecule, and so the difference from distributed multipole models is more in practical technicalities than in philosophy. Hence, it is worth considering the other approximations which are required to move accurate electrostatic potentials into polypeptide modelling, relative to the errors involved in neglecting the electrostatic effects of lone pair and π electron density. These are mainly the issues of transferability—how accurately can the charge distribution calculated for a molecule in one conformation represent its electrostatic interactions in another, and how well can we build up an electrostatic model for a polypeptide from calculations on smaller model molecules?

9.7.1 Variation with Conformation

Most molecular modelling assumes that the charge distribution of the molecule does not change with conformation, i.e. the atomic charges are constant. It has been noted that potential-derived charges are dependent on the conformation used in their derivation (Reynolds *et al.*, 1992, Stouch and Williams, 1993). The same effect is apparent with the more accurate representation of the charge distribution provided by a distributed multipole analysis. If a DMA is performed on a two-residue model polypeptide $(CH_3CO.(NHCH_2CO)_2NHCH_3)$ in the extended conformation, and then the multipoles and atoms transformed to represent the molecule in the alpha-helical conformation, the polarity of the alpha-helix section is significantly overestimated (Price and Stone, 1992).

The r.m.s. error in the electrostatic potential at points on the water-accessible surface is 19 kJ/mol when compared with the values calculated from a DMA of an equivalent wavefunction for the alpha-helical conformation. The contribution to this error from the through space polarisation of one residue by another is relatively small, and can be estimated quite well using polarizability calculations. This observation is supported by the excellent prediction of the potential around the blocked two-residue polypeptide by a model made up from blocked single-residue calculations, with the same torsion angles, which gives an r.m.s. error of only 4 kJ/mol on the water-accessible surface around the alpha-helix conformation. Thus the major change in the charge density with conformation appears to come from the local migration of charge through the bonds with changes in the torsion angles. For example, when the peptide is in the extended conformation, the two amide groups are coplanar and so there is the possibility of hyperconjugation, which is lost on changing to the alpha-helix conformation.

It is chemically reasonable for the charge distribution of an atom to change with conformation, and the calculations suggest that this can have a significant effect on the strength of intermolecular interactions, depending on the nature of conformational change. For example, Koch and Stone (1996) estimated the effect of hydration on rotational barriers for the methyl groups in N-methylacetamide, OH rotation in ethanol and the $cis/trans$ isomerisation of N,N-dimethylacetamide. These calculations showed the errors involved in neglecting the change in the atomic charge densities with conformation on the electrostatic interaction with water. These changes can be represented by conformation-dependent multipole moments expressed in terms of a Fourier series in the dihedral angle. Ignoring this effect produces qualitatively wrong results for the isomerisation reaction, where the change in electronic structure is most significant. Thus, more insight into the electron density variation with conformation of differing functional groups will be required to provide a reliably accurate and efficient scheme for modelling the electrostatic properties of flexible molecules.

9.7.2 Transferability between Molecules

Distributed multipole electrostatic models cannot be used in modelling polypeptides or protein-binding sites if this requires an *ab initio* wavefunction for the entire system. For all such modelling, it is necessary to build up models by transferring the electrostatic model from smaller model molecules. This produces another limitation to the accuracy of the electrostatic model.

The transferability of atomic multipoles was tested by Faerman and Price (1990) by comparing the DMAs of a range of amides and small peptides, seeking a scheme whereby DMA models of polypeptides could be built up by transferring multipoles from smaller molecules. Short-range inductive effects were clearly apparent in the values of the multipoles: for example, the charge component on an amide nitrogen —CO.N.$R_c R_t$ varied from -0.805 ± 0.015 e when both bonded R groups were hydrogen, to -0.547 ± 0.016 e where $R_t = H$, $R_c = CH_3$, and -0.289 ± 0.018 e when both were methyl groups. The higher multipole moments showed the same pattern of being approximately equivalent on atoms which were bonded to the same functional groups, i.e. subject to the same short-range inductive effects. When atomic multipole sets were transferred between molecules, taking account of the bonding environment, the resulting model molecular charge distributions were very close to neutral and gave reasonable predictions of the electrostatic properties.

The conclusion that an atomic charge distribution is only even approximately transferable between molecules if it is bonded to the same functional groups (giving the same short range inductive effects), seems to hold in other studies of the transferability of atomic charge distributions. These include the transferability of atomic charge densities by Atoms-in-Molecules partitioning for hydrocarbons $CH_3(CH_2)_mCH_3$ (Bader et al., 1987), and the Lego approach of Walker and Mezey (1993) to building up approximate molecular charge densities from fragments.

This observation has not been widely applied in determining model potentials. It implies that the smallest model molecule that can be used in building up a polypeptide backbone is a N-acetyl-N'acetamide-blocked residue, and that the more convenient formamide is likely to be misleading. The adequacy of $CH_3.CO.\mathbf{NH.CHR.CO}.NH.CH_3$ as a model molecule for building up polypeptides is supported by calculations by Price et al. (1991) on these molecules, in the alpha-helical conformation, for all the naturally occurring protein residues. In all cases, the net charge on the central residue was less than 0.020 e, except Glu⁻ and Asp⁻ which differed by 0.032 e and 0.024 e, respectively, from the formal net charge. Thus the errors involved in removing these small net charges and building up a polypeptide from the residues are small. Hence, the prospects for building up accurate model charge distributions for large molecules are good, now that we can use moderate sized organic molecules for the fragments which match the short-range inductive effects, provided the effects of conformational changes on the charge distribution are considered.

9.8 OVERVIEW AND FUTURE DIRECTIONS

The most important conclusion from the use of electrostatic models which provide a reasonably accurate representation of *ab initio*-based charge distributions is that the electrostatic forces generally dominate the orientation dependence of the intermolecular potential between neutral molecules. Thus an accurate electrostatic model, plus an approximate model of the molecular shapes, can be used to find the low-energy structures of the system. Fine details of the structure, such as van der Waals contact distances or total energies, obviously require an accurate model for all the contributions to the potential. This can be done using *ab initio* supermolecule or Intermolecular Perturbation Theory calculations. However, such methods are very limited in both the size of system that can be studied and the proportion of the potential energy surface that can be surveyed. Thus such studies may miss less obvious low-energy structures, such as the T-shaped structure of the *s*-tetrazine dimer, which is sufficiently close in energy to the hydrogen-bonded structure that both have been observed experimentally (Haynam et al., 1983). Both structures were found as minima in a scan of the accessible potential energy surface of the *s*-tetrazine dimer with a DMA electrostatic model (Price and Stone, 1987). Thus accurate electrostatic models are providing the starting point potential energy surface for modelling of van der Waals complexes (Wales et al., 1995).

The reason for the empirical success of this simple model is that generally the electrostatic interaction is the most orientation-dependent contribution. The dispersion energy is always attractive, and a crude hard-sphere model is generally successful (to within the few tenths of an Angstrom that may be associated with the repulsion anisotropy) because the repulsive wall rises exponentially. In contrast, the electrostatic potential will change sign frequently on the van der Waals surface of a typical organic molecule, and is significant for all but the saturated hydrocarbons. The complex orientation dependence of the electrostatic contribution results in it generally dominating the orientation dependence of the total potential. Thus a scan of the electrostatic potential energy surface is a valuable preliminary in locating possible low-energy structures, or detecting whether the potential energy surface could be sufficiently flat that a dynamical model must be considered.

Further work is required to quantify other contributions to the intermolecular potential for a more detailed knowledge of the potential energy surface, particularly when detailed structure, absolute energies or dynamical properties are required. There are theories linking the other contributions to the intermolecular potential, particularly dispersion, polarisation and repulsion, to the charge distributions of the isolated molecules which will need to be extended from small molecules to organics, as discussed by Price (1996a).

Accurate distributed multipole models are just one method of representing the molecular charge distribution. Comparable accuracy can be obtained, in principle, by using sufficient non-nuclear point charge sites, or even sufficient terms in the central multipole expansion for orientations where this is valid. Central multipoles are most successful for small, almost spherical molecules, whereas point charges are more appropriate for biological molecules. Distributed multipoles combine the advantages of both multipole expansions and atomic charges to provide a theoretically well-justified model. This approach can be further improved, by introducing the spatial extent of the charge distribution and thus including the effects of interpenetration of the charge distributions by using the recently proposed Gaussian multipoles (Wheatley and Mitchell, 1994).

The main disadvantage of DMAs is the complexity of the resultant model. The importance of using DMAs will be very dependent on the accuracy required, the molecule and the region around the molecule being sampled in the simulation. It seems likely that a major future use of accurate electrostatic models may be to provide a benchmark against which simpler electrostatic models, more suitable for simulations, can be developed.

Thus, whether as a means of locating starting points for high-level detailed calculations, for guiding the development of more approximate model potentials, or as a legitimate alternative to empirical force fields, accurate models for the electrostatic interaction will continue to contribute to our understanding of molecular interactions.

ACKNOWLEDGEMENTS

In a chapter such as this it is impossible to reference all the work that has contributed to electrostatic modelling, or acknowledge all the coworkers who have made my own work in the area so enjoyable. However, specific thanks go to my current graduate students, Robert Apaya, David Coombes and Irene Nobeli for their help with this chapter. Most of this work has been performed with the financial support of the EPSRC.

REFERENCES

Alderton, M. J. (1983) PhD thesis, University of Cambridge.
Allen, F. H., Davies, J. E., Johnson, O. J., Kennard, O., Macrae, C. F., Mitchell, E. M., Mitchell, G. F., Smith, J. M. and Watson, D. G. (1991) *J. Chem. Inf. Comp. Sci.*, **31**, 187.
Almlöf, J., Faegri, K. and Korsell, K. (1982) *J. Comput. Chem.*, **3**, 385.
Amos, R. D. (1985) *Chem. Phys. Lett.*, **113**, 19.
Apaya, R. P. (1996) PhD thesis, University of London.
Apaya R. P., Lucchese, B., Price, S. L. and Vinter, J. G. (1995) *J. Comput. Aided Mol. Design*, **9**, 33.
Bader, R. F. W. (1990) *Atoms in Molecules. A Quantum Theory*, Clarendon Press, Oxford.
Bader, R. F. W., Larouche, A., Gatti, C., Carroll, M. T., MacDougal, P. J. and Wiberg, K. B. (1987) *J. Chem. Phys.*, **87**, 1142.
Battaglia, M. R., Buckingham, A. D. and Williams, J. H. (1981) *Chem. Phys. Lett.*, **78**, 421.
Bauer, G. E. W. and Huiszoon, C. (1982) *Mol. Phys.*, **47**, 565.
Bernstein, F. C., Koetzle, T. F., Williams, G. J. B., Meyer, E. F., Brice, M. D., Rogers, J. R., Kennard, O., Shimanouchi, T., and Tasumi, M. (1977) *J. Mol. Biol.*, **112**, 535.
Buckingham, A. D. and Fowler, P. F. (1985) *Canad. J. Chem.*, **63**, 2018.
Coombes, D. S., Price, S. L., Willock, D. J. and Leslie, M. (1996) *J. Phys. Chem.*, **100**, 7352.
Cooper, D. L. and Stutchbury, N. C. J. (1985) *Chem. Phys. Lett.*, **120**, 167.
Cox, S. R., Hsu, L-Y and Williams, D. E. (1981) *Acta Cryst.*, **A37**, 293.
Cox, S. R. and Williams, D. E. (1981) *J. Comput. Chem.*, **2**, 304.
Desiraju, G. R. and Parthasarathy, R. (1989) *J. Am. Chem. Soc.*, **111**, 8725.
Dykstra, C. E. (1993) *Chem. Rev.*, **93**, 2339.
Faerman, C. H. and Price, S. L. (1990) *J. Am. Chem. Soc.*, **112**, 4915.
Filippini, F. and Gavezzotti, A. (1993) *Acta Cryst.*, **B49**, 868.
Flanagan, K., Walshaw, J., Price, S. L. and Goodfellow, J. M. (1995) *Protein Engineering*, **8**, 109.
Frau, J. and Price, S. L. (1996) *J. Comput. Aided Mol. Design*, **10**, 107.
Gavezzotti, A. (1991) *J. Am. Chem. Soc.*, **113**, 4622.
Gavezzotti, A. (1994) *Acc. Chem. Res.*, **27**, 309.
Gavezzotti, A. and Filippini, G. (1994) *J. Phys. Chem.*, **98**, 4831.
Gray, C. G. and Gubbins, K. E. (1984) *Theory of Molecular Fluids, Vol. 1: Fundamentals*, Clarendon Press, Oxford.
Haynam, C. A., Brumbaugh, D. V. and Levy, D. H. (1983) *J. Chem. Phys.*, **79**, 1581.
Hättig, C. and Heß, B. A. (1994) *Molec. Phys.*, **81**, 813.
Hobza, P. and Sandorfy, C. (1987) *J. Am. Chem. Soc.*, **109**, 1302.
Holden, J. R., Du, Z. and Ammon, H. L. (1993) *J. Comput. Chem.*, **14**, 422.
Hunter, C. A., Singh, J. and Thornton, J. M. (1991) *J. Mol. Biol.*, **218**, 837.
Hurst, G. J. B., Fowler, P. W., Stone, A. J. and Buckingham, A. D. (1986) *Int. J. Quant Chem.*, **29**, 1223.
Karfunkel, H. R. and Gdanitz, R. J. (1992) *J. Comput. Chem.*, **13**, 1171.
Koch, U. and Stone, A. J. (1996) *J. Chem. Soc. Farad. Trans.*, **92**, 1701.
Laidig, K. E. (1993) *J. Phys. Chem.*, **97**, 12760.
Legon, A. C. and Millen, D. J. (1987) *Acc. Chem. Res.*, **20**, 39.
Mitchell, J. B. O., Nandi, C. L., McDonald, I. K., Thornton, J. M. and Price, S. L. (1994) *J. Mol. Biol.*, **239**, 315.
Mitchell, J. B. O., Nandi, C. L., Thornton, J. M., Price, S. L., Singh, J. and Snarey, M. (1993) *J. Chem. Soc. Farad. Trans.*, **89**, 2619.
Mitchell, J. B. O. and Price, S. L. (1989) *Chem. Phys. Lett.*, **154**, 267.
Mitchell, J. B. O. and Price, S. L. (1990) *J. Comput. Chem.*, **11**, 1217.
Momany, F. A. (1978) *J. Phys. Chem.*, **82**, 592.
Mulliken, R. S. (1955) *J. Chem. Phys.*, **23**, 1833.
Murthy, C. S., O'Shea, S. F. and McDonald, I. R. (1983) *Molec. Phys.*, **50**, 531.
Náray-Szabó, G. and Ferenczy, G. G. (1995) *Chem. Rev.*, **95**, 829.
Nobeli, I., Price, S. L., Lommerse, J. P. M. and Taylor, R., in preparation.
Nyburg, S. C. and Faerman, C. H. (1985) *Acta Cryst.*, **B41**, 274.

Perkins, T. D. J., Mills, J. E. J. and Dean, P. M. (1995) *J. Comp. Aided Mol. Design*, **9**, 479.
Pertsin, A. J. and Kitaigorodsky, A. I. (1987) *The Atom–Atom Potential Method. Applications to Organic Molecular Solids*, Springer-Verlag, Berlin.
Price, S. L. (1985) *Chem. Phys. Lett.*, **114**, 359.
Price, S. L. (1987) *Molec. Phys.*, **62**, 45.
Price, S. L. (1996a) *Phil. Mag.*, **73**, 95.
Price, S. L. (1996b) in *Computer Modelling in Inorganic Crystallography*, edited by Catlow, C. R. A., Academic Press, London, in press.
Price, S. L., Andrews, J. S., Murray, C. W. and Amos, R. D. (1992) *J. Amer. Chem. Soc.*, **114**, 8268.
Price, S. L., Faerman, C. H. and Murray, C. W. (1991) *J. Comput. Chem.*, **12**, 1187.
Price, S. L., Harrison, R. J. and Guest, M. F. (1989) *J. Comput. Chem.*, **10**, 552.
Price, S. L., Lo Celso, F., Treichel, J. A., Goodfellow, J. M. and Umrania, Y. (1993) *J. Chem. Soc. Farad. Trans.*, **89**, 3407.
Price, S. L. and Richards, N. J. (1991) *J. Comput. Aided Mol. Des.*, **5**, 41.
Price, S. L. and Stone, A. J. (1983) *Chem. Phys. Lett.*, **98**, 419.
Price, S. L. and Stone, A. J. (1987) *J. Phys. Chem.*, **86**, 2859.
Price, S. L. and Stone, A. J. (1992) *J. Chem. Soc. Farad. Trans.*, **88**, 1755.
Price, S. L., Stone, A. J. and Alderton, M. (1984) *Molec. Phys.*, **52**, 987.
Price, S. L., Stone, A. J., Lucas, J., Rowland, R. S. and Thornley, A. E. (1994) *J. Am. Chem. Soc.*, **116**, 4910.
Pullman, A. and Perahia, D. (1978) *Theor. Chim. Acta*, **48**, 29.
Rauhut, G. and Clark, T. (1993) *J. Comput. Chem.*, **14**, 503.
Reynolds, C. A., Essex, J. W. and Richards, W. G. (1992) *Chem. Phys. Lett.*, **199**, 257.
Ritchie, J. P. (1985) *J. Am. Chem. Soc.*, **107**, 1829.
Ritchie, J. P. and Copenhaver, A. S. (1995) *J. Comput. Chem.*, **16**, 777.
Rodham, D. A., Suzuki, S., Suenram, R. D., Lovas, F. J., Dasgupta, S., Goddard, W. A. and Blake, G. A. (1993) *Nature*, **362**, 735.
Ryan, M. D. (1994) *Modelling the Hydrogen Bond, ACS Symposium Series*, **569**, 36.
Singh, U. C. and Kollman, P. A. (1984) *J. Comput. Chem.*, **5**, 129.
Sokalski, W. A. and Sawaryn, A. (1987) *J. Chem. Phys.*, **87**, 526.
Sokalski, W. A., Keller, D. A., Ornstein, R. L. and Rein, R. (1993) *J. Comput. Chem.*, **14**, 970.
Sokalski, W. A. and Sawaryn, A. (1992) *J. Mol. Struct. (Theochem.)*, **256**, 91.
Spackman, M. A. (1986) *J. Chem. Phys.*, **85**, 6587.
Spackman, M. A. (1996) *J. Comput. Chem.*, **17**, 1.
Stewart, R. F. and Jensen, L. H. (1969) *Z. Kristalogr.*, **128**, 133.
Stoicheff, B. P. (1954) *Can. J. Phys.*, **32**, 339.
Stone, A. J. (1981) *Chem. Phys. Lett.*, **83**, 233.
Stone, A. J. (1991) in *Theoretical Models of Chemical Bonding*, Vol. 4, edited by Maksić, Z. B., Springer-Verlag, Berlin.
Stone, A. J. and Alderton, M. (1985) *Molec. Phys.*, **56**, 1047.
Stone, A. J., Popelier, P. L. and Wales, D. J. (1994) *ORIENT: a program for calculating electrostatic interactions*, Version 3, University of Cambridge.
Stone, A. J. and Price, S. L. (1988) *J. Phys. Chem.*, **92**, 3325.
Stouch, T. and Williams, D. E. (1993) *J. Comput. Chem.*, **14**, 858.
Suzuki, S., Green, P. G., Bumgarner, R. E., Dasgupta, S., Goddard, W. A. and Blake, G. A. (1992) *Science*, **257**, 942.
Taylor, R., Kennard, O. and Versichel, W. (1983) *J. Am. Chem. Soc.*, **105**, 5761.
van der Wenden, E. M., Price, S. L., Apaya, R. P., IJzerman, A. P. and Soudijn, W. (1995) *J. Comput. Aided Mol. Design*, **9**, 44.
Vigné-Maeder, F. and Claverie, P. (1988) *J. Chem. Phys.*, **88**, 4934.
Vinter, J. G. (1994) *J. Comput. Aid. Mol. Design*, **8**, 653.
Wales, D. J., Stone, A. J. and Popelier, P. L. A. (1995) *Chem. Phys. Lett.*, **240**, 89.
Walker, P. D. and Mezey, P. G. (1993) *J. Am. Chem. Soc.*, **115**, 12423.
Weiner, S. J., Kollman, P. A., Nguyen, D. T. and Case, D. A. (1986) *J. Comput. Chem.*, **7**, 230.
Wheatley, R. J. (1993) *Chem. Phys. Lett.*, **208**, 159.
Wheatley, R. J. and Mitchell, J. B. O. (1994) *J. Comput. Chem.*, **15**, 1187.

Wiberg, K. B. and Rablen, P. R. (1993) *J. Comput. Chem.*, **14**, 1504.
Williams, D. E. and Cox, S. R. (1984) *Acta Cryst.*, **B40**, 404.
Williams, D. E. and Weller, R. R. (1983) *J. Am. Chem. Soc.*, **105**, 4143.
Willock, D. J., Price, S. L., Leslie, M. and Catlow, C. R. A. (1995) *J. Comput. Chem.*, **16**, 628.
Williams, D. E. and Starr, T. L. (1977) *Comput. Chem.*, **1**, 173.

10 Protein–ligand Interactions

G. NÁRAY-SZABÓ

Lóránd Eötvös University, Hungary

10.1 INTRODUCTION

Binding of small ligands to protein crevices is one of the key events in a variety of biochemical processes. Some examples are drug action, enzyme-catalysed chemical reactions, immune response and signal transduction. As an answer to the challenge made by the primary importance of protein–ligand interactions, several molecular modelling methods have been developed in the past fifteen years with the goal to describe quantitatively their steric and energetic aspects both from the static and dynamic points of view. For recent reviews see Davis and McCammon (1990), Sharp and Honig (1990), Warshel and Åqvist (1991) and Kollman (1993), while for electrostatic aspects see Náray-Szabó (1996). Efficient methodology, combined with the continuously growing number of known three-dimensional protein structures, have often made it possible to apply computer-aided molecular design to the development of new drugs and diagnostics.

In the vast majority of cases protein–ligand interactions are governed by non-covalent intermolecular forces, especially electrostatics. A qualitative and pictorial rationale is offered by the concept of molecular recognition (Behr, 1994). However, if quantitative estimation of the effects related to protein–ligand association is required, sophisticated models and molecular orbital or force-field methods are needed. In this chapter an overview is given of qualitative and quantitative aspects of protein–ligand interactions with more emphasis on the latter. After discussing the principles and methodology, we present some applications in quantitative structure–activity relationships and computer-aided drug design.

10.2 PRINCIPLES

The simplest yet very effective and general concept to treat protein–ligand interactions is *complementarity*. If we do not consider dynamic aspects and remain in the framework of the rigid-body approximation we may consider the association between the protein (host) and its ligand (guest) as fitting a key into its lock (Fischer, 1894). Complementarity is determined by three major factors: steric, electrostatic and hydrophobic (Náray-Szabó, 1989a; 1993; Náray-Szabó and Ferenczy, 1995). Most important is the steric fit but host and guest must match electrostatically, too, i.e. the interaction between them should be attractive. Hydrophobic complementarity is related to hydration and dehydration of the ligand upon complex formation and can be formulated as the matching between regions of the host and guest that are of similar polarity (have the same ability to bind water molecules) (Gráf et al., 1988). This property is well characterised by the molecular electrostatic field (MEF) and the gradient of the molecular electrostatic potential (MEP). Thus we shall discuss hydrophobic aspects of complementarity in terms of matching of the MEF patterns produced both by the protein and the ligand on the same contact surface.

Let us discuss the above criteria on a more quantitative basis. The host–guest interaction-free energy can be written as follows:

$$\Delta G_{int} = \Delta E_{vac} - T\Delta S_{vac} - (\Delta E_{hydr} - T\Delta S_{hydr}) \qquad (10.1)$$

where ΔE_x is the interaction energy *in vacuo* for $x =$ vac, while for $x =$ hydr it denotes the change in the hydration energy upon association. $T\Delta S_x$ is the corresponding entropy term which is related, for example, to freezing of rotational degrees of freedom upon association for $x =$ vac. This is neglected here since we restrict our treatment to rigid ligands.

ΔE_{vac} can be decomposed into various terms, most of which have a direct physical meaning. According to the decomposition scheme of Morokuma (1971) (see also Kitaura and Morokuma, 1976),

$$\Delta E_{vac} = \Delta E_{es} + \Delta E_{pol} + \Delta E_{ex} + \Delta E_{ct} + \Delta E_{disp} + E_{mix} \qquad (10.2)$$

where the subscripts refer to electrostatic, polarisation, exchange, charge transfer, dispersion and mixed terms, respectively.

ΔE_{ex} and ΔE_{disp} are responsible for the steric fit. The exchange repulsion energy increases abruptly as the non-bonding partners get closer to each other than the sum of their van der Waals radii. On the other hand, ΔE_{disp} represents attractive and non-directional dispersion forces depending on inverse higher powers of the interatomic distance. Its value becomes optimal if the binding crevice is filled by the ligand as perfectly as possible. In an aqueous medium this is explained in terms of density differences between water and the protein, the latter being more dense. Thus interacting atoms may get closer to each other than in the hydrated case (Kollman, 1984). Another explanation of the steric fit is based on entropy considerations. According to the concept of macromolecular crowding (Zimmerman and Minton, 1993) the water-accessible surface of dissolved molecules tends to reduce in order to avoid unfavourable perturbation of water structure around the solute. Combining the above effects it is rationalised that the better the steric complementarity, the larger the gain in free energy upon steric interaction between the protein and its ligand.

Electrostatic complementarity is accounted for by ΔE_{es} in equation (10.2). This term is defined by the following formula:

$$\Delta E_{es} = \int \rho_P(1)\rho_L(1)/r_{12} \, dv_1 \, dv_2 \qquad (10.3)$$

where ρ denotes the sum of electronic and nuclear charge distributions and P and L refer to the protein and the ligand. ΔE_{es} can be written in terms of the protein MEP, V_P,

$$\Delta E_{es} = \int V_P(1)\rho_L(1)\,dv_1 \qquad (10.4)$$

with

$$V_P(1) = \int \rho_P(2)/r_{12}\,dv_2 \qquad (10.5)$$

An intuitive reformulation of equation (10.4) is due to Kollman (1977):

$$\Delta E_{es} \approx C \sum_i V_{Pi} V_{Li} \qquad (10.6)$$

where V_{Pi} and V_{Li} are the MEP values due to the protein and the ligand in a set of appropriately defined reference points, $\{i\}$. Equation (10.6) has no strict physical meaning since the potential product may not have an energy dimension, but it may be assumed that, for example, V_{Li} is proportional to an appropriately defined charge at reference point i. According to equation (10.6) we can define perfect electrostatic complementarity as the requirement to produce MEP values both by P and L that are of opposite sign at any point i belonging to their contact surface (Nakamura et al., 1985).

Using molecular graphics we demonstrate protein–ligand complementarity on the example of binding of the xylitol inhibitor in the active-site cleft of D-xylose isomerase. In Figure 10.1 the enzyme-inhibitor complex is displayed in a solid-surface representation. The steric fit is not perfect, and there are some regions behind the xylitol molecule that remain empty. Such empty spaces are often filled by water molecules that reinforce binding (Náray-Szabó and Nagy, 1989). Figure 10.2 presents electrostatic complementarity of the Asp...His...substrate triad with its protein environment. It is seen that positive and negative potential regions of the triad and its protein surroundings match to a considerable extent.

Hydrophobic complementarity is related to the hydration-free energy term in equation (10.1), $\Delta E_{hydr} - T\Delta S_{hydr}$. The hydration energy of a solute at a given site is approximately proportional to the value of the MEF here (Peinel et al., 1980; Nagy et al., 1987; Náray-Szabó, 1989b). Thus, those parts of the associating molecules that are characterised by small MEF values, i.e. small hydration energy (hydrophobic regions), tend to associate in order to minimise unfavourable entropy effects by expelling water from the contact surface. On the other hand, hydrophilic regions with large MEF values also tend to associate assuming that the local charges, producing the large field, are opposite in sign, i.e. electrostatic complementarity is maintained. We may formulate the *similis simili gaudet* principle (Gráf et al., 1988), stating that molecular regions characterised by similar MEF values tend to associate more strongly than those with dissimilar ones. This principle has been used to explain the specificity of point mutants of trypsin and subtilisin (Náray-Szabó, 1989c). It has been found experimentally that the activation-free energy of the enzymatic reaction is larger for charged–charged and polar–polar pairs than for charged–polar ones. This may be unusual since, as is known, in the gas phase charged–polar (monopole–dipole) interactions are stronger than polar–polar (dipole–dipole) ones.

Though protein–ligand complementarity can be discussed in several cases by assuming that both interacting partners remain rigid upon association, it should be stressed that this is not always true. Sometimes the *hand-and-the-glove* analogy works better, i.e. both the biopolymer and its interacting partner alter their shapes during the binding process. According to the induced-fit model of Koshland (1960), the final conformation of the crevice and the fitting ligand might be essentially different from the initial ones. A precise modelling of this process is possible only by applying molecular mechanics combined with an appropriate geometry optimisation algorithm. Qualitative principles do not work sufficiently, and sophisticated computational methods are needed that yield information on both energetics and geometry changes due to the ligand binding process.

10.3 METHODOLOGY

It is clear that protein–ligand complexes with several thousands of atoms and dozens of thousands of electrons cannot be treated by any available all-atom molecular orbital or density functional methods even using the largest supercomputers. There are at present two essential approaches to overcome this problem. Molecular force fields work with simplified expressions of the molecular energy accounting for its variation upon bond stretching, bending, and torsion, as well as upon changing of non-bonding atom–atom distances. A cross-term describing coupled motions may be also considered. Another possibility is to dissect the protein–ligand complex into an active site where essential changes take place and an environment influencing these events. This allows a simplified mathematical treatment, considerably reducing the computational work.

10.3.1 Molecular Force Fields

The change in the total energy of a protein may be approximated as a sum of the following contributions (see e.g. Dinur and Hagler, 1990; van Gunsteren, 1988):

$$\Delta E = \Delta E_{\text{stretch}} + \Delta E_{\text{bend}} + \Delta E_{\text{tors}} + \Delta E_{\text{cross}} + \Delta E_{\text{nb}} \qquad (10.7)$$

The terms in equation (10.7) are defined as follows

$$\Delta E_{\text{stretch}} = \sum_a k_{ra}(r_a - r_{0a})^2 + \text{ higher terms} \qquad (10.8)$$

where summation runs over all covalent bonds of the protein molecule, and k_{ra} and r_{0a} denote the stretching force constant and equilibrium distance of the bond a:

$$\Delta E_{\text{bend}} = \sum_b k_{\theta b}(\theta_b - \theta_{0b})^2 + \text{ higher terms} \qquad (10.9)$$

In equation (10.9) summation includes all angles formed by covalent bonds having just one common centre and $k_{\theta b}$ and θ_{0b} stand for the bending force constant and the equilibrium value of the bond angle b:

$$\Delta E_{\text{tors}} = \sum_c (V_c^1 \cos \varphi_c + V_c^2 \cos 2\varphi_c + V_c^3 \cos 3\varphi_c) \qquad (10.10)$$

For the torsional energy term summation refers to all torsional angles of the protein defined by two covalent bonds each having just one centre belonging to a third bond. V_c^1 refers to the barrier to rotation around the bond c, while V_c^2 and V_c^3 are parameters for higher-order correction terms that may be put equal to zero in simpler force fields.

Figure 10.1 Steric complementarity between D-xylose isomerase and its xylitol inhibitor. The active-site cleft is represented by the solid surface in green. Figure made using the GRASP program (Nichols *et al.*, 1991) by Ms T. Gérczei (Budapest)

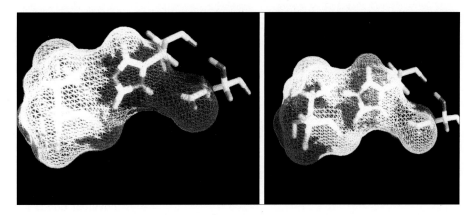

Figure 10.2 Electrostatic complementarity betwen D-xylose isomerase and its xylitol substrate. The MEP of the Asp...His...substrate triad (represented by capped sticks) is displayed on the van der Waals surface (*left*) and complements the potential emerging from the protein (*right*). Dark red: -10, light red: -5, white 0, light blue 5, dark blue 10 kcal/mol. Figure made using the GRASP program (Nichols *et al.*, 1991) by Ms T. Gérczei (Budapest)

Cross-terms involve products corresponding to stretching–bending, stretching–torsion or bending–torsion interactions:

$$\Delta E_{\text{cross}} = \sum_a \sum_b k_{r\theta ab}(r_a - r_{0a})(\theta_b - \theta_{0b}) + \sum_a \sum_c k_{r\varphi ac}(r_a - r_{0a})(\varphi_c - \varphi_{0c})$$
$$+ \sum_b \sum_c k_{\theta\varphi bc}(\theta_b - \theta_{0b})(\varphi_c - \varphi_{0c}) \quad (10.11)$$

The last term in equation (10.7) describes non-bonding interactions:

$$\Delta E_{\text{nb}} = \sum_{a>b}[q_a q_b/(\varepsilon r_{ab}) + (A_{ab}/r_{ab}^n - B_{ab}/r_{ab}^m)] \quad (10.12)$$

The first expression in the square brackets refers to electrostatic and the second to van der Waals interactions. q_a is the net charge on atom a, r_{ab} is the distance between atoms a and b, while ε is the dielectric constant. A_{ab} and B_{ab} are adjustable parameters, and the exponents may be also varied. In most cases $m = 6$, while $n = 9\ldots12$.

Various peptide and protein force fields (ECEPP, Momany et al., 1975; AMBER, Weiner et al., 1984; CHARMM, Brooks et al., 1983; GROMOS, Berendsen et al., 1981; OPLS, Jorgensen and Swenson, 1885) differ mostly in the parameters of equations (10.8)–(10.12) that are obtained from fitting calculations to experimental data or deriving them from precise *ab initio* molecular orbital calculations. Several protein properties, such as torsional angle preferences and basic principles of secondary structure formation, are correctly reproduced. Simulated crystal structures of proteins correspond to experimental ones within 100 pm root-mean square deviation averaged over the backbone atoms (van Gunsteren, 1988).

While the bonding parameters of various molecular force fields are relatively easy to derive, the non-bonding terms, playing a crucial role in protein–ligand binding, are much more problematic, the more so, since calculation of ligand binding energies requires a high degree of accuracy since energy differences in the order of 10 kJ/mol are obtained. It is a general problem that, if comparing empirical force fields, large differences can be observed between non-bonding parameters that represent the same interaction (see e.g. Rullman and van Duijnen, 1990).

The parameter ε, often called dielectric constant, is a number accounting for the effect of protein-core polarisation, counterions partly neutralising ionised surface side-chains and water as a solvent on the gas-phase electrostatics of the protein molecule. Its value may range from 1.4 to 4 in various force fields, while it is also possible to use a distance-dependent dielectric constant, $\varepsilon = r_{ab}$ (Gelin and Karplus, 1979). It should be stressed, however, that neither the protein nor its surrounding can be treated as a dielectric continuum, thus the use of any dielectric constant is only a very rough estimation. This is the main reason for the divergence of estimates of hydrogen-bond strengths by various force fields (Rullman and van Duijnen, 1990). We may state, in general, that non-bonding terms of equation (10.12) have a meaning only as a complete set and not as individual contributions to the total interaction energy with a well-defined physical meaning.

10.3.2 Active Site–environment Partition

Association of a small molecule with a protein can be effectively modelled by partitioning the macromolecule and its surroundings into a ligand-binding region and environment. While the former can be described at a sophisticated level (by *ab initio* molecular orbital methods or molecular dynamics) the latter can be represented much more simply by using a simple force field and a polarisable or non-polarisable medium exerting only electrostatic effects on the ligand. The idea was first formulated by Warshel and Levitt (1976) for an enzymatic reaction, and it was later refined and applied to a variety of systems by several authors (Tapia *et al.*, 1978; Thole and van Duijnen, 1982; Náray-Szabó and Surján, 1983; Bash *et al.*, 1987a; Gao, 1992).

A plausible partitioning scheme is as follows. The central machinery (**C**) is surrounded by a polarisable environment (**P**), which is embedded in a non-polarisable region of the protein (**N**) and water, containing pH-dependent functional groups and counterions, as solvent (**W**). In most cases **C** is treated by some semiempirical or *ab initio* molecular orbital method. However, for the binding of small molecules it is often enough to represent the ligand by a set of appropriate point charges. Region **P** may be considered as an ensemble of polarisable atoms with empirical polarisabilities (Warshel and Levitt, 1976) or by a reaction field coupled to **C** (Tapia *et al.*, 1978; Thole and van Duijnen, 1982). The non-polarisable region, **N**, is represented in most cases merely by its electrostatic effect derived from its atomic monopoles. In several models **P** and **N** are merged and are considered in some cases as a polarisable, in others as a non-polarisable environment. The simplest representation of the **PN** region is by atomic monopoles, including an extra Coulomb term in the Hamiltonian of **C** originating from these monopoles (see e.g. Tapia and Johannin, 1981).

Correct treatment of the biophase, **W**, surrounding the protein, seems at a first glance less important and simpler than any of **C**, **P** or **N**. However, this is not the case as the effect of water as a solvent on ligand binding is crucial (Warshel *et al.*, 1986). It is important to note that polarisation effects in the surroundings reduce pure Coulombic effects in **C**. Accordingly, bare models of the protein, including only the central region while neglecting **P**, **N** and **W**, lead to a serious overestimation of protein electrostatic effects. Region **W** includes acids and bases, too, determining its pH and thus the ionisation state of surface side-chains of the protein. The electrostatic effect of these side-chains may be quite strong. Counterions, always present in the biophase, shield the electrostatic field of the side-chains therefore their effect should also be considered. This is done in simpler models by reducing side-chain net charges, sometimes even to zero. In some molecular dynamics simulations it is possible to consider them explicitly. However, the best treatment seems to be offered by the numerical solution of the discretised Poisson–Boltzmann equation, which will be discussed later.

10.3.3 Reaction Field Theories

A successful method, based on the above active site – environment partition, is the self-consistent reaction field (SCRF) theory of Tapia and Goscinski (1975). **C** may be represented by an effective Hamiltonian

$$\mathbf{H_C} = \mathbf{H_C^0} + \mathbf{H}_{\text{int}} \quad (10.13)$$

where $\mathbf{H_C^0}$ is the Hamiltonian of the unperturbed central region and the interaction operator is defined as follows:

$$\mathbf{H}_{\text{int}} = -\mu_C \langle \mathbf{F_{PNW}} \rangle - \mu_C \langle \mathbf{gM_C} \rangle \quad (10.14)$$

where μ_C is the dipole moment operator, $\langle \mathbf{F_{PNW}} \rangle$ and $\langle \mathbf{gM_C} \rangle$ denote the average fields of the permanent and induced point dipoles representing the environment, $\mathbf{M_C}$ is the expectation value of μ_C, and $\mathbf{g}$ is a response function. The first term in equation (10.14) represents the coupling of $\mathbf{C}$ with the electrostatic field of the environment dipoles. The second term stands for second-order effects: dipoles in $\mathbf{C}$ induce a field in the environment which, in turn, interacts with $\mathbf{C}$.

In the version of the SCRF theory applied to proteins (Tapia *et al.*, 1980) the solvent is replaced by the protein core, a set of polarisable dipoles surrounding $\mathbf{C}$ corresponding to the solute in the original scheme. Both permanent and induced electrostatic fields originating from the environment can be expressed approximately in terms of the wavefunction of $\mathbf{C}$:

$$\langle \mathbf{F_{PNW}} \rangle \approx \mathbf{g}_o \langle \Psi_C | \mu_C | \Psi_C \rangle \tag{10.15}$$

$$\langle \mathbf{gM_C} \rangle \approx \mathbf{g}_i \langle \Psi_C | \mu_C | \Psi_C \rangle \tag{10.16}$$

Combining equations (10.15) and (10.16) we have an effective non-linear eigenvalue problem for Ψ_C:

$$[\mathbf{H}_C^0 - \mu_C (\mathbf{g}_o + \mathbf{g}_i) \langle \Psi_C | \mu_C | \Psi_C \rangle] \Psi_C = E_C^{\text{eff}} \Psi_C \tag{10.17}$$

$\mathbf{g}_o$ and $\mathbf{g}_i$ are the orientation and inductive parts of the response tensor $\mathbf{g}$, which is usually parameterised using empirical data.

The merit of the SCRF model is that it correctly describes polarisation effects of the environment on the central region which may be important in enzymatic reactions. On the other hand, it does not consider charge transfer between $\mathbf{C}$ and the environment. Eventual spurious effects by this approximation can be greatly reduced by an appropriate definition of the central region. Charge-transfer effects, though small in magnitude, may be important for hydrogen-bonded systems. It is therefore advisable to include all atoms involved in hydrogen bonds when defining $\mathbf{C}$. The generalised version of the SCRF theory also allows a formulation at the level of statistical mechanics (Monte Carlo and molecular dynamics methods) (Tapia, 1991).

Since proteins exert their action in the biophase, containing water as the solvent and counterions partly shielding surface-ionised side-chains, the effect of this environment should be adequately treated in the parameterisation of $\mathbf{g}$ in order to obtain quantitative information on realistic models. This problem has been addressed by Warshel and Levitt (1976) and led later to the development of the first consistent model for treating protein/solvent polarisabilities in protein electrostatics, the Protein Dipoles Langevin Dipoles (PDLD) method (Russell and Warshel, 1985; Warshel, 1991). The PDLD model of a protein molecule unifies $\mathbf{P}$ and $\mathbf{N}$, but treats $\mathbf{W}$ separately in order to surmount problems related to the use of effective dielectric constants.

The central region, $\mathbf{C}$, is represented by a set of point charges that can be obtained using a molecular orbital method or another approximation (see e.g. Warshel, 1991) for the adequate molecular model of $\mathbf{C}$. Protein atoms in $\mathbf{P}$ (including $\mathbf{N}$) are described by considering explicitly their net charges and induced dipoles associated with their atomic polarisabilities, while water molecules in region $\mathbf{W}$ are described by the Langevin dipoles model. Thus the total electrostatic energy of the charges in $\mathbf{C}$ is given as follows:

$$E_{\text{es}}^C = E_{\text{Coul}}^{CC} + E_{\text{Coul}}^{CP} + E_{\text{pol}}^{CP} + E_L^{CW} \tag{10.18}$$

The first term in equation (10.18) is the interaction energy between the charges in **C** (in kJ/mol):

$$E^{CC}_{Coul} = 1390 \Sigma_{ij} q_i q_j / r_{ij} \qquad i,j \in \mathbf{C} \tag{10.19}$$

The second term is due to the interaction of the charges in **C** with those in **P**:

$$E^{CP}_{Coul} = 1390 \Sigma_{ij} q_i q_j / r_{ij} \qquad i \in \mathbf{C}, j \in \mathbf{P} \tag{10.20}$$

The polarisation term refers to the energy associated with polarising the induced dipoles of the protein atoms in **P** (in kJ/mol):

$$E^{CP}_{pol} = -695 \Sigma_i \mu_i \mathbf{f}^P_i \tag{10.21}$$

Here $\mathbf{f}^P_i$ is the electrostatic field at atom i of **P** emerging from the permanent charges in **C** and **P** (excluding i itself). The induced dipole moment at atom i can be determined from a self-consistent iterative procedure, but solving the following coupled set of equations:

$$\mu^n_i = \gamma_i \mathbf{f}^n_i \tag{10.22}$$

$$\mathbf{f}^n_i = \Sigma_{j \neq i} q_j \mathbf{r}_{ji} / r^3_{ji} - \Sigma_{j \neq i} [\mu^{n-1}_j - 3(\mathbf{r}_{ji} \mu^{n-1}_j) \mathbf{r}_{ji} / r^2_{ij}] / r^3_{ij} \tag{10.23}$$

where q_i and γ_i are the permanent charge and polarisability of atom i, $\mathbf{r}_{ji} = \mathbf{r}_i - \mathbf{r}_j$. μ^n_i is the dipole induced on atom i, and $\mathbf{f}^n_i$ is the force emerging at atom i in the nth iteration step. The initial field is taken to be equal to that originating from **C** and **P**:

$$\mathbf{f}^0_i = \mathbf{f}^P_i \tag{10.24}$$

The last term in equation (10.18) is the polarisation energy of the Langevin dipoles representing the solvation of the system (in kJ/mol):

$$E^{CW}_L = -695 \Sigma_i \mu_{Li} \mathbf{f}^P_i \tag{10.25}$$

where μ_{Li} are the Langevin dipoles of the protein which are polarised (by changing their magnitudes and directions) and depend both on $\mathbf{f}^P_i$ and contributions from other Langevin dipoles:

$$\mu^{n+1}_{Li} = (\coth X^n_i - 1/X^n_i) \mu_0 \mathbf{f}^n_i / f^n_i \tag{10.26}$$

with

$$X^n_i = C \mu_0 f^n_i / k_B T \tag{10.27}$$

and

$$\mathbf{f}^n_i = \mathbf{f}^P_i + \mathbf{f}^n_{\mu i} \tag{10.28}$$

where $\mathbf{f}^n_{\mu i}$ is the field on the ith dipole emerging from all other dipoles. The iteration starts with $\mathbf{f}^n_{\mu i} = 0$. The parameters C and μ_0 can be fitted to molecular dynamics simulation results obtained by using an explicit all-atom solvent model. Similarly, as in case of the SCRF method, the PDLD method can also be combined with statistical techniques in order to yield a free-energy perturbation model (Warshel, 1991).

It is possible to give a more precise expression for the electrostatic contribution to the ligand-binding energy which is based on the linear response approximation (Lee et al., 1993):

METHODOLOGY

$$\Delta\Delta G_b^{es} = \tfrac{1}{2}\Sigma_i q_i(V_i^{PL} - V_i^{P}) + \tfrac{1}{2}\Sigma_j q_j(V_j^{PL} - V_j^{L}) \quad (10.29)$$

where the first and second summations are over charges of the protein and the ligand, respectively. V^X is the MEP produced by the protein–ligand complex (PL), the protein (P) or the ligand (L). Equation (10.29) allows us to consider conformational changes upon binding because different charge distributions can be used for the complexed and uncomplexed forms of the protein and the ligand. However, the intrinsic conformational energy changes should be added to obtain correct binding-free energy changes.

10.3.4 The Poisson–Boltzmann Equation

In large asymmetric systems such as proteins a simplified description of electrostatics is possible with continuum models where molecules are not treated explicitly. Presently the most popular continuum treatment of proteins involves the solution of the Poisson–Boltzmann equation as proposed by Honig and coworkers (Rashin and Honig, 1985; Gilson and Honig, 1987; Sharp and Honig, 1990; Nicholls et al., 1991). The treatment is based on the Poisson equation, which relates the spatial variation of the MEP, V, to the charge density distribution, ρ, and the dielectric constant, ε:

$$\nabla[\varepsilon(\mathbf{r})\nabla V(\mathbf{r})] + 4\pi\rho(\mathbf{r}) = 0 \quad (10.30)$$

If the polarisability is assumed to be uniform, it can be represented by a single dielectric constant and for point charges equation (10.30) reduces to Coulomb's law. If the polarisability is not uniform, $\varepsilon(\mathbf{r})$ varies in space. This is the case, for example, when each atom is assumed to have a different polarisability.

In order to account for ionic strength effects the mobile ions in **W** should be represented by a mean field approximation. The Boltzmann expression for the concentration of an ion i at point **r** with a charge of q_i is as follows:

$$C_i(\mathbf{r}) = C_i^{BULK} \exp[-q_i V(\mathbf{r})/kT] \quad (10.31)$$

where C_i^{BULK} is the bulk ion concentration. Thus the net charge density for mobile ions is given by

$$\rho_m(\mathbf{r}) = \Sigma_i q_i C_i(\mathbf{r}) \quad (10.32)$$

and replacing the charge density distribution in equation (10.30) by the sum of densities due to mobile ions and the protein, $\rho(\mathbf{r}) \to \rho_m(\mathbf{r}) + \rho_P(\mathbf{r})$ we obtain the Poisson–Boltzmann equation:

$$\nabla[\varepsilon(\mathbf{r})\nabla V(\mathbf{r})] + 4\pi\{\lambda(\mathbf{r})\Sigma_i q_i C_i^{BULK} \exp[-q_i V(\mathbf{r})/kT] + \rho_P(\mathbf{r})\} = 0 \quad (10.33)$$

where $\lambda(\mathbf{r}) = 1$ for ion-accessible regions and 0 otherwise. The advantage of this equation is that it is applicable to arbitrary geometries and non-uniform dielectrics. In cases where the protein is not highly charged the exponential term in equation (10.33) can be linearised, since $q_i V(\mathbf{r})/kT \ll 1$. Thus we obtain

$$\nabla[\varepsilon(\mathbf{r})\nabla V(\mathbf{r})] + 4\pi[-2IV(\mathbf{r})/kT + \rho_P(\mathbf{r})] = 0 \quad (10.34)$$

where $I = \tfrac{1}{2}\Sigma_i q_i^2 C_i^{BULK}$, the ionic strength.

After solving the Poisson–Boltzmann equation, the electrostatic energy expression is obtained via the charging integral approach. This is a thermodynamic coupling process considering an initial state where all the explicit charges and dipoles of the protein (q_{Pi} and μ_{Pi}) are discharged then converted into the fully charged state. The conversion is done by using a parameter $\lambda(=0 \to 1)$ and setting charges and dipoles to λq_{Pi} and $\lambda \mu_{Pi}$. The electrostatic energy is then given by

$$E_{es} = \Sigma_i \int_0^1 V_i(\lambda) q_{Pi}\, d\lambda + \Sigma_i \int_0^1 \mathbf{E}_i(\lambda) \mu_{Pi}\, d\lambda \tag{10.35}$$

where the sum is over all charges and dipoles and V_i and $\mathbf{E}_i$ are the MEP and MEF at the centre of each charge and dipole, respectively. In the simplest case, when all response functions are linear we obtain the well-known expression

$$E_{es} = \tfrac{1}{2}\Sigma_i V_i q_{Pi} + \tfrac{1}{2}\Sigma_i \mathbf{E}_i \mu_{Pi} \tag{10.36}$$

Numerical solution of the Poisson and Poisson–Boltzmann equations is quite complicated since these are non-linear, three-dimensional partial differential equations. In cases where its three-dimensional structure is considered explicitly, the protein is represented by the positions, charges and radii of all its atoms which describe its shape and polarity. The region inside the protein envelope is represented by values of the dielectric constant accounting for electronic and, if appropriate, dipolar polarisability, while the region outside requires the ability to consider the asymmetric distribution of charge, dielectrics and salt concentration. This involves finite difference methods that may lead to numerical errors in the calculations which are, however, not too serious. In the case of protein–ligand interaction energies the statistical and systematic errors due to the discrete grid representation of the protein can be reduced to a range between 1 and 5 per cent of calculated binding energy differences (Shen and Wendoloski, 1996).

10.3.5 Thermodynamic Cycle-perturbation Method

A very precise, though laborious, procedure for the treatment of protein–ligand interactions is the thermodynamic cycle-perturbation method (Tembe and McCammon, 1984; McCammon, 1987). The method has been designed to account for free-energy changes of ligand binding due to small alterations in ligand or protein structure. In order to compute the difference between binding energies of ligands **L** and **L'** to the protein **P** the following thermodynamic processes have to be considered:

$$\mathbf{P} + \mathbf{L} \to \mathbf{PL} \qquad \Delta G_1 \tag{10.37}$$

and

$$\mathbf{P} + \mathbf{L'} \to \mathbf{PL'} \qquad \Delta G_2 \tag{10.38}$$

In principle, the change in free energy, $\Delta\Delta G = \Delta G_2 - \Delta G_1$, can be obtained from molecular dynamics calculations in which processes in equations (10.37) and (10.38) progress sufficiently slowly in the appropriate solvent. Unfortunately, a direct approach is not very useful, except for very simple systems, because each process must be performed slow enough so that the system remains in thermodynamic equilibrium and it is very difficult to ensure this during the simulation. In the thermodynamic cycle-perturbation method the following (non-physical) processes are considered:

$$\mathbf{P} + \mathbf{L} \to \mathbf{P} + \mathbf{L'} \qquad \Delta G_3 \tag{10.39}$$

and
$$\mathbf{PL} \to \mathbf{PL}' \quad \Delta G_4 \quad (10.40)$$

Combining processes in equations (10.37)–(10.40) we obtain a thermodynamic cycle:

$$\begin{array}{ccc} & \Delta G_1 & \\ \mathbf{P} + \mathbf{L} & \longrightarrow & \mathbf{PL} \\ \Delta G_{\text{solv}} \downarrow & & \downarrow \Delta G_{\text{bind}} \\ \mathbf{P} + \mathbf{L}' & \longrightarrow & \mathbf{PL}' \\ & \Delta G_2 & \end{array} \quad (10.41)$$

where the terms ΔG_{solv} and ΔG_{bind} stand for the differences between solvation-free energies and binding energies of **L** and **L'**, and both can be calculated by molecular simulation. Since the changes in the processes in equations (10.39) and (10.40) are, in general, much smaller and more localised than those in processes in equations (10.37) and (10.38), the calculations are greatly simplified. The relative free energy change is now obtained as follows:

$$\Delta \Delta G (= \Delta G_2 - \Delta G_1) = \Delta G_{\text{bind}} - \Delta G_{\text{solv}} \quad (10.42)$$

Shifting the prime from **L'** to **P**, we can apply the very same procedure to cases where the binding energy difference of the same ligand (**L**) to a point mutant of the protein (**P'**) and the wild type (**P**) is to be calculated.

10.4 APPLICATIONS

Quantitative description of protein–ligand interactions plays an important role in enzyme kinetics, molecular immunology, and rational drug design, among others. The methods discussed in Section 10.3 provide efficient tools for such a description and at the present stage of development we may state that in a variety of cases almost quantitative agreement with experiment could be achieved. The high performance of these methods allows us to design even specific ligands, e.g. potent inhibitors, for a given protein. In the following we give a brief overview of some interesting results.

10.4.1 Quantitative Structure–activity Relationships

The high performance of the available computational methods allows us to derive quantitative relationships between molecular structure and ligand-binding energies. The essential difference between this and earlier phenomenological approaches (see e.g. Kubinyi, 1993) is in the adequate molecular model used for the treatment of protein–ligand interaction. Such models are constructed in most cases on the basis of the three-dimensional structure of the protein–ligand complex, obtained by X-ray diffraction of nuclear magnetic resonance methods.

One of the first successful attempts to simulate protein–ligand interactions quantitatively has been published by Blaney et al. (1982b). They considered a thermodynamic cycle to calculate the binding energy differences of thyroid hormone analogues to prealbumin and qualitatively reproduced the experimentally observed relative free energies of association. However, though the predicted trend was correct, the largest difference between calculated and experimental values was as much as 18 kJ/mol.

The Protein Dipoles–Langevin Dipoles method combined with equation (10.30) has been successfully applied to the study of antibody–antigen interactions (Lee et al., 1992). The calculations could reproduce the difference in binding-free energies of phosphorylcholine analogues to the murine myeloma protein, McPC603. Calculating the binding energies for the analogues $(CH_3)_3N^+(CH_2)_nCOO^-$ Lee et al. obtained for $n = 2$ and $3 - 26$ (-23) and -36 (-30) kJ/mol, respectively, in agreement with experiment (in parentheses). The difference in binding energies, 10 kJ/mol, also agrees quite well with the experimental value, 7 kJ/mol.

Bash et al. (1987b) used the free-energy perturbation method to study thermolysin inhibitors with the general formula A–PO$_2$–X–B where A and B are large hydrophobic groups and X is variable. They were able to reproduce the 17 kJ/mol tighter binding of X=NH than X=O as found by experiment. Their calculations were repeated by Shen and Wendoloski (1996), who solved the linearised Poisson–Boltzmann equation and obtained binding energy differences that agree also well with experiment, especially if atomic net charges are used that are fitted to the molecular electrostatic potential outside the van der Waals envelope (see Table 10.1).

The thermodynamic cycle-perturbation method has been successfully applied to a variety of protein–ligand interactions. In general, it accounts both for electrostatic and non-electrostatic effects. Shen and Wendoloski (1995) developed a method, based on empirical solvation entropy, to estimate the non-electrostatic part of the interaction energy. An example where electrostatics plays a major role is the binding of benzamidine derivatives by trypsin (Wong and McCammon, 1986). In order to compare the binding of differently substituted benzamidines to trypsin, the simulation of the protein–ligand complex was extended with simulations of the separated inhibitor or enzyme in water. It has been found that p-F benzamidine binds somewhat less strongly to trypsin than does benzamidine itself ($\Delta\Delta G = 4$ kJ/mol). This is mainly due to solvation effects, since it is less difficult to desolvate benzamidine than its parafluoro derivative ($\Delta G_{solv} = -3$ kJ/mol).

Menziani et al. (1989) studied benzenesulfonamide inhibitors of carbonic anhydrase by

Table 10.1 Calculated binding energy differences (kJ/mol) of the thermolysin inhibitor Cbz–Gly–ϕ(PO$_2$)–X–Leu–Leu using atomic net charges obtained from different procedures (after Shen and Wendoloski, 1996)

Modification	Mulliken charges	MEP fitted charges	Experimental
NH to O	18.4	17.3	17.3
NH' to O'	13.0	11.6	10.8
NH to CH$_2$	3.3	7.5	7.1

the free energy perturbation method. Substituting the benzene ring in the *para* position by chlorine (i.e. mutating the 4 − H substituent to 4 − Cl), they were able to calculate the free energy difference (−4 kJ/mol) in good agreement with experiment (−6 kJ/mol).

HIV protease inhibitors were treated by Reddy *et al.* (1991). They compared the binding-free energy of a heptapeptide and a hexapeptide inhibitor. This rather large change in the molecular structure was calculated to lead to a loss in free energy of 14 ± 5 kJ/mol, in good agreement with experiment (16 ± 5 kJ/mol).

Hirono and Kollman (1990) studied ribonuclease T_1 inhibitors. In agreement with experiment they found a 13 kJ/mol preference for 2'GMP over 2'AMP. This arises from a combination of solvation (29 kJ/mol) and binding energy (42 kJ/mol) differences. The fact that the solvation-free energy difference is positive suggests that a purely hydrophobic site (with no binding energy difference) would prefer adenosine over guanosine binding. Thus for the protein it is easier to achieve adenosine rather than guanosine specificity. The relative free energies of binding of 2'GMP and 2'AMP to the Glu/Gln-46 and Ala/Gln-46 mutants of the enzyme were also qualitatively correct, 13 kJ/mol versus the 10 kJ/mol experimental value (Hirono and Kollman, 1991). However, it has been concluded that if the three-dimensional model of one protein–ligand complex is available, it is reasonable to expect the calculated binding-free energy difference to be an upper bound to the true value, since the system may not reach the correct geometry of the inhibitor whose structure is unknown.

Komeiji *et al.* (1993) studied the relative binding-free energy of tryptophan to the native Trp repressor protein and its Ser/Cys-88 mutant. The calculated values was 6.7 ± 1.3 kJ/mol, in excellent agreement with the experimental value, 7.1 kJ/mol. The Ser-88 side-chain forms a hydrogen bond with tryptophan and it might be expected that a Ser/Cys mutation would weaken this. The authors were able to show that electrostatic effects cancelled to a considerable extent and most likely the larger van der Waals repulsion in the protein–ligand complex of the mutant, where sulfur replaced oxygen, was the origin of the reduced free energy of binding of the mutant.

10.4.2 Computer-aided Drug Design

In the light of the excellent agreement between experimental and calculated binding-free energy differences for a series of similar ligands to the same protein or those for the same ligand to a series of point mutants, it is not surprising that successful *a priori* predictions of ligand-binding energies are also possible. This allows the skilled specialist to design specific ligands that can be later synthesised and passed to preclinical and clinical trials in the long process of drug development. This method gains increasing attention in drug design and the first pharmaceutical products, where it has been used during the development process at small companies such as Auguron and Vertex, will appear on the market soon.

An early qualitative approach is due to Blaney *et al.* (1982a), who designed potent thyroid hormone analogues that bind to prealbumin. Based on geomeric parameters obtained by protein crystallography they analysed the steric complementarity between the protein and its ligand by molecular graphics. They found an empty space in the contact region and predicted that if this space is filled by a bulky substituent, the free energy of binding will increase. Indeed, substituting the molecule by bromine in the appropriate position, they measured the inhibitory power, which increased by a factor of 15 as compared to the parent compound.

Prior to experiment, Rao et al. (1987) quantitatively predicted the effect of the Asn/Ala-155 mutation of the enzyme subtilisin on the binding energy and the catalytic rate of a typical substrate. The results ($\Delta\Delta G_{bind} = 0.4 \pm 3.3$ kJ/mol and $\Delta\Delta G_{cat} = 14 \pm 4$ kJ/mol) were in good agreement with the experimental values, 2 and 16 kJ/mol, respectively. An interesting result from this calculation was that Thr-221 interacts with the oxyanion transition state of the enzymatic reaction. This led to a set of collaborative experimental and theoretical studies and showed that the presence of Thr-221 has a twentyfold contribution to the stabilisation of the transition state relative to alanine (Mizushima et al., 1991).

Merz and Kollman (1989) used free energy simulations to study thermolysin inhibitors. According to their studies the Cbz–Gly–ϕ(PO$_2$)–X–Leu–Leu molecule with X = CH$_2$ binds almost as tightly as the analog with X = NH. This was at variance with the expectations of experimentalists, who confirmed the prediction with subsequent measurements (Morgan et al., 1991).

Cieplak and Kollman (1993) have carried out a study of the JG365 tight binding hydroxyethylene inhibitor of the human immunodeficiency virus protease. They considered which amide groups might be mutated to their ethylene or fluoroethylene isosteres in order to lead to more favourable binding. One should look for locations in the enzyme binding site where the amide group is not interacting too strongly with the protein and thus the binding-free energy is less positive than the free energy of solvation. The authors were able to find a number of promising candidates for JG365 modifications on the basis of their calculations.

REFERENCES

Bash, P. A., Field, M. J. and Karplus, M. (1987a) *J. Am. Chem. Soc.*, **109**, 8192.
Bash, P. A., Singh, U. C., Brown, F. K., Langridge, R. and Kollman, P. A. (1987b) *Science*, **236**, 574.
Behr, J. P. (ed.) (1994) *The Lock-and-Key Principle. The State of the Art—100 Years on*, Wiley, Chichester.
Berendsen, H. J. C., Postma, J. P. M., van Gunsteren, W. F. and Hermans, J. (1981) in *Intermolecular Forces*, edited by Pullman, B., Reidel, Dordrecht, p. 331.
Blaney, J. M., Jorgensen, E. C., Connolly, M. L., Ferrin, T. E., Langridge, R., Oatley, S. J., Burridge, J. M. and Blake, C. C. F. (1982a) *J. Med. Chem.*, **25**, 785.
Blaney, J. M., Weiner, P. K., Dearing, A., Kollman, P. A., Jorgensen, E. C., Oatley, S. J., Burridge, J. M. and Blake, C. C. F. (1982b) *J. Am. Chem. Soc.*, **104**, 6424.
Brooks, B. R., Bruccoleri, R. E., Olafson, B. D., States, D. J., Swaminathan, S. and Karplus, M. (1983) *J. Comput. Chem.*, **4**, 187.
Cieplak, P. and Kollman, P. A. (1993) *J. Comput.-Aided Mol. Des.*, **7**, 291.
Davis, M. E. and McCammon, J. A. (1990) *Chem. Rev.*, **90**, 509.
Dinur, U. and Hagler, A. T. (1990) *J. Comput. Chem.*, **11**, 1234.
Fischer, E. (1894) *Ber. Deutsch. Chem. Ges.*, **27**, 2984.
Gao, J. (1992) *J. Phys. Chem.*, **96**, 537.
Gelin, B. R. and Karplus, M. (1979) *Biochemistry*, **18**, 1256.
Gilson, M. K. and Honig, B. (1987) *Nature*, **330**, 84.
Gráf, L., Jancsó, Á., Szilágyi, L., Hegyi, G., Pintér, K., Náray-Szabó, G., Hepp, J., Medzihradszky, K. and Rutter, J. W. (1988) *Proc. Natl. Acad. Sci. USA*, **85**, 4961.
van Gunsteren, W. F. (1988) *Protein Engng*, **2**, 5.

Hirono, S. and Kollman, P. A. (1990) *J. Mol. Biol.*, **212**, 197.
Hirono, S. and Kollman, P. A. (1991) *Protein Engng*, **4**, 233.
Jorgensen, W. L. and Swenson, C. J. (1985) *J. Am. Chem. Soc.*, **107**, 569.
Kitaura, K. and Morokuma, K. (1976) *Int. J. Quant. Chem.*, **10**, 325.
Kollman, P. A. (1977) *J. Am. Chem. Soc.*, **99**, 4875.
Kollman, P. (1984) in *Crystallography and Drug Action*, edited by Horn, A. S. and De Ranter, C. J., Clarendon Press, Oxford, p. 63.
Kollman, P. A. (1993) *Chem. Rev.*, **93**, 2396.
Komeiji, Y., Uebayashi, M., Someya, J. I. and Yamato, I. (1993) *Protein Engng*, **5**, 759.
Koshland, D. R., Jr (1960) *Proc. Natl. Acad. Sci. USA*, **44**, 98.
Kubinyi, H. (1993) *QSAR: Hansch Analysis and Related Approaches*, VCH, Weinheim, 1993.
Lee, F. S., Chu, Z. T., Bolger, M. B. and Warshel, A. (1992) *Protein Engng*, **5**, 215.
Lee, F. S., Chu, Z. T. and Warshel, A. (1993) *J. Comput. Chem.*, **14**, 161.
McCammon, J. A. (1987) *Science*, **238**, 486.
Menziani, M. C., Reynolds, C. A. and Richards, W. G. (1989) *JCS Chem. Commun.*, 853.
Merz, K. and Kollman, P. A. (1989) *J. Am. Chem. Soc.*, **111**, 5649.
Mizushima, N., Spellmeyer, D., Hirono, S. and Pearlman, D. (1991) *J. Biol. Chem.*, **266**, 11801.
Momany, F. A., McGuire, R. F., Burgess, A. W. and Scheraga, H. A. (1975) *J. Phys. Chem.*, **79**, 2361.
Morgan, B. P., Scholtz, J. M., Ballinger, M. D., Zipkin, I. D. and Bartlett, P. A. (1991) *J. Am. Chem. Soc.*, **113**, 297.
Morokuma, K. (1971) *J. Chem. Phys.*, **55**, 1236.
Nagy, P., Ángyán, J. G., Náray-Szabó, G. and Peinel, G. (1987) *Int. J. Quant. Chem.*, **31**, 927.
Nakamura, H., Komatsu, K., Nakagawa, S. and Umeyama, H. (1985) *J. Mol. Graphics*, **3**, 2.
Náray-Szabó, G. (1989a) *J. Mol. Graphics*, **7**, 76.
Náray-Szabó, G. (1989b) *Int. J. Quant. Chem. Quantum Biol. Symp.*, **16**, 87.
Náray-Szabó, G. (1989c) *Catal. Lett.*, **2**, 185.
Náray-Szabó, G. (1993) *J. Mol. Recogn.*, **6**, 205.
Náray-Szabó, G. (1996) in *Molecular Electrostatic Potentials: Concepts and Applications*, Vol. 3, *Theoretical and Computational Chemistry*, edited by Murray, J. S. and Sen, K., Elsevier, Amsterdam, in press.
Náray-Szabó, G. and Ferenczy, G. G. (1995) *Chem. Rev.*, **95**, 829.
Náray-Szabó, G. and Surján, P. R. (1983) *Chem. Phys. Lett.*, **96**, 499.
Náray-Szabó, G. and Nagy, P. (1989) *Int. J. Quant. Chem.*, **35**, 215.
Nicholls, A., Sharp, K. A. and Honig, B. (1991) *Proteins Struct. Funct. Genet.*, **11**, 281.
Peinel, G., Frischleder, H. and Birnstock, F. (1980) *Theoret. Chim. Acta*, **57**, 245.
Rao, S., Bash, P., Singh, U. C. and Kollman, P. A. (1987) *Nature*, **328**, 551.
Rashin, A. A. and Honig, B. (1985) *J. Phys. Chem.*, **89**, 5588.
Reddy, M. R., Viswanadhan, V. N. and Weinstein, J. N. (1991) *Proc. Natl. Acad. Sci. USA*, **88**, 10287.
Rullman, J. A. C. and van Duijnen, P. T. (1990) *Reports in Molecular Theory*, **1**, 1.
Russell, S. T. and Warshel, A. (1985) *J. Mol. Biol.*, **185**, 389.
Sharp, K. A. and Honig, B. (1990) *Annu. Rev. Biophys. Chem.*, **19**, 301.
Shen, J. and Wendoloski, J. (1995) *Protein Sci.*, **4**, 373.
Shen, J. and Wendoloski, J. (1996) *J. Comput. Chem.*, **17**, 350.
Tapia, O. (1991) in *Theoretical Models of Chemical Bonding*, Vol. 4. *Theoretical Treatment of Large Molecules and Their Interactions*, edited by Maksic, Z.B., Springer-Verlag, Berlin, p. 435.
Tapia, O. and Goscinski, O. (1975) *Mol. Phys.*, **29**, 1653.
Tapia, O. and Johannin, G. (1981) *J. Chem. Phys.*, **75**, 3624.
Tapia, O., Sussman, F. and Poulain, E. (1978) *J. Theor. Biol.*, **71**, 49.
Tapia, O., Lamborelle, C. and Johannin, G. (1980) *Chem. Phys. Lett.*, **72**, 334.
Tembe, B. L. and McCammon, J. A. (1984) *Comput. Chem.*, **8**, 281.
Thole, B. T. and van Duijnen, P. T. (1982) *Chem. Phys.*, **71**, 211.
Warshel, A. (1991) *Computer Modelling of Chemical Reactions in Enzymes and in Solutions*, Wiley, New York.
Warshel, A. and Åqvist, J. (1991) *Annu. Rev. Biophys. Chem.*, **20**, 267.

Warshel, A. and Levitt, M. (1976) *J. Mol. Biol.*, **103**, 227.
Warshel, A., Sussman, F. and King, G. (1986) *Biochemistry*, **25**, 8368.
Weiner, S. J., Kollman, P. A., Case, D. A., Singh, U. C., Ghio, C., Alagona, G., Profeta, S. and Weiner, P. (1984) *J. Am. Chem. Soc.*, **106**, 765.
Wong, C. F. and McCammon, J. A. (1986) *J. Am. Chem. Soc.*, **108**, 3830.
Zimmerman, S. B. and Minton, A. P. (1993) *Annu. Rev. Biophys. Biomol. Struct.*, **22**, 27.

Index of Complexes

Neutral
Ar crystal   26
Ar_2 47
Ar_3 26, 47, 57–61
$Ar–H_2$ 22, 33, 124, 133
$Ar–HF$ 22, 33–34
$Ar–CO_2$ 66
$Ar–NH$ 133
$Ar–NH_3$ 5, 124, 141
$Ar–H_2O$ 5, 35–36, 124
$Ar–C_6H_6$ 110, 115
$Ar_2–HF$ 25, 27, 62–66
$Ar_2–HCl$ 25, 27, 61–66
$Ar_2–CO_2$ 26, 66–67
He_2 16, 21–24, 30, 31, 61, 81–84, 90, 92–93, 98–99
He_3 27, 55, 57, 60–61
$He–HF$ 95, 124–131
$He–CO$ 34, 131–6
$He_2–CO_2$ 66
$He–C_2H_2$ 35, 133
Ne_2 89
Ne_3 55, 57, 61
$(HF)_2$ 5, 87, 97, 101, 124, 136, 164, 167, 169, 171, 174–5, 273, 307, 310
$(HF)_3$ 5, 72–73, 273
$(HF)_n$ 87, 266, 273–86
$HF–ClF$ 310
$FH–N_2$ 310
$FH–H_2O$ 221, 310
$FH–N_2O$ 173, 310
$FH–CO$ 166
$FH–NCH$ 166, 169, 171, 176, 221
$FH–CH_4$ 204–8
$FH–CH_3Cl$ 173
$HF–H_2CO$ 310
$FH–NCCH_3$ 166, 176
$HF–HCCH$ 310
$FH–SiH_4$ 204–8
$(HCl)_2$ 124, 167
$(HCl)_3$ 73
$ClH–OH_2$ 166
$ClH–NH_3$ 173
$ClH–CH_4$ 204–8
$ClH–NCH$ 166, 176
$ClH–NCCH_3$ 166, 176
$ClH–SiH_4$ 204–8
$(H_2O)_2$ 5, 6, 22, 30, 36–37, 84–88, 90, 95–96, 97, 107, 124, 136, 162–4, 167, 169–71, 173–5, 221
$(H_2O)_3$ 5, 26, 37, 68–72, 107
$H_2O–NH_3$ 183–91
$H_2O–HCN$ 192–6
$H_2O–N_2O$ 208–12
$H_2O–NH_2OH$ 185–91
$H_2O–CO_2$ 208–12
$HOH–NCCH_3$ 192–6
$H_2O–C_2N_2$ 192–6
$HOH–HCCH$ 195, 310
$H_2O–C_6H_6$ 322
$H_2O–ACh$ 245–51
$(H_2O)_2–ACh$ 245–51
$(NH_3)_2$ 123, 136–42, 169, 171
$(ND_3)_2$ 141–2
$NH_3–NH_2OH$ 184–91
$NH_3–C_6H_6$ 322
$(HCN)_2$ 310
$(HCN)_n$ 266, 268–74
$(CO_2)_2$ 87
$CO@C_{60}$ 142–8
$(Cl_2)_2$ 310
$CH_3OH–O(CH_3)_2$ 196–204
$CH_3OH–N(CH_3)_3$ 196–204
CH_3OH-pyridine   318
$(CH_3)_2NH–O(CH_3)_2$ 196–204
$(CH_3)_2NH–N(CH_3)_3$ 196–204
$HCONH_2–H_2CO$ 309, 311, 314, 317
$(CH_3Cl)_2$ 320
$(C_6H_6)_2$ 300, 322
$C_6H_6–ACh$ 245–51
$(C_6H_6)_2–ACh$ 245–51
$HCONH_2–C_6H_6$ 322
uracil dimer   303
nucleic acid base pairs   323

s-tetrazine dimer 329

Cations
He–Li$^+$ 95
(HF)$_n$H$^+$ 266, 287
(H$_2$O)$_n$H$^+$ 229, 267
(H$_2$O)$_n$Al$^+$ 233
(H$_2$O)$_n$B$^+$ 233
(H$_2$O)$_n$NO$^+$ 234
(H$_2$O)$_n$(BOH)H$^+$ 233
(H$_2$O)$_n$(NH$_3$)$_m$H$^+$ 228
(H$_2$O)$_n$Cs$^+$ 229
(H$_2$O)$_n$(CH$_3$OH)$_m$H$^+$ 230
H$_2$O–N(CH$_3$)$_4^+$ 247
(H$_2$O)$_m$(N$_2$H$_4$)$_n$H$^+$ 231
(ROH)$_n$H$^+$ 232
(CH$_3$OH)$_n$Na$^+$ 232
(CH$_3$OH)$_n$Cs$^+$ 232
(ROH)$_m$(NH$_3$)$_n$H$^+$ 230
(NH$_3$)$_m$H$^+$ 228
(NH$_3$)$_m$(HCN)$_n$H$^+$ 224–7
(NH$_3$)$_m$(CH$_3$CN)$_n$H$^+$ 228
(NH$_3$)$_m$(CH$_3$CHO)$_n$H$^+$ 228
(NH$_3$)$_m$((CH$_3$)$_3$N)$_n$H$^+$ 228
HCN(NH$_4$)$^+$ 241
HNC(NH$_4$)$^+$ 241
CH$_3$CN(NH$_4$)$^+$ 241
CH$_3$NC(NH$_4$)$^+$ 241
(HCN)CH$_3$(NH$_4$)$^+$ 241
(HNC)CH$_3$(NH$_4$)$^+$ 241
(CH$_3$CN)CH$_3$NH$_4^+$ 241
(CH$_3$NC)CH$_3$NH$_4^+$ 241
(CH$_3$CN)CH$_3$NH$_4^+$ 241
(CH$_3$NC)CH$_3$NH$_4^+$ 241
(CH$_3$CN)CH$_3$NCH$^+$ 241
(CH$_3$NC)CH$_3$NCH$^+$ 241
(CH$_3$COOH)$_n$H$^+$ 231
C$_6$H$_6$-NH$_4^+$ 251
C$_6$H$_6$-N(CH$_3$)$_4^+$ 246–51
C$_6$H$_6$–Na$^+$ 253
C$_6$H$_6$–guanidinium 322
C$_6$H$_6$–imidazolium 322

Anions
Ar–Cl$^-$ 55
Ar$_2$–Cl$^-$ 74–75
(FHF)$^-$ 267, 287, 289–90
(HF)$_{n-1}$F$^-$ 266, 287
(H$_2$O)–OH$^-$ 222
(H$_2$O)$_2$–Cl$^-$ 75
(H$_2$O)$_n$F$^-$ 238
(H$_2$O)$_n$Cl$^-$ 238
(H$_2$O)$_n$Br$^-$ 238
(H$_2$O)$_n$I$^-$ 238

Subject Index

absorption coefficient   120, 133
acetic acid   267
acetoin   244
acetol   244
acetylcholine   245–50
acetylcholinesterase   245
active site   338
additivity approximation   243
adenine   323
adenosine   347
adenosine-phosphate   237
adenosine A_1 receptor   237
adiabatic separation   106, 119
AIM (*see* atoms in molecules)
alpha helix   328–9
AMBER   339
2′AMP   347
angular momentum   107–9, 126, 133, 138
anharmonic effects   36, 87, 162, 176
annealing, simulated   311
anthracene   304
antibody–antigen interactions   346
antisymmetrization   13, 50–1, 92
API (*see* atmospheric pressure ionization)
approximation
   additivity   243
   Born–Oppenheimer   4, 8, 105
   frozen-core   37, 163
   harmonic   162, 165, 168, 176, 221
   linear response   342
   mean field   343
   Mulliken   53
   random phase   28
   rigid-body   335
ATM (*see* Axilrod-Teller-Muto)
atmospheric pressure ionization   230
atomic point charge model   304–6, 310, 316, 330, 340–1
atomic polar tensor   181
atoms in molecules   240–1, 307, 329
aug-cc (*see* basis set, correlation-consistent)
average H-bond energy (*see* interaction energy, per H-bond)

Axilrod-Teller-Muto   27–29, 55, 59, 70–1, 74–5
azabenzenes   302, 305, 308, 313
azahydrocarbons   313–4

barrier
   activation   233
   rotation   328
basis set extension (*see* basis set superposition error, secondary)
basis incompleteness   84–6
basis-set
   correlation-consistent   85–9, 98, 160–3, 169–71, 175, 223, 238, 241–3
   Dunning (*see* correlation-consistent)
   limit   37, 171
   minimal   160
   optimized   25, 30, 37, 88, 160
   Pople type   160
basis set superposition error   xiv, 81–102, 226, 241, 246
   definition   24, 81–2, 168, 220
   need to remove   24, 90–6, 101–2, 269–70
   overcorrection   83, 98, 101, 171, 223–4
   reduction   169–71, 220
   secondary   94
bending frequencies
   intermolecular   62
   shift   188–91, 195–6, 210–2
benzamidine derivatives   346
benzene   245, 299, 301, 304–5
binding energy (*see* interaction energy)
Boltzmann distribution   120
bond functions   30, 37, 85–6, 88, 94, 98, 102
Born–Oppenheimer approximation   4, 8, 105
BSSE (*see* basis set superposition error)

CAD (*see* collision-assisted dissociation)
cAMP   325–6
carbonic anhydrase benzenesulfonamide inhibitors   346
CEPA   93

chain
  H-bonded 266
  infinite 268–71, 275, 282
charge density operator 299
charge-overlap (*see* penetration)
charge transfer 97, 220, 236, 253, 310, 336, 341
  contribution to induction energy 55
  contribution to polarization energy 95
charging integral approach 343
CHARMM 339
chemical Hamiltonian 97–8
chemical ionization 218
CHF (*see* coupled Hartree–Fock)
CI (*see* configuration interaction)
CID (*see* collision-induced dissociation)
Cl–Cl interaction 319–21
Clebsch–Gordon
  coefficients 11, 113
  coupled products 112, 133
clusters 217–258, 265–92
collision-assisted dissociation 228
collision-induced dissociation 228, 230, 244
collocation method 119
complementarity 335–7, 347
configuration interaction 161, 172
continuum models 343
convergence
  with respect to level of theory 159, 162, 173, 222
  with respect to length of chain 268–71, 275, 282–3, 287
  with respect to order of moments 299–303, 307
  (*see also* perturbation theory, electrostatic energy, exchange)
conversion factors xvi
cooperativity (*see also* nonadditivity) 227, 247, 265–92, 285
coordinates
  angular 106
  Cartesian 106
  curvilinear 106–7
  internal 4–5
  normal 106
  space-fixed 108
coordination number 46
core ion 230–1
Coriolis
  coupling 138, 145
  terms 112, 115, 120, 126
coronene 304
correlation
  intramonomer 15–25, 27, 51, 97
  electrostatic 18–19, 309, 316
  exchange 20–1

induction 19–20
  (*see also* exchange)
COSMIC 327
Coulomb energy (*see* electrostatic energy)
counterpoise
  Boys–Bernardi 24, 37, 59, 81–102, 168–71, 220, 226
  local 96–8
  theorem 92–3, 96
  virtuals only 90–6
coupled cluster 16, 18, 20, 22, 51, 57, 161, 163, 172–4, 223
coupled Hartree–Fock 16, 19, 28, 54
coupling between vibrational modes 176
crown ether 234–5
crystals 282, 297, 311–8, 322, 339
  rare gas 56
Cumulative Atomic Multipole Moments 307
cyanoacetylene 267
cyclic structure 164, 268–70, 275–83
cyclosporin 298
cytosine 323

damping
  factors 12
  functions 113
Davidson correction 172
decoupling of vibrational modes 187
defect 267, 287–8
deformation energy 99–100
delocalization effects 243
density functional theory 218, 275–81, 287–91
derivatives, analytical 99, 162, 182–3, 313
deuterium labeling 244
DFT (*see* density functional theory)
dielectric
  constant 339, 343
  continuum 339
diffuse functions 30, 102, 160, 169, 220–2
N,N-dimethylacetamide 328
dinitronaphthalene 312
diphosphate 237
dipole moment 230, 299, 309
  of ammonia dimer 137–41
  changes on complexation 89–90, 95
  of CO 144
  definition 120–2
  of HCN 228
  model 133
dipole polarizability (*see also* polarizability) 89–90
discrete variable representation 116–19, 144
dispersion
  definition 17–18, 48–52
  effects 89, 161

energy   7–10, 14, 220–3, 336
  -induction   49
  magnitude   23, 27, 34, 67, 87
  models   113, 131–2, 138, 252, 329
  nonadditivity of (see also
     nonadditivity)   54–7, 62–7, 71–2
  requirements for calculation of   30, 88, 95,
     97, 298
  (see also exchange)
dissociation energy (see interaction energy)
Distributed Multipole Analysis   307–330
distributed multipoles (see multipole)
DMA (see Distributed Multipole Analysis)
DMAREL   313
DNA   157, 217, 321–2
drug design   326, 347
DVR (see discrete variable representation)

ECEPP   339
effective potentials   4, 6, 46, 72, 235, 238
eigenvalues, negative   209
electron diffraction   277
electrospray ionization   218
electrostatic energy   7–9, 220, 335–6
  absence of   56
  BSSE   95
  convergence of   23
  definition   48, 298, 341–4
  force   297–330
  magnitude   87, 88, 230, 236, 243, 253
  models   113, 252
  relation to nonadditivity   54
electron impact ionization   218, 234
enthalpy
  binding   159, 165–76, 235, 245, 250–2
    comparison with experiment   222,
       227–8, 243
    definition   162, 165, 224
    relation to other properties   204, 241–3
  of evaporation   232
  of sublimation   312–3, 316
entropy   312, 336
  binding   227–8, 235
    comparison with experiment   228
    corrections   238, 250
  solvation   346, 337
equilibrium
  distance   86–9, 159, 162–5, 226, 266, 276
  geometries (see geometry optimization)
ESI (see electrospray ionization)
ethanol   328
Euler angles   106, 108–12, 143
Ewald summation   313
exchange
  convergence of   23
  corrections   10, 27

correlation   20–1, 52, 65, 97
  -Coulomb   8
  decomposition of   63
  definition   49, 53
  deformation   13–17
  -dispersion   14, 21–3, 28, 49–50, 58, 63, 66
  energy   7–10, 14, 49, 66–7, 74–5, 220, 336
  effects   8, 12, 14–17, 48–50
  -induction   14–16, 23, 28, 49, 66
  models   34, 131, 138
  multipole   62
  nonadditivity (see also nonadditivity)   27–9,
     49, 62, 74–5
  overlap   62
extrapolation (see convergence)

FAB (see fast atom bombardment)
fast atom bombardment   218
Fermi golden rule   130
finite difference methods   344
f-functions   86, 171, 220
formic acid   267
free energy   228, 345
  binding   227–8, 235, 237, 251–2,
     336, 347–8
  perturbation model   342
free rotation   109, 192
frequency
  imaginary   162, 221, 224, 276
  shift   165, 267
  (see also vibrational)
frozen-core approximation   37, 163
functions
  bond   30, 37, 85–6, 88, 94, 98, 102
  diffuse   30, 102, 160, 169, 220–2
  f-   86, 171, 220
  g-   223
  Laguerre   113
  Legendre   118, 125, 132
  response   341
F_2 299

GAUSSIAN   181–2, 189, 196, 268
geminals   98
  Gaussian   22, 32, 99
geometry optimization   162, 182, 220–2, 233,
     235, 241, 338
  with counterpoise corrections   24, 99–100
ghost orbitals   90, 96, 220
  (see also basis set superposition error)
GIAO   271, 285, 290
glyme   234–5
2'GMP   347
g-orbitals   223
GROMOS   339
guanine   323

guanosine 347

Hamiltonian
   interaction 9, 49, 92, 97, 340
   Møller–Plesset 15, 51
   nuclear motion 107–10, 112, 115, 138
   radial 144
   trimer 26
   (see also Møller–Plesset, interaction energy, perturbation theory)
harmonic
   approximation 162, 165, 168, 176, 221
   oscillator 106, 115
   (see also vibrational)
Hartree–Fock limit 161
heats of (see enthalpy)
Heitler–London
   exchange 50, 53, 58, 63, 66, 70–2
   wave function 92
   energy 93, 95–8
   theory 14
Hessian matrix 182
hindered rotor 134–5
Hirschfelder partitioning 307
Hirschfelder–Silbey 13
histidine 309
HIV protease 348
   inhibitors 347
HS (see Hirschfelder–Silbey)
hydration energy 336
hydrogen-bonding 87, 157–76, 265–92, 316, 347
   angular dependence 319
   anti- 311
   basis set superposition error in 89
   chemical Hamiltonian calculations 97
   changes in dipole moments caused by 95
   charge-transfer contribution to 341
   CH–Y 239–44, 268–74, 323
   NH–OC 317
   XH–C 239–44
   XH–π 245–53
   electrostatic character of 319, 321, 323
   force fields to simulate 339
   ionic 287–90
   solids 266, 275
   in trimers 68–73
   multipole models of 310
hydrogen chloride 267
hydrophobic factors 336

imidazole 267
indazole 316
induced-fit model 338
induction–correlation 52

induction–dispersion 10, 29, 52, 54, 63
induction energy 7, 8–10, 14, 27, 66–67, 253
   accuracy of calculation 23, 298
   means of calculating 16
   models of 54, 71, 113, 131–2
   nonadditivity of 48–50, 53–6, 62–3, 70
   (see also exchange)
intensity
   IR 135–6, 162, 181–3, 200–1, 266, 271, 284, 289
   Raman 183, 196, 201–203
interaction energy 196, 229, 253
   definition of 4, 81–2, 219–20, 224, 336
   effects of BSSE 86–9
   ionic H-bond 224
   partitioning 99
   per H-bond 269, 276
   relationship with other properties 204–5, 210, 231
   sensitivity to level of theory 158–61, 174, 222–3, 241–2, 291
   water dimer 158
interchange tunneling frequency 141
ion pair 164
ionosphere 217
isotopic substitution 141, 189–90

KERD (see kinetic energy release distribution)
kinetic energy 107–10
kinetic energy release distribution 219, 231

β-lactam antibiotics 326
Laguerre functions 113
Langevin dipoles 342
lattice structure 46
Legendre
   expansion 110–11, 125
   functions 118, 125, 132
Legon–Millen rules 310–11, 322
Lennard–Jones potential 112
linear response approximation 342
localization 90–1
local minimum 246
London theory 7–9, 13
lone pair directionality 317–21

macromolecular crowding 336
magic number 218–19, 229–31
MALDI (see matrix-assisted laser desorption ionization)
many-body effects 6, 45–76, 238
mass spectrometry 218
   electrospray 237
   high-pressure 231, 234, 241, 245
   tandem 230, 232, 234
   time-of-flight 230

matrix
- -assisted laser desorption ionization 218
- Hessian 182
- rare gas 186–91, 193, 205, 209–12, 268, 275
- Wilson GF 106, 182

MBER (*see* molecular beam electric resonance)
mean field approximation 343
METECC 31
methanol 267
N-methyl acetamide 328
MEF (*see* molecular electrostatic field)
MEP (*see* molecular electrostatic potential)
methyl glycolate 244
methyl lactate 244
molecular
- beam electric resonance 184, 192, 209–10
- dynamics 218, 229, 233, 238, 341, 344
- electrostatic field 336, 344
- electrostatic potential 336–7, 343–4
- force fields 338–9
- graphics 337
- matching 326
- mechanics 338
- recognition 335

Møller–Plesset 87, 161, 182, 268
- additivity 243
- advantages of 172–3
- BSSE of 93
- comparison with DFT 275, 287–9
- contribution to binding energy 89, 158, 169, 220, 238
- effect on geometry 162–5, 252
- effect on vibrational frequencies 186
- intermolecular 6, 49
- multibody terms 55–6
- partitioning 9, 15, 51–2
- (*see also* perturbation theory)

MOLPAK 312
moment, multipole 220, 297–330
Monte Carlo 218, 238, 341
- quantum 120
Morokuma decomposition 336
Mulliken
- approximation 53
- charge 231, 305
- spectroscopic convention 183
multiphoton ionization 218
multipole
- anisotropic atomic 308
- moments 12, 297–330
- point 307
- (*see also* exchange)
multipole expansion 7, 11, 252, 308
- central 297–304, 330
- distributed 8, 55, 309–13, 297–330, 306–8

of induction energy 54
murine myeloma protein 346

naphthalene 304
natural energy decomposition 236
neutron diffraction 277, 290
Newton–Raphson minimization 313
m-nitroaniline 314
NMR 277, 346
- chemical shift 266, 273, 285, 290
- shielding constant 271, 273
nonadditivity xiv, xv, 25–30, 45–76, 271, 282
- (*see also* cooperativity)
nonadiabatic effects 4
nonlinear optical coefficient 311
nozzle beam, supersonic 122, 218
nuclear quadrupole splitting 138, 141
nucleic acid bases 309, 323

OPLS 339
orbital exponents 17, 160
ORIENT 308
orthophosphate 237
overlap integral 51
oxohydrocarbons 313–4

Pauli exclusion principle 49, 51
PDLD (*see* Protein Dipoles Langevin Dipoles)
penetration effects 8, 12, 300, 308–10
permutation
- -inversion 122–123
- operator 53
perturbation theory xiv, 3–38
- absence of BSSE in 83, 96, 101
- accuracy 21–5
- applications 31–8, 125, 309–10, 314, 317, 320, 322
- calculation of spectra 129, 131, 134
- convergence 23, 50–1
- double 15–16
- intermolecular 47–8, 298
- many-body 15–16
- nonadditivity 25–31, 48–50
- supermolecular approach, relation to 51–2, 57, 93, 329
- (*see also* Møller–Plesset)

phenanthrene 304
phenylalanine 322
phosphodiesterase III 325–6
phosphorylcholine 346
Poisson–Boltzmann equation 340, 343–4, 346
polarizability 12, 49, 183, 220, 230, 340
- changes of complexation 89–90
- definition 120–2
polarization

polarization (*cont.*)
  energy 9, 14, 27, 95, 220, 252, 298, 336, 342
  expansion 9–13
  function 30, 160
polymers 265–92
polypeptide 328
Pople correction 172
prealbumin 346–7
PROMET 312
Protein Dipoles Langevin Dipoles 341, 346
protein–ligand interactions 335–47
proton
  affinity 173, 228–9
    difference 241
  conductivity 267
  solvated 217
  transfer reaction 101, 164–5, 221, 233, 237, 241, 244
  transport 287
  wire 235
protonated cluster ion 219, 224–32
pyrazine 305
pyrene 304
pyridazine 305
pyridine 305
pyrimidine 305, 310
pyrylene 304

QCISD 173–74, 241, 243
quantitative structure-activity relationships 345–7
quantum Monte Carlo 120

Raman scattering 105
random phase approximation 28
rare gas matrix 186–91, 193, 205, 209–12, 268, 275
Rayleigh–Schrödinger 7, 9, 58
  symmetrized 13–15, 32
reaction-field theory 340–3
  self-consistent 340
reduced resolvent 9
relaxation energy (*see* deformation energy)
response function 341
ribonuclease T_1 inhibitors 347
rigid rotor 106, 108, 114, 133
rigid-body approximation 335
ring diagrams 18
rotational constant 108, 133, 144–5
rotational predissociation lifetime 130
rotor
  hindered 134–5
  rigid 106, 108, 114, 133
RPA (*see* random phase approximation)
R12 98–9

saddle point 162, 192, 276, 287
SAPT (*see* perturbation theory)
SCF deformation energy 51–2, 70, 72
SCRF (*see* self-consistent reaction field)
secondary interaction 185, 317
selection rules 126, 134
self-consistent reaction field 340
separation
  adiabatic 106, 119
similis simili gaudet principle 337
size-consistent 47, 81, 172
  definition 161
size-selective 218, 228, 232
solids 267, 280, 311
  H-bonded 266, 275
solvation shell 46, 219, 228
spin functions, nuclear 123
SRS (*see* Rayleigh–Schrödinger, symmetrized)
stabilization energy (*see* interaction energy)
stacking interactions 322–3
stratosphere 217
stretch 266, 269, 281
stretching frequencies 89, 159, 222, 267
  CH 193–195, 271
  FH 73, 87, 205–8, 280–3, 289
  OH 69, 89, 199–200
  water 186–91, 210–2
  shift 181, 240, 266–7, 291
  (*see also* vibrational)
subtilisin 337, 348
supermolecular approach 6, 219
  BSSE in 81–102
  comparison to SAPT 21–5, 47–8, 298
  nonadditivity 51–2, 56–7, 63
s-tetrazine 305

thermal vibrational energy 162, 168, 176, 221, 246
thermodynamic cycle perturbation method 344–6
thermolysin inhibitors 346, 348
thymine 323
thyroid hormone 346–7
torsional energy 338
transferability 327–9
transition frequency 135–8
transition state 192, 238
  see also saddle point
transition state analog 251
s-triazene 305
triphenylene 304
triple-dipole term 27–29, 55, 59, 70–1, 74–5
troposphere 217
Trp repressor protein 347
trypsin 337, 346
tryptophan 347

Subject Index

tyrosine   322

umbrella inversion   137–41
uncoupled Hartree–Fock   55
units   xvi
uracil   299, 309

van't Hoff plots   219, 228
variational principle   xiv
VCP (*see* counterpoise, virtuals only)
vibration-rotation tunneling   46, 68, 73, 105–48
vibrational (*see also* stretching and bending frequencies)
  (*see also* zero-point energy)
  averaging   137, 280
  modes   181
  predissociation   46, 128
  spectra   182–212

virial coefficient
  second   35
  third   26
VRT (*see* vibration-rotation tunneling)

water   46, 157, 309
  -accessible surface   336
Watson–Crick structure   323
Wilson GF-matrix   106, 182

xylitol inhibitor   337
D-xylose isomerase   337

zero-point energy   221, 238, 246, 269, 276
  BSSE of   89
  contribution to binding   162, 168, 174, 176, 224
  of water dimer   36

DATE DUE			
DUE NOV 1 1 2005			